Early Earth Systems

For Patti

whose love and loyalty
have been a constant
source of strength
for >3×10^1 yr

Exaltare super coelos, Deus,
Et super omnem terram Gloria tua

Psalm 108, v.5

EARLY EARTH SYSTEMS

A Geochemical Approach

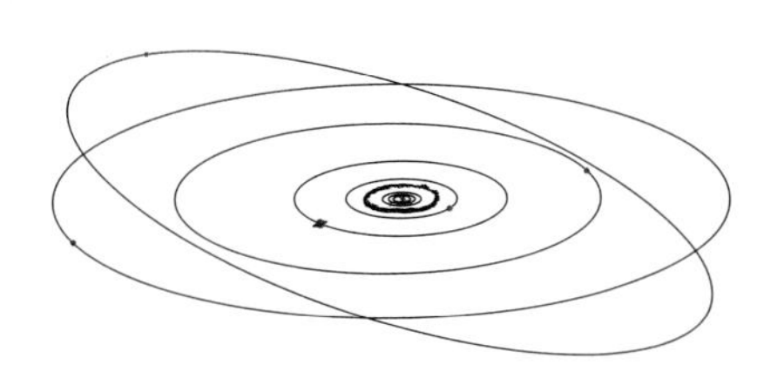

Hugh Rollinson

BLACKWELL PUBLISHING
350 Main Street, Malden, MA 02148-5020, USA
9600 Garsington Road, Oxford OX4 2DQ, UK
550 Swanston Street, Carlton, Victoria 3053, Australia

First published 2007 by Blackwell Publishing Ltd

1 2007

Library of Congress Cataloging-in-Publication Data

Rollinson, Hugh R (Hugh Richard), 1949-
Early Earth systems : a geochemical approach / Hugh Rollinson.
p. cm.
Includes bibliographical references and index.
ISBN-13: 978-1-4051-2255-9 (pbk. : acid-free paper)
ISBN-10: 1-4051-2255-2 (pbk. : acid-free paper)
1. Earth–Origin. 2. Earth–Internal structure. 3. Geochemistry.
4. Historical geology. I. Title
QE500.5.R65 2007
551.7–dc22

2006029667

A catalogue record for this title is available from the British Library.

Set in Trump Mediaeval 9.5/11.5
by NewGen Imaging Systems Pvt Ltd., Chennai, India
Printed and bound in Singapore
by Markono Print Media Pte Ltd

The publisher's policy is to use permanent paper from mills that operate a sustainable forestry policy, and which has been manufactured from pulp processed using acid-free and elementary chlorine-free practices. Furthermore, the publisher ensures that the text paper and cover board used have met acceptable environmental accreditation standards.

For further information on
Blackwell Publishing, visit our website:
www.blackwellpublishing.com

CONTENTS

LIST OF BOXES

PREFACE

The guy on the other end of the phone was offering me a job! A British government funded contract to work in the Geological Survey of Sierra Leone. As a fresh graduate this was an exciting prospect – a whole unexplored Archaean craton at my feet. The next three and a half years were an exciting time. It was the early 1970s when important new discoveries were being made in Archaean rocks in South Africa and the novel ideas of plate tectonics were still being newly applied to early Earth history. My position as a survey geologist provided me with the opportunity to map several greenstone belts and what seemed like an endless tract of granite gneisses and at the same time apply these new ideas in what appeared to be virgin territory. These were 'formative' years which cemented my fascination with the Archaean, an absorption which has remained with me over subsequent decades.

Sierra Leone was followed with a PhD back home in the UK in which I had the opportunity to study Archaean lower crust in the Lewisian Gneisses of northwest Scotland. This was a classic Archaean terrain and a chance to develop my skills where giants of Archaean geology had previously trod and where some of the principles of unraveling complex Archaean terrains were first established. This was also the time when Archaean 'grey gneisses' were being recognized around the world as the metamorphosed equivalents of TTG magmas and the essential building materials of Archaean continental crust. Sure enough the Lewisian too turned out be made of TTG magmas, albeit metamorphosed to amphibolite and granulite grade.

Some years later I took a position in the University of Zimbabwe – another of the world's classic areas of Archaean geology. At the time, Zimbabwe was a relaxed and beautiful country in which to carry out fieldwork. A whole new region of the world to get to grips with, a new craton to live on, and a greenstone belt in the back garden as it were. There were further bonuses. A second 'classic' craton, the Kaapvaal Craton, not far to the south, an Archaean 'orogenic belt', the Limpopo Belt, located between the two, and some great people to work alongside.

More recently I have had the opportunity to work with a team on the Isua Greenstone Belt in west Greenland and to examine, in a region of superb exposure, what is perhaps the most ancient sequence of sediments and lavas preserved from the early Earth. This too was a very fruitful time and turned my thinking to the question of the origin of life, a particular emphasis of that multidisciplinary project. In addition, the great antiquity of these rocks forced me to begin to think about the earliest stages of Earth history and moved me back in time from the Archaean into the Hadean.

At about the same time I was also invited to coordinate a program of research in the Baltic Archaean Shield together with colleagues from the Institute of Precambrian Geology in St Petersburg, Russia. This research took me back to problems of Archaean crustal growth and focused on Archaean sanukitoid magmatism. These rather rare and unusual magmas have provided a new window into Archaean crust and mantle evolution, and an opportunity to work with scientists from a different cultural background, with their particular insights into the early Earth.

And so, what of this place? This desert. Oman has provided me with a different sort of opportunity – a place to read, to think and reflect, and of course to write. That is not to say that the geology is irrelevant – far from it. It is world class - in scientific interest, in access and in exposure, and the rocks of the Oman ophiolite provide great analogues for early Earth processes. But that is another story.

So much for the serendipitous benefits of one's career and the influences of many different geographies on my thinking. There is more, for what I have tried to do in this book is be an advocate for the Earth Systems paradigm and argue for its importance in understanding the early Earth. That is not to say that I have produced a new synthesis, nor a new model for how the Earth has developed in its earliest stages. Instead, what has emerged, in the process of thinking about the early Earth in this 'modern' way, is an agenda. An agenda derived from thinking about the early Earth by using the Earth Systems approach. This agenda identifies what we do not yet know and where we are to go in future research. It is an agenda of questions.

Many people have influenced my thinking over the years – both in their writings and personality. Here, worthy of mention are those who have taken the time to read and comment on chapters – Paul Taylor, Richard Tweedy, Jan Kramers, Ken Collerson, John Tarney, David Catling and Euan Nisbet. In addition Kent Condie and Martin Brasier read and helpfully commented on the entire manuscript. I am grateful to all of them. In addition and not to be forgotten, are generations of students at the Universities of Zimbabwe, Gloucestershire and at the Sultan Qaboos University, Oman whose questions and responses have allowed me to develop and clarify the ideas presented here.

Hugh Rollinson
Sultan Qaboos University
Muscat

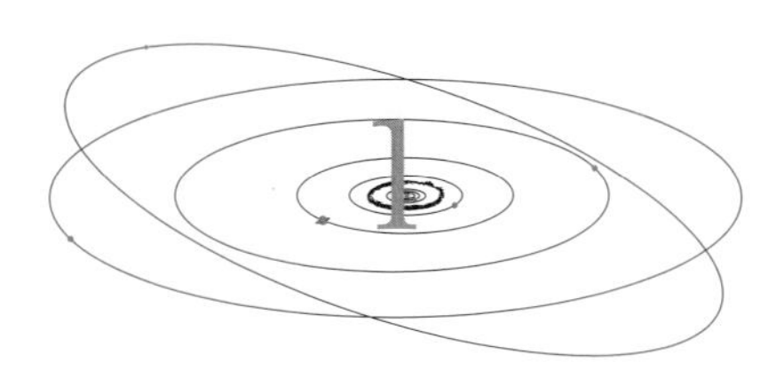

THE EARTH SYSTEM

1.1 Introduction

The Earth is not the planet it used to be. It has changed substantially since it first formed and at the present time is still changing. The quest of this book is to explore the way the Earth was in its earliest history, in particular during the first 2 billion years of its 4.6 billion year life. This knowledge is vitally important, for it provides essential clues in charting how the Earth has arrived at its present physical, chemical, and biological state and leads to an understanding of the large-scale processes which bring about change in our planet.

Over the past decade, there has also been a major change in the way we think about the Earth. The Earth Systems paradigm is now strongly influencing contemporary thinking in the Earth Sciences. Thus, a distinctive feature of this text is an emphasis on the insights which the "systems approach" offers into processes on the early Earth. In particular we will explore the linkages that exist between the different parts of the Earth System. So we will discuss in some detail the interactions that take place between the Earth's crust and mantle; its oceans and atmosphere; its oceans, ocean crust, and mantle; and between the "Earth" and life. These are topics of considerable importance for the modern Earth System, and in seeking to apply this approach to the early Earth, we shall discover how the early Earth System operated.

The findings of planetology imply that during the first 100 Ma of its life the Earth must have experienced a number of extreme changes unlike anything in the recent geological past (see Chapter 2, Section 2.4). These processes shaped the primordial Earth and yielded a "starting point" for the Earth System. What exactly this primordial Earth looked like is one of the subjects for this book and it plunges us directly into some of the biggest scientific questions we might ask. For example, exactly when and how did the Earth form, and is it different from the other "terrestrial" planets? What was the composition of the earliest atmosphere and oceans? Did they come from inside the Earth, or were they acquired from elsewhere within the solar system? To what extent had the Earth melted in its first 100 Ma? What sort of imprint did this leave on the Earth's mantle and can we still see this imprint today? When did the first crust form? What did it look like, and is it still preserved, either within the continents or in the mantle? What were the conditions which permitted life to form, and how exactly and where did life begin? These questions have been asked for a long time, but in recent years we have begun to obtain some answers.

This book also tracks the changes in the Earth from its primordial state to the time at which "modern" processes were established. Whilst this time is not exactly known there is evidence to suggest that some modern processes were established by the end of the Archaean, 2.5 billion years ago, maybe earlier. The change from the primordial Earth toward the modern Earth over a period of 2 billion years requires significant interactions between all the major reservoirs of the Earth system and understanding these requires insights from the Earth Systems way of looking at the Earth.

1.1.1 Earth system science

1.1.1.1 The new paradigm

Over the past two decades Earth Scientists have begun to realize that a reductionist paradigm provides an incomplete picture of the Earth. Segmenting the Earth Sciences into fields such as "igneous petrology" and "carbonate sedimentology" with firm boundaries demarcating these subdisciplines provides inadequate answers to whole Earth problems. The new insights of the past two decades have been prompted by two disparate areas of research. First, comparative planetology arising out of the NASA space missions, has forced us to think about the Earth in the context of our planetary neighbors. This in turn has contributed to a more "global" view of the Earth. At the same time scientists studying climate and the atmosphere have become increasingly aware that they could not fully explain these systems by studying them in isolation. They found that they needed contributions from other parts of the Earth System to understand the parts that they were considering. In particular the scientific elements of the Gaia hypothesis of James Lovelock have been a stimulus to this process (e.g. Lovelock, 1988). These new approaches have caused Earth Scientists to think beyond the traditional "boxes" of their specialist disciplines and to consider the links that exist between their particular "box" and other parts of the Earth System.

Forty years ago, our understanding of how the Earth works was revolutionized by the theory of Plate Tectonics. This theory, however, only described the workings of the solid Earth, with a particular emphasis on the origin of ocean basins and active mountain belts. In contrast, Earth System Science is about much more than simply the solid Earth. Its scope encompasses the whole of the Earth System – the deep Earth and surficial processes; it includes the oceans, the atmosphere, and the Earth's diverse ecosystems. It also offers a new level of integration between the Earth Sciences, (traditionally the "solid" Earth), Environmental Science and Physical Geography (traditionally surface processes and ecosystems), Oceanography, and Atmospheric Science. Furthermore, Earth System Science has implications which reach far beyond the way in which we do science, extending into the realms of environmental management and policy (Midgley, 2001).

Hence the philosophy of Earth System Science is holistic. It argues that we need to explore the Earth System as an integrated whole, rather than simply as a series of separate entities. As stated above, the reductionist approach has now been shown to be wanting, for it is incapable of tackling large-scale issues such as global change. The new paradigm encourages interdisciplinary thinking and an exploration of the processes which *link* the different "boxes" (see Fig. 1.1), in order first to describe, and then to quantify the exchanges that take place. There is a much greater emphasis than before on defining reservoirs, residence times, fluxes, transport mechanisms, and transport rates (Table 1.1). The Earth System Science perspective therefore, sets a new research agenda for Earth Scientists, Environmental Scientists, Atmospheric Scientists, Oceanographers, and Geomorphologists, for suddenly the interesting science is that which is happening at the *boundaries* of the once traditional disciplines.

1.1.1.2 The Gaia hypothesis

One area of thought which has had a profound influence on the evolution of Earth System Science thinking has been the Gaia hypothesis, popularized by James Lovelock (Lovelock, 1979, 1988). Lovelock, a planetary scientist, recognized that the Earth is unique amongst the terrestrial planets in that it possesses an atmospheric blanket, which, in the words of the classical Goldilocks narrative, was not too hot (as is Venus) and not too cold (as is Mars) but "just right" for life to exist. Lovelock, following the work of planetary scientists in the 1950s and 1960s, argued that the uniqueness of the Earth's atmosphere, as evidenced by its disequilibrium state, was because the planet was the host to life. Thus he turned on the head conventional arguments that state that the Earth is the home to life *because* it had a suitable atmosphere. In other words the atmosphere which we have today was produced by and is controlled by "life."

This idea that there is an interaction between living organisms and the Earth has given rise to the concept of "self-regulation" in the Earth System. In his early work Lovelock likened this to the process of homeostasis – the process whereby a living organism regulates its internal environment in order to maintain a stable condition. This caused some misunderstanding of his ideas, for many believed that he was arguing that the Earth *was* a living organism. The scientific community began to back off at this point and a number of quasi-religious groups weighed in, hijacking the Gaian concept, taking it away from its scientific roots in a completely different direction. Sadly, this diversion of the Gaian concept has harmed its scientific credibility. Nevertheless, the ideas promoted by James Lovelock have had an important impact on the development of Earth System Science. Whether or not they are given the Gaian label is relatively unimportant.

Lovelock's great insight was to recognize that life affects the global environment and profoundly affects surficial processes on the Earth. Thus when the Earth System is perturbed, either from outer space or from the interior of the Earth, the Earth's self-regulatory systems come into play, eventually restoring the system to its original conditions. This makes the Earth Surface System resilient to all but the most extreme perturbations. Lovelock and subsequent workers have illustrated this process with a simple model which they called "Daisyworld."

Daisyworld is a computer model of a simplified Earth in which variation in the Earth System is described by one parameter, surface temperature, which in turn is affected by a single property of living matter – its reflectivity to solar radiation – its albedo (Watson & Lovelock, 1983). There are a number of versions of Daisyworld, for it is an evolving model. Here the version of Kump et al. (1999) is used. Daisyworld is a world in which there are vast tracts of white daisies. They grow on grey soil, which is the only other surface feature of this Earth model. Daisyworld is subject to a sun with an increasing luminosity. It might be thought that an increase in solar luminosity would lead to an increase in surface temperature and the daisies would die. What in fact happens in the model is that, in response to the increased surface temperature, the white daisies increase in abundance. The greater proportion of white daisies leads to an increased reflectivity (albedo) of the Earth's surface, leading eventually to a lowering of the surface temperature. Thus Daisyworld is an example of a negative feedback loop (Table 1.1) and illustrates the interconnectedness between life and the Earth in a manner which regulates the Earth's surface temperature. It is therefore a simplified model for how the Earth's climate system works. The important general points are that first, there can be a realistic coupling between life and the physical environment and second that the Earth can behave as a self-regulating system.

1.1.2 A systems approach to the modern Earth

The application of systems theory to the Earth recognizes that rather than separating the Earth into its component parts, there is a focus on the whole. There is a recognition that natural systems are open and interact with other systems outside of themselves and that through this interaction they acquire new properties and evolve over time. Earth System Science views the Earth as a synergistic physical system, governed by complex processes involving the solid Earth (crust, mantle, and core), including its land surface, its atmosphere, hydrosphere (oceans, rivers, and lakes), cryosphere (ice caps, sea-ice, and glaciers), and biosphere. The relevant interactions operate over timescales ranging from milliseconds to billions of years and on spatial scales from microns to millions of kilometers. The Earth

Systems approach not only includes solid Earth processes such as the origin and motion of continents and the creation and destruction of ocean basins, but also the origin and evolution of life and the changing pattern of the Earth's climate through time.

The language of systems is briefly summarized in Table 1.1 and these terms are now increasingly used to describe Earth processes. Earth System Science places a strong emphasis on quantifying processes as a prelude to mathematical modeling. For example, models of the Earth's modern climate system are becoming increasingly mathematically sophisticated with the intention of producing three-dimensional global climate models. Similarly, a major goal of modern geochemistry is to quantify elemental fluxes into and out of major Earth reservoirs.

This type of approach is possible for processes which we can quantify now and in the recent past, but as we go further back in time quantification becomes much more difficult. It is clear that there is a trade off between the timescale of the model and its mathematical complexity. For long timescales, of the type considered in this book, Earth system processes can only be satisfactorily defined by zero-dimensional, box models.

1.1.2.1 Identifying reservoirs and fluxes – box models for the Earth

Box models are diagrams in which the different components, or reservoirs in the Earth system are represented as boxes. These are particularly useful for illustrating processes which need to be viewed over a long timescale. Currently there are a number of box models relevant to the Earth's surface systems, the most detailed of which are being developed by climate modelers. One example is the Grid ENabled Integrated Earth system model (GENIE), which seeks to develop a grid-based computing framework for the Earth System on a timescale of tens of thousands of years, to include the atmosphere, ocean, sea-ice, marine sediments, land surface, vegetation, soil, and ice sheets (http://www.genie.ac.uk/).

The importance of timescale in developing box models is illustrated well by Kump et al. (1999) in their discussion of the carbon cycle. They show that the organic and inorganic carbon cycles operating on timescales of less than a century include the atmosphere, biosphere, the oceans, soil, and marine sediments, but on a longer timescale (thousands to millions of years) the additional reservoir of sedimentary rocks also becomes important, since it contains up to three orders of magnitude more carbon than the Earth's surface reservoirs.

Probably the most comprehensive box model for the whole Earth is the GERM model (Geochemical Earth Reference Model, Staudigel et al. (1998), http://www.earthref.org/), Fig. 1.1. The purpose of this model is to provide a geochemical reference model for the Earth, similar to the Preliminary Reference Earth Model (PREM) used in geophysics (see Chapter 3, Section 3.1). In detail the model

- divides the modern Earth into a complete set of geochemical reservoirs, (Fig. 1.1 and Table 1.2); and will
- provide an internally consistent set of data describing their chemical and isotopic composition; and seeks to
- establish values for the chemical fluxes between these reservoirs over relevant timescales (ca. 1000 yr to 1 billion yr).

It is this type of model which is most relevant to the systems of the early Earth. For at this time period the only processes which can be discerned are those operating over long timescales and the Earth reservoirs are only broadly defined.

1.1.2.2 Quantifying fluxes – geochemical cycles

Where there are exchanges between different reservoirs in a geochemical box model of the Earth, and these can be quantified, then these exchanges may be described in terms of geochemical cycles. The basic conditions for identifying an elementary geochemical cycle are that the "essential reservoirs must be identified, their contents estimated and the fluxes between the reservoirs must be evaluated over a sufficient length of time compared to the relaxation time (homogenization time) of the system" (Albarede, 2003). The basic transport equations for modeling geochemical cycles are given by Lasaga and Berner (1998).

TABLE 1.1 A summary of some of the key terms used in Earth System Science, many of which were first used in the field of chemical engineering.

System	An entity made up of different parts, but which are related. Together the different parts function as a whole. Individual parts are called components
Reservoir	A part of a system, defined in terms of the amount of material contained (either as mass units or volume). Usually one of the "boxes" in a box model
Flux	The amount of energy or matter that is transferred into/out of a reservoir in unit time. A more general term for this process is *Mass Transfer*
Feedback loops	A linkage between two or more components of a system so that there is a self-perpetuating mechanism of change. A set of actions produces automatic reactions within the system. Feedback may be positive, amplifying change, or, negative, diminishing the effects of change.
Steady state	The condition when a system is unchanging in time. A reservoir is in a steady state when the inflow and outflow are equally balanced
Perturbation	A temporary disturbance to a system
Forcing	A long-term, persistent influence on a system bringing about a disturbance to the system
Residence time	The average length of time a substance spends within a reservoir in a steady state with respect to the abundance of that substance

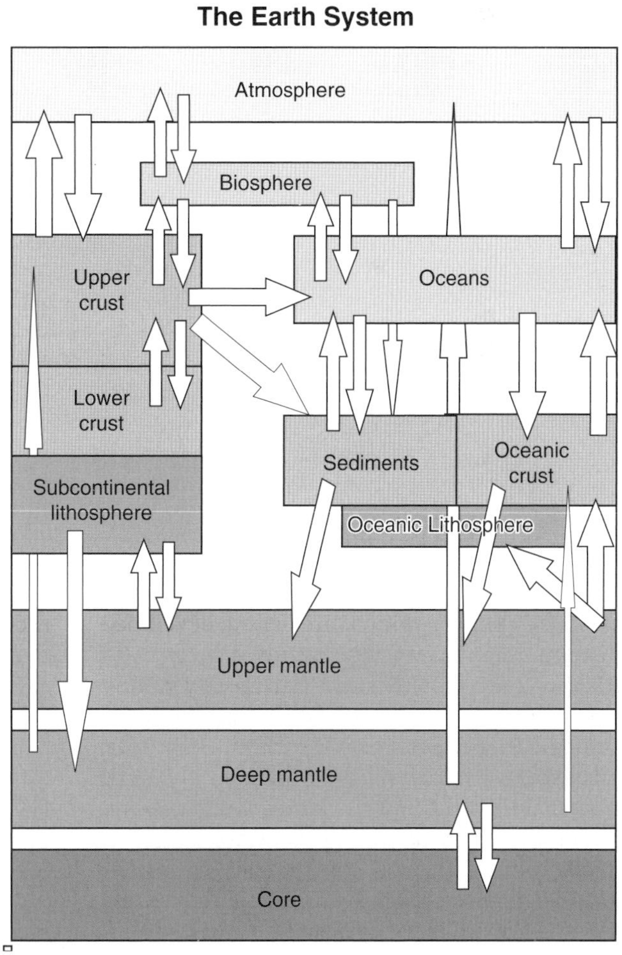

FIGURE 1.1 A box model for the Earth System, showing some of the major reservoirs and the interactions (white arrows) between them.

TABLE 1.2 The principal reservoirs in the GERM Earth model (Source http://www.earthref.org).

Atmosphere	*Stratosphere*				
	Troposphere				
Hydrosphere	*Cryosphere*				
	Lakes				
	Rivers				
	Seawater	Deep water			
		Surface water			
		Hydrothermal vents			
Solid Earth	*Silicate Earth*	Continental crust	Lower crust		
			Middle crust		
			Upper crust		
		Oceanic crust	Mafic crust	Fresh MORB	E-MORB
					N-MORB
				Hydrothermal systems	
				Komatiites	
				Mature oceanic crust	Extrusive rocks
					Dykes
					Gabbro section
					Transition zone
			Marine sediments	Subducted sediment	
			Oceanic plateaux		
			OIB		
		Mantle	Continental mantle		
			Depleted mantle		
			Enriched mantle	EMI	
				EMII	
				FOZO	
				HIMU	
	Core				

This definition of a geochemical cycle demonstrates that recognizing the appropriate scale of the cycle is vital. Particularly important is the timescale. It is clear that on short timescales catastrophic transfers may take place between reservoirs which on a longer timescale would be "smoothed out." As has already been noted, the appropriate choice of timescale is particularly pertinent to a consideration of the carbon cycle (see Kump et al., 1999), for whether or not the sedimentary rock reservoir is included, completely alters the mass balances in the system. The choice of timescale will also vary for the geochemical cycle of different elements in the same set of reservoirs. This is because elements have different residence times in the same reservoir. An unreactive element will have a long residence time, whereas a reactive element will be quickly removed from the reservoir and thus has a short residence time. Defining the length scale of geochemical reservoirs is also an important factor in developing models of geochemical cycles. In some cases linked reservoirs should be merged and at other times separated. This may be linked back to the timescale of a geochemical cycle.

The simplest geochemical cycles are based around the concept of maintaining a steady state to satisfy the constraints of mass balance. More complex cycles incorporate the dynamic response of the system to perturbation (Lasaga & Berner, 1998). In this case, the resultant feedback processes are normally quantified using linear models, but, depending on which geochemical cycle is being considered, more complex, nonlinear responses may be more appropriate.

The key point for an Earth Systems approach to the Earth is that models of geochemical cycles provide the means whereby exchange processes can be quantified. This quantification is element-specific and is a function of the timescale under consideration. It provides the basis for calculating the steady state abundances of particular substances such as oxygen or carbon dioxide in the atmosphere, or sulfate in the ocean. Since these abundances are related to fluxes which relate to biogeochemical and physical processes, we can gain an understanding of the controls on key variables in the Earth system.

An issue of particular interest in the early Earth is the rise of atmospheric oxygen (Chapter 5, Section 5.3). Figure 1.2 shows a box model for the modern steady state oxygen cycle. This model can be used to show how a modern geochemical cycle might be used as the basis for understanding fluxes in the Archaean. For example we can explore how oxygen levels might have been controlled in the early Earth – through say a smaller amount of organic carbon or a greater volume of reducing gases released from the mantle. The model also illustrates the importance of "Earth systems thinking," since it shows how mantle processes can influence the composition of the atmosphere. Other examples of whole-Earth geochemical cycles – for water, carbon and nitrogen – are given in Chapter 5 (Section 5.1). These geochemical cycles are characterized by reservoirs that are large and

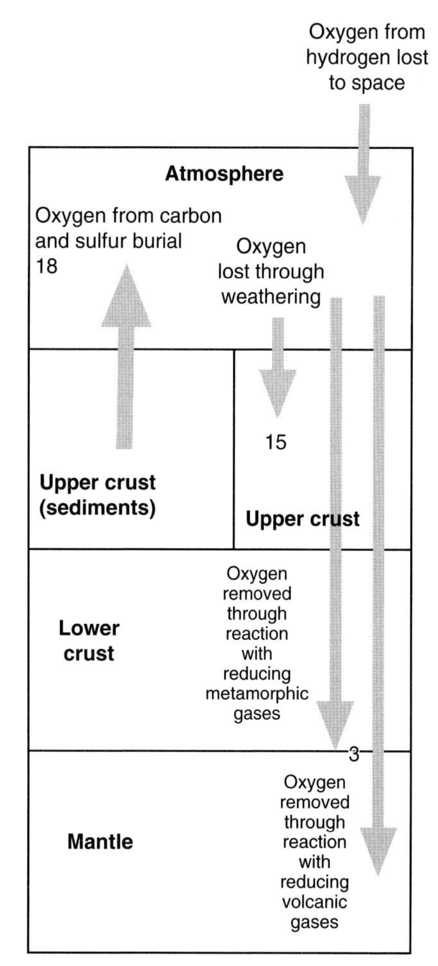

FIGURE 1.2 Box diagram for the modern oxygen cycle (after Catling & Claire, 2005). The direction of the arrows show the contributions to and removal from the atmosphere. The value of the fluxes is in Tmol/yr (10^{12} mol). The principal contributor of oxygen to the atmosphere is the burial of organic carbon and pyrite. Oxygen is lost through continental weathering and through reaction with reducing gases from the deep Earth.

broadly defined, and a timescale of mass transfer which is between 10^6 and 10^9 years.

1.1.3 Early Earth Systems

If the defining character of modern Earth System Science is "the need to study and understand the between-component interactions" (Lawton, 2001), it is now time to enquire how this might be applied to the early Earth. During its first 100 Ma, perhaps even within the first 30 Ma, the Earth differentiated into some of its principal reservoirs – the core, mantle, oceans, and atmosphere (see Chapter 2, Section 2.4, and Chapter 5, Section 5.4). Also at this time a crust formed, initially basaltic in composition (Section 4.5, this volume), followed some time later by a felsic, continental crust (Section 1.4). And then, at some stage, life appeared.

In these very first stages of Earth history it is likely that the interactions between the reservoirs were dynamic and extreme. For example it is possible that there was a strong interaction between the mantle and a primitive atmosphere during the Earth's magma ocean stage whereby the mantle became hydrated (see Chapter 2, Section 2.4.1). These likely interactions demonstrate that in some ways the interconnectedness of the Earth System was *most* apparent, and the Earth Systems approach is most relevant, during the Earth's earliest history, in this case within the first 30–100 Ma.

A central question however, is the extent to which we can access the Earth system at the earliest stage of its history. How much information is there? It is difficult enough grappling with the modern Earth system, to which we have good access. Is it possible to know enough about the early Earth to attempt to describe the early Earth system? It is the claim of this book that such an approach is possible. In the following sections of this chapter we consider the nature of the geological record for the early Earth and examine the data on which much of what is written here rests. Furthermore, modern isotope studies show that the Earth's mantle preserves a remarkable record of its earlier history and that events in the very early development of our planet have left their fingerprint in the modern mantle.

Inevitably we can only understand the early Earth in a qualitative rather than quantitative manner. The best we might hope for is a semi-quantitative understanding and so, in this sense, our understanding of early Earth system will never compete with our understanding of the modern Earth. Nevertheless, even a better qualitative understanding of the large-scale processes and interactions in the early Earth system is vitally important, for many of the basic parameters of the early Earth are, at present, very poorly defined. For example, in the case of the first continental crust, we are not clear on how it formed. In the language of systems we have not even agreed on the inputs and outputs. We are still constructing the box model. The same can be said for most aspects of the early Earth. Compared with our knowledge of modern geochemical cycles, this seems rather elementary – but constructing basic models for the components of the early Earth is a necessary prelude to understanding the earliest Earth system. This book does not therefore provide a sophisticated quantification of geochemical fluxes in the early Earth, but what it does do is set the agenda for such studies in the future. If we can begin to agree on the box models, then in the future we have some hope of beginning to quantify the mass transfers between them. And if we know what the "boxes" are, then we shall know where to look and what to measure. Part of this agenda is to demonstrate that in the Archaean the "solid Earth" has an important contribution to make to our understanding Earth systems processes, contrary to the emphasis of much modern Earth systems science.

Along the way we will discover discrepancies between what observational approaches and theoretical approaches can tell us about the early Earth and also many examples of contested interpretations of the rock record. Controversy of course is the life-blood of science and normally leads to an improved understanding of a topic. However, where controversial interpretations are still unresolved we must tread carefully not to build these data into our models.

This approach requires insights from all branches of the Earth sciences, although the reader will detect a bias toward geochemistry, for this is a particularly powerful tool with which to explore the early Earth and the author's prime area of expertise. An exploration of early Earth Systems is different from the climate modeling of modern Earth System Science with its strong societal implications, but is still of value, for an "Early Earth perspective" raises profound philosophical questions about origins, which we do well to ponder.

1.2 The nature of the early geological record

How can we know about the early history of the Earth? What evidence do we have for this period of Earth history, and how reliable is this evidence? In this section we review the nature of the early geological record, enquire about the types of data that might be extracted from it and assess the quality of the information available. Of course in addition to the geological record, there is also much information to be gained about the early Earth from the planetary record. This is discussed in Chapter 2.

The earliest period of Earth history is subdivided into the *Hadean* and the *Archaean*. Here, the Hadean (literally – the hidden period of Earth history, or for some, the "hellish period"), is taken to be the period of time between the formation of the Earth at about 4.6 billion years and the beginning of the Archaean. However, the beginning of the Archaean (literally – the beginning of the rock record), has not been defined by the International stratigraphic commission (Gradstein et al., 2005). Here, following common usage, it will be taken as 3.8 Ga, and the Archaean as the period of time between 3.8 and 2.5 billion years ago (Fig. 1.3). In this text billions of years are abbreviated to giga-years (Ga – 10^9 years ago) and millions of years to mega-years (Ma – 10^6 years ago).

There are few terrestrial materials older than about 3.8 Ga and so our knowledge of the Hadean, the first 750 Ma of Earth history, depends upon a small number of terrestrial samples, supplemented with inferences from the study of meteorites and planetary materials and deductions based upon the distribution of some radiogenic isotopes in the Earth's mantle. In contrast Archaean rocks abound and regions with rocks dated between 2.5 and 3.8 Ga are found in most major continents. A map showing their distribution is given as Fig. 1.4.

Regions where Archaean rocks are extensively preserved are known as *cratons*. This term

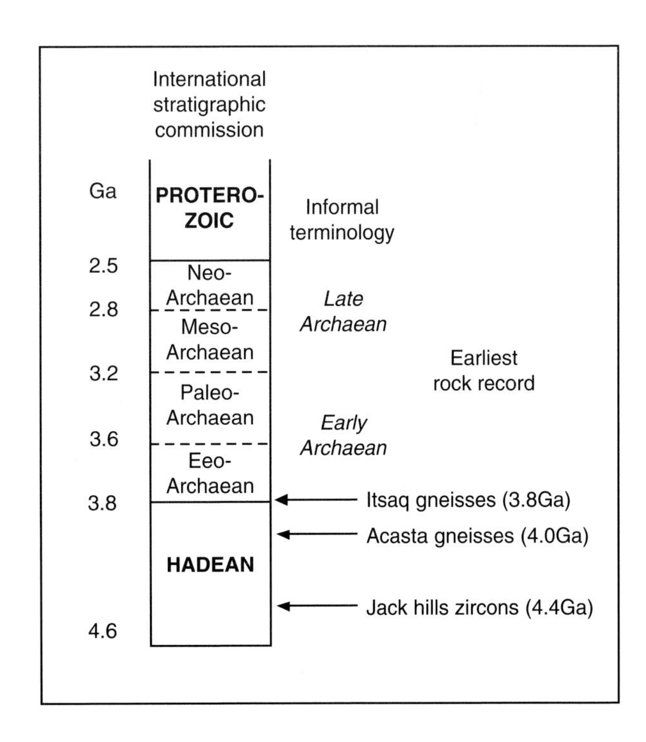

FIGURE 1.3 A geological timescale for the earliest part of Earth history (partially after Gradstein et al., 2005).

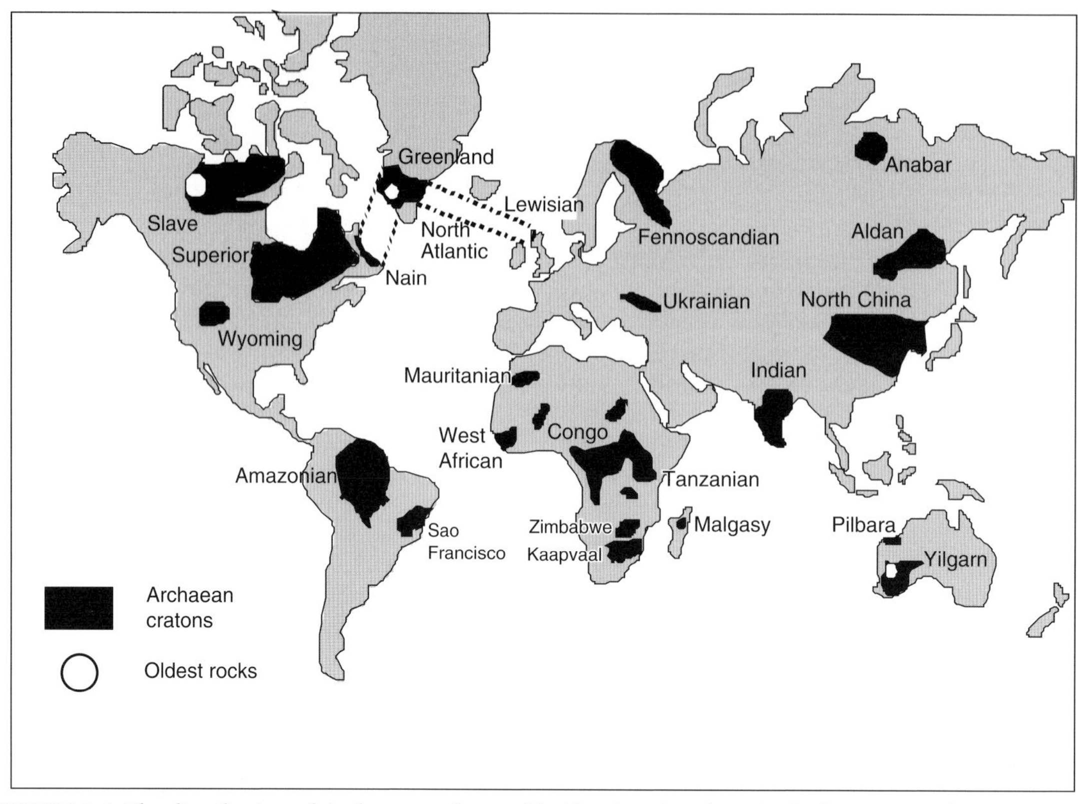

FIGURE 1.4 The distribution of Archaean rocks worldwide, showing the principal cratons and regions where the oldest rocks are preserved. Not all these areas contain exposed Archaean rocks, for some cratons are now partially covered by younger sediment, or have been reworked during later orogenic events. Not shown is the Archaean Enderby Land Craton in Antarctica. The dashed lines indicate the rocks of the North Atlantic Craton now separated by the creation of the Atlantic Ocean.

implies an area of continental crust, normally made up of crystalline rocks, which has been stable for a very long period of time. The term is synonymous with the older term *shield*. The names of some of the major cratons are given in Fig. 1.4 and a typical craton, the Zimbabwe Craton is illustrated in Fig. 1.5.

1.2.1 Types of Archaean crust

Traditionally Archaean crust has been subdivided into two types – granite–greenstone belt terrains and high-grade gneiss terrains. This subdivision reflects a major contrast in both lithological association and metamorphic grade. Typical granite–greenstone belt terrains would be the Zimbabwe or Superior Cratons and a typical high-grade gneiss terrain would be the North Atlantic Craton. In this book a slightly different approach is taken to that previously used in the literature, and Archaean crust will be classified as one of three types of lithological association – Archaean greenstone belts, late Archaean sedimentary basins, and Archaean granite–gneiss terrains.

Of these, late Archaean sedimentary basins are the most "modern" in their form and can be clearly recognized as upper crust. Similarly, the relatively low metamorphic grade of many Archaean greenstone belts indicates that they too represent upper Archaean crust. In contrast, Archaean granite–gneiss terrains more commonly belong to the deeper continental crust and variously represent middle to lower continental crust. There are only a few places

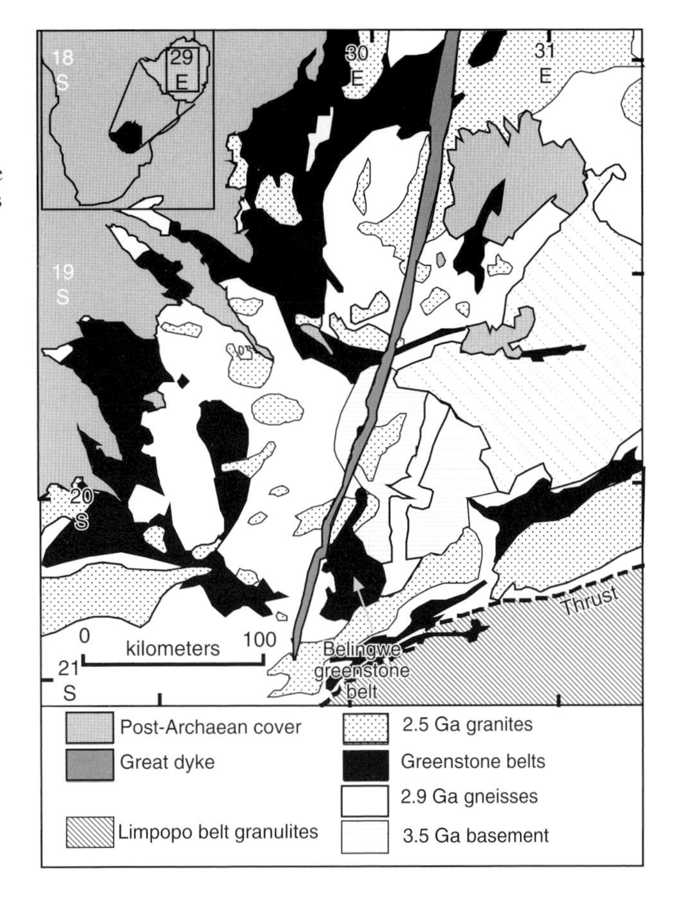

FIGURE 1.5 The central part of the Archaean Zimbabwe Craton, showing the principal geological components: granitoid gneisses of various ages, greenstone belts, younger granite intrusions (2.5 Ga granites) and the late Archaean "Great Dyke." To the south the Archaean Limpopo Belt granulites are thrust onto the Zimbabwe Craton. In the northwest the craton is partially covered by younger sediments. The area of the map within Zimbabwe is identified on the inset map.

in the world where these relationships are made clear, one of which is the Kapuskasing structural zone in the southern part of the Superior province in Canada. This region shows a west to east transition from a shallow level of erosion (<10 km) to a deep level of erosion (20–30 km), over a distance of about 150 km, exposing a cross section through Archaean continental crust. With increasing depth of erosion the lithologies change from the metavolcanic rocks of the Michipicoten greenstone belt, through the midcrustal Wawa granitoid gneisses, into the granulite facies gneisses of the Kapuskasing zone, showing a transition from Archaean upper to lower crust (Percival & West, 1994).

Later in this chapter, each of the major Archaean lithological associations – greenstone belts, late Archaean sedimentary basins, and granite–gneiss terrains – is examined in detail, in order to clarify which type of information each preserves about the early Earth. However, before we turn to this subject it is worth reflecting in more detail on the precise nature of the Archaean geological record.

1.2.2 The Archaean geological record

The Archaean geological record presents particular problems and poses particular challenges. Archaean rocks have commonly experienced long deformation histories. Frequently they are either metamorphosed and/or altered, and recovering their original character requires patient fieldwork, sometimes in very remote locations, and thoughtful geochemistry. When we come to the earliest history of the Earth we find that our knowledge base is very thin and the only information is preserved in single mineral grains – zircons (zirconium silicate grains, $ZrSiO_4$) of Hadean age, protected in younger Archaean sediments (see Section 1.4.3).

A very specific challenge in the Archaean geological record are those rocks which are without modern counterparts. One example is the ultramafic rock komatiite, first discovered as a lava in the Barberton area of South Africa. At the time of their discovery, in the early 1960s, ultramafic lavas were thought to be an impossibility. Over the past 30 years we have learned a great deal about komatiites and they are discussed in detail in Chapter 3 (Section 3.2.1.2). Nevertheless, why they are more common in the early history of the Earth than at the present is an important scientific question.

Similarly, banded iron formation (BIF), a sedimentary rock produced by chemical precipitation, is extremely rare in the Phanerozoic but common in the Archaean and Proterozoic record. Explaining its origin in terms of the atmospheric or ocean chemistry of the early Earth is an important part of recovering the history of early Earth. This is discussed in Chapter 5 (Section 5.4.3.2).

1.2.3 Advances in geochronology

A repeated theme, which will be recognized in the later parts of this chapter, is the importance of U–Pb zircon geochronology in unraveling the early history of the Earth. It is impossible to overemphasize the importance of the recent advances in this field for the study of the early Earth.

Zircon is an important mineral in geochronology for it contains a sufficient quantity of the trace element uranium to provide a workable chronometer and it is a robust "time capsule," which can survive even under extreme conditions. In the past conventional zircon dating methods relied on studying groups of zircon grains, but the method was unable to take account of the complexities that may exist within a single zircon grain. A major breakthrough came with the development, in the early 1980s of a single grain method, using the Sensitive High Resolution Ion Microprobe (SHRIMP). The SHRIMP instrument, developed first at the Australian National University, allowed the measurement of very precise U–Pb ages from within individual zircon crystals (Compston et al., 1982). In fact individual zircon grains, or areas of grains, as small as 20–30 μm across can now be dated with a precision of 1–2 Ma, in rocks as old as 3–4 Ga.

This method is now widely used and is hugely important in stratigraphy of all geological ages. The detailed examination of individual zircon grains now permits the recognition of multiple events in metamorphic rocks; the recognition of different populations of detrital zircons in Archaean clastic sediments, and of inherited, xenocrystic zircons or zircon cores in Archaean granitoids. In fact it is single zircon grains, surviving through several cycles of erosion, sedimentation, and metamorphism, which are the most ancient terrestrial materials so far discovered (see Section 1.4.3).

1.3 Archaean lithological associations

Archaean Cratons contain three different types of lithological association, each of which provides important information about the early stages of the Earth System. These associations are Archaean greenstone belts, late Archaean sedimentary basins, and Archaean granite–gneiss terrains. We will consider each in turn.

1.3.1 Archaean greenstone belts

Archaean greenstone belts are one of the most important primary sources of information about surface processes in the early Earth. Greenstone belts are sequences of volcanic and sedimentary rocks, in varying states of deformation, which occur as sublinear belts, tens to hundreds of kilometers long in Archaean Cratons (Fig. 1.5). They tend to form a striking physiographic and geological contrast to the granitoids plutons and gneisses, which most commonly surround them. Their name comes from the color of the lightly metamorphosed

TEXT BOX 1.1 Radioactive dating methods applicable to the early Earth

1. Isochron calculations

An isochron diagram is an bivariate plot of the ratio of a radioactive parent isotope to a reference stable isotope, plotted on the *x*-axis of the diagram, to the ratio of the radioactive daughter isotope relative to the reference isotope plotted on the *y*-axis, for a suite of cogenetic samples. For example, in the Sm–Nd (samarium–neodymium) isotopic system ($^{147}Sm \rightarrow {}^{143}Nd$), an isochron diagram plots the $^{147}Sm/^{144}Nd$ isotope ratio on the *x*-axis and the $^{143}Nd/^{144}Nd$ isotope ratio on the *y*-axis. In a geochemical system that has been closed to outside interferences, a suite of samples which formed at the same time, but which have slightly different Sm/Nd ratios, will plot on a straight line, the slope of which is proportional to the age of the samples (see Box 1.1 Fig. 1). The age of the sample suite is calculated from the equation

$$t = 1/\lambda \ln(\text{slope} + 1)$$

where t is the age of the rock in years and λ is the decay constant (the decay constant for the Sm–Nd system is $6.54 \times 10^{-11}\ yr^{-1}$). The calculated age has an error, based on the statistical fit of a straight line to the sample points and the statistical errors in the isotopic measurements. The isochron age of an igneous rock therefore, is the time at which the rock crystallized and the isotopic system became isolated from that of neighboring rocks.

Isochron age calculations are commonly made for the Rb–Sr (rubidium–strontium), Sm–Nd (samarium–neodymium), and U–Pb (uranium–lead) radioactive systems. They are most commonly applied to whole-rock systems, that is, a suite of samples thought to have formed at the same time, such as an igneous pluton or a suite of lavas. Isochron age calculations may also be made for a suite of minerals in a rock, in which case they date the time at which the minerals "lost isotopic contact" with each other, that is, became closed systems. This approach can be useful in dating metamorphism.

Isochron age calculations are as good as the assumptions behind them – namely that the samples all formed at the same time and that they have been geochemically undisturbed since that time. The best isochron ages will have errors of a few tens of millions of years on an age of 2000–3000 million years.

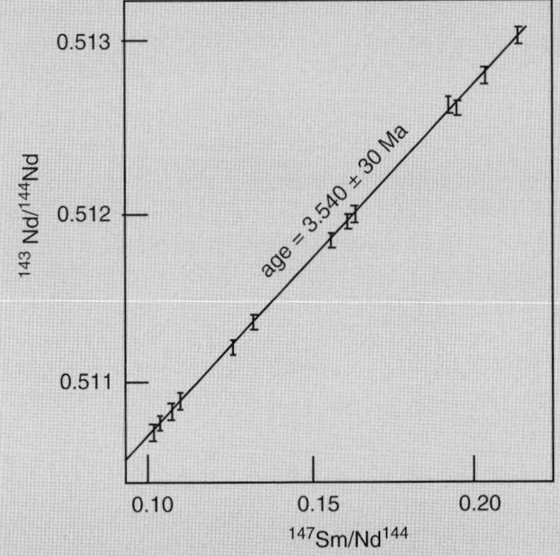

BOX 1.1 FIGURE 1 Isochron diagram for the Sm–Nd system showing a suite of 13 samples which formed at the same time, but which have slightly different Sm/Nd ratios. The calculated age for the 13 samples is 3,540 ± 30 Ma.

2. Model age calculations

Model age calculations are based, as their name implies, on a particular model of mantle isotopic evolution. They are a measure of the time when a particular sample became separated from its mantle source. They are most commonly used for the Nd-isotopic system, but increasingly are also being used in the Hf- (hafnium) isotope system.

A model Nd age is based upon the present-day measured $^{143}Nd/^{144}Nd$ and $^{147}Sm/^{144}Nd$ isotope ratios in a rock sample. The present-day $^{143}Nd/^{144}Nd$ isotope ratio is extrapolated back in time, using the measured $^{147}Sm/^{144}Nd$ isotope ratio, to the time when it was the same as that of the mantle. This is effectively "undoing" the radiogenic ingrowth in that sample of ^{143}Nd over time. One of the most frequently used mantle models is that of the depleted mantle (DM).

Graphically the process is illustrated in Box 1.1 Fig. 2. Normally, however the result is calculated from the measured isotopic ratios in the sample compared to reference ratios for the depleted mantle.

Model Nd-ages are as good as the assumptions they are based on, that is, the mantle model used and the assumption that the sample came from a depleted mantle source. They work quite well for igneous rocks. For sediments they can measure the age of the sedimentary source.

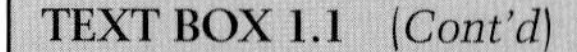

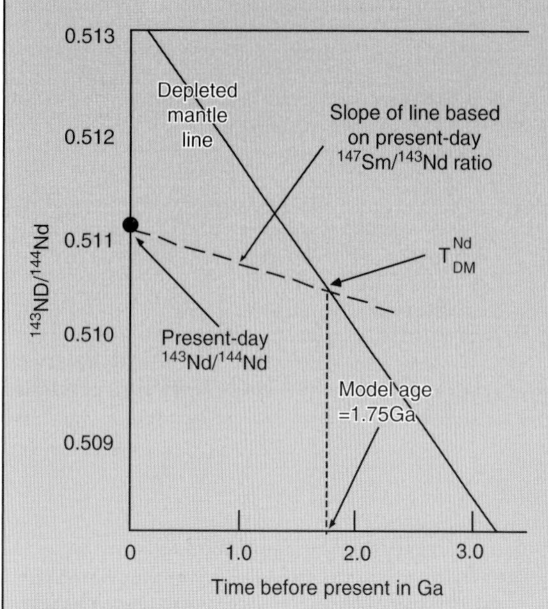

BOX 1.1 FIGURE 2 The evolution of the depleted mantle with respect to $^{143}Nd/^{144}Nd$ over geological time. The model depleted mantle neodymium age (T^{Nd}_{DM}) is the time at which the $^{143}Nd/^{144}Nd$ ratio of the sample is the same at that in the depleted mantle.

3. U–Pb (uranium–lead) zircon ages

There are a number of ways of obtaining U–Pb ages on zircons, but in the last two decades the use of the ion microprobe has become the method of choice. Zircon (zirconium silicate, $ZrSiO_4$) is a common mineral in granitic rocks, often abundant as a detrital mineral in clastic sediments and a rare accessory mineral in mafic rocks and some metamorphic rocks. The frequent occurrence of the mineral zircon, and its resilience to later thermal events that would reset other isotopic systems are the reasons for its frequent use in studies of the early Earth.

Zircons are analyzed in an ion probe using the technique of secondary ion mass spectrometry in which a small area (20 μm in diameter) of a zircon grain is bombarded by high-energy ions and this leads to the ejection, or sputtering, of both neutral and charged species from the zircon grain. Of interest are the charged ions of U and Pb and these are filtered and analyzed according to their mass. Of particular interest are ^{206}Pb, the decay product of ^{238}U and ^{207}Pb, the decay product of ^{235}U.

The results are plotted on what is known as a Concordia diagram (see Box 1.1 Fig. 3), a graph of $^{206}Pb/^{238}U$ and $^{207}Pb/^{235}U$ ratios, showing the locus of agreement of U–Pb ages obtained in the two U-Pb decay systems (^{238}U and ^{235}U). Analyses which plot on the Concordia curve indicate the time of crystallization of the grain. Analyses which plot off the Concordia are interpreted in terms of late Pb-loss from the grain, most normally as the result of a subsequent thermal event which affected the rock.

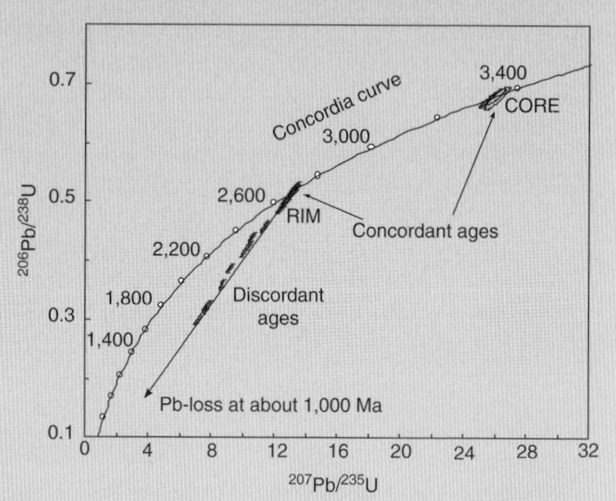

BOX 1.1 FIGURE 3 U–Pb Concordia curve showing the compositions of individual analyses as error ellipses. The data, which are for a single grain, show that the grain has a core which is about 3,350 Ma old and a rim which is about 2,700 Ma old. The rim has experienced subsequent lead loss, giving rise to a series of discordant points on the Concordia diagram. This lead loss reflects a later thermal event at about 1,000 Ma.

U–Pb ages on zircons yield a wealth of data. Granitic igneous rocks can be dated for their time of crystallization, metamorphic rocks for the time of zircon growth, and sediments provide information about their igneous sources. In addition, the crystallization history of individual grains in igneous rocks can yield information about source materials or mixing processes involving older rocks, as illustrated in Box 1.1 Fig. 3. In some cases the precision is excellent, yielding error limits of a few mega-years on zircons 2000 to 3000 Ma old.

basalts which are inevitably present and which were described by an earlier generation of field geologists as "greenstones." Greenstone belts are sometimes known as "supracrustal belts," or as "schist belts," because of their high state of deformation. They were originally important as a source of metals, in particular gold and nickel, but now they are equally valued as the source of scientific knowledge about the early Earth. A map of a typical greenstone belt, the Barberton Belt is given in Fig. 1.6 and its stratigraphy is summarized in Fig. 1.7.

The most comprehensive recent review of greenstone belt geology is by De Wit and Ashwal (1997) who discuss all the major aspects of greenstone belt geology and provide descriptions of greenstone belts from all the major cratons. More than 260 greenstone belts have been recognized world wide (Table 1 of De Wit & Ashwal, 1995), although the details of many are poorly known. Some, however have been studied intensively and have become important sources of information about the early Earth. Some examples of these are listed in Table 1.3.

The greenstone belts of interest here formed between 3.8 and 2.5 Ga, although they are not unique to the Archaean and many Proterozoic and Phanerozoic examples are also known. Within individual cratons greenstone belt ages may vary by as much as 1 billion years (e.g. the Kaapvaal Craton), but normally the time span is much shorter, of the order of only a few hundred million years. Most

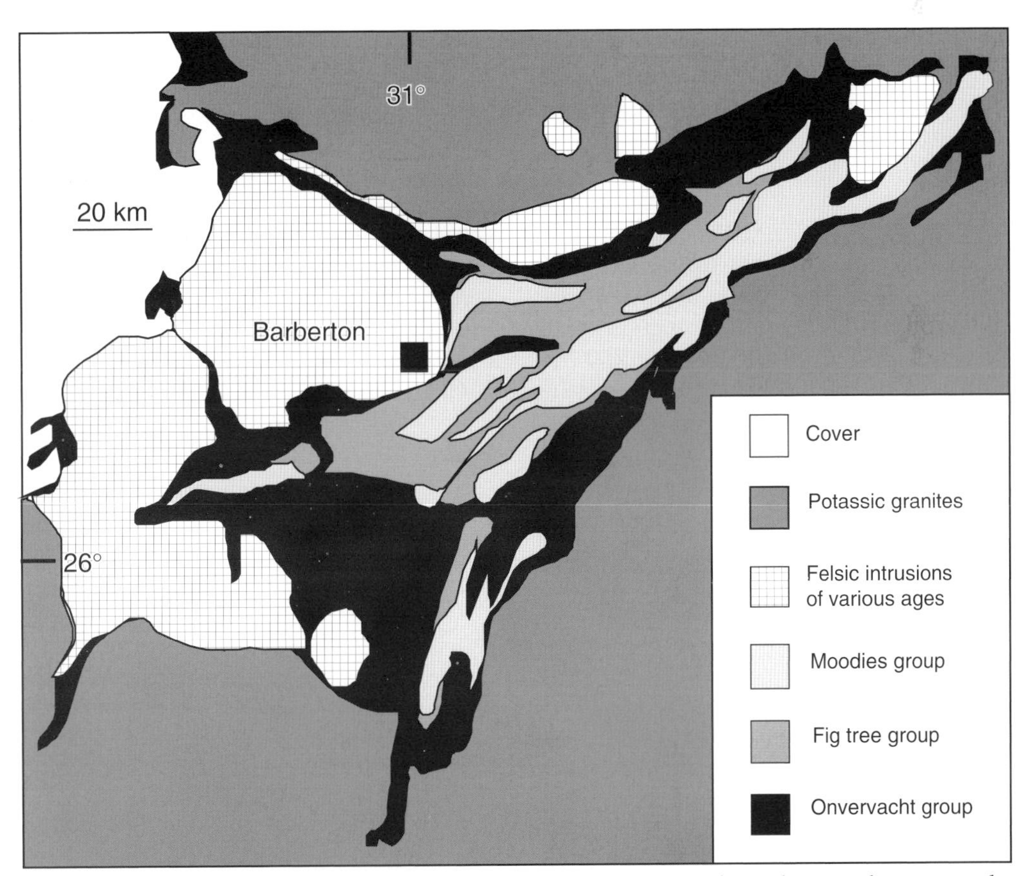

FIGURE 1.6 A geological map of the Barberton greenstone belt in South Africa, showing the main geological components (after De Ronde & De Wit, 1994).

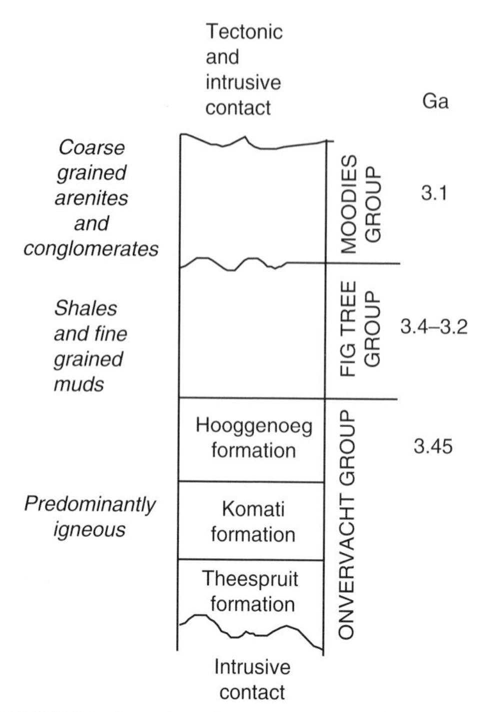

FIGURE 1.7 Stratigraphic succession of the Barberton greenstone belt, South Africa (after De Ronde & De Wit, 1994).

greenstone belts have short histories, less than 100 Ma, many less than 50 Ma, but a few preserve a long history. The Barberton greenstone belt in South Africa, preserves over 300 Ma of Earth history (Figs. 1.6 and 1.7), raising questions about whether there are major time breaks in this succession and if so, how the different parts of the succession might relate to each other.

1.3.1.1 The environment of formation of greenstone belts

Originally all Archaean greenstone belts were thought to have similar stratigraphic successions and represent a similar environment of formation. Now it is known that this is not the case and a large number of different types of greenstone belt have been recognized with differing stratigraphic successions, representing many different environments of formation (see Table 1.4).

However, arriving at this understanding has not been straightforward, for establishing the stratigraphic succession of a greenstone belt is fraught with difficulty. Often critical contacts are obscured and the deformed state of most greenstone belts means that many geological relationships are ambiguous. Seismic and gravity studies of greenstone belts suggest that in many cases true stratigraphic thicknesses have been overestimated by a factor of 2 or 3 (De Wit & Ashwal, 1995). Many fierce debates have taken place over precise stratigraphic relationships in greenstone belts, see for example Myers (2001) and Fedo et al. (2001) on the structure and stratigraphy of the Isua greenstone belt in west Greenland, and the debate between Blenkinsop et al. (1993) and Kusky and Kidd (1992) over whether or not

TABLE 1.3 Examples of greenstone belts used in this text as important sources of information about the early Earth.

Greenstone belt	Country	Craton	Age (Ga)	Features
Abitibi	Canada	Superior	2.7	Very large/diverse stratigraphies/exploration driven by mining
Belingwe	Zimbabwe	Zimbabwe	2.7	Unconformable on the craton/very well preserved komatiites/study driven by academic interests
Barberton	South Africa	Kaapvaal	3.45	One of the first to be studied in detail/ study driven by academic and mining interests
Kalgoorli	Australia	Yilgarn	2.7	Abundant komatiites/exploration driven by nickel mining interests
Isua	Greenland	North Atlantic	3.8	Oldest greenstone belt/may preserve evidence for ancient life/initial interest was driven by mining/more recently by academic interest

TABLE 1.4 Key lithological associations recognized in Archaean greenstone belts (modified from Eriksson et al., 1994).

Tectonic setting	Principal lithologies	Depositional system	Basin configuration	Tectonic elements
Volcanic basin	Mafic/ultramafic volcanic rocks	Individual volcanoes; shallow water, chemical-biogenic, evaporitic;	Not preserved	May be thrust bound at site of emplacement
Arc – Forearc	Conglomerate–sandstone–mudrock/turbidite/ironstone/chert/tuff	Fan delta and submarine fan or ramp	Elongate between volcano plutonic and metasedimentary/ophiolitic elements	Bounded by thrust complex on one side; may be deformed into a lithotectonic complex
Arc – Interarc	Calc–alkaline volcanic/plutonic/mafic volcanic; pyroclastic	Individual volcanoes; subaerial and subaqueous fans; chemical-biogenic, evaporitic, and pelagic	Outer margins pass to forearc and backarc elements	Local structures around tectonic elements
Cratonic extensional	Sedimentary and bimodal mafic/felsic volcanics	Fluvial and fan delta or submarine fan; shallow to deep marine shelf; subaerial and subaqueous volcanic centers	Elongate between cratonic margins or active and remnant arcs	Fixed position listric or back-stepping normal faults; intrabasinal transfer faults
Stable Shelf	Chemical-plagic +/− siliclastic	Chemical and siliciclastic platform or fluviodeltaic; tidal and delta channels	Regionally or interregionally tabular	Not seen; thermal relaxation and eustatic change
Compressional foreland	Conglomerate–sandstone–mudrock/turbidite/ironstone/volcanics	Alluvial fan, fluviodeltaic; shallow marine; submarine fan or submarine ramp	Elongate, wedge shaped against, between or on thrust sheets	Extra- and intra basinal thrust faults; may be divided into separate compartments
Collisional Graben	Breccia–conglomerate-sandstone–mudrock/mafic–felsic volcanic	Alluvial fan, fluvial, fan delta; shallow to deep marine; lacustrine; in volcanic basins subaerial to subaqueous fan	Elongate, wedge shaped between bounding faults; enclosed or open to marine basin	Longitudinal, strike-slip or oblique slip faults and various intra basinal faults

there is a major thrust in the Belingwe greenstone belt succession. Discussions of this type may be entertaining (at a distance) but are not trivial, for these competing observations have major implications for a correct understanding of Archaean tectonic processes.

Detailed zircon chronologies from some individual greenstone belts now show that they have very long histories, with significant chronological gaps. A good example is the Barberton greenstone belt (see Fig. 1.7). When these age differences are associated with differing lithological assemblages it raises the possibility that individual greenstone belts may be composed of accreted fragments which have formed in diverse tectonic settings. A related issue, the subject of ongoing debate, is the extent to which greenstone belts can be correlated across a given craton. One school of thought sees greenstone belt terrains as accreted fragments of crust of diverse provenance and age, whereas there are others who believe that greenstone belts represent fragments of a former, stratigraphically contiguous depositional basin, now disrupted through deformation. This can be seen in the contrasting interpretations of the greenstone belts of the Zimbabwe Craton, where Wilson et al. (1995) have proposed a 700 km, craton-wide correlation of greenstone belts, whilst Dirks et al. (2002) argued that individual greenstone belts are made up of unrelated thrust slices and that craton-wide correlations are impossible.

1.3.1.2 Greenstone belt sedimentary sequences

Eriksson et al. (1994) showed that the original tectonic setting of Archaean greenstone belts can be recovered by the application of sequence stratigraphy. The underlying assumption of this approach is that Archaean sediments can be treated in exactly the same manner as modern sediments and so can be used as indicators of former tectonic, climatic, and depositional settings. In contrast to previous attempts to correlate stratigraphic successions in greenstone belts using lithological contrasts, sequence stratigraphy divides sedimentary successions into genetic packages bounded by unconformities rather than lithological boundaries. In the absence of fossils in the Archaean this process is assisted by precise U–Pb zircon geochronology.

Eriksson et al. (1994) defined seven dominant lithological associations, which they used as a basis for defining the different tectonic environments of greenstone belt formation in the Archaean (Table 1.4). What is clear from their analysis is that tectonic environments between 3.8 and 2.5 Ga were remarkably similar to those in the modern Earth. This study also emphasizes the fact that greenstone belts do not represent a single geotectonic setting, but instead reflect a wide range of tectonic environments. This implies that individual greenstone belts may contain sedimentary and volcanic rocks formed in multiple tectonic settings. Thus, the Barberton Greenstone Belt evolved from a volcanic basin to a volcanic arc, to a compressional foreland basin over a period of 300 Ma. At the scale of a whole craton, greenstone belts in the Pilbara Craton preserve a huge diversity of tectonic environments over 725 Ma of geological history (Eriksson et al., 1994), although more recently Smithies et al. (2005) have suggest that for 300 Ma plume activity dominated this craton.

1.3.1.3 Igneous rocks found in greenstone belts

The most common igneous rocks in greenstone belts are basalts. Frequently they occur as pillow lavas, lightly metamorphosed to chloritic schists. Also present are komatiites (ultramafic lavas), showing a wide range of unusual volcanic textures. Boninites, andesites and rhyolites are also present, but are generally volumetrically subordinate to basalts (De Wit & Ashwal, 1995).

The environment of formation of basaltic rocks is often determined from their trace element chemistry. There have been a large number of chemical studies of the chemistry of greenstone belt basalts claiming a variety of geotectonic settings for their origins. These studies may be correct, but there is a fundamental problem with Archaean tholeiitic basalt chemistry: there is no exact modern analogue of these basalts, so precise comparisons are difficult (Arndt et al., 1997; see Chapter 3, Section 3.2.1.1) not least because the mantle has evolved chemically through time. There are claims that some Archaean basalts represent former ocean floor, or intracratonic, within-plate basalts, or are ophiolitic remnants (Arndt et al., 1997). In addition, some Archaean basalts and komatiites may be the product of plume magmatism which formed oceanic plateaux (Puchtel et al., 1997). Andesite-rhyolite and boninitic lavas are most plausibly the product of arc magmatism. All these claims are possible, and taken together with the nature of the interbedded sediments, the chemistry of volcanic rocks can be used to indicate probable environments of deposition.

1.3.1.4 What can greenstone belts tell us about the early Earth system?

Despite their deformation state and metamorphism some Archaean greenstone belts contain regions of low strain and low grade metamorphism which provide critical information about the early Earth. One of the most intensively studied of these is the Belingwe greenstone belt in Zimbabwe (Fig. 1.5).

Most of what we know about the Earth's early mantle comes from the study of greenstone belt basalts and ultramafic lavas (Chapter 3, Section 3.2.1). These rocks formed as direct mantle melts and provide our best "window" into the Archaean mantle. In the Belingwe greenstone belt there are remarkably fresh komatiites (ultramafic melts) from 2.7 Ga (Nisbet et al., 1987). These rocks preserve many details of their original volcanic features, including very fresh olivine, with inclusions of silicate glass. The geochemistry and textures of this particular suite of komatiites have been used to provide much tighter constraints on Archaean mantle characteristics than were previously available from more altered samples elsewhere in the world.

Greenstone belt sediments also provide important clues to the nature of Earth surface processes. In addition the greenstone belt sedimentary record contains a wealth of information about Archaean weathering processes in the form of erosion surfaces and paleosols. Chemical sediments, such as evaporites, limestones, and cherts provide information about ocean chemistry (Chapter 5, Section 5.4). The significance of the very common banded ironstones is more enigmatic. They may reflect hydrothermal input into the oceans. In this context there has been much debate as to whether the oldest known greenstone belt, the 3.8 Ga Isua Greenstone Belt in West Greenland preserves genuine sedimentary rocks or whether they are actually tectonized igneous rocks (Nutman et al., 1984). There is now good evidence that at least some of the disputed "sediments" at Isua are sedimentary and water-lain in origin (Fedo et al., 2001). This is important for it is possible that, in the absence of an effective climatic greenhouse, the reduced solar luminosity early in Earth history would have given rise to a frozen Earth (Sagan & Mullen, 1972).

More controversial is whether there is evidence for former life in greenstone sediments. At present the much publicized carbon isotope evidence for life in the 3.8 Ga Isua succession (Rosing, 1999) is under severe scrutiny. This is discussed in some detail in Chapter 6 (Section 6.3.2.1). In contrast stromatolite fossils have been documented in greenstone belt sediments from about 3.5 Ga onwards, and sulfur isotope evidence from the Belingwe Greenstone Belt demonstrates that there were complex algal communities by 2.7 Ga (Grassineau et al., 2001 – see Chapter 6, Section 6.3).

The diversity of lithological "packages" identified in Table 1.4 after Eriksson et al. (1994) demonstrates that greenstone belts preserve rocks from a wide variety of tectonic settings. It is clear therefore that greenstone belts preserve important information about tectonic processes in the Archaean. The powerful combination of precise zircon geochronology, igneous geochemistry, and structural mapping in greenstone belts such as Abitibi in Canada has demonstrated that "modern-style" tectonic processes such as arc accretion and the assembly of oceanic plateau fragments were taking place at 2.7 Ga (Ludden & Hubert, 1986; Desrochers et al., 1993).

1.3.2 Late Archaean sedimentary basins

There are only a few Archaean cratons which preserve late Archaean sedimentary basins, which are unconformable on the craton. These sedimentary basins are very different from greenstone belts, for they are not strongly deformed, nor do they have the elongate shape and the abundance of basalts frequently associated with greenstone belts. Basins of this type have previously been described as "late Archaean cover successions" (Nisbet, 1987). In the past, their relatively undeformed state has misled workers into thinking that they were Proterozoic, rather than Archaean in age, although now their Archaean age is well established. The unconformable relationship between late Archaean sedimentary basins and the greenstone belts and gneisses of the underlying craton indicates that basins of this type formed either after the craton stabilized, or during the later stages of craton formation. This unconformable relationship is also the reason for their importance, for they provide unique insights into the processes operating within and on stable continental crust during the late Archaean.

The relatively small number of late Archaean basins may reflect a low preservation potential, but may also indicate that relatively few cratonic nuclei had stabilized before the end of the Archaean. Examples of late Archaean basins include the (3.1–2.7 Ga) Witwatersrand basin in South Africa (Robb & Meyer, 1995), and its correlative the 3.0–2.67 Ga Pongola Supergroup in Swaziland (Walraven & Pape, 1994; Gold & von Veh, 1995), both unconformable upon the Kaapvaal Craton. In western Australia the 2.77–2.71 Ga Fortescue Group is unconformable upon older Archaean rocks of the Pilbara Craton (Blake et al., 2004).

1.3.2.1 The Witwatersrand Basin

The best known late Archaean sedimentary basin is the Witwatersrand Basin, for it hosts the world's richest gold province and is also an important source of uranium. For these reasons the geology of this basin is known in considerable detail. The Witwatersrand basin contains a 7 km-thick succession of terrigenous sediments and volcanic rocks, formed between 3.074 and 2.714 Ga ago.

There are three principal components to the stratigraphy (Robb & Meyer, 1995) – the 3.074 Ga Dominion Group, a 2 km-thick sequence of lavas and volcaniclastic rocks which rest unconformably on the gneisses and greenstone belts of the Kaapvaal Craton. This is overlain by the clastic sediments of the 2.9 Ga West Rand and the 2.9–2.76 Ga Central Rand Groups, which together make up the Witwatersrand Supergroup. Unconformably overlying the basin is another 5 km of lavas and sediments of the 2.7 Ga Ventersdorp Supergroup (see Fig. 1.8).

Much of the gold in the Witwatersrand basin is located in conglomerates within the Central Rand Group. These are part of a succession of predominantly coarse clastic

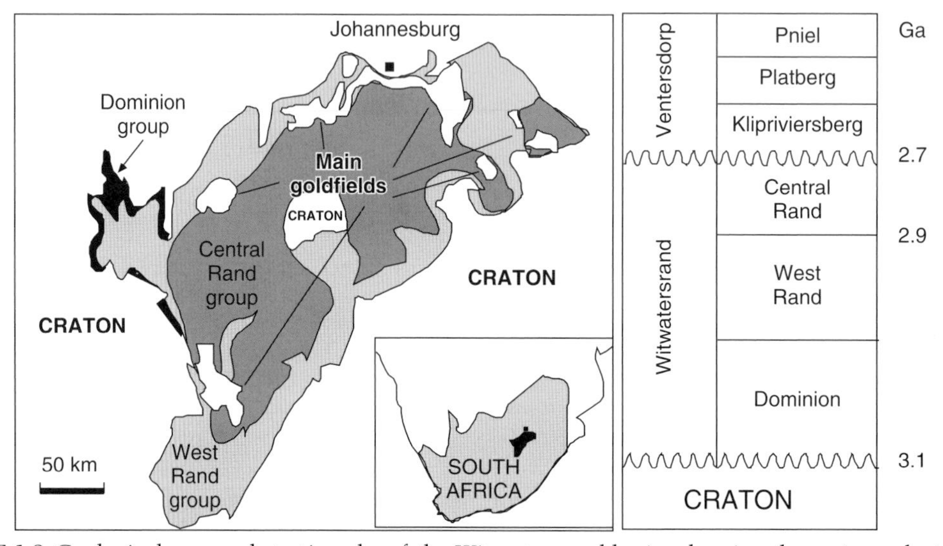

FIGURE 1.8 Geological map and stratigraphy of the Witwatersrand basin, showing the main geological units and the location of the main goldfields.

sediments, which represent a braided fan/delta system (Buck & Minter, 1985) and the gold formed as paleoplacer deposits within this environment.

Although the Witwatersrand basin was in existence for at least 360 Ma, sedimentation within this basin was episodic and there were long periods when there was no deposition. Robb and Meyer (1995) used detrital zircons to identify the provenance of the sediments and showed that they were deposited contemporaneously with pulses of granite magmatism elsewhere in the craton. They suggest that sedimentation in this basin represents a complex response to a variety of plate tectonic linked processes operating on and within the Kaapvaal Craton. Catuneanu (2001) following Coward et al. (1995) suggested that the basin developed as a retroarc foreland system in response to late Archaean crustal thickening during the collision of the Kaapvaal and Zimbabwe Cratons to the north. However, it is difficult to see how such a model can adequately explain the 360 Ma history of the basin, and more recently Kosticin and Krapez (2004) proposed, on the basis of detrital zircon data, that the West Rand Group formed in a passive margin setting which switched to a retroarc basin in Central Rand Group times.

1.3.2.2 What can late Archaean sedimentary basins tell us about the early Earth system?

Sedimentary basins develop as a response to the interaction between sediment supply and crustal subsidence. In detail this reflects the complex interplay of tectonism, magmatism, eustasy and weathering rates, which are themselves a function of paleoclimate (Eriksson et al., 2001). In the late Archaean probably a number of these factors were different from the modern. For example there is evidence that climatic regimes were different and so it is likely that weathering rates and sediment supply were also different (Chapter 5, Section 5.3.2.1). In addition, many believe that the mantle was hotter at this time. If this is the case then tectonic processes and subsidence rates would also be different from those of modern sedimentary basins.

Sediments in late Archaean sedimentary basins can provide some answers to these questions, with one proviso, best illustrated by a long-running debate which has been taking place between workers in the Witwatersrand Basin. The argument is about whether or not gold in the Witwatersrand Basin is primary and detrital in origin, or secondary and hydrothermal in origin. In essence this is a debate about the extent to which primary data on early Earth surface processes can be extracted from ancient sediments. As is often the case with scientific controversies of this type, the answer is that both schools of thought are partially correct, and it is most probable that the gold was deposited in detrital form, but has subsequently been modified by hydrothermal fluids (Frimmel, 2005). Therefore identifying the primary characteristics of the Witswatersrand sediments requires some care.

Given this caveat, the Witwatersrand sediments can provide some information about the weathering environment during the late Archaean. Not least in importance is the unusually high abundance of accumulated gold. The presence of detrital grains of pyrite and uraninite, normally oxidized in the modern weathering environment, suggests that the atmosphere was low in oxygen (Frimmel, 2005). Climatic indicators for the late Archaean also come from torroidal shaped gold grains and faceted pebbles in the Witwatersrand Basin indicating that the surrounding landscape experienced eolian deflation as part of the gold concentration process (Minter, 1999). In addition, in the Pongola Supergroup, Gold and von Veh (1995) report the presence of diamictites of glacial origin and paleosols. The Witwatersrand sediments also contain important information about the early biosphere, for the gold is often associated with bituminous carbon (Gray et al., 1998). This was interpreted by many authors as the in situ preservation of algal mat structures. Gray et al. (1998), however, in their recent reexamination of this topic conclude that the carbon originated as immature algal kerogen, but show that that it was subsequently mobilized by hydrothermal fluids to its present location.

At a very different scale, late Archaean sedimentary basins also have the potential to provide important information about late Archaean tectonic processes, in particular the processes of epicontinental basin development in the late Archaean. At present however, we are probably faced with more questions than answers. For example, is the smaller size of the Witwatersrand basin (wavelength 490 km) relative to modern intracratonic basins, and the long timescale of its evolution (360 Ma) compared with the modern, typical of Archaean basin processes? If so what does this mean? Some partial answers are provided by a recent study of the Fortescue Group, in the Pilbara Craton, in which Blake et al. (2004) showed that sedimentation rates in late Archaean fault-bounded epicontinental basins are similar to those in modern examples, indicating that the substrate to the basin was of equivalent thickness and rigidity to younger counterparts. They also showed that the eruption rate of flood basalts in the same sequence was similar to that of more recent flood basalt provinces.

1.3.3 Archaean granite–gneiss terrains

Granitic gneisses are the most abundant rock type of Archaean cratons. In many cratons they are the dominant rock type and in some they are the only rock type. In detail, however, they take many forms. Some occur as relatively undeformed plutons, whereas others are highly deformed banded gneisses, some metamorphosed at granulite grade. Granite gneisses may form the basement to greenstone belt sequences, but granitic plutons also intrude greenstone belt sequences. Compositionally they are variable, but many belong to the magmatic suite **t**onalite- **t**rondhjemite-**g**ranodiorite (a suite of low potassium, high-silica granitoids, dominated by the minerals quartz and plagioclase feldspar), giving rise to the acronym TTG which is frequently used to describe the composition of Archaean granitoids. Archaean granite–gneiss terrains have also been termed "grey-gneiss terrains" because of their grey appearance in the field. Where they are metamorphosed to amphibolite facies and granulite facies they tend to be known as "Archaean high-grade gneiss terrains." Some of the better known granite–gneiss terrains are listed in Table 1.5.

The apparent lithological simplicity of Archaean granite–gneiss terrains belies their geological complexity. This fact was recognized in the pioneering work of John Sutton and Janet Watson, in the early 1950s, in the Lewisian Complex of northwest Scotland. Their careful fieldwork, subsequently supported by geochronological studies, showed that it is possible to disentangle a billion years

TABLE 1.5 Examples of important Archaean granite–gneiss terrains discussed in the text.

Craton	Country	Ages (Ga)	Features
North Atlantic (Lewisian Gneiss)	UK	2.5–2.95	The first Archaean granite–gneiss terrain to be chronologically segmented as a result of structural mapping
North Atlantic (Godhabsfjord area)	Greenland	2.5–3.8	The first Archaean granite–gneiss terrain to be identified as segmented into terranes, on the basis of zircon geochronology
Zimbabwe	Zimbabwe	2.5–3.8	Some evidence to support concentric growth around a central ancient nucleus
Kapuskasing	Canada	2.5–2.8	An exposed Archaean crustal cross section from upper to lower continental crust
Limpopo Belt	Zimbabwe, Botswana, South Africa	2.7–2.9	The north and south marginal zones are the lower crustal sections of the Zimbabwe and Kaapvaal Cratons respectively, thrust onto their respective cratons

of Earth history by detailed structural mapping (Sutton & Watson, 1951). This methodology was subsequently successfully applied to many other Archaean granite–gneiss terranes, revealing spectacular complexities in what had previously been mapped just as "crystalline basement." Further advances were made with the advent of precise isotopic dating techniques allowing distinct age provinces to be identified within Archaean cratons.

Granite–gneiss terrains of the upper and middle crust are frequently associated with greenstone belts (Fig. 1.5). In contrast, the felsic granulite gneiss belts of the lower crust host a slightly different rock assemblage, and contain lenses, usually less than 1 km long and a few hundred meters wide, of basalt, a variety of sedimentary rock types and layered ultramafic–mafic–anorthositic intrusions, all metamorphosed to granulite grade. It is possible that the supracrustal rocks are also fragments of greenstone belts, infolded deep into the crust, but this is by no means certain and their origin is the subject of some debate.

1.3.3.1 The granite–gneisses of the Zimbabwe Craton – a typical granite–greenstone belt association

The greenstone belts of the Zimbabwe Craton are enclosed by a variety of granitic gneisses of various ages (Fig. 1.5). In the center of the craton there are 3.5 Ga gneisses, which are associated with an early generation of greenstone belts. These are surrounded by 2.9 Ga gneisses which are broadly coeval with a second generation of greenstone belts, but which form the basement to the majority of greenstone belts which formed at 2.7 Ga (Wilson et al., 1995). The youngest granitoid rocks of the craton are a ubiquitous suite of potassic granites which intruded both gneisses and greenstone belts at about 2.5 Ga. Although on a map they appear to dominate the outcrop, they are for the most part, kilometer-thick, sheet-like intrusions located entirely in the upper crust. Southeast of the craton, and separated by a major thrust zone are the high-grade gneisses of the Limpopo granulite belt, whose age (2.7–2.9 Ga) and metamorphic grade suggest that these rocks are equivalent to the lower crust of the craton, now thrust up to a higher structural level (Rollinson & Blenkinsop, 1995). The relationships between the different types of granitoid gneisses in this craton demonstrate that granite–gneiss terrains may have a long history, which in the case of the Zimbabwe Craton is in excess of 1 billion years. Furthermore, the geographic distribution of gneisses of various ages may suggest that the Zimbabwe Craton grew outwards laterally, from the nucleus of ancient gneisses (Wilson et al., 1995).

1.3.3.2 The granite–gneiss terrain of the Gothåbsfjord region of west Greenland – a typical high-grade gneiss terrain

The granitic gneisses of the Gothåbsfjord region of west Greenland have become something of a geological classic, for the combination of their superb, ice-cleaned exposures and the detailed mapping of the Geological Survey of Denmark and Greenland has made them one of the best known terrains of their kind. The main rock types are granitoid gneisses of intrusive origin with compositions in the tonalite–trondhjemite–granodiorite range, and a small amount of mafic and ultramafic rocks of supracrustal origin. All rock types are metamorphosed to amphibolite or granulite grade. Three decades of geochronological study have defined a number of different age provinces within this region and show that these granitoid gneisses preserve a long history. The application of detailed zircon geochronology led to the recognition that the Gothåbsfjord region is an assembly of separate terranes, of differing age, subtly different lithological character and of different metamorphic grade, separated from one another by mylonite zones (Friend et al., 1988). As more has become known of this remote area, so the number of distinct terrains has increased. Recently, Friend and Nutman (2005) identified six different terrains in this region, to which they gave characteristically tongue-twisting Greenlandic names (Fig. 1.9). It is thought that these terranes were assembled during the late Archaean, during the period 2.95 to 2.70 Ga ago and their assembly reflects the deep crustal record of continent-continent collision in the late Archaean (Friend & Nutman, 2005).

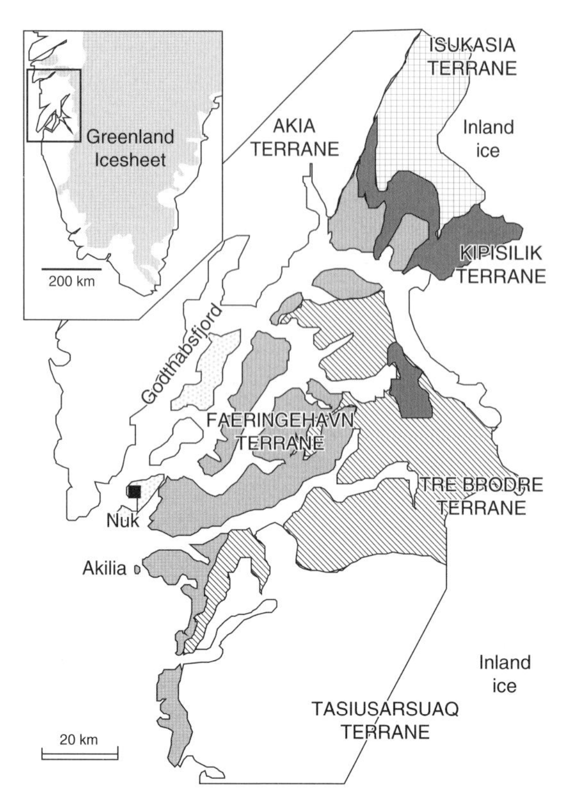

FIGURE 1.9 The granite–gneiss terrane of the Gothåbsfjord region of west Greenland, showing the 2.97–3.22 Ga Akia, >3.60 Ga Isukasia, 2.96–3.075 Ga Kipisilik, >3.60 Ga Faeringehavn, ca. 2.825 Ga Tre Brodre and the 2.82–2.92 Ga Tasiusarsuaq terranes. The map also shows the location of Akilia Island (Nutman et al., 1996; Friend & Nutman, 2005). The Isukasia terrane is thought to be at the lowest structural level, overlain by the Kapisilik, the Faeringehavn, and the Tre Brodre (highest structural level) terrains. This map is simplified from Friend and Nutman (2005).

1.3.3.3 What can Archaean granite–gneiss terrains tell us about the early Earth system?

Because of their relatively indestructible nature Archaean granite–gneiss terrains are the host to all other Archaean rock associations, including the oldest terrestrial materials (see the following section). In their own right, however, they are our principal source of information about the processes of continent formation, for they provide the basis for our understanding of how and when the continents formed, and how they have developed through time.

The precise mechanism of continental growth is a matter of debate. Whether the continents grew principally by lateral or by vertical accretion is not entirely clear. This raises the question of the extent to which modern processes can be used to explain Archaean crustal growth. For there are some regions of Archaean crust whose composition and origin are best explained by modern style, continental margin magmatism. Examples include the Minto Block in Canada (Stern et al., 1994) and the northern marginal zone of the Limpopo Belt in Zimbabwe (Berger & Rollinson, 1997). These are issues which will be developed further in Chapter 4.

1.4 The oldest rocks

Rocks older than about 3.8 Ga are extremely rare and are known in outcrop from only a few localities worldwide. Here we describe three localities, from which much of our knowledge of the pre-3.8 Ga Earth has been gained. These localities are marked in Fig. 1.4.

1.4.1 The Gothåbsfjord region of west Greenland – the most extensive area of ancient rocks

The Gothåbsfjord region of west Greenland provides the most extensive outcrop of early Archaean rocks currently known (Nutman et al., 1996). These are collectively known as the Itsaq gneisses and are located in the Isukasia and Faeringehavn terranes (Fig. 1.9) and comprise a number of geological units, which outcrop over an area of 2,000 sq. km between Greenland's west coast and the inland ice-sheet. These units have ages between 3,600 and ca. 3,900 Ma.

1.4.1.1 The Isua Greenstone Belt

One of the most important areas of ancient rocks is the Isua Greenstone Belt (or supracrustal belt, as it is often called) (Appel et al., 1998), located in the northern part of the Isukasia terrane (Fig. 1.9). The Isua Greenstone Belt comprises a sequence of metabasalts, ultramafic rocks, clastic, and chemical sediments, with ages between 3.7 and 3.8 Ga (Nutman et al., 1996). The most ancient materials so far recorded from Isua are mineral inclusions in pyrite grains from within the iron formation. These have ages of around 3.86 Ga and one population gives ages of 4.31 ± 0.06 Ga (Smith et al., 2005). The outcrop now defines a linear belt folded into a horse-shoe shape, although internally the greenstone belt is probably a series of stacked thrust sheets of different provenance and age (Myers, 2001; Rollinson, 2002). The Isua Greenstone Belt contains a record of the Earth's oldest mafic magmatism, possible evidence of early life (Rosing, 1999), and chemical and clastic sediments containing information about the early atmosphere and oceans. Interestingly, however, absent from the Isua succession is evidence for the existence of continental crust (Kamber et al., 2005).

1.4.1.2 The Amîtsoq Gneisses

The first clues of very ancient crust in the Gothåbsfjord region of west Greenland came, in the early 1970s, from the isotopic dating of felsic gneisses close to the town of Nûk (Moorbath et al., 1972). Refinements in geochronological methods, particularly with the development of precise, single-zircon U–Pb dating have confirmed the great antiquity of these rocks and showed that the gneiss precursors have intrusion ages of between 3.65 and 3.87 Ga. These felsic gneisses, initially known as the Amîtsoq Gneisses, have now been included in the broader group of gneisses termed the Itsaq Gneisses (Nutman et al., 1996). One of the oldest areas within the Itsaq Gneisses is to the south of the Isua Greenstone Belt, where the rocks have zircon ages of between 3.78 and 3.81 Ga. However, these gneisses include fragments of mafic and ultramafic rocks which must be even older than 3.81 Ga. These mafic and ultramafic rocks could be older than the rocks of the Isua Greenstone belt and could therefore be some of the oldest rocks on Earth.

1.4.1.3 The Akilia association

One group of Gneisses, originally designated as part of the Amîtsoq Gneisses is found on Akilia Island, a small island about 1 km across, located in the outer part of Gothåbsfjord, close to the town of Nûk (Fig. 1.9). Akilia Island has become famous in the scientific literature because Amîtsoq Gneisses from this locality have been dated at 3.87 Ga, and yet include older lithologies which may be of sedimentary origin (Mojzsis et al., 1996). It is in these older lithologies that graphite has been found which has an alleged biogenic isotopic signature. This claim, which has profound implications for the timing of the beginning of life on Earth, has been much contested and is a subject to which we will return in Chapter 6 of this book.

1.4.2 The Acasta Gneisses, northwestern Canada – the most ancient rocks

The Acasta Gneisses are located on the western margin of the Slave Craton in northwestern Canada. Their great antiquity was first reported by Bowring et al. (1989) who described two samples containing zircons with crystallization ages of 3.96 Ga. Detailed SHRIMP studies of zircons from relatively undeformed samples (Bowring & Williams, 1999) indicate that these rocks are in fact 4.03 Ga old and contain even older inherited cores with ages up to 4.2 Ga (Iizuka et al., 2006).

The Acasta Gneisses outcrop over an area of about 40 sq. km around the Acasta River, on the edge of the Canadian Shield. They comprise a heterogeneous assemblage of strongly foliated and banded tonalitic and granodioritic gneisses (Bowring et al., 1990). They include lenses of amphibolite and, less commonly, metasedimentary and ultramafic inclusions and preserve a complex history of intrusion and deformation. These gneisses now form the basement to the Proterozoic rocks of the Wopmay Orogen and are exposed in a structural high, within this orogen. They possess a Proterozoic mineral fabric indicating that they were also metamorphosed during the Proterozoic. U–Pb zircon ages for the gneiss complex range from 3.6 to >4.0 Ga, and the oldest known material is a xenocrystic zircon in a tonalitic gneiss, which has an age of 4.2 Ga.

Whilst the great age of the Acasta Gneisses is not in doubt, they have been the subject of some controversy. Bowring and Housh (1995) calculated initial ε_{Nd} values for a range of samples from this area and found that they are extremely heterogeneous. This observation, coupled with the ancient zircon ages for these samples, has profound implications for the

Nd-isotopic evolution of the Earth and implies a very early differentiation event in the mantle and the formation of very early continental crust, which has subsequently been destroyed. Moorbath et al. (1997), however, have challenged this interpretation of the Nd-isotopic data and find no support for the early (pre-3.8 Ga) differentiation event in the mantle, claimed by Bowring and Housh (1995) – see Chapter 3, Section 3.2.3.1.

1.4.3 Zircons from western Australia – the most ancient terrestrial materials

The oldest known terrestrial materials are 4.1–4.4 Ga detrital zircons from the Mount Narryer area of western Australia (Froude et al., 1983). The first locality to be shown to host extremely ancient zircons (4.1–4.2 Ga) was a sequence of highly metamorphosed quartzites, conglomerates, and pelites from Mount Narryer, at the northwestern edge of the Yilgarn Craton in western Australia. Subsequent discoveries, in a less metamorphosed sequence of sediments from the Jack Hills, some 60 km away identified even older zircons, some 4.28 Ga old (Compston & Pidgeon, 1986). Recently, Wilde et al. (2001) reported a zircon grain with an age of 4.404 ± 0.008 Ga (Fig. 1.10). To date this is the oldest known terrestrial material. Supporting evidence for very ancient crust in this area comes from Nelson et al. (2000) who reported zircon xenocrysts with ages of up to 4.18 Ga from felsic gneisses in the same region. Taken together these observations indicate that the rocks of this part of the Pilbara Craton (mostly with ages between 3.3 and 3.7 Ga) were derived from substantially older parental materials, which appear now to have been destroyed.

The composition of the original crust in which these ancient zircons crystallized has been the subject of some debate. However, evidence from rare mineral inclusions preserved within the zircon grains, and from the trace element signature of the zircons themselves, shows that it is likely that they formed in granitic crust (Maas et al., 1992; Peck et al., 2001). Recently it has been shown that their Ti-content is consistent with crystallization temperatures of about 700 °C, indistinguishable from present-day zircon growth temperatures in hydrous granitoid magmas (Watson & Harrison, 2005). In detail, some zircon crystals record a number of different discrete age events within the same grain (4.3, 4.2, 4.15 Ga, see Fig. 1.10), indicating that they have experienced complex thermal histories. This history

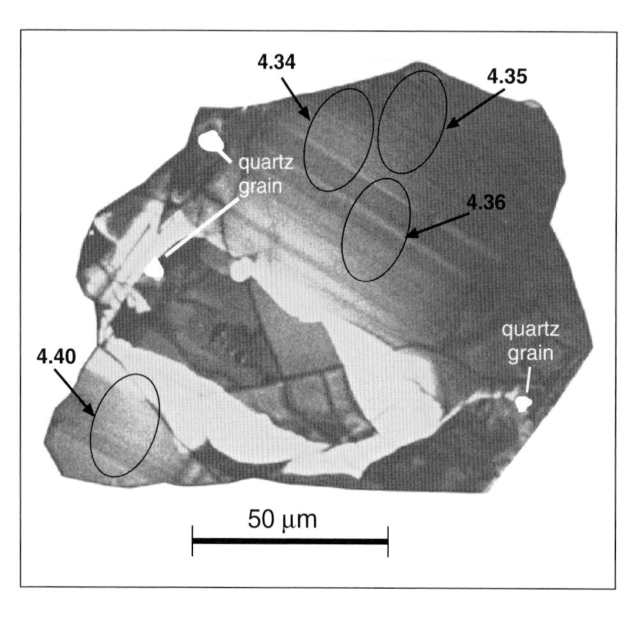

FIGURE 1.10 Cathodoluminescence image of a detritol zircon grain from the Jack Hills, western Australia, showing U–Pb ion microprobe (SHRIMP) ages for the different parts of the grain in Ga. One part of this grain (bottom left) gave an age of 4.404 ± 0.008 Ga (after Wilde et al., 2001). Photograph courtesy of John Valley.

most likely indicates the reworking and metamorphism of already formed crust, which in turn, indicates that the crust was sufficiently thick for metamorphism to take place, and therefore represents a continental, rather than oceanic setting (Peck et al., 2001).

Not surprisingly, this most ancient part of the terrestrial record has been the subject of intense scientific interest, and a wide range of isotopic studies have been made on the Jack Hills zircons, in a quest to find out as much as possible about the earliest state of our planet. For example, oxygen isotope studies have been used to suggest the presence of oceans on the Earth at 4.4 Ga (Mojzsis et al., 2001; Wilde et al., 2001) a topic which will be discussed later in this book (Chapter 5). Hf-isotope studies offer the opportunity to understand the timing of mantle differentiation and continental crust formation in the very early Earth (Amelin et al., 1999; Bizzarro et al., 2003), although at present this discussion is clouded by uncertainties surrounding the decay constant of ^{176}Lu (see Chapter 3, Section 3.2.3.2). The decay product of now-extinct isotopes, no longer preserved even within the earliest part of the rock-record have also been identified in 4.1–4.2 zircons from the Jack Hills and offer the potential to date very early metamorphic events within the early Earth's now destroyed proto-crust (Turner et al., 2004).

1.5 How much do we really know about the early Earth?

In the preceding sections of this chapter we have reviewed our state of knowledge about the early Earth. In summary, we might conclude that there is a reasonable amount of evidence on which to develop an understanding of the early Earth system. Nevertheless, there are three caveats to this statement which need to be discussed.

First, the geological record is incomplete. This is particularly true for the first 750 Ma of Earth history, about which we know very little. As was shown in the preceding section, there is very little rock record preserved from before 3.8 Ga. The dynamic nature of our modern planet has destroyed any evidence of its beginning. This is also clear from observing the Moon and our planetary neighbors. Like them, early in its history the Earth would have suffered from the impacting of asteroids and comets, and yet the record of these very early events has largely not been preserved on Earth. In a similar way, observation of other planets provides ideas about the nature of the Earth's earliest crust and oceans. In fact, the evidence we do have indicates something of a paradox, for comparisons with other planets might suggest that the Earth's first crust was basaltic in composition, and yet the oldest known terrestrial materials are zircons, which are thought to be derived from now destroyed felsic rocks.

Second, it is possible that the early geological record has a preservation bias. A haunting question which should lie in the back of the mind of all who study the early Earth is, "Is what we see now, representative of what was there then?" Is there something about Archaean and older rocks which is unusual, and which has caused these particular rocks to be preserved rather than destroyed? A related question is whether or not something special happened in the history of the Earth, at about 3.8 Ga, which gave rise to the preserved rock record. The answer to both these questions must be a qualified "yes," for it is very likely that, given normal erosion processes, those rocks which formed at the Earth's surface are more likely to be destroyed than deep crustal plutonic rocks. So for example De Wit and Ashwal (1995) have proposed that the comparative rarity of Archaean andesites is a preservation issue, for andesites typically form subaerial strato-volcanoes, which are highly susceptible to erosion and nonpreservation. In a similar way Morgan (1985) and Pollack (1997) have argued that continental crust is most likely to be preserved when it is underlain by a thick lithospheric root, with low radiogenic heat production, insulating it from the asthenospheric mantle.

Finally, it is important to revisit a central tenet of modern geological interpretation – the principle of uniformitarianism. Originally formulated by James Hutton in the eighteenth century the principle of uniformitarianism states that the physical and chemical processes which shape the modern world also operated in a similar manner in the past, and so can be used to explain past events. Here uniformitarianism will be adopted as a working model and several of the chapters begin with a description of the modern Earth System. There are, however, three provisos to this position.

First, the earliest stage of Earth history must have been substantially different from the modern, not least because it was subject to *planetary* processes. The processes of impacting, accretion, and the related melting are not part of what are normally included within uniformitarian thinking.

Second, there are rocks-types which are unique to the Archaean, implying that the Archaean Earth must have somehow been different. Examples of these rock-types include the ultramafic lavas komatiites, sedimentary BIFs found in greenstone belts, and calcic anothosites – typically found in the deeper crust (De Wit & Ashwal, 1997).

Third, it will be argued in this book that many of the principal Earth reservoirs have experienced gradual changes over time. These are not changes which are cyclical, rather they are unidirectional, on a timescale of billions of years. This type of long-term, unidirectional change is known as secular change. It is seen, for example, in the chemical evolution of the atmosphere and the oceans and probably also took place in the mantle.

However, despite these caveats, modern research is showing increasingly that many Archaean rocks appear to have formed through "modern-type" processes. For example, in their review of greenstone belt sedimentation Eriksson et al. (1994) found that there are many similarities between Archaean and modern sedimentary environments. They state "lithological associations in Archaean greenstone belts do not appear to be unique to that time period. This implies that Phanerozoic analogues are generally suitable to define Archaean depositional environments." Similarly, De Wit and Ashwal (1995) in their survey of Archaean greenstone belts find that "at present it is difficult to get away from the fact that most geological observations in greenstone belts indicate similar surface and near-surface processes as are deduced from a variety of Phanerozoic orogenic belts." Thus it would seem as though a uniformitarian approach can be used as a working hypothesis in the Archaean, with the understanding that in some instances, some aspects of the early Earth were different from today. Given this premise, it is important to enquire at which point in time modern-style tectonic processes began to operate on Earth.

This book therefore will look at the early Earth with modern eyes, and will use our understanding of modern Earth processes to interpret the early Earth system. Only when this is patently not explaining the observed facts, or in situations where there does not seem to be a modern analogue, will other types of process be invoked.

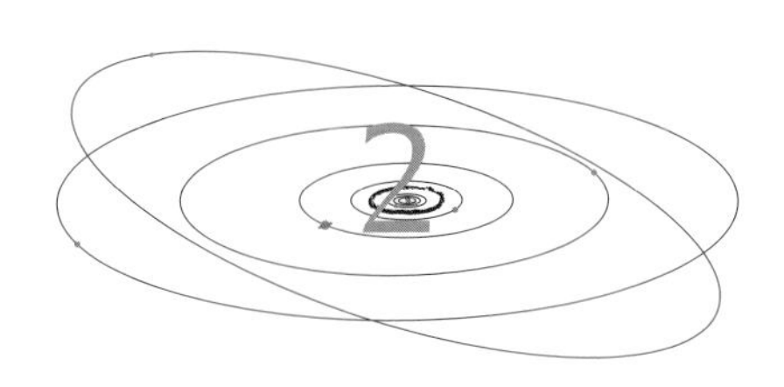

THE ORIGIN AND DIFFERENTIATION OF THE EARTH

The origin and differentiation of the Earth – the big picture

The Universe was already 9 billion years old when our solar system was born at 4.567 Ga. The compression and collapse of a gas cloud in the interstellar medium gave rise to a flattened disk of gas and dust rotating around an otherwise nondescript, medium-sized star. It was in this rotating disk of gas and dust – the primitive solar nebula – that our planetary system was formed.

The original building blocks of the Earth are thought to be preserved in a group of primitive meteorites known as the carbonaceous chondrites. These contain inclusions rich in calcium–aluminum minerals which formed at high temperature within 10^4–10^5 years of the formation of the solar system. Also present are chondrules, olivine-rich spheroidal melt droplets, a few millimeters in diameter, which formed within the first 4 Ma of Solar System history.

The formation of the Earth can be explained by the "standard model of planetary formation," in which dust fragments, including the types described above from meteorites, accumulate through the process of accretion – first, into kilometer-sized planetesimals over a timescale of 10^4 years, and then into planetary embryos – up to 4,000 km in diameter – over an interval of 10^6 years. The final stage of planetary accretion took place over 10^7–10^8 years and involved collisions between a relatively small number of large planetary bodies, giving rise to "giant impacts." These late, large-magnitude impacts are thought to have had a profound influence on the earliest history of the Earth System.

The latest of these involved an oblique collision between the proto-Earth and a body the size of Mars, at about 30 Ma after the formation of the solar system. This impact generated a huge amount of thermal energy so that a significant portion of the Earth was vaporized. This vapor coalesced around the Earth and cooled to form the Moon. A consequence of this impact is that a significant proportion of the Earth's mass would have been molten, creating what has become known as a magma ocean.

Similarly, the Earth's core is thought to have formed during the early stages of accretion, perhaps by as early as 10 Ma after the formation of the solar system. Geochemical constraints require the core to have also formed within a deep magma ocean, with liquid metal separating from a silicate melt at depths as great as 1,000 km.

Although the circumstantial evidence for the existence of a terrestrial magma ocean is strong, independent geochemical evidence has been hard to find. Recently however, the first geochemical clues of mineral fractionation with a deep molten mantle have been found, supporting the terrestrial magma ocean concept.

It was in this earliest Earth System that there was the strongest interaction between the different Earth reservoirs. There were intense, dynamic interactions between core, mantle, proto-ocean, and atmosphere. In addition there was probably an early basaltic crust, now long since lost by recycling into the mantle.

2.1 The origin and early history of the Universe

Enquiring into the origin of the Universe is to pose one of the most fundamental and difficult scientific questions of all. It also brings us into a very rapidly advancing and exciting field of science. In recent decades, cosmology has advanced extremely rapidly, and in the last few years has begun to provide the first self-consistent answers to age-old questions (Silk, 1994; Longair, 1998; Rees, 2000).

In many ways, the question of the origin of the Universe is far beyond the scope of this book. Nevertheless, in the context of understanding the origin of the Earth a brief review of the cosmological background is provided here. This is in fact necessary, for some of the fundamental features of the Earth system can only be understood within a framework of cosmology and cosmochemistry. There are two specific areas that are important within Earth System Science. The first is that the elemental abundances in the early solar system are a function of its setting within the galaxy and this "cosmic geography" ultimately determines the "raw materials" of the terrestrial planets. A second reason is that a planetary view of the Earth provides insight into the ultimate Earth system, in which there was profound, dynamic interaction between what are now the different components of the modern Earth system.

2.1.1 The Big Bang theory

The prevailing theory of the origin and evolution of the Universe is the Big Bang theory. According to this theory, about 14 billion years ago the Universe expanded to its present enormous volume from an initial volume, which was effectively zero. Three sets of observations have profoundly shaped the way in which we think about our Universe and led to the Big Bang theory. First was the discovery that our Universe is expanding. Second, there were predictions about the abundances of the light elements H, He, and Li in the Universe, and third was the discovery made by Penzias and Wilson (1965) that our part of the Universe is filled with microwave radiation.

2.1.2 Evidence for the Big Bang theory

2.1.2.1 An expanding Universe

Hubble (1929) discovered that there is a simple, linear relationship between the distance to a remote galaxy and the cosmological redshift – the redshift in the spectral lines from that galaxy (Hubble, 1929). Hubble's observations showed that the greater the distance to a galaxy, the greater the redshift in its spectral lines. These measurements strongly indicated that galaxies appear to be moving away from us with speeds proportional to their distance. The net effect of this motion is that, as time goes on, the galaxies are getting further and further apart. A very important consequence of these observations is that at some point in the past all matter must have been concentrated in one place. Astronomers define this point in time as the beginning of the Universe. At this time all the matter of the Universe was concentrated in an infinitely small volume with a state of infinite density.

2.1.2.2 The abundance of the light elements

The Big Bang theory predicts that the early universe was very hot. In the early stages of the formation of the Universe the light elements H (and its isotope deuterium), He, and Li were formed during the "Big Bang nucleosynthesis." The deuterium found today in the interstellar medium could only have formed at the beginning of the Universe (Songaila et al., 1994). Calculations based upon the initial ratio of protons and neutrons suggest that if the Big Bang theory is correct then about 24% of the ordinary matter in the Universe will be He and the rest hydrogen (Schramm & Turner, 1998). This value is in good agreement with recent observations indicating that the Big Bang theory passes one of its key "tests."

2.1.2.3 The Cosmic Microwave Background

If the early Universe was extremely hot, it is possible that, even today, the remnants of this initial fireball might be detected. Support for this hypothesis came from the discovery by Penzias and Wilson (1965) of what came to be known as the Cosmic Microwave Background. This discovery coincided with the work of theoretical physicists who showed that if the

Universe began with a hot Big Bang, then the Universe should be filled with electromagnetic radiation cooled from the early fireball to a temperature of around a few Kelvin. In subsequent years a large number of direct measurements of the Cosmic Microwave Background at different wavelengths yielded an intensity–wavelength plot which had the characteristics of black body radiation at 2.73 K. This is the remnant of the initial fireball of the Big Bang.

2.1.3 The current state of the Big Bang theory

The Big Bang theory allows us to pose (and answer) a number of fundamental questions about the Universe. These include questions about the nature of matter and energy in the Universe, about its speed of expansion and whether this is changing with time, about the age of the Universe, and about its ultimate fate – whether or not it will continue to expand. Cosmologists have debated these issues for decades, but some recent exciting results have begun to provide some significant answers.

Central in this discussion was the balance between an expanding Universe from the initial impulse of the Big Bang and the counteracting influence of the gravitational attraction of matter in the Universe. In the 1990s astronomers found a way of measuring the expansion rate of the Universe at different times in its history and found, to their great surprise, that the Universe is expanding faster now than it did earlier in its history (Reiss et al., 2001). This observation led to the recognition of what has become known as "dark energy," a force that appears to counteract the effects of gravity (Bahcall et al., 1999). Understanding the nature and origin of dark energy is one of the outstanding problems of modern cosmology.

More recently, in 2002 the Wilkinson microwave anisotropy probe (WMAP) was launched to map the fine detail of the Cosmic Microwave Background. After a year of mapping the first results were published in 2003 (Bennett et al., 2003), with some far-reaching conclusions. A best-fit cosmological model for the Cosmic Microwave Background is satisfied by a Universe composed of only 4.4% of ordinary baryonic matter (protons and neutrons). The rest is made up of nonbaryonic matter, identified as dark matter (22% of the Universe), that is, particles, which can only be detected by their gravitational effect, and dark energy, making up 73% of the Universe (Fig. 2.1). They found that their estimate of the total mass–energy of the Universe was in good agreement with earlier results from high altitude balloon experiments (e.g. de Bernadis et al., 2000) signifying that the geometry of the Universe is flat. These data strongly support the theory of the inflation of the Universe – that is, an extremely rapid expansion of the Universe shortly after the Big Bang.

In its present form, standard Big-Bang cosmology is a very open subject, for it contains three important elements – dark matter, dark energy, and inflation – about which there is

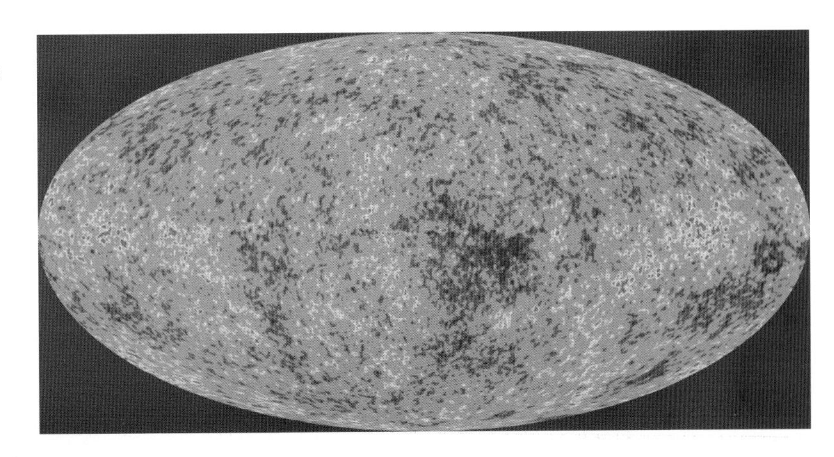

FIGURE 2.1 A WMAP image of the infant universe showing 14 billion year old temperature fluctuations (shown as differences in grey shading) that correspond to the seeds of galaxies (from http://map.gsfc.nasa.gov/m_or.html, courtesy of NASA/WMAP Science team).

no known mechanism, and therefore may be subject to significant shifts in understanding in the future.

The best-fit cosmological model for the Cosmic Microwave Background from the WMAP also precisely constrains the age of the Universe. This is a welcome contribution to what has been a comparatively inexact science, with two irreconcilable camps preferring either a "young" 10 Ga or "old" 20 Ga age for the Universe. A number of different approaches had been used such as the calculation of velocity–distance relationships measured from galaxies, the application of radioactive dating methods to old stars (Cayrel et al., 2001), and relationships between the luminosity and mass of a star. The recent WMAP result indicates that the age of the Universe is a precise 13.7 ± 0.2 Ga (Bennett et al., 2003). This is in good agreement with an earlier less precise result of 13.4 ± 1.6 Ga based on the rate of expansion of the Universe (Lineweaver, 1999). However, all results in this field are model-dependent and may be the subject of change in the future.

2.1.4 The early history of the Universe

Current thinking in cosmology indicates that the history of the Universe may be described in the following stages (Fig. 2.2):

An initial singularity. At the beginning of the Universe, 13.7 billion years ago, all matter was in one place at a single instant; this event in cosmological parlance is known as a "singularity." This term describes the inference that an infinitely large amount of matter is gathered at a single point in space-time.

The Big Bang. At the Big Bang there was a huge expansion of matter, an expansion that has continued ever since.

Inflation. Between 10^{-50} and 10^{-30} s after the Big Bang there was a particularly rapid expansion of the Universe. This process is known as the inflation of the Universe and represents the first burst of growth of the Universe. During inflation the part of the Universe that we see today expanded by a factor of 10^{60}. When the Universe was only one second old temperatures were of the order of 10 billion degrees. At that point the universe was permeated with radiation and subatomic particles. When the Universe was a few minutes old and temperatures were about 1 million degrees protons and neutrons formed atomic nuclei leading to the production of primordial hydrogen and helium ions.

An opaque Universe After 100,000 yr conditions in the Universe were similar to those inside

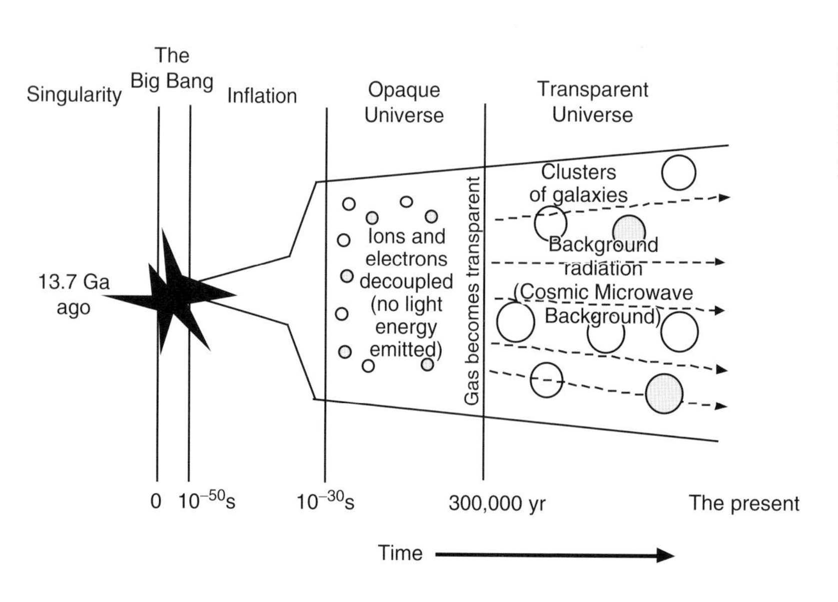

FIGURE 2.2 The origin, inflation, and expansion of the Universe from the initial singularity to the present.

the sun today. An almost uniform plasma of electrons, hydrogen, and helium ions filled the Universe. At this time the free electrons acted as a block to photons – generated from the light energy generated in the Big Bang, and prevented them escaping, rendering the early Universe opaque.

A transparent Universe. After 300,000 yr temperatures dropped to 4,500 K and gave rise to the formation of atomic matter, and atoms of hydrogen, helium, and deuterium were formed. Because electrons were removed from the plasma through the formation of atoms, radiation streamed out and the Universe became transparent. Initially the Universe contained abundant ultraviolet- and X-rays, now cooled down to microwave wavelengths. This is what is recorded as the Cosmic Background radiation.

The present Universe. As the universe continues to expand the initial radiation will appear to be derived from a much cooler body. Hence today the Cosmic Background radiation is 2.73 degrees above absolute zero.

It is a long journey from the formation of the Universe at 13.7 Ga ago to the formation of the Earth at about 4.57 Ga. This journey is the subject of the next sections of this chapter, and in them we shall consider the relationship between the Earth and its host – the solar system. In so doing we shall discuss the processes which have led to star formation, to the formation of the chemical elements, the condensation of the solar system, and ultimately to a model for planetary accretion and hence the Earth.

2.2 Star Formation

In this section we consider the processes which led from the Big Bang to the formation of stars and galaxies.

2.2.1 The anisotropy of the Universe

As cosmologists began to accumulate measurements of the Cosmic Background radiation at the edge of the Universe they were impressed by the uniformity of the results. However, many theorists predicted that in detail the results should not be uniform, leading to a search for microscale variability in the cosmic background radiation. This was first discovered using a differential microwave radiometer on the COBE space probe and demonstrated that the Cosmic Microwave background was very slightly variable on the scale of one part in 100,000. A better resolution of this temperature fluctuation was obtained by the WMAP (Bennett et al., 2003). These results are shown in Fig. 2.1 and provide a detailed map of the temperature fluctuations in the Universe at the beginning of time, demonstrating that matter and energy were not evenly distributed in the very early Universe.

This unevenness in the distribution of matter and energy has been described as the "lumpiness," or more correctly, the anisotropy of the early Universe. Elsewhere it is stated that the Universe is homogeneous or isotropic. This is not a contradiction, rather a difference of scale. On a large scale the Universe is homogeneous and everywhere has a temperature of 2.73 K at its edge. On a fine scale however, microvariations in temperature have been measured.

The significance of this variation in the intensity of the Cosmic Microwave Background is that it shows how matter and energy were distributed when the Universe was still very young. It is thought that these early inhomogeneities subsequently developed into the regions in the present Universe where there is matter, that is, galaxies and galaxy clusters, and those regions from which matter is, absent – space. The early inhomogeneous distribution of matter also reflects an inhomogeneous distribution of density, and it is these initial density differences that gave rise to small differences in gravitational forces which began to draw matter together.

2.2.2 The formation of galaxies

Primordial gas clouds, composed of hydrogen and helium, are thought to be the beginnings of galaxies and represent the first large-scale structures to form in the evolving Universe. These huge gas clouds with masses 10^{15}–10^{16} greater than that of the sun (that is, 10^4–10^5 times greater than our own galaxy, the Milky

Way), formed due to gravitational forces working against the expansion of the Universe.

Just as primordial gas clouds formed through weak, small scale density fluctuations in the Universe, so also a similar process led to their fragmentation into molecular gas clouds, with dimensions greater than 10^{14} km, and masses a million times greater than the sun. This process took place in the first few hundred million years of the Universe when it was more compact than at the present time. It was a chaotic, turbulent process, in which strong gravitational forces hurled matter together at supersonic speeds. Fragments of the gas cloud, compressed by the effects of density and by shock waves, collided into each other forming the basis of what we now recognize as galaxies.

2.2.3 Galaxies

Today galaxies are huge concentrations of stars, that range in size from 80,000 to 150,000 light years in diameter. (A light year is the distance traveled at the speed of light in a year). Within this huge volume, galaxies contain very large numbers of individual stars – ca. 100 billion. There are at least 10 billion galaxies in the Universe.

There are two main groups of galaxies, spiral galaxies and elliptical galaxies. Spiral galaxies (see Fig. 2.3) are disk-shaped, contain abundant dust and gas and are home to regions of active star formation. Elliptical galaxies are spheroidal in shape, contain very little dust and gas, are not host to regions of star formation and so contain mainly old stars. Our own galaxy, the Milky Way, is of the spiral type. The spiral arms are thought to have been produced by density waves which sweep through the galaxy. The Sun is located in the outskirts of the Milky Way galaxy, between two spiral arms. It is these arms that form the Milky Way appearance at night.

In addition to the two main groups of galaxies, there are a number of other galaxy types. Of these Quasars (quasi-stellar objects) are galaxies with very energetic nuclei. They form small, very bright objects which emit as much as a 100 times the radiation of a normal galaxy. They only exist at immense distances and are thought to represent an early stage of

FIGURE 2.3 A typical spiral galaxy – the Whirlpool Galaxy, M51 and companion galaxy (from the HubbleSite picture gallery http://hubblesite.org/, Credit NASA, http://www.nasa.gov/, ESA, http://www.spacetelescope.org/, S. Beckwith (STSci), http://www.stsci.edu/, and the Hubble Heritage team (STSci/AURA), http://heritage.stsci.edu (http://www.stsci. edu/and http://www.aura-astronomy.org/)).

formation of large galaxies, formed, in part, through galactic collisions which took place early in the history of the Universe.

2.2.4 The process of star formation

As large molecular clouds fragment and collapse, star formation can take place. This process is triggered by density inhomogeneities in the gas cloud, producing regions which become gravitationally unstable and contract. Some stars appear to have formed in groups and are geometrically arranged, implying that a particular "event" might have initiated their genesis. Such an event could be caused by shock waves propagated from supernovae explosions or by density waves within the arms of a spiral galaxy. The process of gravitational collapse within a star leads to an increase in the temperature and pressure of the gas at its center, igniting thermonuclear fusion and turning hydrogen into helium. The radiation from this process leads to an outward pressure preventing further collapse.

The key features of stars which are of interest to astronomers are their mass, their luminosity, their surface temperature, and their distance from us. These parameters are used to classify stars and place them into an evolutionary sequence. A widely used classification diagram, based on optical data for stars, is the Hertzsprung–Russell Diagram (the H–R diagram) which is a plot of stellar luminosity versus effective surface temperature. The luminosity of a star is a function of its radius and effective temperature. The surface temperature is determined from its color and is vastly different from temperatures in the core of the star. For example, our sun has a surface temperature of about 5,700 K but a core temperature of 14 million K.

Data from nearby stars measured by the Hipparcos satellite and displayed on a Hertzsprung–Russell Diagram show four main groups of stars (Fig. 2.4). These are the following:

Main sequence stars, which make up 90% of all stars, define a curved trend across the center of the diagram, showing a relationship between mass and luminosity, such that large stars have a high luminosity and a high surface temperature and small stars have a lower

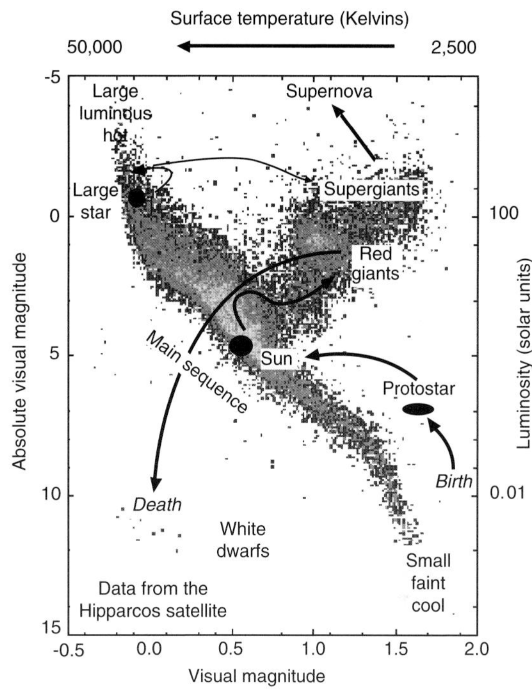

FIGURE 2.4 Hertzsprung–Russel diagram for the classification of stars. The data in this diagram are taken from the Hipparcos satellite which measured the properties of more than 10,000 nearby stars. Key variables are the relationship between the luminosity of the star (relative to the sun) or absolute visual magnitude, and its color index – the inverse of the surface temperature. The data in this diagram show the relative importance of the different types of star. The arrows show a typical birth to death cycle of a small star (lower part of diagram) and of a large star with a mass 25 times that of the sun (upper part of diagram).

luminosity and surface temperature. Hot, large stars burn up quickly and have a short life span whereas smaller cooler stars are more long-lived. For example a star in the top left of the diagram will last for about 10 million years before going supernova, whereas a star similar to our Sun will probably last 10 billion years.

Red giants are stars which are cooler, but more luminous than stars on the main trend. These are stars which have exhausted their supply of hydrogen and are now burning helium. This process causes them to expand. *Super giants* are similarly stars which are more luminous but much cooler than main sequence stars.

White Dwarfs are stars which are less luminous than main trend stars. These stars have exhausted H and He in their core and can be thought of as dead stars radiating away their energy.

2.2.5 The evolutionary cycle of stars

The cycle of birth–life–death for a star may be charted on a Hertzsprung–Russell diagram as illustrated in Fig. 2.4. Most stars evolve onto the main sequence and then into the Dwarf stage as follows (Lewis, 2004):

- An interstellar cloud of hydrogen and helium gas and dust is compressed. The contraction leads to an increase in density and a lowering of transparency and temperature, and simple molecules form.
- Continued contraction, now driven by gravitational collapse causes temperatures to rise. Molecules break down to atoms, ions, and subatomic particles. The highest temperatures are in the center of the mass, and it is here that nuclear burning begins giving birth to a protostar (Fig. 2.4).
- The newly-formed star subsequently evolves to be part of the main sequence of stars. The precise position on the main sequence depends upon the mass of hydrogen present in the star. At this point in its life temperatures in the core of the star are as high as 10^7K. Nuclear fusion takes place within the center of the star and helium is formed. At higher temperatures, above 10^8K, He reacts through nuclear fusion to form heavier elements such as carbon and oxygen (see Section 2.2.6.1). It is at this stage of stellar evolution that planetary formation may take place, although these processes are entirely unrelated to the high pressures and temperatures that ignite the fusion.
- Evidence from open star clusters shows that with aging stars migrate to the right on a H–R diagram and eventually convert into a Red Giant (Fig. 2.4). At this stage two separate thermonuclear processes are taking place in both the core and the shell. Hydrogen burning in the shell leads to massive expansion of up to 100 times the diameter of the original star.
- The final stage of stellar evolution depends upon star mass. A star smaller than the sun is unable to burn He and so ends up as the core of a Red Giant, which quickly lose its outer layers to space and fades away as a White Dwarf (Fig. 2.4). In larger stars, as the core collapses, temperatures and pressures become high enough for He to ignite and be converted to C and O in a supernova, and there is a huge explosion marking the end of their life (Fig. 2.4). Within the dense supernovae, neutron stars may form as the extremely dense endpoint in the life of a star.

2.2.6 The origin of the elements

Stars shine because nuclear reactions take place in their core. When these reactions take place there is a slight lowering of the mass of the nuclei undergoing fusion which is liberated as energy. It is the fusion process which gives rise to the formation of the chemical elements.

2.2.6.1 The processes of element formation

Our understanding of the process of element formation is based upon the elemental composition of stars, deduced from optical spectra, and from the theoretical calculations and experimental observations of nuclear physics. From these different lines of reasoning it has become clear that nucleosynthesis took place in a variety of different environments. Although the principal processes of nucleosynthesis take place in stars and supernovae, it is now also recognized that the nucleosynthesis of some light elements happened during the Big Bang, and to a lesser extent through the interactions between cosmic rays and matter in interstellar space.

Cosmological nucleosynthesis. The elements H, and its isotope D, He, and Li were created in the first few moments of the Big Bang. These are the essential ingredients of the cosmos and the starting composition for all other elements. The ratio of He/H, in terms of the number of atoms, is about 25% as a consequence of this event, and although some additional He has been created in stellar nucleosynthesis (see below) the ratio in the Universe as a whole has remained essentially unchanged since the beginning of time.

Stellar nucleosynthesis. Elements with atomic masses up to that of iron (^{56}Fe) are created in stars through a variety of different reactions, taking place over a wide range of temperatures.

Hydrogen burning and helium production. Hydrogen burns in the core of a star to form ^{4}He through either the proton–proton chain reaction, which takes place at 5×10^{6} K or at higher temperatures ($> 20 \times 10^{6}$ K) through the carbon cycle (the C–N–O cycle) in which carbon acts as a nuclear catalyst in the production of He. This process is also known as the quiescent burning phase of a star and is a slow process which takes billions of years and covers much of the life of a star. Our sun is currently in this phase.

Helium burning to form carbon and oxygen. As the hydrogen in a star is used up, the star contracts and its temperature rises to greater than 10^{8} K. At this stage nuclear reactions take place which permit the synthesis of the elements carbon, nitrogen, and oxygen, from helium. ^{12}C forms from ^{4}He, through what is known as the triple alpha reaction, and when sufficient ^{12}C is present, further reaction leads to the formation of ^{16}O.

Carbon and oxygen burning. When the helium is almost completely consumed then the carbon and oxygen can be transformed into elements with masses up to that of silicon. This takes place after the stellar core has contracted further and increased in temperature. In detail carbon fusion reactions (^{12}C) lead to the formation of ^{24}Mg, ^{23}Na, and ^{20}Ne at about 6×10^{8} K. Oxygen fusion reactions lead to the formation of ^{32}S, ^{31}P, ^{31}S, and ^{28}Si at about 10^{9} K.

Silicon burning. As carbon and oxygen burning proceeds the stellar core becomes enriched in Si and is at temperatures of about 10^{9} K. At these temperatures nuclear reactions in the stellar core induced by photons lead to the formation of elements with masses up to that of iron. Elements heavier than Si cannot be formed by the process of nuclear fusion because beyond this point the nuclear reactions cease to be an energy source. The most energetic fusion reaction is hydrogen burning; a lesser amount of energy is produced by He-burning, even less from C and O, and progressively less until Fe is reached at which point no energy is released at all.

Explosive burning in a supernova. In contrast to the nuclear reactions thus far described, the formation of elements beyond the mass 56 (Fe) consumes energy. This process is that of neutron capture and involves the absorption of neutrons by the atomic nucleus. Hence heavier elements such as silver, gold, or lead can only be formed in a highly energetic environment within a star, such as found in a supernova explosion, because only in this environment is sufficient energy released to allow the energy-inefficient process of heavy-element formation to take place. Supernovae explosions are the endpoint of large stars and represent the violent collapse of a Fe-rich stellar core, during which neutrons, produced in the core collapse, are captured by other nuclei. They are built into heavy nuclei, through rapid neutron capture, up to the elements Th and U. Hence, the reason that the heavy elements are so rare is because the process by which they are formed is rare – approximately only one in a million stars is massive enough to go supernovae.

2.2.6.2 The interpretation of solar elemental abundances

Solar element abundances (Anders & Grevesse, 1989) are plotted against increasing atomic numbers in Fig. 2.5. These data were obtained by the optical analysis of the solar spectrum. Abundances are measured relative to 1 million silicon atoms on a logarithmic scale.

There are three important observations to be made from this graph (Fig. 2.5):

First, the graph has an overall smooth trend from light to heavy elements, indicating that solar abundances are greatest for light elements and least for heavy elements. This is consistent with the discussion above, in which H and He abundances date from the Big Bang, and are the starting material from which all other elements have been built. It is also consistent with the burning of He to form C and O, and the burning of C and O to form successively heavier elements.

Second, superimposed upon the smooth trend is a smaller scale irregularity, such that

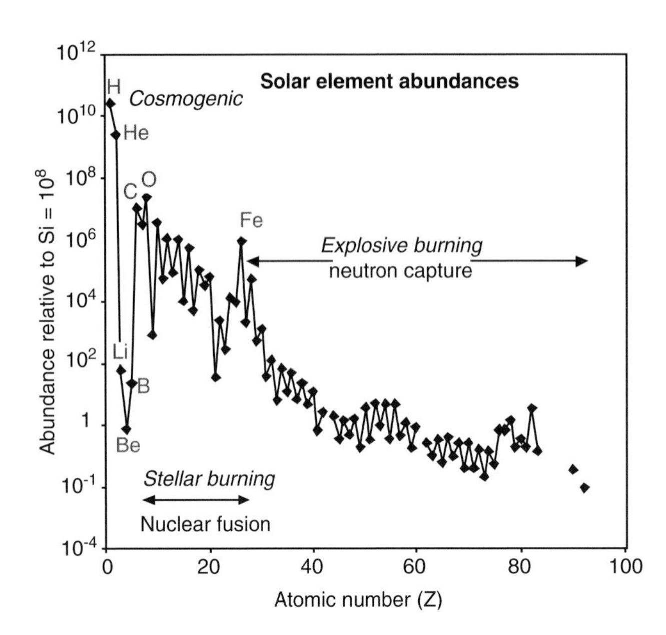

FIGURE 2.5 Abundances of elements in the sun, by atomic number, relative to solar Si = 10^6 (Data from Anders & Grevesse, 1989).

elements with even atomic numbers have higher abundances than those with odd atomic numbers. This can be explained by the greater stability of atomic nuclei with paired neutrons. Thus elements with even atomic numbers have the greater nuclear stability and therefore the greater abundance.

Third, some elements have anomalous abundances. Hydrogen and helium and iron have anomalously high concentrations. H and He have been discussed already. In the case of Fe this relates to the high binding energy and associated stability for Fe. The elements lithium, boron, and beryllium have anomalously low concentrations for they are not produced in stellar nucleosynthesis, as already discussed.

2.3 The condensation of the Solar System

The generally accepted model for the origin of the Solar System is that the Sun formed in a similar manner to any other medium angular momentum star, by the collapse of a region of gas in the interstellar medium. Compression associated with this collapse gives rise to an increase in temperature within the nebula and a small part (2–10%) of the gas condenses and forms a flattened disk of dust surrounding the star. This rotating disc of gas and dust is known as the primitive solar nebula, and it is from this that the planets and other bodies of the solar system formed. In this section we shall consider the processes which led to the formation of planets from the solar nebula, concentrating in particular on the formation of the Earth.

Our knowledge of the state of the primitive solar nebula in its early stages of development comes from a number of sources. These include the observations and calculations made by astronomers, from the principles of elemental behavior in planetary systems and from the study of meteorites and other extraterrestrial materials.

2.3.1 Evidence from astronomical observations

Astronomical observations of molecular clouds and young stellar objects provide the basis for our understanding of the early solar system (Cameron, 1995; Alexander et al., 2001). The first stage in this process is when a fragment of an interstellar molecular cloud collapses to form a disk-like nebula, or protoplanetary disk. This process normally takes between 10^6 and 10^7 years, although if triggered by interstellar shock waves it may be as short as 10^5 years.

During gravitational collapse within the molecular cloud, the adiabatic compression of gas from the interstellar fragment leads to the formation of an extremely hot core of gas, the central protostar. Most commonly, this central star becomes surrounded by a disk of condensed matter, which will become the planetary system. Once such a disk is formed all matter subsequently accreting to the star is processed through this disk. As the gravitational collapse of the molecular cloud continues adiabatic compression leads to the heating of the cloud, and the highest temperatures in the solar nebula are reached during this early stage of its development.

2.3.1.1 *Solar nebula evolution*

Astronomers recognize four classes (0 to III) of young stellar objects which can be used to track the development of the early solar system history. These are as follows. Class 0 objects are deeply embedded in their parental molecular cloud with most of their circumstellar mass still in the cloud rather than in the disk. A particular feature of this stage of solar nebula evolution is the presence of narrow bipolar jets of high velocity gas, which are ejected outward from center of the disc, perpendicular to the midplane. Excellent Hubble space telescope images of these jets have been observed on the young star HH30 (Boss, 1998). These jets continue into the later stages of nebular evolution and are a source of thermal energy in the otherwise cooling nebula. This stage of solar nebula evolution lasts about 10^4 years.

Class I objects are characterized by the build up of material in the disk followed by a burst of accretion onto the star. It is during this phase that the protostar acquires its main mass and grows rapidly over a timescale of 10^3–10^5 yr.

Class II objects are T Tauri stars, young stars with a cool surface but high luminosity, which lie above the Main Sequence trend on a H–R diagram, and are identified as protostars on Fig. 2.4. These stars retain well-developed dusty discs which typically have a mass of 0.02 solar masses. This stage may last from 10^5 to several million years. During this stage of solar nebula development, solar winds directed radially outward inhibit further accretion to the sun, and planetesimals and planets begin to form (Alexander et al., 2001). At this stage the nebula has lost most of its original (adiabatic) heat, facilitated by its disk-shape, and the condensation of water to ice is at the orbit of Jupiter.

Class III objects are weak line T Tauri stars (T Tauri stars in which the characteristic emission lines are only weakly observed in their optical spectra) and have little or no evidence of a disk. At this stage of solar nebula evolution, which may last between 3 and 30 Ma, the sun has formed, and the material of the nebula is being dissipated by solar winds in the inner part. In the outer part of the nebula material is dissipated by photo-evaporation caused by UV radiation from the solar wind. A positive pressure gradient near the inner edge of the nebula facilitates planetesimal formation.

Temperatures within the nebula vary over time and within the midplane of the disc decrease with distance from the core as shown in Fig. 2.6. They are also strongly dependent on the mass of the nebula, expressed as a fraction of the solar mass. The model in Fig. 2.6 is calculated for nebula masses of 0.02 and 0.04 solar masses (Boss, 1998).

A much cited example of a solar nebula is the main sequence star Beta Pictoris, which shows a central core, with a disc of matter rotating around the core (Fig. 2.7). Beta Pictoris is thought to be a young star (10–20 Ma old) showing the early stages of planetary formation and is "a somewhat messy planetary system caught in formation" (Artymowicz, 1997). It has a large (>1,000 AU) dust-dominated, thin disk surrounding the central star. The dust is of a similar composition to that of the solar system and contains Mg olivines and pyroxenes. There is no evidence of ice. Scattered light, from within the disk, analyzed using infrared thermal imaging has identified particles of all sizes – from the micron and millimeter scale. This distribution of dust particles has been used to infer the presence of

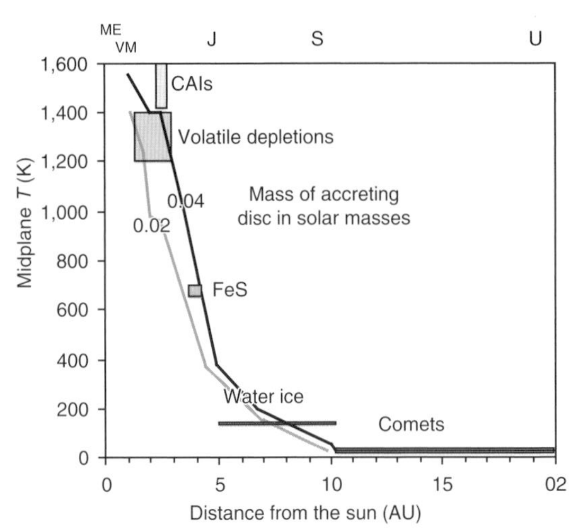

FIGURE 2.6 Midplane solar nebular temperatures (K) calculated for 0.04 and 0.02 solar masses in the accreting disk, and estimated temperatures from meteorites and comets, plotted against distance from the sun, expressed both as astronomical units (1 AU = Earth–Sun distance) (bottom) and planetary distance (top). The temperatures presented here are indicative only, as they are dependent upon the size of the disc and the thermal model used (after Boss, 1998).

planetesimal belts around the star (Okamoto et al., 2004; Telesco et al. 2005).

2.3.2 Evidence from cosmochemistry

A rather different approach to understanding the condensation of the solar nebula came from the work of the geochemist V.M. Goldschmidt carried out in the 1920s. Goldschmidt proposed, what has now become, a widely used geochemical classification of the elements. This work was in part based upon the study of meteorites, and so his classification is very relevant to the understanding of planetary processes.

Chemical elements display different chemical affinities, explained largely by their differing electronegativities, and may be classified into *lithophile elements* – those with an affinity for silicates and oxygen, *chalcophile elements* – those with an affinity for sulfur, *siderophile elements* – those with an affinity for metallic iron, and *atmophile elements* – those with an affinity for the gaseous atmosphere. Initially Goldschmidt's classification of the elements was helpful in understanding two rather different subjects – the differences

FIGURE 2.7 Beta Pictoris imaged in 1993, using the 2.2 m telescope of the Hawaii University at Mauna Kea. Kalas and Jewitt (1995) demonstrated that the disk is asymmetric, a feature that is often the result of gravitational perturbations from planets, although not thought to be the case in this study.

TEXT BOX 2.1 A geochemical classification of the elements

Geochemists have their own particular way of classifying the chemical elements as found in rocks and minerals.

Major elements are those chemical elements which make up the principal part of rocks. These are the elements which make up the main rock-forming minerals. They are normally reported as metals, but, since oxygen is also an important part of the Earth, by convention they are presented as percentages by weight of the metal oxide. They tend to be listed in the order – SiO_2, TiO_2, Al_2O_3, FeO, MnO, MgO, CaO, Na_2O, K_2O, and P_2O_5. Sometimes structurally bound water (listed as H_2O) may also make up several percentages by weight of the rock. Major element analyses should sum to about 100 wt%. A typical set of major element analyses is given in Chapter 3, Table 3.1.

Trace elements are those elements which do not normally make up rock-forming minerals in their own right, and are present at the part per million level (ppm) or lower. Over the past 50 years methods of determining the concentrations of trace elements in rocks have improved dramatically, so that determinations with precision at the parts per billion (ppb, 10^{-9}) level are not uncommon. A typical set of trace element analyses is given in Chapter 3, Table 3.2.

Minor elements are those elements which have concentrations between 1.0 and 0.1 wt%. They include elements such as Ti, Mn, K, and P. These elements may behave either as a major element or a trace element. For example, K may be present as several wt% in granite and may be part of a major rock-forming mineral (K-feldspar). In a basalt however, K may be present only at the ppm level, substituting for Na in plagioclase feldspar.

A geochemical classification of trace elements according to their position in the periodic table

It is useful in geochemistry to recognize particular groups of elements on the basis of their position in the periodic table. This is because their chemical similarities lead us to expect some similarity in their geochemical behavior in natural systems. In this text, reference will be made to:

- the *noble gases* (rare gases or inert gases) Ne, Ar, Kr, and Xe;
- the lanthanide series, or the *rare Earth elements (REE)*, as they are more normally known – elements 57 to 71 – La, Ce, Pr, Nd, (Pm), Sm, Eu, Gd, Tb, Dy, Ho, Er, Tm, Yb, and Lu;
- the *platinum group elements (PGE)*, Ru, Rh, Pd (elements 44–46), Os, Ir, and Pt (elements 76–78) and sometimes including Au (79);
- the *transition metals*, that is, the first transition series, Sc to Zn, elements 21 to 30.

A geochemical classification of trace elements according to their behavior during partial melting

During partial melting trace element behavior is governed by the preference of a particular element for its host (a mineral phase) or the melt phase. Elements which tend to become part of the melt are known as *incompatible elements*, and those which prefer to remain in the mineral phase are known as *compatible elements*. In detail this is governed by relationships between the particular element and the structure of the relevant mineral phase. The degree of incompatibility of a specific element in a particular mineral phase is expressed as the mineral–melt partition coefficient. Incompatible elements have mineral–melt partition coefficients < 1.0 and highly incompatible elements have partition coefficients $\ll 1.0$, whereas compatible elements have partition coefficients of > 1.0.

A geochemical classification of trace elements according to their ionic charge and size ratio

Small, highly charged ions behave differently from large ions with a low charge during geochemical processes. Small highly charged ions include metals such as Ti, Hf, Nb, and Zr and are known as the *high field strength elements (HFSE)*. These are elements which tend to be immobile when hydrous fluids react with a rock. Examples of larger ions carrying a low charge are Ba, Sr, and K. These are known as the large ion lithophile elements (LILE) or *low field strength elements*. These elements tend to be mobile in hydrous fluids.

A geochemical classification of the elements based upon their electronegativities

Lithophile elements are those which have a preference for a silicate host, whereas chalcophile elements have an affinity for sulfur and so will most frequently be found in sulfides. Siderophile elements are those which will partition preferentially into a metallic iron phase and so are enriched in the Earth's core and in iron meteorites. Atmophile elements prefer the gaseous phases of the Earth atmosphere. This classification is discussed more fully in Chapter 2, Section 2.3.2.

A cosmochemical classification of the elements based upon the solar condensation sequence

During the condensation of a solar nebula different mineral phases condense as the nebular temperature decreases. It is an understanding of this process which has led to an appreciation of the mineralogy of meteorites. Phases which condense at high temperatures (1850–1400 K) are known as *refractory*, whereas phases which cool at lower temperatures are know as *volatile*. Highly volatile phases condense below 640 K (see Section 2.3.2.1).

between the major meteorite groups (Section 2.3.3.1) and the differentiation of the Earth.

However, Goldschmidt's scheme only relates to the condensation of major elements into mineral phases. As the solar nebula hypothesis gained credence, it became clear that there are other element groupings which relate to the condensation of a high-temperature solar gas. These are: the *refractory elements* – those which formed above the condensation temperature of the Mg silicates and Fe–Ni metal, at 1,300–1,400 K, the *moderately volatile elements* – those formed in the range 1,300–670 K, and the *volatile elements* – those that formed below the condensation temperature of FeS, at 670 K (Larimer, 1988). Palme and O'Neill (2003) have proposed a cosmochemical/geochemical classification of the elements based on these two elemental groupings (Table 2.1).

2.3.2.1 The solar condensation sequence

The processes of vaporization and condensation are of major importance in the solar nebula. In order to systematize this process, Grossman (1972) used thermodynamic equilibria to calculate the composition of phases in equilibrium with a gas with cosmic element concentrations, at a pressure of 10^{-3} atm, and as a function of temperature. This work has subsequently been developed by others and is systematized in Lewis (2004). It is worth noting that in detail it is likely that the assumption of equilibrium conditions will not always apply to the condensation sequence (Wood, 1988). Nevertheless, observations of this type are invaluable in providing a first pass at interpreting the processes of planetary formation.

The solar condensation sequence may be described as a series of steps which describe the formation of phases, and subsequent reactions between phases, during the condensation of the solar gas (Lewis, 2004). These steps explain the main sequence of mineral phases forming in a solar nebula in relation to temperature and distance from the sun (Fig. 2.6). The temperatures indicated are taken from the adiabatic curve of Lewis (2004):

1 Formation of the refractory siderophiles. (The metals W, Os, Ir, and Re – although in reality concentrations are so low that these phases do not nucleate.)

2 Formation of refractory oxides (ca. 1,700 K). (Al, Ca, and Ti oxides such as corundum, spinel, perovskite, and some silicates. The phases also include the REEs and U and Th.)

3 Formation of iron–nickel metal (ca. 1450 K) (also included are the minor elements Co, Cu, Au, Pt, Ag and may include the nonmetals P, N, C).

4 Formation of magnesium silicates (ca. 1,420 K) (the principal components are olivine and Mg–pyroxene).

5 Formation of alkali metal silicates (ca. 1,020 K) (the major component is plagioclase feldspar).

TABLE 2.1 A cosmochemical and geochemical classification of the elements based upon their Lithophile/siderophile/chalcophile affinities and their refractory or volatile character (after Palme and O'Neill, 2003).

	Lithophile (silicate)	Siderophile and chalcophile (metal and sulfide)
Refractory (T_c *1,850–1,400 K*)	Al, Ca, Ti, Be, Ba, Sc, V, Sr, Y, Zr, Nb, Ba, REE, Hf, Ta, Th, U, Pu	Mo, Ru, W, Re, Os, Ir, Pt
Main component (T_c *1,350–1,250 K*)	Mg, Si, Cr, Li	Fe, Ni, Co, Pd
Moderately volatile (T_c *1,230–640 K*)	Mn, P, Na, B, Rb, K, F, Zn	Au, As, Cu, Ag, Ga, Sb, Ge, Sn, Se, Te, S
Highly volatile ($T_c < 640$ *K*)	Cl, Br, I, Cs, Tl, H, C, N, O, He, Ne, Ar, Kr, Xe	In, Bi, Pb, Hg

Condensation temperature T_c is at a pressure of 10^{-4} bars.

6 Formation of the moderately volatile chalcophiles at 670 K (FeS, with Zn, Pb, and As).
7 Formation of silicates with mineral-bound OH (ca. 430 K; this group includes the hydrated silicates – the amphibole tremolite, serpentine, and chlorite).
8 Formation of ice minerals (ca. 140 K) (to include water ice, solid hydrates of ammonia, methane, and rare gases).
9 A residue made up of permanent gases (gases which under natural conditions will not condense are H_2, He, and Ne).

This condensation scheme provides a valuable framework for understanding the mechanisms behind the formation of the different components found in primitive meteorites and a basis for understanding the differentiation of the Earth. Both of these topics are discussed in subsequent sections of this chapter.

2.3.3 Evidence from Meteorites

Meteorites are extremely important to our understanding of solar system evolution, because, in their most primitive form, they are our most ancient samples of the solar system. As such they provide valuable information about the condensation of the solar nebula from which our solar system formed. Whilst they represent, to date, our most abundant sample of extraterrestrial material, we have no idea how representative this is of the solar nebula material as a whole.

Meteorites, as rocks produced in a solar nebula, have formed through a range of processes, some of which are quite different from those observed on Earth. Thus while igneous differentiation processes and metamorphism are recognized in meteorites, there also other processes operating which are not observed in terrestrial rocks. These include evaporation and condensation events related to melting in a gas-rich medium (as discussed in Section 2.3.2), impacting events, and metal-silicate fractionation.

2.3.3.1 Types of meteorite

Meteorites may be subdivided into two main categories – unmelted meteorites, that is those which come from a parent body which has not been fractionated since its aggregation early in the history of the solar system, and melted, or differentiated meteorites (Fig. 2.8). Unmelted meteorites are stony meteorites, the chondrites, and are made up of the same silicate minerals that are found on Earth. Melted meteorites are of three types. They include some stony meteorites (the achondrites), the iron meteorites, whose composition is dominated by a metallic iron–nickel alloy, and stony-iron meteorites, meteorites which are made up of approximately equal proportions of

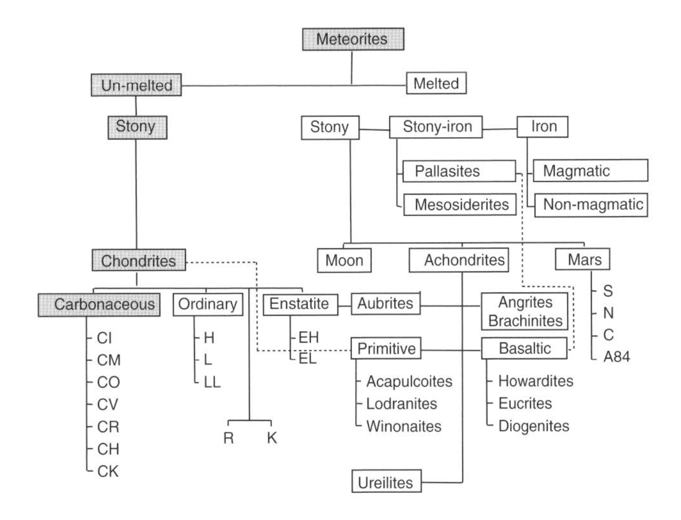

FIGURE 2.8 Meteorite classification from the Natural History Museum of London's *Catalog of Meteorites, Fifth Edition* (Grady, 2000).

silicate minerals and iron–nickel metal (Grady, 2000). Melted meteorites have experienced chemical differentiation during their history and so do not preserve the most primitive matter of the solar system. For this reason much of the discussion which follows will concentrate on the unmelted, chondritic meteorites.

Chondrites. Chondrites are stony meteorites and are the most abundant meteorite type (87% of all meteorites). Their radiometric ages are around 4.56 Ga and these ages are thought to define the time when the solar system formed. Chemically their element abundance patterns, apart from the very light and/or volatile elements, are the same as that of the sun and other stars, and for this reason they are thought to represent undifferentiated cosmic matter. Chondrites therefore are thought to represent the most primitive material in the solar system. They are the "stuff" from which all other rocky materials were built.

Chondrites are ultramafic in composition and contain the minerals olivine, pyroxene, and metallic iron. They are composed of three main components (see Fig. 2.9), each of which represents a different component of primitive solar nebula material:

- Chondrules – spheroidal ultramafic melt droplets a millimeter or so in diameter, which tend to dominate the texture of their host and from which chondrites take their name.
- CAI's – refractory inclusions, or Ca–Al-Inclusions, up to 2 cm across, enriched in Si-poor, Ca–Al-rich minerals. The most abundant source of CAIs is the Allende meteorite, which fell in 1969.
- Matrix – porous, fine grained mineral matter that fills the space between the chondrules and CAIs.

Chondrites are subdivided into carbonaceous (C), ordinary (O), and enstatite (E) varieties (Fig. 2.8). Carbonaceous chondrites are volatile rich and contain abundant carbon in their matrix. Because they have a high volatile content they are thought to be the most primitive of all chondrites. Within this group there are a number of varieties named after type specimens designated CI, CM, CV, etc. An earlier classification used C1 to C3. CI chondrites are the most primitive meteorites within the carbonaceous chondrite groups and the most primitive of all meteorite types. They are the least chemically fractionated and have the highest volatile content. Ordinary chondrites, as their name implies are the most abundant

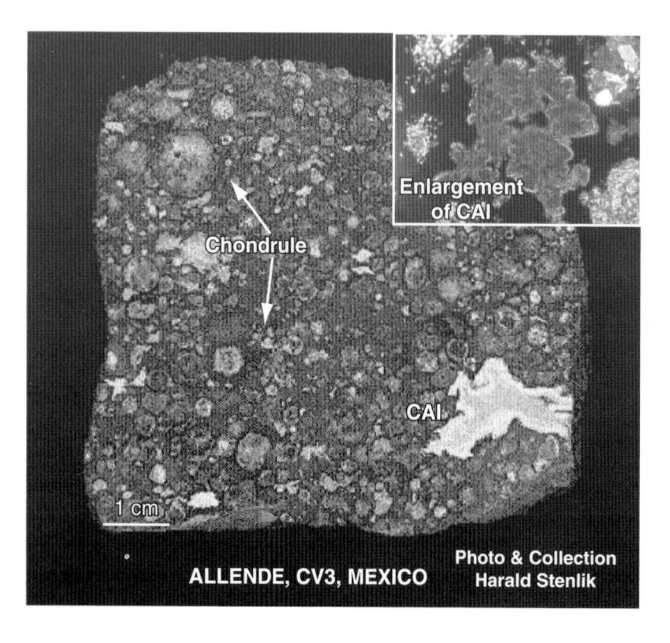

FIGURE 2.9 A slice of the Allende meteorite showing rounded chondrules, a large white CAI and a dark (fine grained) matrix. The inset at the top right shows an enlargement of a CAI in thin section (from http://www.meteorite.com/gallery/allende.htm; photo Harald Stehlik).

of all chondrites. These are subdivided on the basis of their iron content. Enstatite chondrites are highly reduced meteorites and more siliceous so that enstatite rather than olivine is the dominant silicate phase. They are thought to have formed under different redox conditions from other chondrites, probably in a different region of the solar nebula (Kallemeyn & Wasson, 1986).

Achondrites. Achondrites are stony meteorites formed by the melting of their parent body. They are differentiated meteorites which have lost their original metal content. Generally they do not contain chondrules. There are a number of different categories of achondrite representing melted chondrites, basaltic igneous rocks, and planetary regolith breccias.

Iron meteorites. Iron meteorites are thought to be derived from the segregated metallic iron cores of small planetary bodies, which were originally a few tens to hundreds of kilometers in diameter. They demonstrate that metal-silicate fractionation was a fundamental process during the evolution of the solar nebula. Mineralogically they are composed of the minerals kamacite (Fe–Ni metal with a low $< 7\%$ Ni content), and taenite (Fe–Ni metal with a high Ni content, 20–50%). Iron meteorites are subdivided into magmatic irons, iron meteorites that have solidified by fractional crystallization from a melt, and nonmagmatic irons, iron meteorites which do not seem to have completely melted. There is also a chemical classification based upon the concentration of Ge and Ga (Wasson, 1985).

Stony irons. Stony-iron meteorites are those which contain equal proportions of silicate minerals and metallic iron. Pallasites are made up of olivine and Fe–Ni metal and are thought to represent samples from the core–mantle boundary of their parent body. Mesosiderites are brecciated mixtures of silicates and Fe–Ni metal.

Meteorites from other planetary bodies. A small number of lunar meteorites are known. These include anorthositic regolith breccias, basalt, and gabbro. Martian meteorites are identified from their noble gas and oxygen isotope compositions and are known as SNCs (pronounced "snicks") after the three common types – shergottites, nakhlites, and chassignites. Probably the most famous Martian meteorite ALH84001 is in a class of its own and is the only sample to date of very primitive Martian crust with a crystallization age of ca. 4.5 Ga. This meteorite was thought at one time to be the host to fossilized bacteria (see Chapter 6, Section 6.1.3.4).

2.3.3.2 The primitive matter of the solar nebula

Chondrites, the most primitive of all meteorites, formed in dynamic energetic, dust-rich zones in the solar nebula. In this environment, dust/gas ratios were constantly changing, temperatures fluctuated through 1,000 K, with multiple cycles of melting, evaporation, condensation, and aggregation. In addition there were influxes of matter from the interstellar dust and the periodic removal of batches of chondritic material to small planetesimals. In this section we explore how the most primitive materials of the solar system were formed and what they can tell us about processes during the condensation of the solar nebula. These materials include chondrules, refractory inclusions (CAIs), and amoeboid olivine aggregates (AOAs), the oldest component parts of chondritic meteorites.

An approach which has been very fruitful in identifying different regions within the solar nebula environment is the study of oxygen isotopes. Clayton et al. (1976) showed that, on a three isotope $\delta^{17}O$–$\delta^{18}O$ plot, the different classes of meteorites defined distinct groups (Fig. 2.10b). On the basis of these distributions Clayton et al. (1976) made two very important observations. First, that chondritic meteorites have very different oxygen isotope ratios from terrestrial rocks. Second, that chondritic meteorites record the fact that a number of different processes took place within the solar nebula (Fig. 2.10). More recently, Clayton and Mayeda (1999) described in detail the oxygen isotope chemistry of carbonaceous chondrites and showed that they lie on a series of mixing lines (Fig. 2.10). Anhydrous minerals from CAIs define the carbonaceous chondrite anhydrous mixing line (CCAM) with a strongly ^{16}O-enriched end-member. Other chondrites define hydration–alteration mixing lines along which there is mixing between a low-temperature end-member, interpreted to be a ^{16}O-poor fluid, and one of a number of anhydrous starting compositions.

2.3.3.2.1 Chondrules

Chondrules are the principal constituent of many chondritic meteorites (Fig. 2.9) and their formation represents a major, pervasive, high-temperature process in early solar system history (see Zanda (2004) for a recent review). They are made up of silicate, metal, sulfide, and glass phases and in detail show a wide variation in chondrule composition, extending from iron-poor to iron-rich and silica-poor to silica-rich varieties. Some chondrules are composite and show high temperature rims on older cores (Kring, 1991).

There are two possible explanations for the chemical variability of chondules. One emphasizes variations in the mix of precursor solids. In this model compositionally different chondrules reflect different starting materials. Alternatively, chondrules vary in composition because of the chondrule-forming process, and record a reaction between chondrules and the ambient gases. A recent experimental study by Cohen et al. (2004) showed that chondrules do in fact vary in composition as a result of open

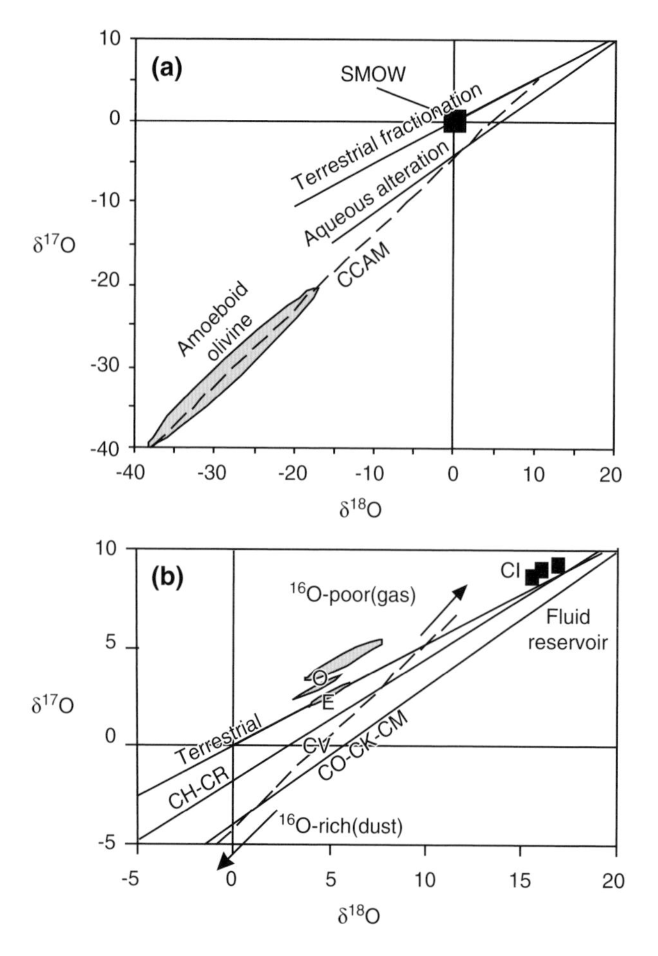

FIGURE 2.10 (a) An oxygen isotope plot for chondritic meteorites showing the two main meteorite trends, relative to the terrestrial mass-dependent fractionation (MDF) trend (which passes through the oxygen isotope standard SMOW). The CCAM line (dashed) represents a mixing line defined by separated minerals from refractory CAI inclusions. Amoeboid olivines plot at the lower ^{16}O-rich end of this line. The aqueous alteration trend is shown in more detail in (b) below. (b) A detail of (a) showing the compositional ranges for the different carbonaceous chondrite groups (CI, CH, CR, CO, CM, CK, CV) and ordinary (O) and enstatite (E) chondrites. The steep carbonaceous chondrite anhydrous mineral line (dashed) is shown to be a mixture of ^{16}O-rich dust and ^{16}O-poor gas. The different carbonaceous chondrites are shown to be the product of the hydrous alteration (^{16}O-poor, fluid reservoir) of two different (CH–CR and CO–CK–CM) anhydrous components (^{16}O-rich). Data from Clayton et al. (1976), Clayton and Mayeda (1999), Fagan et al. (2004) (see Text Box 5.1, Chapter 5).

system processes, implying that it is the nebular environment rather than a difference in starting material that is the principal control on chondrule formation. It is likely therefore, that different chondrule types formed in regions of different chondrule density, which in turn experienced different degrees of evaporative loss.

The precise nature of the chondrule-forming event, the "chondrule factory," is also the subject of some discussion. Some workers favor the formation of chondrules by the direct condensation of the solar nebula gas as a melt. Alternatively, chondrules could be the residues of evaporation. The presence of relict olivine grains in some chondrules (Jones, 1996) argues against a simple condensation model and may imply multiple condensation-evaporation events.

A further important and yet incompletely resolved matter is the nature of the heating event in which chondrules formed. The experimental study of Cohen et al. (2004) shows that chondrules probably formed in a matter of minutes to hours. This implies a flash heating process (Boss, 1998), such as might be caused by shock-heating or lightning strikes in the solar nebula. If this is the case then chondrule melting temperatures represent transient temperatures rather than the ambient temperature of the solar nebula.

2.3.3.2.2 Ca–Al-rich inclusions

CAIs in chondritic meteorites contain the refractory minerals melilite, spinel, pyroxene, plagioclase, and perovskite. CAIs formed in a high-temperature part of the solar nebula, where temperatures were above the formation temperature of forsterite (>1,444 K). CAI's are commonly found in carbonaceous chondrites, and more rarely in ordinary and enstatite chondrites. They are also thought to have formed through evaporation and condensation processes in the solar nebula and may be the product of transitory heating events (Wood, 1988). Oxygen isotope studies show that CAIs are characterized by an extreme enrichment in ^{16}O indicating that they formed in an unusual ^{16}O-rich environment in the solar nebula, the origin of which is not resolved (Nittler, 2003). Mixing trends on an $\delta^{17}O - \delta^{18}O$ diagram suggest that there are at least three different primordial oxygen isotope reservoirs recorded in the formation of CAIs (Fig. 2.10b, Young & Russell, 1998; Clayton & Mayeda, 1999). An important recent observation is that CAIs are now known to have formed over a very short time interval (0.05–0.5 Ma) (Bizzarro et al. 2004; Wood, 2004). Wood (2004) has used this short timescale as constraint to infer that they must have formed in young stellar objects with short histories, that is, in stellar objects of Class 0 or Class I, rather than in the longer T Tauri stage (see Section 2.3.1.1).

2.3.3.2.3 Amoeboid olivine aggregates

AOAs are irregular objects also found in chondrites and are up to 1 mm long. They comprise granular olivine, intergrown with the refractory phases diopside, anorthite, and spinel, and contain grains of Fe–Ni metal. The presence of olivine indicates that these aggregates formed in the solar nebula at lower temperatures than CAIs, at below the forsterite condensation temperature. However, oxygen isotope studies by Fagan et al. (2004) show that AOAs have an identical ^{16}O enrichment to that found in CAIs (Fig. 2.10). This implies that AOAs and CAIs formed from the same parcel of gas in the same nebular region, but their mineralogical differences suggest that they formed over different temperature intervals (Fagan et al., 2004).

2.3.3.2.4 The relationship between chondrules, CAIs, and AOAs

The contrasting mineralogy of CAIs and chondrules indicates that they formed at different nebular temperatures and perhaps in different nebular environments. This is confirmed by oxygen isotope studies which show that there are significant oxygen isotope compositional differences between chondrules and refractory CIAs and AOAs. Chondrules from C and O chondrites also have different oxygen isotope compositions and appear to have formed in different nebular environments. In contrast, CAIs from all chondrite types and AOAs are isotopically similar and share the same range of light-isotopic oxygen rich compositions (Guan et al., 2000).

If AOAs and chondrules originally formed in different nebular environments, how they subsequently came together in chondrites is a problem. CAIs and AOAs, must have formed in a restricted region within the solar nebula and were subsequently dispersed and mixed with chondrules. In part the relationships may be explained in terms of the timing of their formation, as is being revealed by modern high precision isotope measurements. For example, recent Pb isotope studies of chondrules in carbonaceous chondrites show that they formed over a period of about 4 Ma, between 4,563 and 4,567 Ma. CAIs on the other hand are known from ^{26}Al studies (see Text Box 2.3) to have formed slightly earlier than, or at the same time as, chondrules but over a short time interval of < 0.6 Ma, at 4,567 Ma (Amelin et al., 2002, 2004; Bizzarro et al., 2004; Haack et al., 2004). In fact increasingly precise age measurements of chondrules and CAIs suggest multiple melting events within very short time intervals (Krot et al., 2005; Young et al., 2005).

These results suggest that CAIs and chondrules formed in a dynamic solar nebula setting, in which a number of different processes were operating, which we now categorize as chondrule formation. It is likely that large oxygen isotope fluctuations were taking place at this time with O-isotopes compositions varying on a scale of months to years.

2.3.3.3 *Presolar matter*

The study of presolar matter is one of the growing areas of modern space science. Presolar grains are of great scientific importance because they bring stellar material into the laboratory where it can be directly analyzed and provide much more stringent tests of models for stellar evolution and nucleosynthesis than can be provided by the remote sensing of stellar dust from observational astronomy. It is thought that presolar grains formed in red giant stars, of a type known as asymptotic giant branch (AGB) stars, and in supernovae (Nittler, 2003).

Presolar grains have been identified in both chondritic meteorites, and in interplanetary dust particles (IDPs) collected from the Earth's stratosphere. They are identified by their rare gas isotopic signature which is different from that which characterizes our solar system. This unusual isotopic signature implies their genesis in a different stellar environment, outside of the solar system. This previous history means that these grains formed before the solar system began, and traversed interstellar space prior to their incorporation into the solar nebula. "Each grain is a frozen piece of a single star which ended its life before the formation of the solar system" (Nittler, 2003). The grains are normally extremely small, about 1 μm, and are only found in meteorites if they are subjected to extensive acid dissolution – a process which has been likened to burning a haystack in order to find the needle. Even then, their extrasolar character can only be recognized after detailed isotopic examination by ion probe.

2.3.3.3.1 *Presolar grains in meteorite residues*

Presolar grains are found in carbonaceous chondrites and include graphite spherules, organic carbon, nano-diamond, silicon carbide, and a range of other metal carbides. More rarely there are grains of silicon nitride and the oxides corundum, spinel, and the phase hibonite (a Ca–Al-oxide) (Nittler, 2003; Zinner et al., 2003). Phases such as corundum, hibonite, and spinel are phases which are expected to condense first in a gas of solar composition. The presence of these phases as presolar grains indicates that their stellar sources have a similar condensation sequence to that of the solar nebulae outlined in Section 2.3.2.1. In detail, presolar grains tend to have very variable isotopic compositions implying that they have been derived from many different stellar environments (Zinner et al., 2003).

2.3.3.3.2 *Interplanetary dust particles*

IDPs collected in the Earth's stratosphere are of multiple origins. Solar system objects come in anhydrous and hydrated forms. The anhydrous, chondritic-porous group are thought to be the most primitive and least-processed of all matter in the solar system and are probably of cometary origin. Hydrated IDPs, on the other hand, are linked to an asteroid origin (Keller et al., 2004).

IDPs of presolar origin are also abundant in stratospheric dust and include silicate-rich and carbon-rich grains. Currently the only silicates observed are forsteritic olivine and a glass embedded with metal and sulfide (GEMS) (Messenger et al., 2003). Some organic molecules have a D- and N-isotopic signature which can only be explained by their origin in a very low-temperature molecular cloud (Keller et al., 2004) and provide important insights into the nucleosynthetic processes which take place within this environment.

2.3.4 Planetary formation

The final stage in the condensation of a solar nebula is that of planetary formation. Planets are thought to form in a protoplanetary disk of the type observed around young stellar objects such as Beta Pictoris (Section 2.3.1.1). The lifetime of such dust-laden protoplanetary disks is of the order of 10^7 years. Until recently we knew very little about processes operating in such protoplanetary disks, and much of what we did know was drawn from the study of the solar system. However, the recent discovery of planetary systems around other stars suggest that our solar system may not be characteristic of planetary systems in general and many new advances can be expected in this field.

2.3.4.1 The standard model of planetary formation

Wetherill (1990) has provided the standard model of planetary formation, based upon the planetesimal hypothesis. This model states that planets grow within a circumstellar disk, via pairwise accretion of smaller bodies known as planetesimals. It should be noted that the process of planet formation is a fundamentally different process from that of star formation. Stellar formation begins with the process of gas condensation, whereas planetary formation begins with the accumulation of solid bodies, and gas accretion takes place only at a late stage in some of the larger planets (Lissauer, 1993).

In order to be successful, it is necessary that the standard model of planetary formation should explain the main features of the solar system, in particular the division of the planets into three main groups – the terrestrial planets, the giant planets and the outer icy planets (Table 2.2). In detail the standard model should also explain the following properties of the solar system (Lissauer, 1993):

- the coplanar nature and spacing of the planetary orbits;
- the presence of cometary reservoirs in the Kuiper belt at ca. 40 AU, and beyond in the Oort cloud, at 10000 AU;

TABLE 2.2 Physical properties of the planets showing the three groups of planets. The Asteroids lie between Mars and Jupiter at 2.7 AU.

		The terrestrial planets				The giant planets		The outer icy planets		
Body	**Sun**	**Mercury**	**Venus**	**Earth**	**Mars**	**Jupiter**	**Saturn**	**Uranus**	**Neptune**	**Pluto**
Increasing mean distance from Sun (Earth to Sun = 1.0 AU)										
Distance from the sun (AU)	0	0.39	0.72	1	1.52	5.2	9.55	19.2	30.1	39.5
Mean density (Terrestrial planets > Jovian planets)										
Actual Density (g cm^{-3})	1.41	5.43	5.25	5.52	3.95	1.33	0.69	1.29	1.64	2.03
Radius (Terrestrial planets < Jovian planets)										
Radius (Earth = 1.0)	109	0.38	0.95	1	0.53	11	9	4	4	0.18

- the existence of satellite systems for most planets;
- the variation in planetary masses, from the relatively small terrestrial planets to the large gaseous planets;
- the concentration of the majority of the angular momentum of the solar system in the planets, despite these representing only a small fraction of its total mass;
- the variable bulk compositions of the dense, refractory, terrestrial planets, the low density giant planets, and the icy outer planets;
- the existence of the asteroid belt and meteorites;
- the record of dense cratering on most solid planetary and satellite surfaces.

The processes of cosmochemistry constrain planetary materials to be one of three types – gas (hydrogen and helium – the most abundant elements of the Universe), ice (water, methane, ammonia, nitrogen – the next most abundant elements in nucleosynthesis), and rock – principally made up of Mg–Fe silicates (Stevenson, 2004). If the standard model of planetary formation is applied to the formation of the Earth and the terrestrial planets, then temperatures in the solar nebula in the vicinity of the Earth's orbit would have been between 500 and 800 K. At these temperatures rock – Fe–Mg silicates, and Fe–Ni metal would condense but not water ice. Micron-sized particles of these minerals would have grown in series of stages into the present configuration of planetary bodies as outlined below and summarized in Table 2.3.

2.3.4.1.1 Stage 1. The formation of small (1–10 km) planetesimals

The first stage of planetary formation involves the accumulation of micron-sized dust particles into kilometer-sized, 10^{12}–10^{18} g, planetesimals, over a timescale of 10^4 years. Small grains in the solar nebula settle toward the midplane of the nebula, where they are buoyant in the radial gas pressure, and accumulate largely through nongravitational, electromagnetic forces, such as weak Van der Waal's binding energies. Recent observations of protoplanetary disks has confirmed the existence of these materials in the "terrestrial region" of the disk (Van Boekel et al., 2004).

2.3.4.1.2 Stage 2. Gravitational accumulation of planetesimals into planetary embryos

When planetesimals grow to radii of between 0.1 and 10 km the nongravitational accumulation growth stage come to an end. At this stage the relative velocities of these larger bodies is their key property and subsequent processes become dominated by mutual gravitational perturbations. Hence there is a transition from nongravitational accumulation to gravitational accumulation. It is through this process that planetary embryos, or protoplanets as they are also called, up to 4,000 km in size (10^{26}–10^{27} g, Mercury- or Mars-sized), are formed. Our understanding of protoplanet formation has been greatly aided by mathematical simulations, a technique pioneered by Wetherill (1990). A more recent simulation by Weidenschilling et al. (1997) found that the distribution of planetesimals within the orbits of Mercury to Mars reduced to 22 planetary embryos, with masses of $> 10^{26}$ g, over 10^6 years. Together these bodies represented 90% of the total mass. The accretion process ceases when the supply of suitably sized, locally available material is exhausted.

TABLE 2.3 Stages of planetary formation according to the "standard model" (Wetherill, 1990), applied to the terrestrial planets.

Stage	Final mass (g)	Time taken (yr)	Main processes
1. Accretion of dust-sized particles into planetesimals.	10^{12}–10^{18}	10^4	Nongravitational accumulation; particles coalesce through electrostatic forces
2. Accretion of planetesimals into planetary embryos	10^{26}–10^{27}	10^6	Gravitational accretion aided by runaway growth
3. Accretion of planetary embryos into planets	10^{27}–10^{28}	10^7–10^8	Giant impacts

2.3.4.1.3 *Stage 3. From embryos to planets*

The final stage of planetary accretion is the accumulation of a few tens of planetary embryos into a small number of full-sized planets, 10^{27}–10^{28} g, over a timescale of 10^7–10^8 years. There are two processes which appear to be important at this stage. The first is the very high probability of giant impacts between planetary embryos. The importance of giant impacts was first postulated by Wetherill (1990) on the basis of his numerical simulation experiments. Giant impacting will almost inevitably lead to planet-scale melting and in some cases to planetary disintegration (Asphaug et al., 2006). The second key process in the final assembly of planets is the dispersion of the gaseous component of the solar nebula. This is thought to have happened on the time scale of 10^6–10^7 years (Canup, 2004). The loss of the gaseous component would reduce the velocity damping on small objects, which would in turn reduce their ability to damp the velocities of the planetary embryos, leading to collisions, and, as indicated above, giant impacts. The noble gas geochemistry of the Earth's mantle indicates that the Earth has retained a memory of the solar nebular gas (see Chapter 5, Section 5.2.1.4). This must have been acquired in the first 10^6–10^7 years of its accretion history.

2.3.4.2 *The formation of giant (gaseous) planets*

According to the standard model, the giant gaseous, or Jovian-type planets form in a different manner from the terrestrial planets. They have a rocky core, a gaseous outer layer, and may have a middle layer rich in ice, possible because of their greater distance from the sun and lower nebular temperature (Fig. 2.6). Their outer gaseous layer is thought to be captured gas from the solar nebula.

Models for the formation of the giant planets suggest that a rocky planetary embryo of about ten Earth masses can form rapidly, within 10^6 years. Once this embryo is established these massive planetary embryos accumulate two Earth masses of solar nebular gas over 10^7 yr (Kortenkamp et al., 2001).

More recently our understanding of the nature of the giant gaseous planets has been challenged by the discovery of planetary systems around other stars (Lissauer, 2002). Many of these "extrasolar" planets are giant gaseous planets, but with rather unusual properties, for many are less than 0.1 AU away from their parent star, unlike Jupiter's 5 AU. In part this is an observational bias given that large giant planets close to their parent stars are more readily detectable. Observations of extrasolar, Jupiter-like planets have led to an alternative model for their formation, whereby they form by disc instability, through the rapid (100 years) gravitational collapse of clumps of dust and gas in the protoplanetary disk (Boss, 2000). At present however, the consensus seems to favor the "rocky core" nucleation model (Stevenson, 2004).

2.4 EARTH DIFFERENTIATION – THE FIRST EARTH SYSTEM

It is most likely that the Earth formed according to the standard model of planetary accretion, as described in the previous section. Given this assumption, the earliest Earth system is likely to have had the following properties:

- *A chondritic composition.* The Earth formed from the primitive materials of the solar nebula, which, apart from the volatile elements, is identical to that of primitive chondritic meteorites. It is likely, therefore, that the initial bulk composition of the Earth was chondritic.
- *The Earth experienced one or more giant impacts.* Given that giant impacting is an essential part of planetary accretion models for the terrestrial planets, leading to planet-wide melting, it is probable that the Earth experienced one or more planet-wide melting events during its earliest history.
- *Formation of a terrestrial magma ocean.* Extensive melting as a result of giant impacts would lead to the formation of a terrestrial magma ocean during the Earth's late accretion history.
- *Extensive melting may have facilitated core formation.* It is probable that during its magma

ocean stage the Earth differentiated into an outer silicate mantle and an inner metallic core.
- *The Moon formed through a giant impact.* It is now thought likely that the Moon formed as a result of a giant impact event. The timing of this event, its contribution to terrestrial melting, and the bulk composition of the Earth are important subjects of current research.

However, the standard planetary accretion model also raises a number of questions about the early Earth. For example:
- *How do we define the age of the Earth?* If the Earth formed by accretion, over a time period of 10–100 million years, at what stage of this process do we attribute the age of the Earth?
- *To what extent were the planetesimals and protoplanets, from which the Earth was built, already differentiated?* Early models for the formation of the Earth assumed that it formed from undifferentiated starting materials. This is not necessarily the case and the Earth may have formed from protoplanets which already possessed their own metallic core.
- *What sort of early crust formed on Earth?* And is it still preserved? This topic is discussed more fully in Chapter 3 (Section 3.2.3.5).

Our purpose in this section is to seek an understanding of the earliest differentiation of the Earth and how the Earth acquired its large-scale layered structure of core, mantle, and atmosphere-oceans. A major result of "planetary thinking" about the early Earth is that we now recognize that several large-scale features of the Earth may all be linked to giant impacting events. These include the formation of the Moon, the existence of a magma ocean, and the separation of the core. Whether or not these are all the result of a single event or many is the subject of some discussion (Kleine et al., 2004).

2.4.1 The formation of the Moon

The standard model for the formation of the Moon is the Giant Impact hypothesis (Stevenson, 1987). This model proposes that a planetesimal, the mass of Mars (about 15% the mass of the Earth), collided with the proto-Earth, at a time after core formation. The impact generated a huge amount of thermal energy so that the impactor and part of the Earth were vaporized, and some of this material coalesced in an orbit around the Earth to form the Moon. On cooling it is widely thought that both the Earth and the Moon went through magma ocean stages (see Section 2.4.3).

2.4.1.1 Modeling the Giant Impact

Particular insights into the Giant Impact process have been gained through the application of numerical modeling. Melosh (2001) points out, however, that such an approach is not for the faint hearted, and that "it has taken squadrons of physicists nearly 50 years to come up with computers and three-dimensional computer codes that can adequately treat the effects of impacts and explosions under relatively simple conditions." Over the past decade or so, numerical solutions to the Giant Impact event have cycled through the more and less probable. At present a probable (and mathematically sophisticated) solution is proposed by Canup and Asphaug (2001) in which they find that an impactor, the mass of Mars (with a mass of $\sim 6 \times 10^{26}$ g), collided at an oblique angle with an almost fully formed Earth to produce an iron-poor Moon. The impactor has been named "Theia," after the mother of the Greek goddess of the Moon (Halliday, 2000).

In detail, an impact of this type, starting with two bodies both with iron cores and silicate mantles, would result in a planet surrounded by a disk of former mantle material. It is from this disk that an iron-depleted Moon would accrete (Canup, 2004). Models of this type are proving highly successful inasmuch as they have the capacity to simultaneously explain the masses of the Earth and Moon, the Earth–Moon angular momentum, and the unusual chemical composition of the Moon.

2.4.1.2 Geochemical tests of the Giant Impact model

Returned lunar basalt samples presented a few surprises when their chemical compositions were first determined, for their chemistry turned out to be rather different from their terrestrial counterparts. Particular chemical features of lunar basalts are their strong depletion in highly volatile elements (Fig. 2.11), and modest depletion of moderately volatile elements, relative to the Earth's mantle.

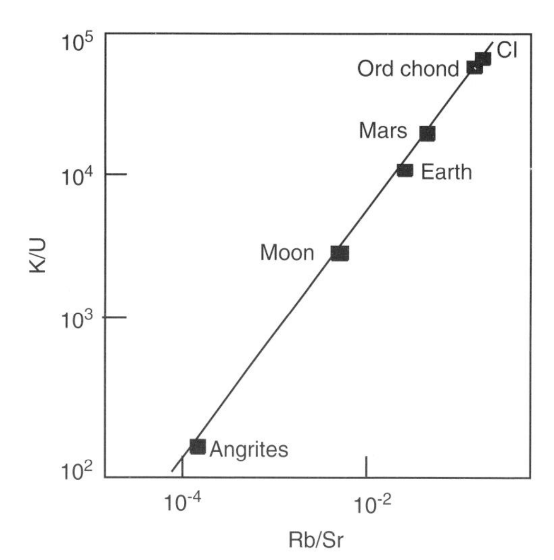

FIGURE 2.11 The ratio of volatile (K, Rb) to refractory elements (U, Sr) in planetary and solar system objects (after Halliday & Porcelli, 2001). The relationships show the volatile depleted nature of the Moon relative to the Earth and the Moon and the Earth relative to primitive CI chondrites.

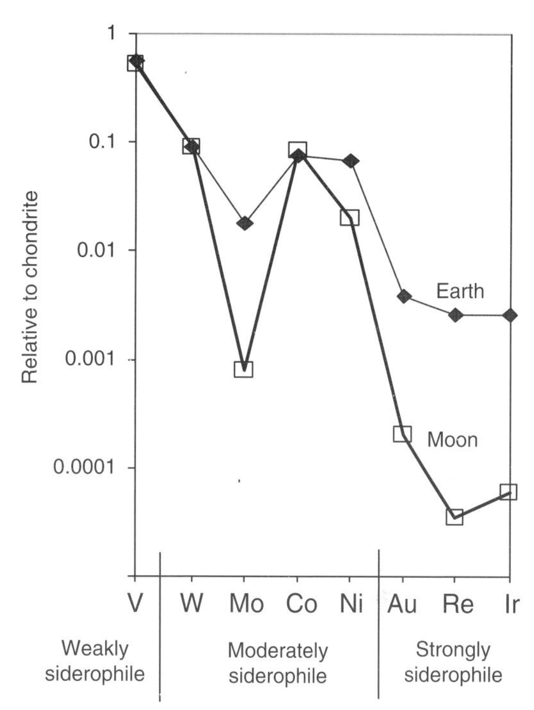

FIGURE 2.12 Depletion factors for siderophile elements in the Earth's mantle and the Moon relative to C1 chondrite, using the median Earth and Moon values from the compilation of Kramers (1998) (see Text Box 2.2).

The Moon also has an unusually low iron content. Typical inner solar system objects contain about 30% iron – CI chondrites, for example contain about 36% iron by mass – whereas the Moon contains only 13 wt%. Some of this iron is probably located in the lunar core, although this will only account for a small amount of iron, since the lunar core is small (3–4% of its mass) compared to that of the Earth. In contrast the more refractory lithophile elements such as Al are more abundant on the Moon given the thick aluminous lunar crust (Taylor, 1987).

Jones and Hood (1990) approached the problem of the lunar composition by assuming that it had the same composition as the Earth's mantle. They showed that if this were the case then it would also have had a large metallic core. However, the Moon does not have a large metallic core and by this logic cannot be made of the same material as the terrestrial mantle. A solution to this paradox was proposed by Kramers (1998) who showed that siderophile element patterns for the silicate Moon are similar to the terrestrial pattern for some elements but are more strongly depleted for others (Fig. 2.12). This model suggests that the Moon formed from an already differentiated body, but was depleted further during lunar core formation (Kramers, 1998).

Very high precision oxygen isotope studies show that, within the limits of analytical error, the Earth and the Moon have identical isotopic compositions. Oxygen isotope studies of meteorites summarized in Fig. 2.10, show that the inner solar system was isotopically heterogeneous, and so the Earth–Moon result implies that the proto-Earth and the impactor Theia must have accreted at a similar distance from the sun and grew from a similar mix of materials (Wiechert et al., 2001). One implication of this finding is that the chemical differences between the Earth and Moon described above may be of a secondary origin, in the sense that they were produced during the differentiation of the Earth, Theia, and the Moon.

2.4.1.3 Consequences of a Giant Impact for the Earth

The likely implications of the Giant Impact hypothesis for the early Earth are that:

- About 30% of the planet's mass would be raised to temperatures greater than 7,000 K, so that the Earth would be surrounded by a silicate vapor atmosphere. In addition much of the planet would have been in a molten state.
- The Earth's atmosphere would have been volatilized, either at impact or during the magma ocean stage, although in the case of an oblique impact, the atmosphere may have condensed back again.
- It is also possible that, if the impactor had a metallic core of its own, this was added to the Earth's core immediately following the impact (Benz & Cameron, 1990).

The Earth however is a dynamic planet, with a convecting mantle, and identifying evidence for these events in the modern Earth's mantle has proved an elusive task, as will be shown below.

2.4.1.4 The timing of the impact

The oldest rock on the Moon is a clast of ferroan anorthosite in an impact breccia which has a relatively imprecise Sm–Nd isochron age of 4,562 ± 68 Ma (Alibert et al., 1994). Tungsten isotope measurements (see Section 2.4.2.5) on lunar rocks indicate that the Moon formed about 30 Ma after the formation of the solar system (Schoenberg et al., 2002). Halliday (2004) suggested that the Moon is younger and formed > 44 Ma to > 54 Ma after the beginning of the solar system, but the consensus view is that the timescale of planetary accretion was relatively short and that the Moon formed within the first 30 Ma of the life of the solar system (Kleine et al., 2002; 2005b; Yin et al., 2002). In fact the tungsten isotope data are now sufficiently good to allow us to discriminate between competing impactor models (Jacobsen, 2005) Kleine et al., 2004 and confirm the model of Canup and Asphaug (2001) in which a Mars-sized impactor collided with an almost fully formed Earth. Thus the process of terrestrial accretion had ceased by about 30 Ma and was in fact terminated by the Moon-forming impact.

2.4.2 Core formation

Geophysical evidence shows that the Earth has a liquid Fe–Ni–S-alloy outer core with a thickness of 2,260 km, and a solid Fe–Ni-alloy inner core, with a radius of 1,215 km. Temperatures at the top of the core are thought to be in the range 3,500–4,500 K and rise to 5,000–6,000 K at the center of the Earth. There is a mismatch of a few percent between the measured density of the Earth's outer core and that predicted for an iron–nickel alloy at high pressure. This "core density deficit," as it is called, is thought to indicate the presence of other elements as impurities in the core in addition to iron and nickel, and this could be the principal reason why the outer core is molten. The impurities act as a form of "antifreeze" in the liquid metal (Stevenson, 1990).

At the present day the Earth's inner core continues to grow at the expense of the outer core and this process generates heat in the modern Earth. Estimates of this heat flux are not well constrained and vary from 2 TW to 10 TW (Labrosse, 2002). The Earth's magnetic field is thought to be driven by a geomagnetic dynamo, which is thought to be governed by thermal convection in the outer core. The driver for this thermal convection is not well understood and could be radioactive heating (if potassium is present in the core), the cooling of the core, or the latent heat arising from crystallization at the inner/outer core boundary. Weak paleointensity measurements of the Earth's magnetic field have been recorded in rocks as old as 3.5 Ga (Yoshihara & Hamano, 2004).

2.4.2.1 Impurities in the Earth's core

Birch (1952) demonstrated that the Earth's liquid outer core is 10% less dense than expected if it were made up of an iron–nickel alloy. For this reason one or more light elements are thought to be present in the core. Although a core density deficit of 10% has been widely used in many geochemical studies this value has recently been revised downwards by Anderson and Isaak (2002) and is now thought to be 5.4%.

A suitable light element diluent must be able to form an alloy with Fe at core pressures,

TEXT BOX 2.2 Geochemical multielement diagrams

Geochemists use a variety of trace element diagrams in which to display multielement data. These take a variety of forms but have some features in common. Typically a multielement diagram will display a range of trace elements equally spaced along the *x*-axis of the diagram and these will be arranged in a specific order according to the particular logic of the diagram. The *y*-axis will record the abundances of those elements, and because the abundances are very variable, they are displayed on a logarithmic scale. Further, the abundances are normalized to the concentrations present in a particular reference material. These reference materials vary but include chondritic meteorites and an estimate of the composition of the Earth's primitive mantle. In other words, the composition of the rock is compared to that of the primitive, undifferentiated Earth. Three different types of multielement diagrams will be discussed here.

Rare Earth element (REE) diagrams

The REEs are the lanthanides, elements 57 to 71 (La to Lu) in the periodic table. They are of particular interest because geochemically they are very similar. All carry a 3+ charge and they show decreasing ionic radii from La to Lu. They are expected therefore to behave as a coherent group and to show smooth, systematic changes in geochemical behavior through the series. Hence REE diagrams are plotted to show the elements of the group in order of increasing atomic numbers (from La (57) to Lu (71)) from left to right. REE concentration in rocks are normalized to the concentrations in chondritic meteorites and there are a number of recommended chondrite concentrations in use. One widely used set of chondritic normalization values is that of McDonough and Sun (1995) reported in Chapter 3, Table 3.2. The purpose of this normalization is to compare samples with the compositions of the original starting material of the Earth.

Of interest is the extent to which the REE vary within the group. Where there are trends of increasing or decreasing abundances within the group the origin of this "fractionation" is of interest and comparisons are made between the light (lower atomic number) and heavy (higher atomic number) members of the group (Box 2.2 Fig.1). Where single elements have abundances which are atypical, the origin of these "anomalies" is of interest.

REE plots have been widely used to understand magmatic processes in igneous rocks, but have also been useful in unraveling sedimentary processes and geochemical exchanges in seawater (see Chapter 5, Section 5.4.2).

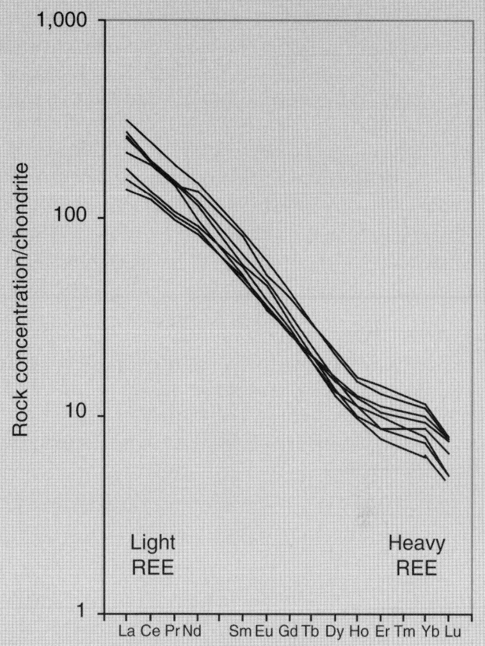

BOX 2.2 FIGURE 1 An REE plot for granitoids from the Baltic Shield, showing the REEs in order of their atomic number on the *x*-axis and concentrations, normalized relative to abundances in chondritic meteorites, shown on the *y*-axis, using a log scale. The graph shows that in these rocks all the REE have higher concentrations than in chondritic meteorites (all values > 1.0) but that the light REE (low atomic number, at the left) have much higher abundances than the heavy REE (higher atomic number, at the right). The reason for this "fractionation" is important to determine.

Multi-trace element diagrams

Multielement diagrams are quite variable and care needs to be taken to identify which set of normalizing values are being used. Clearly this will vary according to the purpose of the diagram. Multielement diagrams have also been known as "spider diagrams", because the apparently chaotic abundance patterns that they sometimes generate resemble the track of a spider walking through ink!

(a) Primitive mantle-normalized plots

A primitive mantle normalized trace element diagram appears to use a rather random set of trace elements.

The series will often start with elements such as Cs or Rb and end with elements such as Y or Yb, although there is no definitive element suite for these diagrams. In this case the elements are not ordered according to atomic number, instead they are arranged in order of decreasing incompatibility during mantle melting. In other words those elements with the greatest preference for the melt phase during mantle melting are plotted to the left of the graph and those with a lesser affinity for the melt phase to the right (Box 2.2 Fig.2). These preferences are determined from laboratory measurements of the partition coefficients of the respective elements in the main mantle minerals, weighted according to the proportion of those minerals in the mantle. These abundances are normalized to the concentrations of those elements estimated to have been present in the Earth's primitive mantle – that is the mantle as it was after core formation but before the formation of the continental crust. A table of the various estimates of these abundances is given in Chapter 3, Table 3.2. Hence the purpose of this normalization procedure is to see how the sample being investigated differs from the primitive mantle from which it was originally derived.

(b) Mid Ocean Ridge Basalt (MORB)-normalized diagrams

An alternative normalization scheme is to use the composition of average midocean ridge basalts (MORB). MORB-normalization is often used for volcanic arc rocks and altered basalts. In this case the elements are arranged slightly differently and are ordered according to their mobility in a hydrous fluid so that mobile, low field strength elements are plotted to the left of the diagram and immobile, high field strength elements to the right (Box 2.2 Fig. 3).

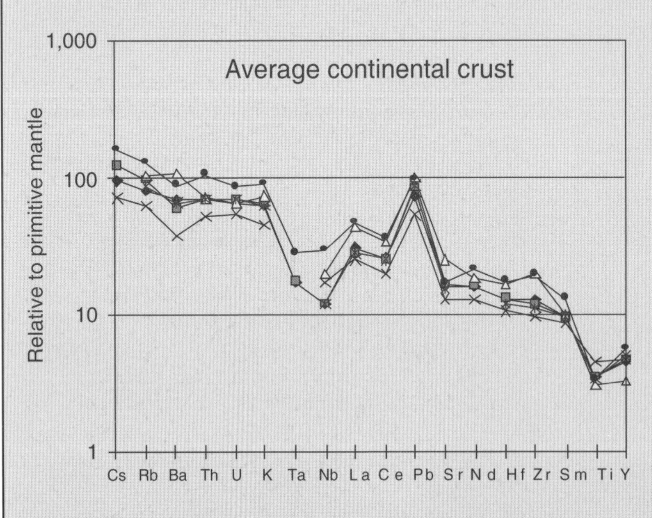

BOX 2.2 FIGURE 2 A mantle-normalized multielement plot for selected trace elements in average continental crust (five different averages). The elements are arranged in order of decreasing incompatibility during mantle melting. The plot shows that all the elements shown are concentrated in the continental crust relative to the primitive mantle but that the highly incompatible elements are highly concentrated (×100) compared with elements such as Ti and Y (×3). Understanding this fractionation is important. Also of interest are the anomalies for Pb and Ta–Nb. For further discussion see Chapter 4, Section 4.2.2.3.

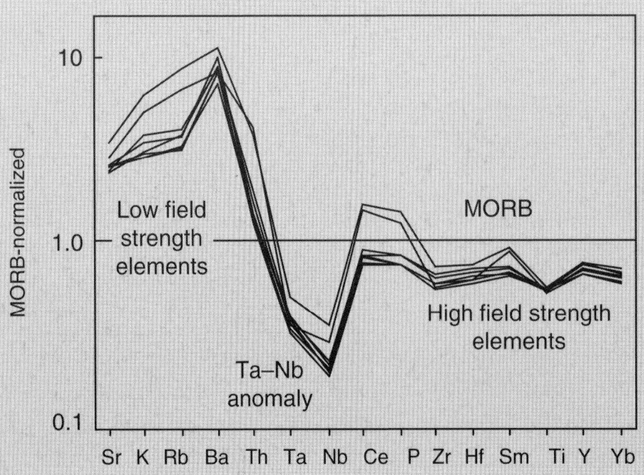

BOX 2.2 FIGURE 3 A MORB-normalized plot for typical arc magmas. This normalization procedure highlights the differences between fluid-mobile elements (the low field strength elements) and immobile elements (the high field strength elements). Fluid-mobile elements have higher concentrations than average MORB, whereas the immobile elements have (in this case) lower concentrations than in average MORB. For further discussion see Chapter 4, Section 4.1.2.

TEXT BOX 2.2 *(Cont'd)*

Siderophile element diagrams

The siderophile elements are those which have a strong affinity for metallic iron. These are the elements therefore that preferentially partition into the Earth's core during planetary formation. Understanding the distribution of the siderophile element concentration in the Earth's mantle can provide important clues about the origin of the Earth's core. There are a range of different siderophile element diagrams using different groups of siderophile elements. The features that they have in common are that they order the siderophile elements according to their siderophile affinity. This is normally in order of increasing siderophile nature from left to right. Concentrations are normalized according to abundances in chondritic meteorites (Box 2.2 Fig. 4) (see Section 2.4).

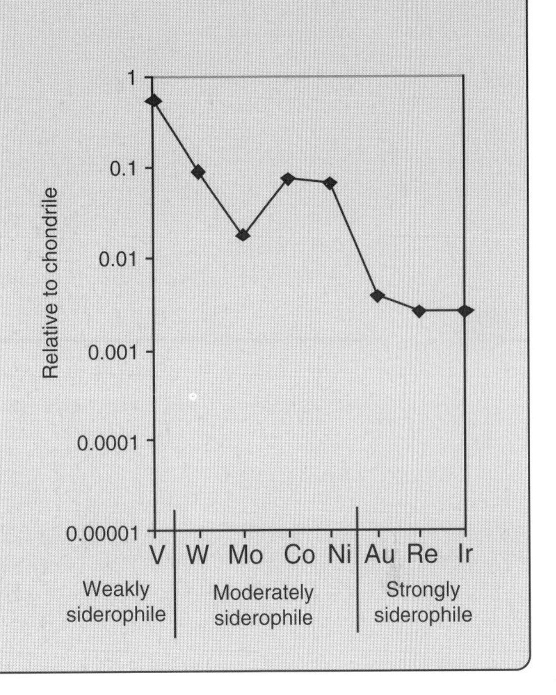

BOX 2.2 FIGURE 4 A chondrite normalized plot for siderophile elements in the Earth's mantle. The elements are arranged in order of increasing siderophile affinity from left to right and show decreasing element abundances with increasing siderophile character, commensurate with core formation.

partition into the core in sufficient quantity, and form an alloy that matches the known seismic properties and density of the outer core. Identifying precisely the appropriate light elements in the core is likely to provide important clues to the process of core formation, as discussed below (Section 2.4.2.4.2). Over the years a large number of elements have been proposed as candidates for the light element contribution in the core, the current contenders being O, Si, and S (Williams & Knittle, 1997). The key lines of evidence in their support are summarized in Table 2.4.

2.4.2.2 The mechanism of core formation

The principal constraint on the formation of the Earth's core is that the Earth must have been sufficiently molten for metal droplets to separate and sink through a silicate melt (Walter & Tronnes, 2004). Experimental studies show that molten iron alloys cannot migrate along the grain boundaries of solid silicates and silicate-metal separation only effectively takes place in a melt. Given this starting point a large number of questions arise about the details of core formation. For example:

- Had the planetary bodies from which the Earth accreted already melted and so acquired their own cores?
- In other words, was the equilibration between silicate and melt primarily a low pressure process (in small bodies – planetesimals or protoplanets) or a high pressure-high temperature process in an almost fully accreted Earth?
- Did chemical equilibrium exist between the silicate and metallic components of the early Earth during core formation?
- For example, if small bodies with metallic cores merged during accretion did they do so without equilibration with the silicate mantles, or was there large-scale melting and rehomogenization?
- To what extent did the Moon-forming impact drive the process of core formation?

Some consensus is now emerging on the process of core formation through an exploration of the likely heat sources for the melting during which silicate and metal can separate, from geochemical constraints based upon the siderophile elements, and from our knowledge of the timing of core formation,

TABLE 2.4 Competing models for the likely light element diluents in a Fe–Ni-rich outer core.

Light element	Approximate %	Evidence	Reference
Sulfur	<6 wt%	FeS forms eutectic properties with Fe, S is commonly found in Fe-meteorites as FeS (Troilite) 2–15 wt%	O'Neill et al. (1998)
	2–15 wt%	Layering absent in outer core	Helffrich and Kaneshima (2004)
	>1 % (atomic) (*Inner core*)	Experimental study of Fe–S system near the pure iron end-member	Li et al. (2001)
Oxygen	25–30 mol% Oxygen	High solubility of O in Fe metal. FeO at core pressure is miscible with Fe. O may be essential for compositional convection May imply an oxidized chondritic starting material May be a product of reaction with the mantle through time	Abe et al. (1999) Alfe et al. (2002)
	<6 ± 1 wt%	Layering absent in outer core	Helffrich and Kaneshima (2004)
Silicon	<0.01 wt%	If at low pressure and fO_2	Gessmann et al. (2001)
	<0.1 wt%	High pressure, low fO_2	Malavergne et al. (2004)
	Up to 17%	High pressure – only permissible if very low fO_2	
	5–7%	High pressure – progressive oxidation during accretion	Wade and Wood (2005)
Potassium	250 ppm (could account for 20% of the heat production of the core)	K and Na concentrations in the silicate Earth ~20% less than in C1 chondrites. Either, lost during accretion or, due to K extraction into a S-rich O-rich Fe liquid in the Earth's core	Gessmann and Wood (2002)
Carbon	ca. 1%	Stable as Fe_3C	Wood (1993)

based principally upon the study of tungsten isotopes.

2.4.2.3 Constraints on core formation

2.4.2.3.1 Thermal constraints

Probably the most important source of heat during the early stages of planetary accretion was from the decay of short lived isotopes, in particular ^{26}Al (Walter & Tronnes, 2004). Baker et al. (2005) recently reported isotopic evidence for planetismal melting and differentiation very early in the history of the solar system, triggered by ^{26}Al decay.

Later in the history of planetary accretion, impact melting became a more important source of thermal energy, leading to the creation of magma oceans. On the Earth, a particularly important impacting event – the formation of the Moon – probably led to widespread melting at about 30 Ma. An extreme impact melting event, or events, would lead to extensive melting, homogenization of discrete metal and silicate phases present in the initial accretionary components, and the formation of a magma ocean in which silicate and metallic phases separate, leading to the formation of the Earth's core and mantle.

2.4.2.3.2 Geochemical constraints

The siderophile elements provide an important control on the processes of core formation. Siderophile elements are those elements which have a chemical affinity for metallic iron (Section 2.3.2) and so might be expected to concentrate in the core. This is in fact what is observed, and siderophile element concentrations in the mantle are depleted relative to concentrations in chondritic meteorites, with the "missing" component presumed to be in the core.

However, when calculations are made to compare the measured siderophile element depletion in the mantle with the expected values, there is a mismatch. Using metal–silicate partition coefficients measured at low pressures and temperatures, it was found that the depletion is not as great as is expected (Fig. 2.13). For some elements the depletion is several orders of magnitude less than that expected after core formation. This problem, first noted by Ringwood (1966) for the elements Ni and Co, and subsequently extended to other siderophile elements, has become known as the "excess siderophile problem."

2.4.2.4 Models of core formation

Explanations of the excess siderophile element problem have led to a number of different models of core formation. For example it is possible that the partitioning of siderophile elements between silicate melt and metal alloy was not an equilibrium process. Second, the assumption that silicate–metal equilibration took place at low temperatures and pressures may be incorrect, and high-pressure partition coefficients may be more appropriate. A third possibility is that core formation was complete before the end of accretion and that some siderophile elements were added late in the accretion process of Earth, after the separation of the core and mantle. In this case the mantle was "topped up" with a late addition of siderophile elements.

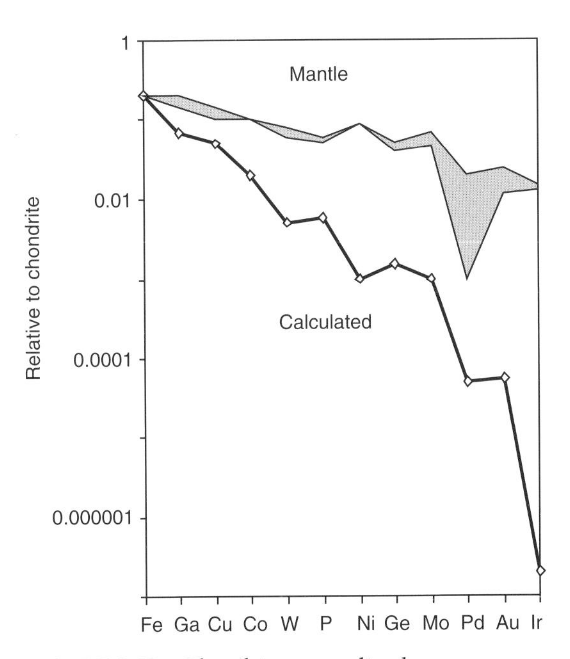

FIGURE 2.13 Chondrite normalized concentrations of siderophile elements in the Earth's mantle relative to those expected in the mantle if all the iron in the Earth's core had equilibrated with silicate at low pressures and temperature. The siderophile elements are arranged such that their siderophile nature increases from left to right (after Rama Murthy and Karato, 1997).

2.4.2.4.1 The low pressure core segregation model

Azbel et al. (1993) and Kramers (1998) proposed that core formation began when the Earth was only 10% formed and took place in a shallow (6 GPa) magma ocean. In this model, protoplanets had already formed cores, which, on impacting, merged without fully equilibrating with the silicate mantle. Hence siderophile element extraction from the mantle was an inefficient process, incompletely scavenging the siderophile elements. This solution to the excess siderophile element problem implicitly accepts that there was not a simple equilibrium relationship between the metallic and silicate components of the Earth during core formation. Kramers (1998) modeled the distribution of siderophile elements on Earth and in the Moon and showed that there are large differences in the extent of depletion from the modest depletion of weakly siderophile elements such as V, to a greater depletion of the moderately siderophile elements (W, Co, Ni, and Mo) to the extensive depletion of the highly siderophile elements (Au, Re, Ir) (Fig. 2.12). This pattern is broadly consistent with low pressure silicate–metal partition coefficients for the siderophile elements and is the pattern expected in the Earth's mantle as a consequence of inefficient metal extraction during core formation.

However, recent modeling by Canup (2004) shows that when there is impacting between bodies with existing metallic cores there will be a general homogenization of silicate and melt and the metallic cores will reequilibrate within a magma ocean.

2.4.2.4.2 The "late veneer" model

The "late veneer" solution to the excess siderophile element problem proposes that there was an initial major accretion event during which the core formed and siderophile elements were removed from the mantle to the core. This was followed by a later stage of accretion during which more oxidizing material was added to the Earth, making up about 7% of the Earth's mass. This contributed to the moderately siderophile element budget of the mantle. Finally, a "late veneer" was added to the Earth representing about 1% of the Earth's mass. Most of the mantle's highly siderophile element content was added at this stage (Newsom & Jones, 1990). The principal geochemical evidence in support of this model is that although the degree of depletion of the highly siderophile elements in the mantle is extreme (0.5% chondritic), their relative abundances are still chondritic (O'Neill et al., 1995). More recent measurements by Becker et al. (2006) find that some highly siderophile elements in the primitive upper mantle (PUM) had ratios that are chondritic, whereas others are suprachondritic. Similar abundances have been found in lunar impact melt rocks. These authors argue that the siderophile element distributions cannot be explained simply by silicate metal equilibration. Instead they suggest that the highly siderophile element abundances reflect a late impacting event which affected both the Earth and the Moon and that on Earth a very early, highly siderophile element-enriched crust has been reworked into the modern mantle.

2.4.2.4.3 The high pressure core segregation model

Currently, the favored solution to the siderophile element problem, and the most popular model of core formation, is that silicate–metal equilibration took place at high pressures, in a deep magma ocean. Support for this model comes from very high pressure experimental studies of trace element partitioning, which show that metal–silicate partition coefficients are greatly reduced at elevated temperatures and pressures. For example Li and Agee (2001) show that at high pressures and temperatures Ni and Co become less siderophile in nature. Similarly, Chabot and Agee (2003) show that V, Cr, and Mn depletions in the mantle can be modeled in a high-temperature magma ocean. A number of recent studies suggest that the mantle has an equilibrium siderophile element signature for liquid metal–liquid silicate partitioning at 30–60 GPa and $> 2,000$ K, at an oxygen fugacity about 2 log units below that of the iron–wustite buffer (Chabot et al., 2005).

Experimental results of this type require the existence of a magma ocean as much as 1,000 km deep in which the metal–silicate separation took place. Most recent models of core formation (e.g. Rubie et al., 2003) require the percolation of liquid metal through a silicate magma ocean as droplets of "metal rainfall." Rubie et al. (2003) calculate a droplet diameter of about 1 cm and a settling velocity of about 0.5 $m^{-1}s$. Subsequently the metal droplets accumulate at depth into a metal layer and descend still deeper as diapirs. Precisely at which point silicate metal equilibration is established is an uncertainty in the deep magma ocean model (Fig. 2.14).

The deep magma ocean model is also successful in explaining the partitioning of light elements into the Earth's core. Gessmann et al. (2001) and Malavergne et al. (2004) have shown that at high pressures silicon solubility increases in liquid iron–alloy, although the oxygen fugacity is an important control. Wade and Wood (2005) propose a model in which the oxygen fugacity of the core increases during accretion by two log units and in which the Si content is between 5 and 7%.

2.4.2.5 The timing of core formation

Although a number of isotopic systems are potentially useful as chronometers of core formation the Hf–W isotope system is the one which is regarded as the most robust (Jacobsen, 2005). The merits of the Hf–W chronometer in establishing the time of core formation are that the isotope ^{182}Hf decays to ^{182}W with a half-life of 9 Ma, so that after

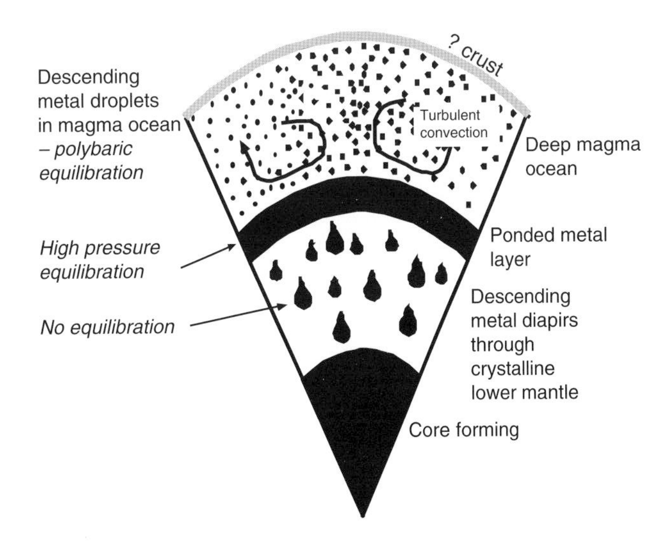

FIGURE 2.14 A generalized model for silicate–metal separation in a deep, turbulent, convecting terrestrial magma ocean. Descending metal droplets equilibrate with silicate melt and pond at the base of the magma ocean. Diapirs of metal descend to the growing core through a crystalline or partially molten lower mantle (after Rubie et al., 2003).

50 Ma the Hf–W chronometer is dead. This means that the Hf–W isotopic system is very useful in dating events that took place within the first 50 Ma of the life of the solar system. In addition the elements Hf and W display very different geochemical behavior making this isotope system particularly useful for dating core formation, for Hf is a lithophile element, with a tendency to remain in the silicate mantle whereas W is a siderophile element under highly reducing conditions, with a preference to concentrate in the Earth's metallic core (see Text Box 2.3).

Measuring W–isotope ratios is currently one of the most challenging tasks in isotope geochemistry, and over the past few years there has been great debate over whether or not the Earth has a chondritic W–isotope ratio. Recently agreement has been reached that the bulk silicate Earth (BSE) has an elevated Hf/W ratio relative to chondrites and that Earth's mantle has an excess ^{182}W relative to chondrites, indicating that the Earth's core formed during the lifetime of ^{182}Hf (Schoenberg et al., 2002). How early, is still a matter of dispute, but Yin et al. (2002) and Jacobsen (2005) provide evidence that the bulk of the metal silicate separation was complete before 30 Ma, a result which is consistent with the rapid growth of the terrestrial planets, that is, within about 10 Ma of the formation of the solar system.

A possible alternative to the above model is that the Earth's core grew gradually through the coalescing of the metal cores of planetesimals without fully equilibrating with the Earth's mantle. In this case the core formation age would be average accretion age of the two objects and would indicate a core segregation age younger than that of the time of Earth accretion (Jacobsen, 2005). These competing models can only be evaluated by further computer modeling of planetary accretion processes.

2.4.3 Was there a terrestrial magma ocean?

The central dilemma over the likely existence of a terrestrial magma ocean is that whilst there is abundant circumstantial evidence from the origin of the Earth's core and the existence of the Moon, there is no direct geochemical evidence in the Earth's mantle that such a feature ever existed.

Further lines of reasoning which have been used to infer the existence of a terrestrial magma ocean are:

- the blanketing effect of a proto-atmosphere in which it contributed to the trapping of thermal energy released from impacting (Sasaki, 1990);
- the heat generated during metal-silicate separation leading to core formation;
- the contribution from short-lived radioactive isotopes to the overall thermal budget of the new-born Earth.

TEXT BOX 2.3 Short-lived radioactive isotopes

Radioactive isotopes with short half-lives of a few million years are important in helping us to understand the history of meteorites and the earliest stages of Earth formation. The basic principles here are that if the lifetime of an isotope is short, it can only have a memory of the very earliest part of Earth history, when it was alive. Further, if there is a fractionation event affecting the parent and daughter elements during the lifetime of the isotope, this can be identified in the measured isotope ratios. There are three isotopes which are important in studies of the early Earth – ^{26}Al, ^{182}W, and ^{142}Nd and they are discussed below in order of increasing half-life.

Aluminum-26 (^{26}Al)

The radioactive isotope ^{26}Al decays to ^{26}Mg with a very short half-life of 7.2×10^5 years, and so ^{26}Al becomes extinct within a few million years of its formation. Some have likened this isotope system to the "second-hand" on the clock of the Universe. ^{26}Al was produced by explosions in supernovae early in the history of the Universe. The discovery of the excesses of the daughter isotope ^{26}Mg in very early solar system objects such as CAIs indicates that they contained ^{26}Al at the time of their formation and so must have formed within a very short time of the supernova explosion (see Section 2.3.3.2).

Tungsten-182 (^{182}W)

The Hf–W isotope system (^{182}Hf→ ^{182}W) has a half-life of 9 Ma and so ^{182}Hf becomes extinct after about 50 Ma. The Hf–W chronometer, therefore, is an important tool in understanding processes which took place within the first 50 Ma of the life of the solar system.

The elements Hf and W are geochemically very different. Hf is a lithophile element, with a tendency to remain in the silicate mantle whereas W is a siderophile element with a preference for the Earth's metallic core. Hence the Hf–W isotope system is particularly useful in dating the time of core formation. It is important to ascertain whether or not there was a separation of Hf and W whilst ^{182}Hf was still live, that is, during the first 50 Ma of Earth history. Such a separation would imply that W was removed to the core, whilst Hf remained in the mantle. The logic here is that chondritic meteorites are the same material as that from which the primitive Earth was made and so should show the same Hf–isotope ratios. However, if there was a fractionation event (e.g. during core formation), whilst ^{182}Hf was alive, then the isotope ratios will be different. If such a fractionation event can be identified, then we know it must have taken place within about 50 Ma of the formation of the solar system and so provides evidence of early core formation (before 50 Ma).

Such an event can be detected from isotopic measurements of ^{182}W, measured as the ratio ^{182}W/^{183}W, compared to the ratio in primitive chondritic meteorites, and expressed in the terminology of ε-^{182}W. It is now known that Earth's mantle has an excess ^{182}W relative to chondrites, indicating that the Earth's core formed during the lifetime of ^{182}Hf (Schoenberg et al., 2002) (see Section 2.4.2.5).

Neodymium-142 (^{142}Nd)

^{142}Nd (not to be confused with the more common isotope ^{143}Nd) is the product of ^{146}Sm decay and has a half-life of 103 Ma (see Text Box 3.2, Chapter 3). ^{142}Nd therefore became extinct within the first 400–500 Ma of Earth history. Sm and Nd are both lithophile elements, and so ^{142}Nd has the potential to identify very early events in the Earth's mantle, events which took place within the first 500 Ma of Earth history. Taking chondritic meteorites as the reference point, positive or negative deviations of ^{142}Nd from the chondritic value, as expressed using the epsilon notation as ε-^{142}Nd, can be used to assess whether there was a Sm–Nd fractionation event during the first few hundred Ma of Earth history.

There are many indications that such an anomaly exists (Chapter 3, Section 3.2.3.1), the most important of which is that described by Boyet and Carlson (2005), who showed that the chondritic ϵ-^{142}Nd value is 20 ppm lower than that for terrestrial rocks, implying a ubiquitous ^{142}Nd anomaly in the Earth's mantle and a very early differentiation event in the Earth.

Whilst these thermal arguments presented here do not constitute proof of melting of the early Earth, they are persuasive. Newsom and Jones (1990) in their discussion of the evidence concluded that "there are many good reasons to believe that the early Earth was very hot. Some would go as far as to suggest that a terrestrial magma ocean was unavoidable." Similarly, Abe (1997) argued that "from a theoretical point of view it seems difficult to accrete the Earth without making some type of magma ocean."

TABLE 2.5 Magma ocean scenarios (after Abe, 1997).

Scenario	Deep or shallow magma ocean	Differentiation
1. The Earth forms in a solar nebula, blanketing effect is provided by solar gases	Deep and shallow	Shallow turbulent magma ocean with no fractionation. Differentiation takes place at lower mantle pressures and is affected by high-pressure minerals such as Mg- and Ca–perovskite
2. The Earth forms in "gas-free" space, blanketing effect of a steam atmosphere	Shallow	Upper mantle differentiation. Differentiated upper mantle material is buried in the lower mantle during the accretionary growth of the Earth
3. Giant impact	Deep and shallow	Chemical differentiation in the lower mantle is less likely due to rapid cooling, unless there is a thick transient atmosphere. The upper mantle is differentiated

2.4.3.1 Magma ocean – definitions and scenarios

The term magma ocean describes a layer, either at the surface of a planet or within its interior, which is completely or partially molten. In the case of the early Earth there are several possible scenarios. First, a distinction must be made between "transient" and "sustained" magma oceans. Transient magma oceans are those formed deep in the Earth from a single large impact which cool and solidify within a fairly short period of time. Sustained magma oceans are those surface magma oceans maintained by the thermal blanketing effect of a proto atmosphere. Second, there are "deep" and "shallow" magma oceans, and the processes of differentiation will be different in each. In a deep magma ocean differentiation will be controlled by the fractionation of high-pressure minerals such as perovskite, whereas in a shallow magma ocean it will be by olivine or garnet. Each will leave a different chemical signature in the residual melt phase. Third, it is important to identify the melt fraction in a magma ocean. "Soft" magma oceans are characterized by a high melt fraction and low viscosity, whereas a "hard" magma ocean will have a lower melt fraction and high viscosity. A soft magma ocean will experience turbulent convection such that chemical fractionation does not take place. A hard magma ocean on the other hand is more likely to become chemically fractionated, crystallizing from bottom up. Table 2.5 shows some likely magma ocean scenarios (from Abe, 1997).

2.4.3.2 Controls on magma ocean formation

The two key controls on the formation of a magma ocean are the thermal input and the melting temperature of the planetary material. The principal thermal controls are the presence of a blanketing atmosphere and the rate of delivery and size of the impactors. The blanketing atmosphere may be derived from the gases of the solar nebula, outgassed steam from the Earth or created by the impacting event itself (Table 2.5). Thus, even without impacting a thick thermal blanket can induce melting at the Earth's surface. If surface temperatures exceed 2,100 K then melting will extend deep into the Earth, into the lower mantle (Abe, 1997).

The melting temperature of the Earth's proto-mantle is governed by its initial composition. Most experimental models assume a peridotitic composition for the Earth's mantle with liquidus temperatures of about 2500 °C at 25 GPa. However, if the Earth was undifferentiated and chondritic then the melting temperature might be several hundred degrees lower (Ohtani et al., 1986; Agee, 1997). Another possibility is that the proto-mantle was originally hydrous. Again, this would significantly depress the mantle solidus (Inoue et al., 2000). A hydrous mantle could have formed through the interaction of a steam-rich atmosphere with an early crust and the mixing of that crust back into the mantle (Sasaki, 1990). Since the solubility of water increases with increasing pressure it would be possible to produce a hydrous deep mantle in which a significant quantity of water was stored (Righter et al., 1997).

2.4.3.3 Geochemical evidence for a terrestrial magma ocean

As already noted, one of the paradoxes of the terrestrial magma ocean is that whilst there are strong theoretical grounds for believing that it once existed, there is little evidence preserved in the Earth's mantle which might constitute proof of such a process. In fact, until recently, the evidence seemed to be to the contrary. The present-day mantle does not possess the physical properties of a once molten material (Takahashi & Scarfe, 1985; Ito & Takahashi, 1987), indicating that if it ever was molten, the record of such an event had been destroyed, in perhaps a turbulent early mantle (Tonks & Melosh, 1990; Solomatov, 2000).

A central test in this discussion is whether or not the magma ocean fractionated and the record of such a process is preserved in the Earth's mantle. The standard accretion model for the Earth implies a chondritic composition for the BSE (see Section 2.4.4.1, below). However, when plotted on a Mg/Si versus Al/Si ratio diagram the Earth is less siliceous than chondrite (Fig. 2.16). This problem can be resolved if the mantle was originally chondritic, but then fractionated in a magma ocean into a silica-poor, olivine-rich, peridotitic upper mantle and a silica-rich perovskite-rich lower mantle (Agee & Walker, 1988). Figure 2.15 shows the compositions of the upper mantle phase majorite and the lower mantle phase Mg–perovskite relative to a chondritic bulk Earth composition and a peridotitic upper mantle in Mg/Si–Al/Si space. It is clear from this plot that it is not possible to produce the field of mantle peridotite compositions by majorite fractionation from a chondritic melt. On the other hand it is possible to produce a "parental" peridotite by fractionating 10–20 wt% perovskite from a chondrite starting composition, as proposed by Ito and Takahashi (1987) and Agee and Walker (1988).

However, it is only recently that there has been convincing trace element and isotopic geochemical evidence to support the notion of a magma ocean. The recent ^{142}Nd isotopic study of Boyet and Carlson (2005) has shown convincingly that the Earth experienced a major differentiation event within 30 Ma of

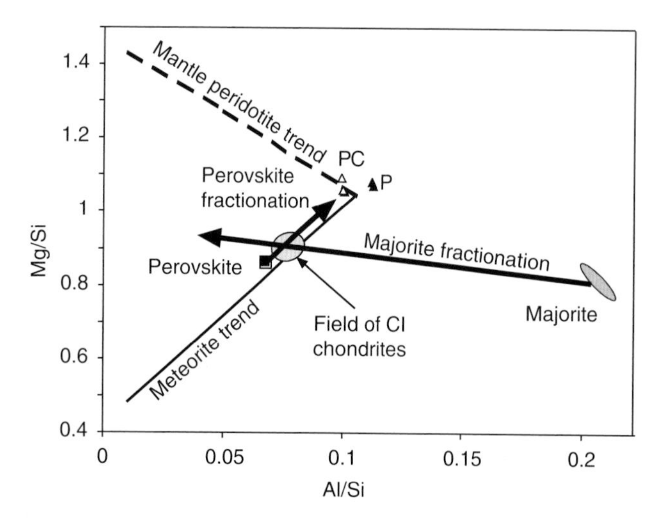

FIGURE 2.15 Mg/Si versus Al/Si (wt%) plot for CI chondrites, majorite, and perovskite. The black arrows show fractionation trends for majorite and perovskite and demonstrate that mantle peridotites cannot be produced from a chondritic starting composition by majorite fractionation but can be by perovskite fractionation. (PC – peridotite-chondrite, P – pyrolite model silicate Earth compositions). Perovskite and majorite data from Ohtani et al. (1986), Ito and Takahashi (1987), and Inoue et al. (2000).

the formation of the solar system (see Chapter 3, Section 3.2.3.1). In addition, high-pressure trace element studies have been found to support the perovskite fractionation hypothesis (Caro et al., 2005; Corgne et al., 2005). Similarly, Walter and Tronnes (2004) reported a number of nonchondritic elemental ratios in the PUM, which might be the consequence of mantle differentiation. These studies show that the fractionation of a mixture of Mg–perovskite, Ca–perovskite, and ferropericlase can explain most of the nonchondritic nature of the upper mantle, and provide important geochemical "clues" to the former existence of a terrestrial magma ocean.

2.4.4 A chondritic model for the silicate Earth

The most primitive of the chondritic meteorites are thought to represent the "raw material" from which the Earth and other terrestrial planets were accreted. These most primitive chondrites, the CI – carbonaceous chondrites (see Section 2.3.3.1), are identified by their high volatile content (they may contain up to 30 wt% of H_2O, S, and C). They lack evidence of thermal processing after their accretion and so can be treated as the least altered condensates of the solar nebula. This view is supported by a plot of element concentrations in CI chondrites against concentrations in the solar photosphere. There is a strong 1 : 1 correlation for all elements except the gaseous elements (H, He, N, O, and the inert gases) and carbon which are depleted in the chondrite.

However, when estimates of the composition of the silicate Earth are compared with primitive chondrite compositions there are a number of important differences. These may be explained in two ways. One possibility is that the primitive Earth was constructed of a chondrite other than CI chondrite, an "Earth chondrite," as is argued by Drake and Righter (2002). In this model there is no single class of meteorites that equates to an Earth composition. Alternatively, the processing of nebular material during planetary formation gave rise to fractionations which are now apparent in the Earth. In this latter model the differences between primitive chondrites and the Earth's mantle may be attributed to fractionation processes that took place within the solar nebula during chondrite formation, fractionations that took place during the accretion process and planetary formation, and fractionations that took place during early Earth differentiation, during the magma ocean stage, and core formation. Each of these possible processes is discussed below.

2.4.4.1 Solar nebula fractionations

A candidate for fractionation during solar nebula differentiation is the element silicon. Hart and Zindler (1986) showed that on a Mg/Si versus Al/Si plot mantle peridotites and chondritic meteorites plot on two quite

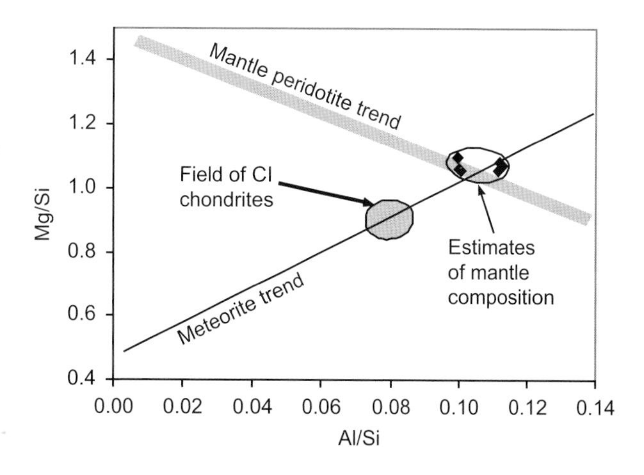

FIGURE 2.16 Mg/Si versus Al/Si plot for meteorites and mantle peridotite samples. The elliptical shaded area is the field for CI chondrites. The diamonds are estimates of mantle composition (see Chapter 3, Table 3.1), located, as expected, at the intersection between the mantle trend and the meteorite trend.

different trends (Fig. 2.16). These trends intersect at a point with a lower Mg/Si and higher Al/Si ratio than most mantle peridotites, but with higher Mg/Si and Al/Si than CI meteorites, suggesting that the Earth is Si-deficient relative to CI chondrites. The implication of this observation is that different chondritic bodies had slightly different compositions, probably related to the preferential accretion or loss of forsterite (high Mg/Si) in the solar nebula and that the Earth formed from a Si-deficient "chondritic" parent.

An alternative view, and one that has some experimental support, is that Si was partitioned into the Earth's core during core formation (Gessmann et al., 2001). If the Earth's metal core contains 7% Si then the bulk Earth and carbonaceous chondrites have the same Mg/Si ratios (Palme & O'Neill, 2003).

Further support for solar nebular fractionation comes from the compilation of Brown and Mussett (1993) who showed that the terrestrial planets have densities, which when corrected for their different internal pressure, vary significantly from that of chondritic meteorites. In fact Mars with an uncompressed density of ca. 3.7 $Mg.m^{-3}$ is the only planet which lies in the chondritic range of 3.4–3.9 $Mg.m^{-3}$. Earth, Venus, and Mercury are denser than chondrites and the Earth's Moon is less dense. These compositional differences are thought to reflect differences in the Fe/Si ratio between the different planets, which in turn reflects the fractionation of Fe from Si in relation to proximity to the Sun, during the condensation of the solar nebula, further supporting the view that the silica-depletion relative to chondrite took place in the solar nebula.

2.4.4.2 Elements fractionated during Earth accretion

The ratios of refractory to volatile lithophile elements are very different between chondritic meteorites and the silicate Earth. These fractionations are thought to have occurred during Earth accretion.

Refractory elements are present in meteorites, the Earth, and the other rocky planets in cosmic abundances and their elemental and isotopic ratios are effectively constant. In contrast, nonrefractory elements show marked variations in absolute and relative abundances with respect to each other and with respect to the refractory elements. This is seen in the extreme volatile loss of the elements H, He, C, N, and the noble gases. In addition the alkali elements (K, Na, Rb, and Cs) are depleted in the Earth's mantle relative to chondritic meteorites (Fig. 2.11). Concentrations of K and Na are approximately 20% less than those in primitive C1 chondrites, although it is possible that there is a small amount of K in the Earth's core (Gessmann & Wood, 2002).

2.4.4.3 Elements fractionated during core formation

Siderophile element concentrations in the Earth's mantle are depleted relative to chondrites (Figs. 2.12 and 2.13), most probably as a consequence of metal–silicate equilibration during core formation. Os-isotope ratios ascribed to the PUM are also different from ratios found in primitive carbonaceous chondrites (Meisel et al., 2001), also possibly a consequence of core formation.

2.4.4.4 Summary

The extent to which particular elements and isotopic ratios might be expected to be chondritic in the Earth's mantle is governed by the chemical behavior of that element during the early stages of Earth formation, as shown in Table 2.6.

2.4.5 The accretion history of the Earth

Defining the "age" of the Earth is a difficult task, for the age of the Earth may mean a number of different things. Harper and Jacobsen (1996) have shown that the age of the Earth may be defined as either the time the solar system began (T_0) or as the time at which accretion ended (T_E). However, both approaches are problematical, for the solar system began before the Earth was formed, and yet the end of accretion is very difficult to define, since the accretion process had a long tail. They suggest that a useful compromise is to define a mean accretion age for the Earth, as the time when 64% of the planetary mass had accreted. Here the main events in the accretion

TABLE 2.6 Summary of elemental behavior in the silicate Earth relative to chondrites.

Element group	Concentration relative to chondrite	Fractionation process
Light gases – H, He, N, and C	depleted	during planetary accretion
Inert gases	depleted	during planetary accretion
Volatile lithophile elements – Mg, Si, Fe, O, Ni, Na, K, Rb, Cs, S, Cu, and Pb	depleted	during planetary accretion
Silicon	depleted	during differentiation of solar nebula or in the core
Refractory lithophile elements – Ca, Al, Ti, Sc, Sr, Ba, Zr, Mo, REE, Hf, Th, U	slightly enriched	excluded from the core and so ca. 1.6 times chondrite
Weakly siderophile – V	weakly depleted	core formation
Moderately siderophile elements – W, Co, Ni, and Mo	moderately depleted	core formation
Highly siderophile elements – Au, Re, Os, PGE	strongly depleted	core formation

history of the Earth are identified. A summary table is given as Table 2.7.

The formation of the earliest matter in the solar system.
The oldest matter of the solar system is found as refractory inclusions (CAIs) and chondrules in carbonaceous chondrite meteorites. The oldest refractory inclusions from the Allende meteorite have been dated at 4,567 Ma (Amelin et al., 2002) and formed over an interval of less than a million years. T_0 therefore is 4,567 Ma. The oldest chondrules formed at the same time, at 4,567 Ma but their period of formation lasted longer, until about 4,563 Ma (Amelin et al., 2004; Haack et al, 2004).

The timescale of accretion. Increasingly evidence suggests that the accretion of the Earth was rapid and was mostly formed after 10 Ma and fully formed after 30 Ma (Jacobsen, 2005).

The formation of the Earth's core. After a period of some controversy, the most recent (and self-consistent) Hf-isotope data suggest an "early" date for the formation of the core. Given that the core grew over a period of time, dating the time of formation of the core is the same problem as dating the time of formation of the Earth. Halliday (2004) suggests that the mean accretion age of the core is about 11 Ma and that core separation was complete by 30 Ma. It is possible that core formation is, on average, earlier than accretion.

The formation of the Moon. The oldest rocks on the Moon are dated at 4,562 ± 68 Ma. However, Hf isotope evidence suggests a formation age of about 30 Ma after the formation of the solar system by 4,537 Ma. This event is thought to mark the end of accretion.

The accretion of a late veneer. Whether or not a late veneer was added to the Earth towards the end of accretion is not clear. The principal evidence comes from the elevated siderophile element chemistry of the mantle. If there was a late veneer, it had to happen after core formation. At present the evidence from the trace element chemistry is ambiguous, because these data can also be explained by the formation of the core at high pressures and temperatures in a magma ocean, or by continuous core formation with decreasing metal input. Currently, the best evidence for a late veneer comes from Os–isotope evidence, where there is a clear mismatch between the composition of the PUM and chondrite. However, even this is uncertain, as is discussed in the next chapter (Chapter 3, Section 3.2.3.4), for

Table 2.7 A chronology for the accretion of the Earth.

Event	Time (Ma)	Time from T_0 (Ma)	References
Formation of the solar system (T_0)	4567.2 ± 0.6 (4569.5 ± 0.2)	0	Amelin et al. (2002) Baker et al. (2005)
CAI formation	4567	0	Amelin et al. (2002) Bizzarro et al. (2004) Krot et al. (2005)
Chondrule formation	4567–4563	4	Amelin et al. (2002, 2004) Bizzarro et al. (2004) Haack et al, (2004) Krot et al. (2005)
Core formation (*Earth 64% formed*)	4,556 4,537	11 30	Yin et al. (2002) Jacobsen (2005)
End of core formation			
End of main growth stage	4,557	10	Jacobsen (2005)
End of accretion	4,537	30	
Differentiation of the mantle – predates formation of Moon *(see Chapter 3, Section 3.2.3.1)*	> 4,537	< 30	Boyet and Carlson (2005)
Moon formation	4,537	30	Schoenberg et al. (2002)
?? Late Veneer			Becker et al. (2005)
Oldest terrestrial materials *(see Chapter 1, Section 1.4.3)*	4,404	163	Wilde et al. (2001)
Late Heavy Bombardment *(see Chapter 6, Section 6.3.1)*	3,800–3,900	770–670	Kring and Cohen (2002)

there are a number of different models for the evolution of Os–isotopes in the Earth's mantle over time.

A late heavy bombardment. There is now good evidence that the Moon experienced a period of intense impacting at about 3.8–3.9 Ga, significantly after the normally accepted end of Earth accretion (Kring & Cohen, 2002). By inference the Earth must have experienced this same event, although attempts to find geochemical evidence of this event in the Earth's oldest sediments at Isua have been disappointing (Frei & Rosing, 2005).

Isotopic accretion modeling. As already discussed, the Earth accreted over a period of time and so defining the "age of the Earth" is not a particularly meaningful exercise. It may, however, be even more serious than this, and may be conceptually wrong, because, by its very nature, the accretion process is an open system and thereby violates one of the fundamental assumptions of geochronology (Hofmann, 2003). The recognition of this problem, that accretion is a disequilibrium process, is giving rise to a new approach to the chronology of accretion – that of isotopic accretion modeling, whereby isotopic ratios are interpreted as part of a dynamic accretion process (Kramers & Tolstikhin, 1997; Halliday, 2004). Such an approach has to make some assumptions about the nature of the unseen accreting material, its similarity or otherwise to the already formed Earth, about the degree of mixing, and about the amount of material lost in the accretion process. Although more complex, isotopic accretion modeling will become a more appropriate approach for the dynamic accretion of the Earth.

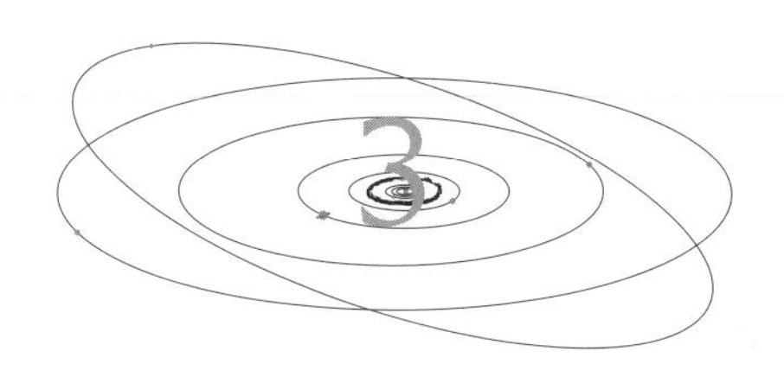

THE EVOLUTION OF THE EARTH'S MANTLE

The evolution of the Earth's mantle – the big picture

The Earth's mantle is peridotitic in composition and is significantly depleted in silica relative to primitive chondrites. Seismological evidence shows that the mantle is layered and can be divided into an upper and lower mantle, separated by a transition zone at 400–660 km depth. Above the transition zone the mantle is dominated by olivine and orthopyroxene with minor garnet and clinopyroxene. The lower mantle is made up of phases Mg- and Ca-perovskite and magnesiowustite. Seismic velocity contrasts between the upper and lower mantle are thought to reflect the phase transformations between the two and are not related to differences in bulk chemical composition. The lower mantle is separated from the outer core by the D″ layer, a hot thermal boundary layer of enigmatic composition.

For many years it was thought that the lower and upper mantle were distinct geochemical sources with minimal mass transfer between the two. On this basis it had been argued that the mantle convects as two separate layers. This is now known not to be the case, for seismic tomographic studies show that subducting slabs penetrate through the mantle transition zone deep into the lower mantle. There is now a strong case for whole-mantle convection, with significant mass transfer from the upper to the lower mantle. Return flow from the lower to the upper mantle may be represented by "superplumes" beneath the Pacific Ocean and Africa. Estimates of the mass flow out of the mantle through basalt melting and into the mantle via subduction indicate that the whole mantle could have been replaced over the past 2–3 Ga.

Calculations based on radioactive heat production show that the Archaean mantle was hotter than the modern mantle. High mantle potential temperatures calculated from the ultramafic lavas – komatiites, common in the Archaean – led to the assumption that the Archaean mantle was substantially hotter than the modern mantle. However, the recent proposal that komatiites are the product of cooler, wet mantle melting weakens the argument for a very hot Archaean mantle, and there are now good grounds for arguing that the temperature of the Earth's mantle has declined by only 100–200°C since the mid–late Archaean.

A study of the radiogenic isotope memory of the Earth's mantle clearly shows that the mantle is not an independent part of the Earth system, nor has it been for a long time. But rather, it records a history of the extraction and recycling of both basaltic and continental crust. Because of the relatively slow mixing rates and rates of diffusion, compositional heterogeneities within the mantle produced by these processes may be preserved for >1.0 Ga so that significant parts of the Earth's prehistory can be seen in recent mantle melts. Mantle melts from early Earth history (basalts and komatiites), therefore, have the potential to record very early mantle heterogeneities.

It is likely that there is no primitive, undifferentiated mantle still preserved within the Earth's mantle. Nd-isotopes indicate that the mantle experienced a major differentiation event, perhaps as early as 30 Ma after the formation of the solar system, in which a Fe-rich basaltic crust formed on a magma ocean and was subsequently buried in the lower mantle and became chemically isolated. Oceanic basalt extraction and recycling can be traced back as far as 4.2 Ga, from Os-isotopes, but the extraction of the continental crust only began on a large scale after about 3.0 Ga. Hence the onset of modern mantle convection is constrained to between 4.2 and 3.0 Ga. The recycling of continent-derived sediment back into the mantle did not begin until about 2.0 Ga.

There are two schools of thought about the structure of the mantle. One sees the mantle as chemically stratified, the other as heterogeneous with "blobs" of enriched mantle randomly distributed throughout a matrix of depleted mantle (the plum pudding model). A layered mantle might be explained either by the deeply subducted oceanic crust now located in the lower mantle, or from the date of the primary differentiation of the Earth in a magma ocean. The heterogeneous "plum pudding" mantle is thought to have been produced through the incomplete mixing of deeply subducted oceanic lithosphere over geological time. Distinguishing between these different explanations of mantle heterogeneity is an important area for future research.

The Earth's mantle is the most important reservoir within the Earth system. It has a thickness of about 2,900 km and makes up 67% of the Earth by mass (83% by volume) and has an actual mass of 4.3×10^{27} g. Our knowledge of the mantle, however, is indirect, for it is concealed and for the most part inaccessible. Nevertheless, evidences of mantle activity are all around us, as described by Harte (1983),

> *The Earth's mantle might be likened to some pantheon of gods. Magnificent displays of power are locally manifested by volcanoes and earthquakes, whilst on the other hand an occult influent is detectable everywhere as a contribution to the heat flow processes at the Earth's surface. Any study of major crustal processes involving magmatism, tectonism, and metamorphism inevitably ponders the power and processes of the mantle.*

Many decades of effort by geochemists and geophysicists have yielded a vast wealth of information about the mantle. These data allow us to build a picture of the mantle as it is now and track its evolution back through geological time.

In the first part of this chapter we shall examine the structure and composition of the modern mantle in order to establish how it works. In so doing we will find tantalizing clues which relate to the mantle's earlier history. It is these clues that we shall explore in the second part of this chapter and use to identify the nature and chemical evolution of the Archaean mantle. These data are then used in the third part of the chapter to constrain models for the Archaean mantle.

A knowledge of how the Archaean mantle system works will become foundational in subsequent chapters of this book which deal with the origin of the continents, oceans, and atmosphere, for each of these parts of the Earth system was ultimately derived from the Earth's earliest mantle.

3.1 Understanding the mantle

In this first section we examine the nature of the modern Earth's mantle. We examine its physical structure, its chemical composition, and the processes operating within it. This approach is necessary to a study of the early Earth, for not only does this provide an insight into how our planet works but also shows that the modern mantle contains a history of much earlier events. We will find that the parts of the present-day mantle are in fact very old and contain an important memory of earlier Earth events.

Initially we shall explore ways in which we can obtain information about the modern Earth's mantle. This will then provide a firm foundation for an understanding of the composition of and processes in the Earth's mantle as we find it today.

3.1.1 Constraints on the nature of the modern mantle

Before we consider in detail *what* we know about the Earth's mantle, first we review *how* we know about the mantle. Here we review the evidence from seismology, heat flow measurements, and mineral physics, and

TEXT BOX 3.1 Glossary of terms relating to the Earth's mantle

adiabatic	An adiabatic process is one where no heat is transferred either in or out of the system. In the mantle melting takes place as a "package" of mantle, with a given potential temperature, rises. Melting and heating occur because of the process of pressure release
asthenosphere	The asthenospheric mantle is that part of the mantle which is semiplastic and makes up the convecting upper mantle. The asthenosphere underlies the rigid nonconvecting *lithosphere*
bulk silicate Earth	The bulk silicate Earth is the composition of the silicate Earth immediately after core formation, before the formation of the continental crust. Hence the bulk silicate Earth composition (BSE) represents the original homogeneous Earth. The composition of the BSE is the same as that of the *primitive mantle*
compatible element	An element which during mantle melting remains in the unmelted host, rather than migrating into the melt
depleted mantle	Mantle peridotite from which a basaltic melt has been extracted. This rock would normally be a *harzburgite*
dunite	A peridotite made up predominantly of the mineral olivine
eclogite	A metamorphic rock, often found in the mantle, made up of the minerals (red) magnesian garnet and (green) sodic clinopyroxene. Many eclogites are metamorphosed basalts
fertile mantle	Mantle peridotite from which a basaltic melt has not been extracted. This rock would normally be a *lherzolite*
harzburgite	A peridotite made up predominantly of the minerals olivine and orthopyroxene. Harzburgites are mantle peridotites from which a basaltic melt has been extracted and so represent *depleted mantle*
incompatible element	An element which, during mantle melting, migrates into the melt phase, from its original host. There are degrees of incompatibility such that some elements show a very strong preference for the melt phase. These are the highly incompatible elements
lherzolite	A peridotite made up of the minerals olivine and orthopyroxene with smaller amounts of clinopyroxene and an aluminum rich mineral such as plagioclase, spinel or garnet. This gives rise to the rock types plagioclase-lherzolite, spinel-lherzolite etc. Lherzolites represent unmelted or *fertile mantle*
lithosphere	The mantle lithosphere is that cool, rigid part of the upper-mantle which forms rigid tectonic plates. It is underlaid by the semiplastic *asthenosphere*. There are differences in structure and composition between the suboceanic lithosphere and the subcontinental lithosphere
kimberlite	Kimberlite is an *ultramafic* rock famous for hosting diamond and is therefore inferred to have been produced deep in the mantle. Kimberlites also bring to the surface *mantle xenoliths* – mantle fragments
mafic rock	A rock, such as basalt, which is rich in magnesium (ma) and iron (-fic)
magnesiowustite	The mineral (Mg, Fe)O. Also known as ferropericlase. A significant component of the lower mantle
majorite garnet	A high-pressure Mg-silicate mineral found in the mantle transition zone (below 400 km)
normative olivine content	The amount of olivine present in a rock as calculated from the major element chemistry of the rock using a specific set of rules – the (CIPW) normative calculation. The amount of normative olivine is not necessarily the same as that observed in the rock. The normative scheme is useful if the rock is very fine grained or altered
ophiolite	A sequence of rocks which include, from top to bottom, pillowed basalts, sheeted dykes, layered gabbros, and mantle peridotites – which are thought to represent former ocean floor
partition coefficient	The coefficient which describes the distribution of a particular element between two coexisting phases. An example of a mineral–melt distribution coefficient

TEXT BOX 3.1 (*Cont'd*)

	would be the distribution of the element Ni between the mineral olivine and a coexisting basaltic melt. Mineral–melt partition coefficients >1.0 imply that the element has a preference for the mineral phase; these are the *compatible elements*. A coefficient <1.0 implies that the element prefers the melt phase; these are the *incompatible elements*.
peridotite	The principal rock-type of the upper mantle comprising the minerals olivine and pyroxene. There are a number of different types of peridotite – see *dunite*, *lherzolite*, and *harzburgite*
perovskite	Magnesian-perovskite (Mg, Fe)SiO_3 is a major mineral in the lower mantle. Ca-perovskite also occurs in the lower mantle
post-perovskite phase	The phase identified in the deep mantle, close to the core–mantle boundary, which represents the very deep transformation of the mineral *perovskite*
primitive mantle	The composition of the Earth's original mantle, after core formation but before crust formation. It is synonymous with *bulk silicate Earth*
serpentinite	An *ultramafic* rock made up of the mineral serpentine, normally produced by the hydrous alteration of an olivine rich rock such as *peridotite*
solidus	The mantle solidus is the melting curve for the mantle. At a particular depth the solidus defines the temperature at which the mantle begins to melt. The mantle solidus is governed by its water content, and so the wet-solidus is at a lower temperature than the dry solidus
rhenium depletion model age	The time at which melt was extracted from a mantle peridotite. More precisely it is a model age calculated relative to the $^{187}Os/^{188}Os$ ratio of the BSE, assuming that the Re/Os ratio of the sample is zero. Note, however, that BSE reference values vary
ringwoodite	The high-pressure mineral γ-Mg_2SiO_4 (a form of olivine) found in the deep part of the mantle transition zone. Ringwoodite can contain water in its lattice
ultramafic	A rock, such as peridotite which is dominated by dark, iron–magnesium-rich minerals such as olivine and pyroxene
wadsleyite	The high-pressure mineral β-Mg_2SiO_4 (a form of olivine but with a spinel structure) found in the mantle transition zone below 400 km. Wadsleyite can contain water in its lattice
xenolith	A mantle xenolith is a fragment of the mantle brought to the surface in a magmatic rock such as a basalt or *kimberlite*

the petrological evidence from ophiolites, mantle xenoliths, and from mantle melting experiments that allow us to understand the nature of the modern mantle.

3.1.1.1 Evidence from seismology

The science of seismology has profoundly influenced the way we think about the interior of the Earth. From the first estimate of velocity–depth curves for the Earth, obtained in the early part of the twentieth century a clear image has now emerged of the layered nature of the interior of the Earth. The synthesis of a huge amount of data derived from seismic body-wave travel times for P-waves (longitudinal waves) and S- waves (transverse-waves) has provided a profile for the Earth of changing seismic velocities V_p and V_s with depth. This profile shows major discontinuities between the crust and mantle, mantle and core, and the inner and outer core. In addition less extreme velocity contrasts have allowed the identification of the mantle transition zone, located between 410 km and about 660 km. This zone is marked by a rapid increase in seismic wave velocity and marks the transition between the upper and lower mantle (Fig. 3.1).

At the core–mantle boundary (2,740–2,890 km) there is a further sharp increase in

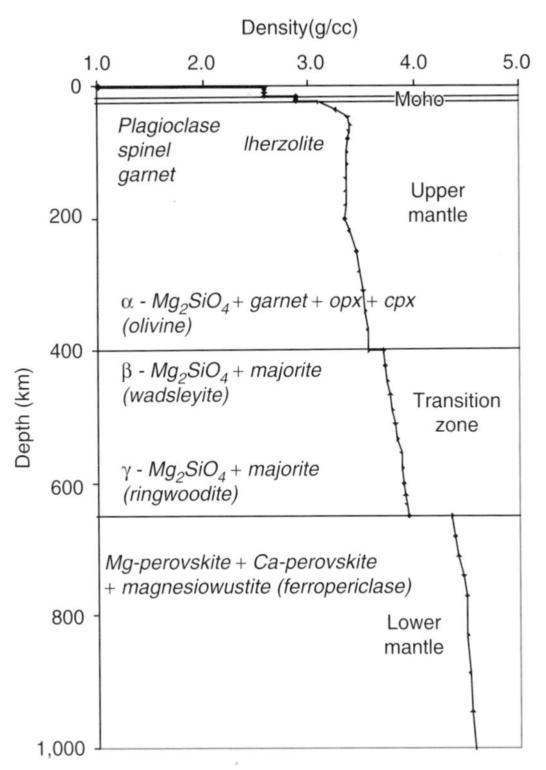

FIGURE 3.1 Density–depth profile for the top 1,000 km of the Earth's mantle showing the main mineral phases present in the different layers. (Depth density data from Montagner & Anderson, 1989.)

wave velocity. This is the D″ (D-double prime) layer, a hot thermal boundary layer (TBL) produced by the transport of heat from the Earth's core to its lowermost mantle. The D″ layer is also a mechanical boundary layer (MBL) for convection in the lower mantle (Panning & Romanowicz, 2004). The D″ layer is about 300 km thick and has an approximate mass of 2×10^{26}g – that is about 5% of the mass of the mantle. In detail the nature of this particular layer in the mantle is poorly understood but of considerable interest. The D″ layer is variable in thickness, contains discrete discontinuities, and is thought to contain lateral chemical heterogeneities making it significantly different in composition from the overlying mantle. It has been suggested that it is the ultimate resting place for subducted slabs, a "slab graveyard" (Russell et al., 1998), although other workers have suggested that the layer is extremely ancient and formed shortly after accretion (Tolstikhin & Hofmann, 2005). What is generally agreed is that the D″ layer plays an important role in the generation of mantle plumes.

The densities of the different layers within the Earth can be also be inferred by combining information from P- and S-body waves with data from surface wave oscillation periods. This then permits a density–depth profile to be constructed for the Earth, the mantle section of which is shown in Fig. 3.1. These data have been progressively improved and refined into a reference model for the Earth showing the average depth–velocity–density structure of the Earth. These data were first presented as a Preliminary Reference Earth Model (PREM – Dziewonski & Anderson, 1981), the mantle part of which was subsequently refined by Montagner and Anderson (1989).

More recent advances in our understanding of the finer detail of mantle structure have come through the development of *global seismic tomography*. Seismic tomography is a way of making a "body scan" of the Earth's interior and is analogous to CAT-scans used in medicine to examine the human body. This method depends upon very large databases comprising large numbers of earthquake waves (both surface and body waves) traversing a particular segment of the Earth's mantle. These data represent a dense mesh of waves which provide the basis for 3-D imaging of seismic velocities within the mantle. In detail, the region of interest within the mantle is divided into sectors 100–200 km on a side, and for each sector the wave velocities are mapped as either fast or slow by inverting the travel time data. This provides an image of the mantle in which relative velocities can be mapped.

Cold regions in the mantle tend to be more rigid and less compressible so that seismic waves travel faster in these regions. In contrast, warm regions are less rigid and less dense and seismic waves travel more slowly. In this way warm/cold, more rigid/less rigid domains in the mantle can be identified and mapped. Tomographic imaging of this type has been used to examine the thermal structure of the Earth's mantle as far down as the core–mantle boundary. These result are presented as velocity/temperature maps of the Earth and are presented either as "slices" through the Earth along a particular line of section or as maps of the mantle for a given depth. The poorest resolution is at about 700 km depth meaning that the upper-mantle to lower mantle boundary is not well resolved, and in some cases the base of the subcontinental lithosphere is not well resolved. At shallower mantle levels seismic tomography has been used to map flow directions in upper mantle utilizing the different wave velocities associated with the different crystallographic orientations of olivine (Fig. 3.2).

3.1.1.2 Mineral phases in the mantle – evidence from mineral physics

The mineralogy of the mantle changes with depth as pressure increases. In the case of the shallow upper-mantle the mineralogy of the mantle is known from mantle xenoliths (see Section 3.1.1.5). The mineralogy of the deeper mantle is known from rare mineral inclusions in diamond (see Section 3.1.1.5), from experimental studies using multianvil and diamond anvil techniques, and from theoretical studies of the bulk mantle composition constrained by the V_p–V_s- and density data for the deep mantle.

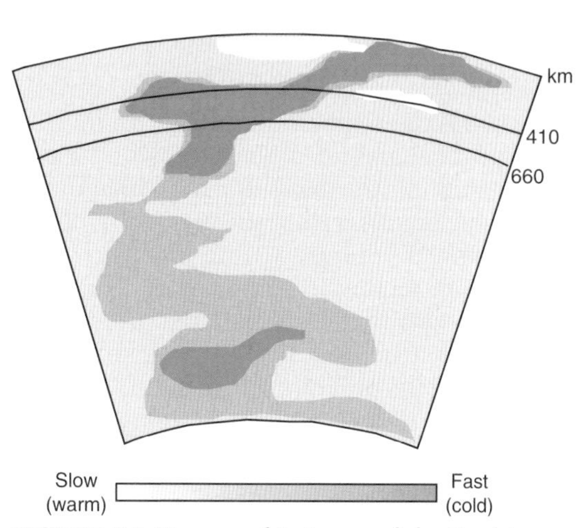

FIGURE 3.2 Tomographic image of the Earth's mantle beneath the Japanese Arc, down to the core–mantle boundary showing the distribution of slow and fast seismic waves. The wave velocity distribution also reflects temperature distribution and shows the penetration of a cold subducting slab through the transition zone into the lower mantle (after Fukao et al., 2001).

Of particular significance are the density increases which take place at the upper and lower boundaries of the mantle transition zone, at the 410 km and 660 km discontinuities. In the past it has been argued that the large density increase which takes place at 660 km depth reflects a change in the bulk composition of the mantle with depth. However, the present consensus is that the contrasts can be accommodated simply by phase changes in the mantle mineralogy. (This debate has huge consequences for whether or not the mantle is chemically layered, and is an important factor in the current debate about the nature of mantle convection).

The upper part of the upper mantle (see Fig. 3.1) is made up of the mineral assemblage olivine (60%) + orthopyroxene (30%) + clinopyroxene (5%) + an aluminous mineral (5%). At the shallowest depths the aluminous phase is plagioclase, but this changes to spinel at about 30 km depth (ca. 0.1 GPa) and from spinel to garnet ca. 75 km (ca. 0.2 GPa). Thus the upper mantle is made up of plagioclase lherzolite down to about 30 km, is spinel lherzolite from 30 to 75 km, and is garnet lherzolite at depths greater than about 75 km. The precise transitions between the aluminous phases are a function of mantle temperature, bulk composition, and pressure.

Between about 300 and 460 km depth, pyroxene begins to dissolve in garnet to form the higher density phase – majorite garnet. At about 400 km (14 GPa) olivine also undergoes a structural change from its low pressure α-form to the more dense β-Mg_2SiO_4 (wadsleyite – see Fig. 3.1). Both these phase changes represent a significant increase in density in the mantle. In the case of the gradual pyroxene to majorite transition there is an increase in density of about 10%, whereas in the case of the more rapid α- to β-Mg_2SiO_4, there is an increase in density of about 8%.

At about 500 km depth, within the mantle transition zone, olivine undergoes a further phase change from β-Mg_2SiO_4 to spinel structured Mg_2SiO_4 – γ-Mg_2SiO_4 (ringwoodite – Fig. 3.1).

This is a more gradual transition than the α- to β-olivine transition and represents only a small (2%) density increase. γ-olivine has a structure similar to that of spinel.

At 660 km depth there is a marked increase in seismic velocities and mantle density (Fig. 3.1). This discontinuity defines the lower boundary of the mantle transition zone. Here γ-Mg_2SiO_4 dissociates to magnesian-perovskite $(Mg, Fe)SiO_3$ and magnesiowustite – or ferropericlase – $(Mg, Fe)O$, and majorite-garnet breaks down to form the phase Ca-perovskite ($CaSiO_3$). Hama and Suito (2001) used the thermoelastic properties of these three phases to calculate that the lower mantle contains 77–83% (molar) magnesian-perovskite, 13–18% magnesiowustite, and 4–5% Ca-perovskite. Magnesian-perovskite is therefore the dominant mineral of the lower mantle and so the most abundant mineral on the planet.

In the region of the core–mantle boundary Mg-perovskite transforms to a "post-perovskite phase" (Oganov & Ono, 2004). This phase is likely to be the dominant phase of the D″ layer and its presence may explain many of the unusual physical properties of this layer. This phase transition is exothermic and so this discovery also has important implications for the generation of mantle plumes at the core–mantle boundary.

3.1.1.3 *Thermal structure*

Heat flow measured in the Earth's crust represents heat which has been transported from below by conduction. This is a composite of heat produced *in* the crust by radioactive decay and heat *transferred from below* from the mantle. Heat production from radioactive decay comes from isotopes of the elements potassium (^{40}K), uranium (^{235}U, ^{238}U), and thorium (^{232}Th). Calculations based upon average heat producing element concentrations indicate that almost 90% of the present heat outflow of the Earth (ca. 41×10^{12}W, Brown & Mussett, 1993) comes from the mantle. Most of this heat is lost through the oceanic crust. There is a problem, however, because geochemical estimates of heat producing element concentrations in the mantle cannot explain the heat outflow of the Earth. This observation requires that a contribution to mantle heat flow comes from elsewhere. These heat sources most probably include the solidification of the inner core and heat from the primordial energy of accretion.

Heat flow data provide important constraints on mantle models. For example, combined with the heat producing element content of crust and mantle rocks, and the physical properties of mantle minerals, they can be used to constrain the nature of thermal convection in the mantle. In addition, variations in the mantle contribution to crustal heat flow between the continents and oceans have been used to make inferences about the nature of the different types of mantle underlying continental crust and oceanic crust (Section 3.1.2 and Chapter 4, Section 4.3.1.2). Furthermore, heat flow data, combined with bathymetric measurements, rates of sea-floor subsidence, and the depth of seismic discontinuities are all a function of mantle temperature and can be used to estimate relative, lateral variations in mantle temperature (Anderson, 2000).

3.1.1.4 *Ophiolites and orogenic lherzolites*

Petrology also provides useful information about the Earth's mantle and the largest mantle samples are found in the lower sections of ophiolite complexes and in orogenic lherzolites. Ophiolite complexes represent obducted fragments of the ocean floor, now exposed as thrust slices on continental crust. Mantle rocks are frequently present as part of the ophiolite succession, underlying basaltic pillow lavas, sheeted dykes and layered gabbros. In the case of the late Cretaceous ophiolite of Oman (the Sumail Ophiolite), over 10,000 sq km of mantle rocks are exposed (Fig. 3.3). However, even in this huge area, only the uppermost 10–15 km of the mantle is represented.

Nicolas (1989) divided ophiolite complexes into two main types – the harzburgite type (e.g. Oman; Bay of Islands, Newfoundland; Zambales, Philippines) and the lherzolite type (e.g. Trinity, California; Lanzo, Italy). He proposed that the harzburgite-type of ophiolite represents oceanic crust created at a fast spreading ridge and a mantle sequence which

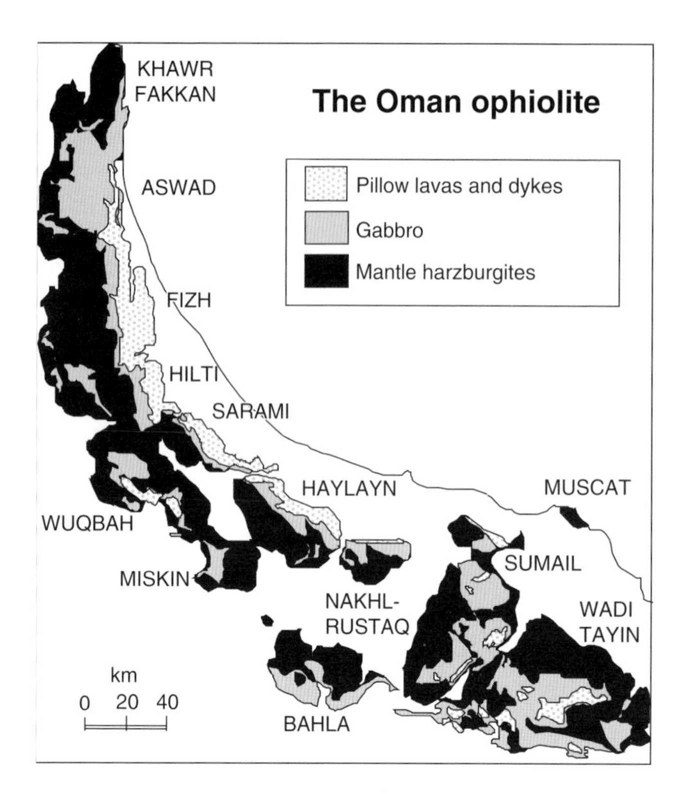

FIGURE 3.3 The Oman ophiolite showing the different crustal blocks. The mantle section is shown in black (after Nicolas et al., 2000).

has experienced a high degree of partial melting, whereas the lherzolite-type represents crust formed at a slower spreading ridge and a mantle section which experienced a lower degree of partial melting.

A further class of mantle samples are variously described in the literature as peridotite massifs, Alpine peridotites, and orogenic lherzolites. They are thought to be slices of mantle peridotite emplaced into or onto the continental crust during continental collision but they cannot be strictly identified as ophiolite fragments.

Ophiolitic mantle therefore provides a great deal of insight into the uppermost part of the oceanic mantle and in particular into the processes of melt accumulation and melt migration beneath a midocean ridge. In recent years new insights have been gained into the process of melt reaction whereby melts derived at depth are modified in composition through reaction with shallow mantle.

3.1.1.5 Mantle xenoliths

If geophysics provides a large-scale picture of the Earth's mantle, then xenoliths provide the detail. This is typical of much of the Earth Sciences where we have to marry together large- and small-scale observations. In the case of the Earth's mantle, both views are necessary, and yet bringing them together is not without its difficulties.

Mantle xenoliths are fragments of the Earth's mantle, usually a few centimeters across, brought rapidly from great depths during volcanism to arrive at the surface in their unmodified state. They represent a very important source of information about the mantle. They are most commonly found in kimberlites and alkali basalts. The most abundant xenoliths are from kimberlites, which sample the deep mantle beneath ancient continental crust. Alkali basalts sample shallower mantle from beneath both continents and oceans. As will be seen below, the mineralogical and chemical diversity displayed in mantle xenoliths adds considerable complexity (and realism) to the geophysical models of the mantle discussed above in Section 3.1.1.1.

3.1.1.5.1 Kimberlite xenoliths

Harte (1983) subdivided the xenoliths found in kimberlite into peridotites, eclogites, and a mica-amphibole-rich suite termed the MARID (**M**ica-**A**mphibole-**R**utile-**I**lmenite-**D**iopside) suite. The peridotite group includes garnet lherzolites, garnet harzburgite, and dunite. These samples are sometime divided into fertile and infertile (or depleted) varieties indicating their relative richness or poverty in basaltic components such as CaO and Al_2O_3. Typically garnet lherzolite is fertile mantle and dunite is infertile. The origin of the eclogite suite is much debated and may be either ancient subducted oceanic lithosphere or subducted oceanic lithosphere with a felsic melt fraction removed. The MARID suite represents samples which are enriched in elements such as K, Rb, Ti, Ba, Sr, and Zr. This process of metasomatism implies a reaction between peridotite and a migrating melt and often results in the presence of additional phases such as amphibole or mica.

The depth of origin and formation temperature can be calculated for mantle xenoliths using mineral–chemical equilibria between the coexisting xenolith phases (Fig. 3.4). In a series of innovative studies on xenoliths from kimberlites in northern Lesotho, Boyd and Nixon showed that when a sufficient variety in temperature and pressure is recorded from the xenoliths of a single kimberlite pipe then it is possible to reconstruct a lithological and thermal log of the underlying mantle, at the time of kimberlite emplacement (e.g. Boyd, 1973; Boyd & Nixon, 1975). Studies of this type show that a typical mantle profile was a layer of "cold" coarse peridotites underlaid by "hot" deformed peridotite penetrated with melt (Harte, 1983). Reconstructions of this type led to the first serious understanding of the nature of the mantle beneath old continental crust.

3.1.1.5.2 Alkali basalt xenoliths

Xenoliths in alkali basalts are most commonly spinel-bearing peridotites. They have been divided into two main groups (Frey & Prinz, 1978). Primitive xenoliths have chemical compositions which suggest that they have had a melt extracted from them. Menzies (1983) showed that this "depleted" character is similar in alkali basalt xenoliths beneath both the oceans and the continents and proposed that a depleted mantle layer underlies both

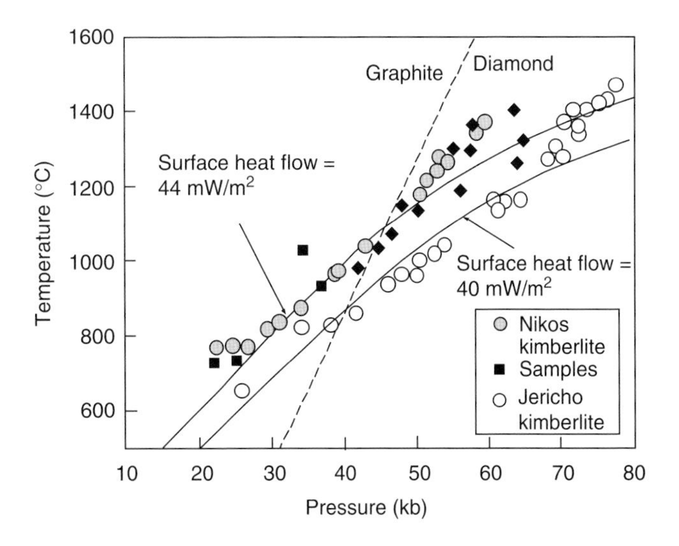

FIGURE 3.4 Mantle pressure and temperature conditions, calculated from kimberlite xenoliths from the Nikos and Jericho kimberlites in the Slave Province of the Canadian shield, indicate that the xenoliths last equilibrated at depths between 60 and >200 km in the upper mantle. These xenoliths represent samples of Archaean subcontinental lithosphere. The data can be fitted to one of two conductive paleogeotherms that correspond to surface heat flows of either 40 mW/m^2 (Jericho kimberlite in the center of the craton) or 44 mW/m^2 (Nikos kimberlite, marginal to the craton). Some xenoliths equilibrated in the diamond stability field (from Schmidberger & Francis, 1999).

the oceans and continents. Other xenoliths, formed in the upper part of the spinel lherzolite field, represent disrupted mantle fragments permeated by melts and illustrate the complexity of some regions of the upper mantle, when examined at this level of detail. Xenoliths in potassic, ultramafic magmas from the Ontong Java oceanic Plateau have been found to come from much deeper in the mantle. Collerson et al. (2000) reported majorite, calcium- and magnesium-perovskite bearing xenoliths in these rocks, which come from the base of the transition zone.

Just as ultramafic xenoliths from kimberlites can be used to build a picture of the deeper subcontinental mantle so also can alkali basalts xenoliths be used to image shallower mantle – from 30 to 75 km depth. In this way Sen (1983) and Sen et al. (1993) and Yang et al. (1998) have used xenoliths to constrain the processes of partial melting within the ocean lithosphere (see Section 3.1.4.3).

3.1.1.5.3 Diamond inclusions

Diamond forms a strong and chemically inert container which permits samples of the deep mantle to be brought to the surface without reequilibration. Until recently it was thought that diamond inclusions were derived from relatively shallow mantle. However, Moore and Gurney (1985) identified majorite garnet inclusions in diamonds from South Africa. More recently Stachel et al. (2000) showed that garnets from Guinea contain a variable majorite component indicating their derivation from between 250 and 440 km depth. The deepest mantle inclusions include the phases periclase, ferropericlase, Mg-perovskite, and Ca-perovskite and have been widely reported including samples from Brazil (Harte & Harris, 1994) and from Guinea (Stachel et al., 2000). These diamonds come from a range of depths down to about 1,700 km – deep in the lower mantle – indicating a very deep origin for some kimberlites (Hayman et al., 2005).

3.1.1.6 Mantle melts as windows into the mantle

Many experimental studies have shown that basalt is the melting product of mantle peridotite. This means that even though the Earth's mantle is inaccessible, its internal composition can be known from its melt products. Basaltic melts then provide an important "window" into the Earth's mantle. Furthermore, because basaltic rocks are preserved within the continental crust from as far back as 3.8–3.9 Ga, they provide a window into the Earth's mantle through most of geological time.

Whilst inferences can be made about variations in the mineralogy of the mantle with depth from basalts as the mantle melts, it is the chemical composition of basalts that is the most powerful source of information about the mantle. In particular, a detailed examination of the trace element and isotopic characteristics of basalts has revealed a number of different basalt types, whose differences can only be explained if they come from chemically different domains within the Earth's mantle. Basalt chemistry, therefore, is the prime source of information about chemical heterogeneity within the Earth's mantle. This heterogeneity is a very important property of the mantle and one which can only be understood when the history of the Earth's mantle is considered over geological time.

Figure 3.5a shows the difference in trace element concentrations for an average mido-cean ridge basalt and an average oceanic island basalt. The ocean island basalt (OIB) is strongly enriched in certain trace elements relative to the ocean ridge basalt. In a similar way Fig. 3.5b shows the isotopic differences between a number of different mantle end-member compositions, thought to be responsible for the range of basaltic compositions observed in Nd-isotope-Sr-isotope space (Hart, 1988).

Of course observations of this type provide only a broad picture of the Earth's mantle, for it is likely that the process of basalt melting takes place over a range of depth intervals and that erupted magmas "drain" the melt products from a broad area of the mantle. Nevertheless basalt chemistry is a powerful tool in identifying geochemically contrasting reservoirs within the Earth's mantle and provides the basis for identifying compositional heterogeneities within the mantle.

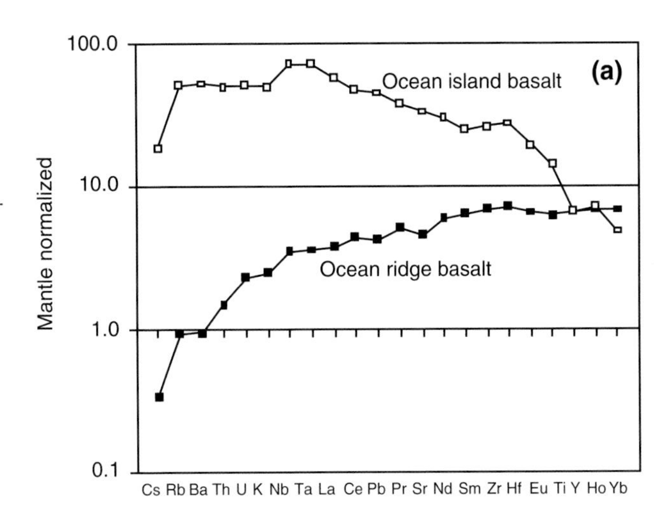

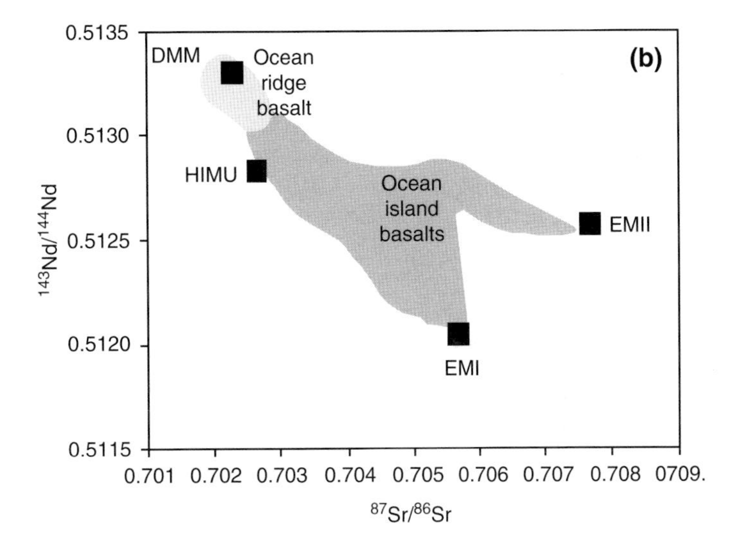

FIGURE 3.5 (a) Primitive mantle-normalized trace element concentrations in average MORB and average OIB (see Text Box 2.2). (b) $^{143}Nd/^{144}Nd$ versus $^{87}Sr/^{86}Sr$ isotopic characteristics of ocean ridge basalt and OIB showing the mantle end-member compositions identified by Hart (1988) – DMM depleted MORB mantle, HIMU – OIB source with high U/Pb ratio, EMI and EMII – enriched ocean-island basalt sources.

3.1.1.7 Summary

Figure 3.6 illustrates the main contributions of geophysical data, ophiolite complexes, xenoliths, diamond inclusions, and mantle melts to our current understanding of the mantle. What is clear from this summary is that the extent to which we understand the mineralogy and composition of the Earth's mantle decreases with depth, and our only detailed knowledge is of the upper mantle.

3.1.2 The composition of the Earth's mantle

Our knowledge of the composition of the Earth's mantle is constrained by both theory and observation. It is widely accepted that the bulk composition of the Earth is chondritic (see Chapter 2, Section 2.4.4). Hence, the broad compositional parameters of the mantle are set by this model. In addition, the composition of the upper mantle is constrained from chemical measurements made on xenoliths and from mantle rocks in ophiolites. Together these observations indicate that the mantle is ultramafic in composition and that the *upper* mantle is made up of peridotite. Further gross compositional constraints are provided by seismic and density data, and from peridotite melting experiments. Figure 3.7 shows the nomenclature for ultramafic rocks, expressed as proportions of olivine, orthopyroxene, and clinopyroxene, and illustrates the range of

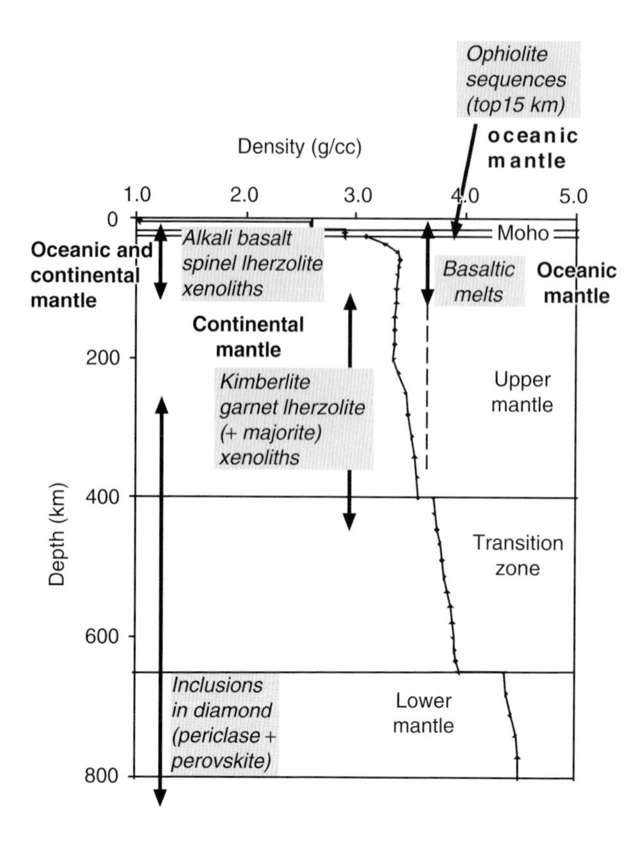

FIGURE 3.6 A summary diagram showing the relative contributions of ophiolites, mantle xenoliths, diamond inclusions, and basaltic melts to our knowledge of the nature of the upper and lower mantle.

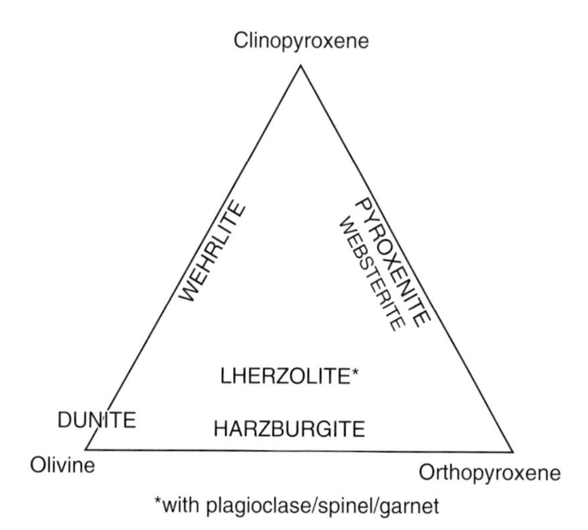

FIGURE 3.7 Classification of ultramafic rocks showing the main rock types found in the upper mantle.

rock types found as xenoliths and in the mantle section of many ophiolite complexes.

Mantle rocks may be classified into those which are relatively enriched in the elements Ca, Al, Ti, and Na and those which are not. Lherzolites are "enriched" peridotites and are thought to be mantle rocks which have not melted and are known as "fertile" mantle – mantle from which a basaltic melt has not been extracted. Typically they contain a few percent of clinopyroxene and an aluminum-rich mineral (plagioclase, spinel, or garnet) in addition to the statutory olivine and orthopyroxene. Mantle which has experienced melt extraction is known as "depleted" mantle and has lower concentrations of the elements Ca, Al, Ti, and Na. This type of mantle is represented by the rock types harzburgite (peridotite without clinopyroxene and an aluminous mineral) and dunite. The relationship between a fertile mantle lherzolite, depleted mantle harzburgite/ dunite and a basaltic melt is

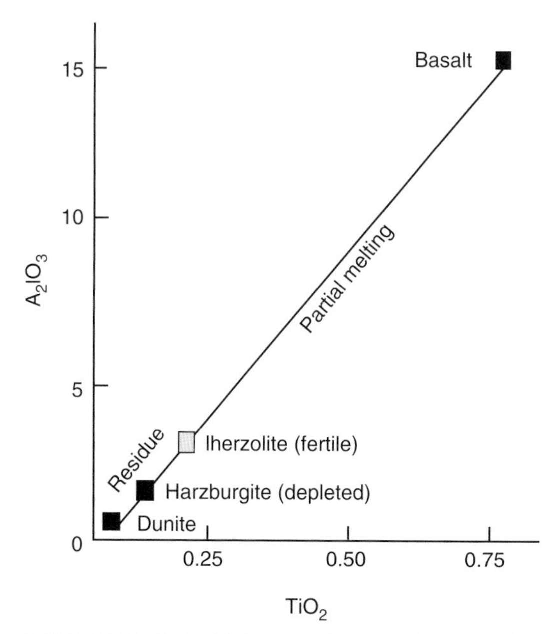

FIGURE 3.8 TiO_2–Al_2O_3 plot showing how fertile lherzolite melts to produce basalt (enriched in TiO_2 and Al_2O_3), leaving a harzburgite residue (depleted in TiO_2 and Al_2O_3), after Brown and Mussett (1993).

illustrated on the TiO_2–Al_2O_3 plot shown in Fig. 3.8.

3.1.2.1 The major element composition of the mantle

Estimates of the chemical composition of Earth's mantle normally refer to the composition of the Earth's mantle as it existed immediately after core formation but before the extraction of the continental crust. This composition is known as the bulk silicate Earth (BSE) or the Primitive mantle and is an important reference composition for the study of the mantle.

Estimating the composition of the Bulk Silicate Earth is strongly dependent upon an appropriate model for the origin of the Earth and accordingly involves a number of assumptions. These are, that for nonvolatile elements, the Earth has a CI chondritic composition, the mantle is homogeneous, that the upper and lower mantle have the same composition, and that the Earth's core is a Fe–Ni alloy with the addition of 10–15% light elements (Allegre et al., 1995).

3.1.2.1.1 Chondritic model

A chondritic model for the Earth's mantle is usually based upon the composition of CI chondrites, adjusted for the loss of volatile elements and for the separation of the siderophile elements into the core. This leads to a mantle composition which is enriched in refractory lithophile elements by about 1.5 times the CI chondrite value. There are however difficulties with the chondritic model because CI chondrites and the Earth have different Mg/Si ratios (Fig. 3.9), as was discussed in Chapter 2 (Section 2.4.4). Some authors believe that this difference is original and dates from processes within the solar nebula indicating different evolutionary histories between the two. If this is true then CI chondrites are not an appropriate starting composition for the composition of the bulk Earth and alternative models have to be considered.

3.1.2.1.2 Peridotite-chondrite models

Rather than assuming a composition for the mantle, its composition may be estimated by directly measuring the composition of mantle samples. The biggest difficulty here is

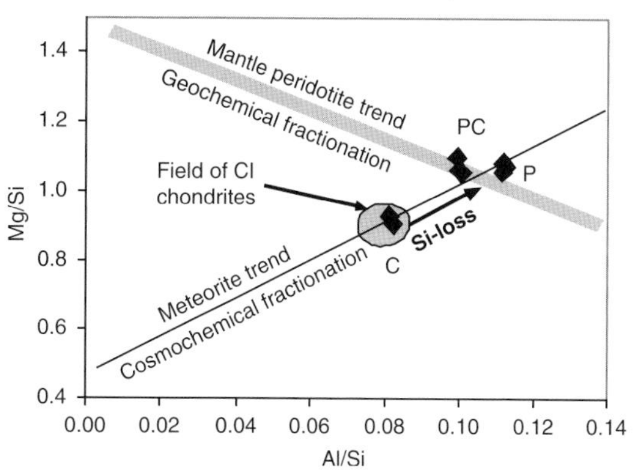

FIGURE 3.9 The geochemical/ cosmochemical fractionation diagram for the Earth showing the various estimates of primitive mantle composition discussed in the text. The elliptical shaded area is the field for CI chondrites and the analyses in Table 3.1 are plotted as weight ratios, for the chondritic model (C), the pyrolite model (P) and the peridotite–chondrite model (PC).

identifying the most appropriate samples. In a novel approach Jagoutz et al. (1979), and Hart and Zindler (1986) showed that on a Mg/Si versus Al/Si plot, mantle peridotites and chondritic meteorites plot on two quite different trends (Fig. 3.9). These trends indicate two very different chemical processes. In peridotites the geochemical fractionation is a consequence of melt extraction, whereas in chondrites it is related to processes operating in the solar nebula (See Chapter 2, Section 2.4.4.1). The two trends intersect at a point with a lower Mg/Si and higher Al/Si ratio than most mantle peridotites, but with higher Mg/Si and Al/Si than CI chondrites, suggesting that the Earth is Si-deficient relative to CI chondrites. This method permits an independent estimate of the composition of a BSE composition prior to mantle differentiation.

Allegre et al. (1995) challenged the approach of Jagoutz et al. (1979) and Hart and Zindler (1986) by suggesting that the Earth's apparent Si deficiency is not primary, but that Si entered the Earth's core during the earliest stages of planetary differentiation (see Williams & Knittle, 1997 for a different view). They proposed an alternative means of calculating mantle compositions without dependence upon Si-ratios, although their result for major elements is very similar to that of Jagoutz et al. (1979), and Hart and Zindler (1986) – see Table 3.1.

3.1.2.1.3 Pyrolite models

Pyrolite models for the mantle are based upon the assumption that the composition of the mantle can be calculated by "reconstituting" the mantle from depleted mantle residues plus basaltic melt. The method was first proposed by Ringwood (1962), who showed that his model agreed well with heat flow data, and with the composition of carbonaceous chondrites minus a core component (Ringwood, 1966). More recently McDonough and Sun (1995) have used a similar approach. These authors examined a suite of fertile mantle peridotites,

TABLE 3.1 Major element compositions of the Earth's mantle as calculated by the different models discussed in the text.

	Chondrite model		Pyrolite model		Peridotite–chondrite model		
	CI Chondrite volatiles and siderophile elts removed (Hart & Zindler, 1986)	Bulk silicate Earth based on CI Carbonaceous Chondrite – Taylor and McLennan (1985)	McDonough and Sun (1995)	Pyrolite (Ringwood, 1991)	Allegre et al. (1995)	Jagoutz et al. (1979)	LOSIMAG (CI bulk silicate Earth, but with low Si and Mg) – Hart and Zindler (1986)
SiO_2	49.52	49.90	45.00	44.76	46.12	45.15	45.96
Al_2O_3	3.56	3.65	4.45	4.46	4.09	3.97	4.06
FeO	7.14	8.00	8.05	8.43	7.49	7.82	7.54
MgO	35.68	35.15	37.80	37.23	37.77	38.31	37.78
CaO	2.82	2.90	3.55	3.60	3.23	3.50	3.21
TiO_2	0.16	0.16	0.20	0.21	0.18	0.22	0.18
Cr_2O_3	0.41	0.44	0.38	0.43	0.38	0.46	0.47
Na_2O	0.29	0.34	0.36	0.61	0.36	0.33	0.33
K_2O	0.03	0.02	0.03	0.03	0.03	0.03	0.03
Weight ratios							
Al/Si	0.081	0.083	0.112	0.113	0.100	0.100	0.100
Mg/Si	0.930	0.909	1.084	1.073	1.057	1.095	1.061
Atomic ratio							
mg#	0.90	0.89	0.89	0.89	0.90	0.90	0.90

screened to eliminate samples which had experienced melt extraction and melt enrichment but found that even the most apparently fertile peridotites have experienced some melt loss. Their estimated composition for the silicate Earth is given in Table 3.1.

3.1.2.1.4 Composition of the lower mantle

A variety of compositions have been proposed for the lower mantle including almost pure perovskite, chondrite, and pyrolite. However, most models of the Earth assume that the upper and lower mantle have the same composition. Recent attempts to directly estimate the composition of the lower mantle have used best-fit curves of the thermoelastic properties of the Earth to a PREM model mantle made up of the phases Mg-perovskite, Ca-perovskite, and magnesiowustite. The recent calculations by Li and Zhang (2005) indicate a pyrolitic composition for the lower mantle, with Mg/Si atomic ratios between 1.29 and 1.39, slightly higher than those for the pyrolite models in Table 3.1 (Mg/Si_{atomic} = 1.24–1.25).

3.1.2.2 The trace element composition of the Mantle

A number of estimates have been made of the trace element content of the Earth's mantle. These are based on trace element concentrations in midocean ridge basalts, OIBs (see Fig. 3.5a), arc basalts, and continental basalts, and these are explored more fully later in this chapter.

In addition estimates of the composition of the trace element content of the Bulk Silicate Earth/ Primitive mantle have been made by a number of authors. These are summarized in Table 3.2.

3.1.2.3 The volatile content of the mantle

The presence of volatiles in the Earth mantle is not only of petrological interest but has profound implications for the evolution of the Earth's atmosphere. In fact "the role of volatiles is so central to the thermal, petrological, climatological and biological evolution of the Earth that one cannot properly address the Earth as an evolving coupled system without the careful integration of volatiles" (Pollack, 1997). Hence this initial discussion of the volatile content of the mantle is a topic to which we will return later, in Chapter 5 (Section 5.1), when we consider the origin of the Earth's atmosphere and oceans.

The two volatiles central to processes in the modern Earth are water and carbon. Water has a profound effect on the properties of the Earth's mantle. In particular it influences the melting temperature of the mantle and its viscosity. Water, as an agent of hydrolytic weakening is thought to be the primary driver behind the longevity of plate tectonics on the Earth (Regenauer-Lieb and Kohl, 2003). The presence of water in the Earth's mantle also has a marked effect on mantle melting. Asimow and Langmuir (2003) recently showed that wet melting at a midocean ridge (see Section 3.1.4, below) increases the total melt production, giving rise to thicker crust and extends the depth of the melt column into the mantle.

The exact amount of water present in the Earth's mantle is the subject of some debate, but recent estimates suggest that it is in the range 6–9 × 10^{24} g (Ohtani, 2005).

Estimates of the carbon content of the Earth indicate that it too is a major component of the Earth's volatile budget. Tingle (1998) estimated that the primordial carbon content of the Earth was between 10^{24} and 10^{25} g. This is many orders of magnitude greater than the total carbon at the Earth's surface, and even allowing for carbon stored in the Earth's core implies that the mantle is a huge carbon reservoir. Precisely locating this carbon, however, is more difficult and this topic is discussed more fully in Chapter 5, Section 5.1.2.

3.1.3 Lateral variations in the Earth's upper mantle

The upper mantle can be divided into a strong upper layer – the lithosphere and a weaker, deeper layer – the asthenosphere. The lithosphere reacts to stress as a brittle solid and forms the rigid plates of the Earth's surface. In contrast the asthenosphere is a relatively weak layer which deforms by creep and is the zone within the upper mantle in which convection takes place. The thickness of these two layers varies greatly between the continents and the oceans giving rise to significant

TABLE 3.2 Trace element concentrations in the Primitive Mantle in ppm, arranged by atomic number.

Z		Jagoutz et al. (1979)	Hofmann (1998)	Sun and McDonough (1989)	McDonough et al. (1992)	McDonough and Sun (1995)	CI chondrite values of McDonough and Sun (1995)
3	Li	2.070		1.600		1.600	1.500
4	Be					0.068	0.025
15	P			95		90	1080
19	K	260	258	250	240	240	550
21	Sc	17.000	14.880		17.300	16.200	5.920
22	Ti	1300	1084	1300		1205	440
23	V	77				82	56
24	Cr	3140				2625	2650
27	Co	105	104			105	500
28	Ni	2110	2080			1960	10500
29	Cu	28	28			30	120
30	Zn	50				55	310
31	Ga	3.000				4.000	9.200
37	Rb	0.810	0.535	0.635	0.635	0.600	2.300
38	Sr	28.000	18.210	21.100	21.100	19.900	7.250
39	Y	4.600	3.940	4.550	4.550	4.300	1.570
40	Zr	11.000	9.714	11.200	11.200	10.500	3.820
41	Nb	0.900	0.618	0.713	0.713	0.658	0.240
50	Sn		0.150	0.170		0.130	1.650
55	Cs		0.027	0.032	0.023	0.021	0.190
56	Ba	6.900	6.049	6.989	6.990	6.600	2.410
57	La	0.630	0.614	0.687	0.708	0.648	0.237
58	Ce		1.601	1.775	1.833	1.675	0.613
59	Pr		0.242	0.276	0.278	0.254	0.093
60	Nd		1.189	1.354	1.366	1.250	0.457
62	Sm	0.380	0.387	0.444	0.444	0.406	0.148
63	Eu	0.150	0.146	0.168	0.168	0.154	0.056
64	Gd		0.513	0.596	0.595	0.544	0.199
65	Tb		0.094	0.108	0.108	0.099	0.036
66	Dy		0.638	0.737	0.737	0.674	0.246
67	Ho		0.142	0.164	0.163	0.149	0.055
68	Er		0.417	0.480	0.479	0.438	0.160
69	Tm		0.064	0.074	0.074	0.068	0.025
70	Yb	0.420	0.414	0.493	0.481	0.441	0.161
71	Lu	0.064	0.064	0.074	0.074	0.068	0.025
72	Hf	0.350	0.268	0.309	0.309	0.283	0.103
73	Ta	0.040	0.035	0.041	0.041	0.037	0.014
82	Pb		0.175	0.185		0.150	2.470
90	Th	0.094	0.081	0.085	0.084	0.080	0.029
92	U	0.026	0.020	0.021	0.021	0.020	0.007

lateral variations in the upper 200–300 km of the Earth's upper mantle.

3.1.3.1 Suboceanic lithosphere

The distinction between oceanic lithosphere and asthenosphere is defined thermally. Oceanic lithosphere is created at a ridge, then cools by conduction and increases in thickness until it reaches an age of 70 Ma, at which point it has reached a maximum thickness of 120 km. In contrast, the asthenosphere is a well-mixed, convecting adiabatic layer. Apart from the uppermost part of the mantle which is compositionally modified by melt extraction, the asthenosphere converts to lithosphere by cooling, with no compositional change.

Between the oceanic lithosphere and asthenosphere is a TBL, about 80 km thick, in which there is small-scale convection (Fig. 3.10). This small-scale convection removes material from the base of the conductively cooling layer and replaces it with hotter material and so maintains a constant lithospheric thickness (White, 1988).

3.1.3.2 Subcontinental lithosphere

In contrast to the oceanic lithosphere, the subcontinental lithosphere is thicker, more chemically differentiated, old, and cold.

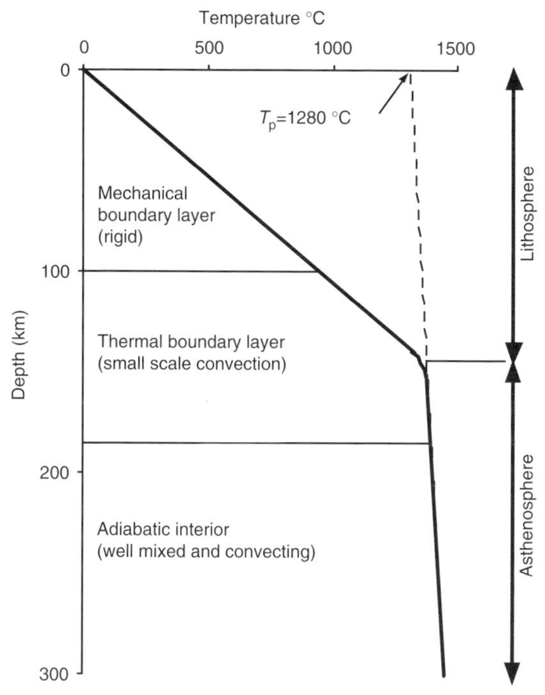

FIGURE 3.10 The thermal structure of old oceanic lithosphere over asthenosphere with a (normal) potential temperature of 1,280°C (the term potential temperature is defined in Section 3.1.4.3). The curve shows the horizontally averaged equilibrium thermal structure of oceanic lithosphere and asthenosphere (after White, 1988). Also shown are the rigid MBL, the thermal boundary layer, and the adiabatic interior of the upper mantle. The lithosphere is sometimes referred to as the "conductive lid" of the mantle, as opposed to the convecting interior (asthenosphere).

Richardson and colleagues working at MIT in the 1980s showed that diamond inclusions in mantle xenoliths from beneath the >3.0 Ga Kaapvaal Craton in South Africa were extremely ancient. This discovery demonstrated for the first time a link between the age of the continental crust and its underlying lithospheric mantle (Richardson et al., 1984, 1985).

Over the past two decades this work has been significantly extended so that now it is widely accepted that Archaean subcontinental lithosphere is thicker, older, colder, more chemically depleted, less dense, and has higher seismic P- and S-wave velocities than its Phanerozoic counterpart. More recently it has been argued that Proterozoic lithosphere is intermediate in composition between Archaean and Phanerozoic (see Table 3.3 for a summary).

More generally it has become clear that the subcontinental lithosphere is a domain within the mantle that is isolated from the rest of the mantle and so has resisted homogenization. In the case of Archaean subcontinental lithosphere, these regions preserve mantle domains which have been isolated since the early part of Earth history and provide an important "memory" of early Earth processes.

3.1.3.2.1 The structure of the subcontinental lithosphere

A detailed seismological examination of the continental lithosphere using the techniques of seismic tomography has provided a much clearer picture of the architecture of the subcontinental mantle. At a global scale, travel-time residuals deduced from surface-wave tomography permit the broad delineation of the mantle beneath Precambrian shields. Results of this type confirm that Archaean cratons have a lithospheric "keel" extending to 250–300 km depth, a feature which is absent from the lithosphere beneath younger crust (James & Fouch, 2002), see Fig. 3.11. At the local scale seismic sections show sharp velocity gradients over distances of about 20 km, which equate to differences between ancient and modern mantle lithosphere (Poupinet et al., 2003).

Two rather different approaches show that Archaean subcontinental mantle is cooler

TABLE 3.3 Principal differences between Archaean, Proterozoic, and Phanerozoic subcontinental lithospheric mantle (after Griffin et al., 2003).

	Archaean lithosphere	Proterozoic lithosphere	Phanerozoic lithosphere
Thickness (depth to lithosphere-asthenosphere boundary)	180–240 km	150–180 km	100–140 km
Compositional variation	Dominated by depleted harzburgite and lherzolite	Fertile and metasomatized lherzolite	Fertile and metasomatized lherzolite
Average density (g/cc)	3.31	3.34	3.36–3.37
Thermal gradient	35–40 $mW.m^{-2}$	40–45 $mW.m^{-2}$	50–55 $mW.m^{-2}$
Calculated Vp @ 100 km, 700°C	8.18	8.05	7.85

than Phanerozoic mantle (Table 3.3). Surface heat flow measurements, corrected for heat production in the crust show that average heat flow measurements in Archaean cratons are lower than those for Proterozoic cratons (Nyblade & Pollack, 1993; although Jaupart & Mareschal, 1999 offer an alternative view). Similarly, paleo-geotherms calculated using mineral thermometers and barometers also show that Archaean subcontinental mantle is cooler than modern subcontinental mantle.

Detailed studies of mantle xenoliths combining age determinations with the results of mineral barometry allow age-depth profiles to be constructed for some kimberlite pipes (Fig. 3.11). The results of such studies do not show any obvious age-depth relationship which is puzzling, maybe reflecting melt infiltration

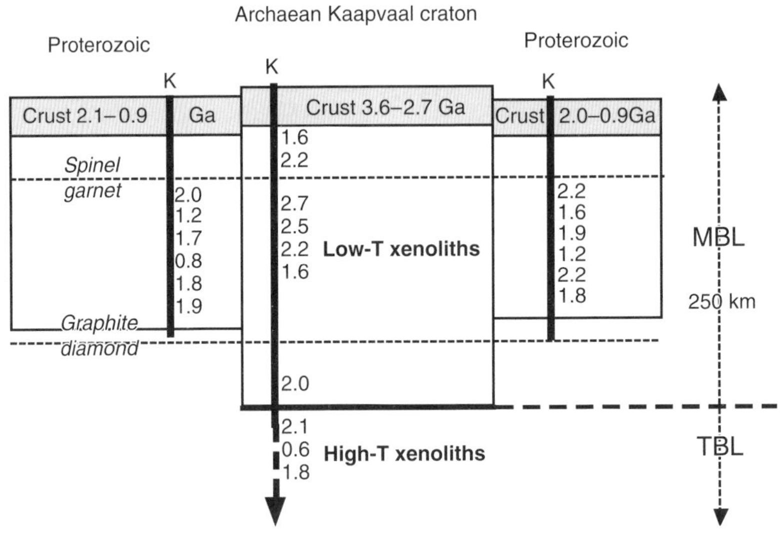

FIGURE 3.11 Schematic section through southern Africa showing the relationship between crust and subcontinental lithosphere. The spinel–garnet and graphite–diamond phase changes are shown. The heavy black lines marked "K" indicate kimberlite pipes and the numbers to their side are the rhenium depletion ages of mantle xenoliths (in Ga), plotted according to depth. The xenoliths are subdivided into those showing low-temperature and high-temperature deformation textures; this distinction also defines the difference between the MBL and the thermal boundary layer (TBL) (after Pearson, 1999).

into older continental lithosphere (Pearson, 1999; Pearson et al., 2002).

The growing consensus that ancient crust is always underlain by ancient lithospheric mantle has been challenged recently by Wu et al. (2003). These authors report an unusual absence of xenoliths with Archaean ages beneath the Archaean North China Craton. They propose that in some cases, therefore, ancient subcontinental mantle can be removed from beneath ancient continental crust by delamination – a process which has previously been postulated but never demonstrated. The subject of mantle delamination is discussed more fully in Chapter 5, Section 5.5.2.

3.1.3.2.2 *The major element composition of the subcontinental lithosphere*

The principal chemical feature of the subcontinental lithosphere is one of strong chemical depletion. This is evident when xenolith compositions are plotted relative to both the Bulk Silicate Earth and the oceanic lithosphere (Fig. 3.12). There are also compositional differences between Phanerozoic and Archaean subcontinental lithosphere as is illustrated by the $CaO–Al_2O_3$ plot in Fig. 3.12 (Griffin et al., 2003). Archaean lithosphere is strongly depleted relative to Proterozoic lithosphere (Fig. 3.12, Table 3.4). On a Mg/Si versus Al/Si diagram subcontinental mantle compositions plot close to the mantle trend (Fig. 3.13), but again the extreme position of average Archaean lithosphere on this plot demonstrates its strongly depleted nature. The compositional differences between ancient and modern mantle give rise to the enhanced buoyancy of ancient mantle, which in part, accounts for its preservation.

3.1.3.2.3 *The trace element composition of the subcontinental lithosphere*

In contrast to the *depleted major-element* character of Archaean subcontinental lithosphere it is often *enriched in trace elements*, relative to a midocean ridge basalt mantle source (Richardson et al., 1985; Jordan, 1988). A resolution of this apparent paradox can be found in the timing of the two events. Major element depletion is thought to have taken place during the early formation of the subcontinental lithosphere whereas the trace element enrichment reflects later melt infiltration.

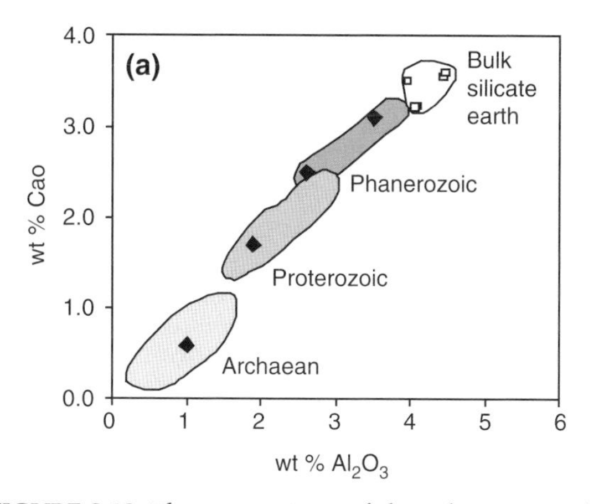

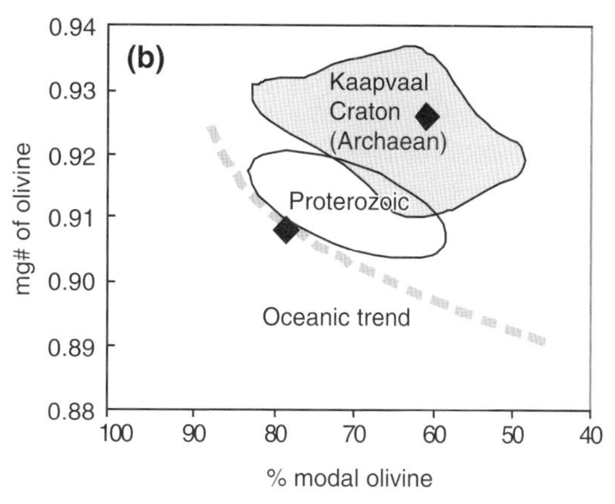

FIGURE 3.12 The composition of the subcontinental lithosphere. (a) Differences in composition between Archaean, Proterozoic, and Phanerozoic subcontinental lithosphere (after Griffin et al., 2003), compared to estimates of the BSE (from Table 3.1). The diamonds represent the average compositions given in Table 3.4. (b) Differences in composition between oceanic and continental lithosphere, as expressed in % modal olivine and olivine mg#; oceanic peridotites define the trend shown as the dashed line whereas garnet peridotites from the Kaapvaal Craton, South Africa – illustrative of Archaean subcontinental lithosphere – plot in the shaded field. The diamonds are the averages of the two fields (after Boyd, 1989). Also shown is the field for Proterozoic subcontinental lithosphere (ellipse) after Griffin et al. (2003).

TABLE 3.4 Compositions of the subcontinental mantle (after Griffin et al., 2003).

	Subcontinental lithospheric mantle (SCLM)			
	Archaean mean garnet peridotite	Proterozoic mean garnet peridotite	Proterozoic mean spinel peridotite	Phanerozoic mean garnet peridotite
SiO_2	**45.70**	**44.60**	**44.50**	**44.40**
Al_2O_3	0.99	1.90	3.50	2.60
FeO	6.40	7.90	8.00	8.20
MgO	45.50	42.60	39.18	41.10
CaO	0.59	1.70	3.10	2.50
TiO_2	0.04	0.07	0.14	0.09
Cr_2O_3	0.28	0.40	0.40	0.40
Na_2O	0.07	0.12	0.24	0.18
K_2O	nd	nd	nd	nd
Weight ratios				
Al/Si	0.025	0.048	0.089	0.066
Mg/Si	1.285	1.232	1.136	1.194
Atomic ratio				
mg#	0.93	0.91	0.90	0.90

Clearly these two events may be widely separated in time which is why it is difficult to measure the true age of the subcontinental lithosphere (Fig. 3.11).

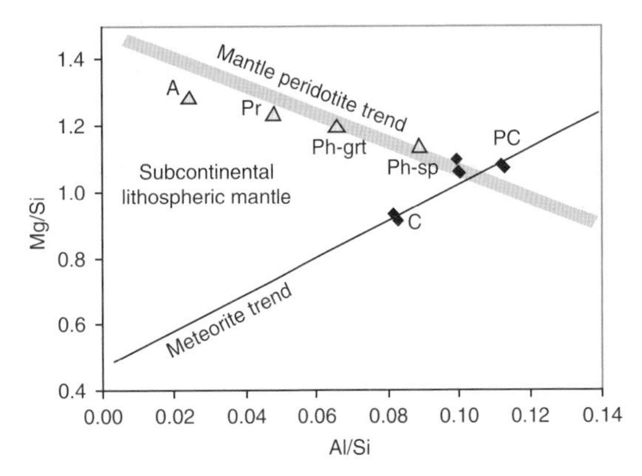

FIGURE 3.13 The composition of the subcontinental lithosphere plotted on a mantle composition Mg/Si–Al/Si weight ratio diagram. Labeled triangles: A, Archaean lithosphere; Pr, Proterozoic lithosphere; Ph-sp, Phanerozoic spinel lherzolite; Ph-grt, Phanerozoic garnet lherzolite (data from Griffin et al., 2003, see Table 3.4). Other symbols as in Fig. 3.9.

3.1.3.3 Understanding the subcontinental lithosphere

One way to understand the origin of the Phanerozoic subcontinental lithosphere is to examine the mantle in modern arc environments, for it is here that new continental crust is being formed. By inference, it is also where new subcontinental mantle is created. There are two key features to the subarc mantle.

First, the subarc environment is the site of major chemical fluxes into and out of the mantle (Chapter 4, Section 4.1). The subarc mantle experiences an input from the subducting slab, which in turn triggers partial melting within the wedge leading to the formation of the primary magmas of volcanic arcs. A principal constituent of the mass flux into the mantle wedge is water from the slab, leading to the view that the subarc mantle is hydrated (and oxidized) – see Fig. 4.1. In detail it is likely that the cool part of the mantle wedge beneath the forearc is underlain by strongly hydrated mantle in the form of serpentinite, whereas beneath the hotter, volcanic zone of the backarc, hydrous mantle phases are unstable and so fluids partition into mantle melts (Hyndman & Peacock, 2003). Schurr et al. (2003) have recently imaged melt-rich zones beneath the Andean arc, confirming what had previously been inferred from geochemistry.

A second feature of the subarc mantle is that it is the site of convective flow. In the mantle wedge cold mantle from beneath the overriding plate travels into the mantle wedge "corner" and is dragged deeper by the downgoing slab. Combining these two features – flow within the subarc mantle and chemical fluxes from the slab – leads to a dynamic view of the formation of the subcontinental lithosphere.

If therefore, the modern subarc mantle is the site where Phanerozoic subcontinental lithosphere is created, we are still left with a large number of questions about the earlier history of the subcontinental lithosphere. Why for example is the Archaean subcontinental lithosphere so different in composition, heat production and thickness from more recent subcontinental mantle? What different processes were operating early in Earth history which are recorded in this mantle domain? Is there a link with komatiite extraction, as suggested by Boyd (1989), or with the extraction of basaltic melts? Or, is there a close link between the formation of this type of mantle and the overlying continental crust? We will return to these issues when we discuss the origin of the continental crust in Chapter 4 (Section 4.5.1).

3.1.4 Melting of the Earth's mantle

In many regions of the Earth, basaltic melts are our primary window into the Earth's mantle. Melt production is also a major mechanism of heat loss for the Earth. For this reason an understanding of melting processes is vital for correctly reading the compositional and thermal state of the Earth's mantle. Furthermore, as will become clear later in this chapter, this is even more important in the early history of the Earth, when the most common samples of the Earth's mantle are its melt products.

3.1.4.1 The process of partial melting

Much of our knowledge of mantle melting comes from insights gained from studies of midocean ridge basalts. Experimental studies indicate that the Earth's mantle solidus is at 1,100°C at the Earth's surface, and since the mantle at midocean ridges is hotter than that, the mantle will melt when brought near to the surface at an oceanic ridge. Even hotter mantle will melt at greater depths (Fig. 3.14).

The principal mechanism driving basaltic volcanism is the pressure-release melting of passively upwelling mantle. Pressure release melting, or adiabatic melting as it is known, is where melting takes place without heat gain or heat loss. Because the asthenospheric adiabat has a lower temperature gradient (1–2°C/kb) than that of the dry mantle solidus (ca. 12°C/kb) a package of ascending asthenospheric mantle will begin to melt as it crosses the mantle solidus (Fig. 3.14). Typically the melting region beneath an ocean ridge is triangular in shape and may extend to a depth of 200 km, and be several hundred km wide. Ascending asthenosphere that rises directly beneath a midocean ridge spreading center will experience the most extensive degree of melting and will follow a flow line such that it forms the upper part of the suboceanic lithosphere. In contrast, asthenosphere which rises away from a spreading ridge will experience a smaller fraction of melting and will ultimately form the lower portion of the suboceanic lithosphere (Fig. 3.14).

On a microscale, melting takes place along grain boundaries, where the minerals which enter the melt are in contact with each other. Melt collects at "triple points" between grains, and when more than 1 or 2% melt is collected then the melt can migrate away from the zone of partial melting. A consequence of this whole process is that the melting residues, which form the oceanic lithosphere are vertically stratified and are more depleted at the top than the bottom. This sequence of melt-depleted residues was termed by Plank and Langmuir (1992) the Residual Mantle Column.

3.1.4.2 Controls on mantle melting

The exact composition of a mantle melt is a function of the source mineralogy, the mantle composition and volatile content, and the degree of melting, as is illustrated in Fig. 3.15. However, a major question in the study of the upper mantle is whether we ever see the direct melting products of the mantle, or whether what is produced in the melting zone of the

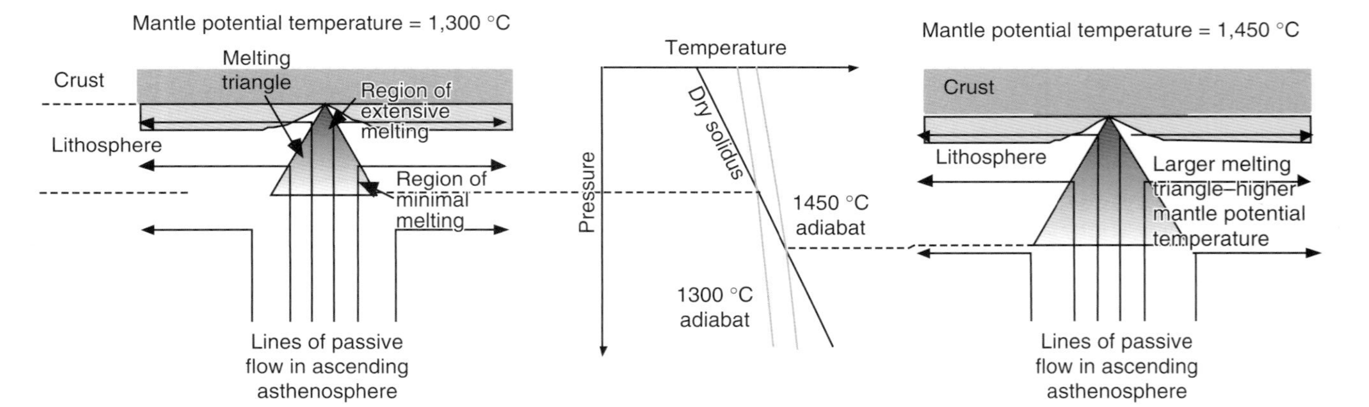

FIGURE 3.14 Model for mantle melting beneath a midocean ridge. The pressure–temperature diagram (center) shows the intersection of mantle adiabats for 1,300°C and 1,450°C with the dry mantle solidus. The diagrams right and left show the melting triangles for mantle potential temperatures of 1,450 and 1,300°C respectively. The higher the mantle potential temperature the deeper and more extensive is the melting, resulting in a thicker ocean crust and lithosphere (after Asimow & Langmuir, 2003).

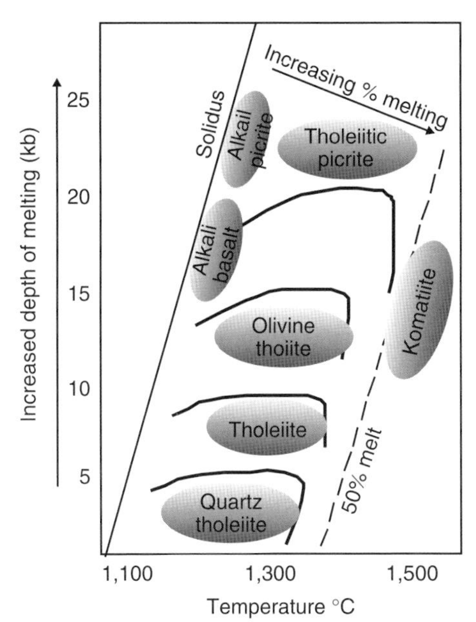

FIGURE 3.15 Pressure–temperature diagram showing the results of experimental studies which illustrate the variable effects of the depth of melting, temperature of melting, and melt fraction on the composition of melts of an enriched lherzolite source (after Jaques & Green, 1980). The melting interval shown is between 0 and 50%; the curved black lines are the percentage of normative olivine in the melt, which increases with depth of melting. Studies of this type were central to the primary magma debate in the 1980s.

mantle is somehow transformed before it erupts at the surface. This knowledge is an important prerequisite to using mantle melts to characterise their source.

Between the 1960s and the 1980s the subject of whether midocean ridge basalts are primary, that is, unmodified mantle melts, or nonprimary, that is, have experienced compositional change prior to eruption was intensely debated. One school of thought (e.g. Presnall & Hoover, 1984, 1986) argued that the least fractionated midocean ridge basalts are primary melts produced at 8–10 kb pressure. The alternative view was that all midocean ridge basalts are derived by olivine fractionation from a picritic melt, generated at pressures of 15–30 kb (Elthon, 1986; O'Hara, 1968). These latter, picritic melts have were fractionated in a crustal magma chamber prior to eruption.

3.1.4.3 *The melt column*

More recently it has become clear that both schools of thought were wrong. These early ideas were based upon the concept of batch melting, where the mantle melts in one location and then a batch of melt moves, virtually unmodified, to the surface. More recent studies have shown that fractional melting is important in the genesis of midocean ridge basalts and that the final product of melting is the accumulation of small melt increments from depth to the base of the oceanic crust. Mantle melting, therefore, is a polybaric process, and it is important to distinguish between the instantaneous melt composition and the aggregated melt composition. The appreciation of the polybaric nature of mantle melting beneath a midocean ridge led to the concept of the integrated melt column (McKenzie & Bickle, 1988) in which the final melt is the integrated composition of many melts produced at different depths.

Developing this argument, Klein and Langmuir (1987) showed that the greater the depth of melting, the greater the amount of melt produced. This means that, in the case of the oceanic lithosphere, there is a relationship between the depth of melting and the thickness of the ocean crust. This line of reasoning was extended by Bown and White (1994) who showed that there is also a relationship between spreading rate, ocean crust thickness, and basalt chemistry, indicating that the thickness of the oceanic crust relates to amount of melt produced during upwelling beneath the ridge. These studies have led to the concept of *melt thickness* – the integrated volume of melt produced in the melt column.

3.1.4.3.1 *Mantle potential temperature*

Melt thickness is also controlled by melting temperature. This is because hot mantle will intersect the mantle solidus at a greater depth than cooler mantle, and as already established, melt thickness is a function of the depth of the melt column. However, since mantle temperatures increase with depth there needs to be a way of assessing mantle temperatures independent of depth. To overcome this confusion the concept of mantle potential temperature is used to compare the heat content of mantle materials at different depths and account for their different compressibilities. The term potential temperature is the temperature a fluid mass would have, if expanded or compressed at some reference temperature, normally taken to be the temperature at the Earth's surface (McKenzie & Bickle, 1988). Potential temperature is also expressed as the temperature of the solid mantle adiabat projected to surface pressure (Fig. 3.10).

Melt thickness therefore can be expressed as a function of the mantle potential temperature. A potential temperature of 1,280°C equates to a melt thickness of about 7 km (normal ocean floor), whereas a mantle potential temperature of 1,480°C equates to a melt thickness of about 27 km (see Fig. 3.14). Clearly, these principles are important when considering melting processes in the early Earth, since many geoscientists believe that mantle temperatures were hotter in the Archaean (Section 3.2.3).

3.1.4.4 Melt extraction from the mantle

Initially melt will move within the melting zone as small droplets of melt. These will tend to migrate upwards from the melting zone and will collect together to move as larger droplets and patches of melt ultimately representing 8–20% of the original lherzolite host. As the melts reach the surface they will tend to move along channel-ways in the mantle whose orientation is related to mantle flow and upwelling beneath a spreading ridge (Braun & Kelemen, 2002). Kelemen et al. (2000) proposed a pattern of melt channels resembling an inverted bush (Fig. 3.16) in which a large number of small melt channels converge to produce a smaller number of wider "melt-highways" focusing melt beneath a midocean ridge. This model of melt movement solves the problem of a melting zone hundreds of kilometers wide (Section 3.1.4.1), coupled to a focused zone of crust formation, only a few kilometers wide.

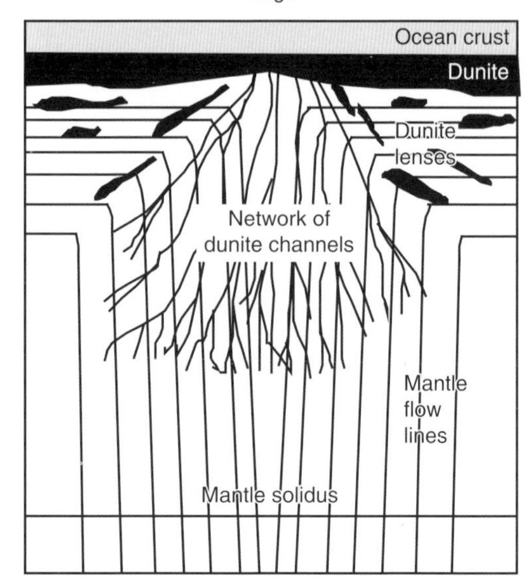

FIGURE 3.16 Network of coalescing dunite channels in adiabatically upwelling mantle beneath a fast-spreading midocean ridge (after Kelemen et al., 2000).

3.1.4.5 Melt reaction and the upper mantle as a chemical filter

One of the most striking features of midocean ridge basalts (MORB) is that they are chemically homogeneous and any model for their origin must account for this feature. This of course was one of the strong arguments for MORB as a primary magma (Section 3.1.4.2). However, those who opposed this view argued, equally plausibly, that the uppermost mantle acted as a filter, controlling the density of erupted melts. In this model, the dense primary magma picrite, produced by deep mantle melting, crystallizes olivine in a crustal magma chamber thereby reducing its density and enabling it to erupt. A thick layer of dunite is left in the magma chamber.

Although this idea has now been largely superceded by the polybaric melting model a new possibility has recently emerged whereby the uppermost mantle may behave as a *chemical* filter. Midocean ridge basalt (MORB) is silica saturated and is in equilibrium with orthopyroxene at pressures greater than about 8 kb. At the lower pressures commensurate with the base of the oceanic crust MORB melts are undersaturated in silica and have the capacity to dissolve orthopyroxene (Braun & Kelemen, 2002). There is field evidence from the mantle section of some ophiolites that this process happens, leaving a residue of dunite and thereby creating a zone of increased porosity (Morgan & Liang, 2003). In this way the upper part of the mantle acts as a chemical filter which transforms melts on their upward pathway by a process of chemical reaction. Support for this process is given by Coogan et al. (2003) who show that the trace element composition of clinopyroxenes in the lower oceanic crust of the Cyprus ophiolite permits the calculation of the melts with which they equilibrated. These melts are significantly different from and more varied in composition than the erupted basaltic compositions in the same fragment of ocean floor, strongly suggesting that the erupted melt compositions have been homogenized in the lower oceanic crust prior to eruption.

3.1.5 Mantle convection, mantle plumes and layering in the mantle

One of the most fundamental, contemporary debates about the nature of the Earth's mantle centers on the subject of mantle convection. There are two conflicting views which are commonly described as "layered convection" and "whole-mantle convection" (see Fig. 3.17). The *layered convection* model is championed by geochemists who prefer to see the mantle as two separate convecting layers. In this model the upper and lower mantle are geochemically isolated from each other and convect separately. *Whole mantle convection* is advocated by geophysicists, who that believe there is evidence for a significant exchange of mass between the upper and lower mantle. Resolving the conflict between these two views is not easy because mantle convection cannot be directly observed. Rather it has to be inferred from seismological and geochemical observations, and from experimental studies in mineral physics and fluid dynamics.

A second debate, which has arisen independently, but which has a direct bearing on the layered mantle debate, is a discussion about the existence and origin of mantle plumes. Again there are two schools of thought, one arguing for a deep source for mantle plumes from the core-mantle boundary, the other stating that mantle plumes must come from the upper mantle (Fig. 3.17). A deep origin for mantle plumes would support whole mantle convection.

These two debates address central questions about the nature of convection within the Earth's mantle. First, are there two layers of convection (upper mantle – lower mantle) or one? Second, are there two separate modes of convection – plate tectonics and plumes?

The resolution of these questions is of great importance for our understanding of the early Earth, for they have a direct bearing on models of Earth differentiation and patterns of mantle convection in the early Earth.

3.1.5.1 What is the nature of the 660 km discontinuity?

In the early days of plate tectonics it was widely believed that convection in the mantle was confined to the upper mantle, above the 660 km discontinuity. This view was supported by the observation that there was no seismicity in the lower mantle, indicating that subducting

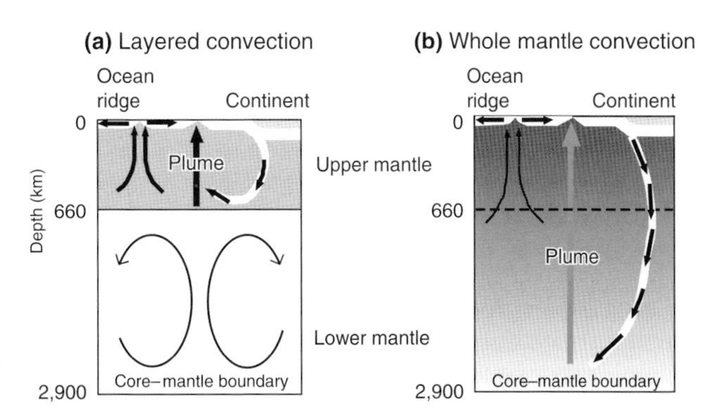

FIGURE 3.17 Schematic diagram illustrating the two different models of mantle convection. (a) A two-layer mantle in which the upper mantle is chemically isolated from the primitive, lower mantle. In this model the plume source is in the upper mantle and subducting slabs do not penetrate the 660 km upper mantle–lower mantle boundary. (b) The whole mantle model in which there is mass exchange between the upper and lower mantle. In this case subduction penetrates the 660 km discontinuity and plumes are sourced in the D″ layer, at the core–mantle boundary.

slabs did not penetrate through the upper mantle–lower mantle boundary. However, this view has been challenged in recent years with the improved resolution of seismic tomographic imaging. The most damning evidence comes from tomographic images showing cold slabs penetrating through the 660 km discontinuity into the lower mantle. Some of these images show subducted slabs descending into the lower mantle (Fig. 3.2). An important consequence of this observation is that there must also be a compensating flow in the reverse direction, from lower to upper mantle.

The strong separation between upper and lower mantle was also believed to result from a compositional difference between upper and lower mantle. Support for this view was found in the geochemical difference between OIB and MORB sources (Fig. 3.5), thought to represent the deep and shallow mantle layers. These differences were thought to primarily reflect the extraction of the Earth's continental crust from the upper mantle over the past 3.5 Ga. However there is no support from mineral physics for a compositional difference between the upper and lower mantle. The 660 km discontinuity in the mantle is marked by an increase in seismic wave velocities which reflect a phase change from the mineral assemblage γ-Mg_2SiO_4 + majorite of the upper mantle to the assemblage Mg-perovskite + Ca-perovskite + magnesiowustite in the lower mantle (Section 3.1.1.2, Fig. 3.1). This phase change is now regarded as isochemical, and the Fe/Mg ratio of the upper mantle is the same as that of the lower mantle. Such an observation is significant because it had been previously argued that long-term differentiation above and below the 660 km discontinuity is difficult to achieve unless there is a density and compositional difference between the two layers.

3.1.5.2 Geochemical arguments for a two-layer mantle

One of the fundamental difficulties with the whole mantle convection model is that geochemical heterogeneities cannot be sustained over geological timescales. And yet, observations from trace element and isotope geochemistry show that the mantle is and has been heterogeneous over geological time and comprises a number of distinct chemical domains.

3.1.5.2.1 The radiogenic argon argument

One of the strongest and most widely used arguments in support of chemically distinct upper and lower mantle reservoirs, isolated throughout much of geological time, comes from the study of argon isotopes. In brief the argument is as follows. The isotope ^{40}Ar (radiogenic argon) is produced by the radioactive decay of ^{40}K. Once formed, argon, unlike the lighter rare gases, is of sufficiently high mass not to be lost from the Earth into space. This means that if we know the K-content of the bulk Earth we can calculate the total mass of ^{40}Ar produced through geological time. This argon is now distributed between the atmosphere, the continental crust and the upper and lower mantle reservoirs (Fig. 3.18). Direct measurement of ^{40}Ar in the Earth's atmosphere and in rock samples tells us that about 45% of the Earth's total argon is in the atmosphere and 5% is in the continental crust. The remaining argon therefore is in the Earth's mantle. However, if the whole mantle has upper mantle potassium concentrations then only about 15% of the Earth's argon can be accounted for. It is argued therefore that the Earth's lower mantle has a higher potassium content than the upper mantle and contains the balance of the mantle argon.

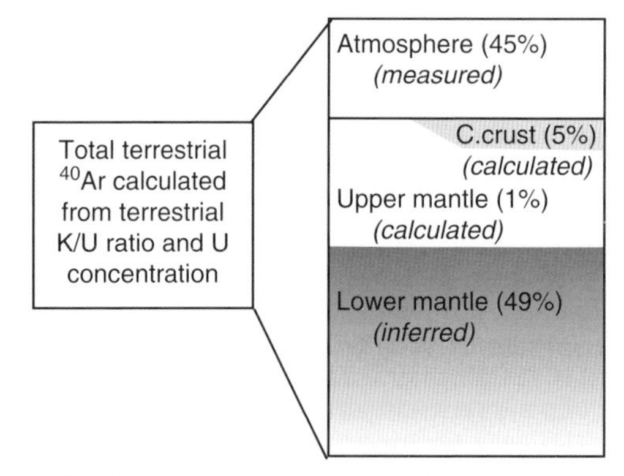

FIGURE 3.18 Present-day terrestrial ^{40}Ar distribution showing the very large difference between upper and lower mantle.

Mass balance calculations of this type lead to the conclusion that the ^{40}Ar concentration of the lower mantle is as much as 50 times greater than that of the upper mantle (Davies, 1999). Arguments based upon the isotope ^{36}Ar, inferred from $^{40}Ar/^{36}Ar$ ratios, suggest an even greater difference. This large disparity between the upper and lower mantle is normally explained by postulating that the Earth's upper mantle is extensively degassed and that its original argon content is now in the atmosphere, whereas the lower mantle is not degassed and it contains its original primordial argon (and potassium) content. The long half-life of ^{40}K (1,250 Ma) requires that the upper mantle was degassed early in Earth's history and that the upper and lower mantle reservoirs have remained chemically separate for the major part of Earth history. The corollary to this is that there has been minimal mass transfer between the upper and lower mantle.

The difference in $^{40}Ar/^{36}Ar$ ratios between MORB and OIB is of the same order as that expected between upper and lower mantle, respectively, leading to the widely held view that the MORB source is in the upper mantle and that the OIB source is in the lower mantle (see Table 3.5).

However, as the two-layer model for the mantle has come under increased scrutiny, so have the premises on which the radiogenic argon argument is based. There are two pertinent issues. First, the total argon inventory of the Earth will be different if the Earth has less potassium than originally assumed by Allegre et al. (1986a), or if some ^{40}Ar is either located in the core or has been lost from the Earth (Davies, 1999). Of these possibilities serious concern has been expressed over the high K/U ratio assumed for the bulk Earth and much lower values have been proposed (Albarede, 1998; Trieloff et al., 2003) – see Table 3.5. Second, the concept of a degassed upper mantle and an un-degassed lower mantle does not sit easily with what we know of the early history of the Earth, when with high internal temperatures extensive degassing must have taken place.

TABLE 3.5 Argon isotopic compositions and K/U ratios in MORB, OIB and the atmosphere.

	MORB	OIB	Atmosphere
Original estimates – Allegre et al. (1986a)			
$^{40}Ar/^{36}Ar$	16,700	390	295
K/U	12,700	12,700	
Recent estimates – Trieloff et al. (2003); Albarede (1998)			
$^{40}Ar/^{36}Ar$	32,000 ± 4,000	8,000	295
K/U	6,200	6,200	

3.1.5.2.2 Helium isotopic ratios

The helium isotope argument for a two-layer mantle is very similar to that for argon isotopes, except that He has not accumulated in the atmosphere because it easily escapes into space. The ratio $^{3}He/^{4}He$ is the ratio of a primordial, cosmogenic isotope ^{3}He to a radiogenic isotope ^{4}He. ^{4}He is produced by the radioactive decay of U and Th and has increased with time. For this reason the terrestrial $^{3}He/^{4}He$ ratio has decreased with time. Using similar reasoning for argon isotopes, measured differences in $^{3}He/^{4}He$ between OIBs and midocean ridge basalts have been used to infer chemical differences between upper and lower mantle reservoirs. In turn these arguments have been used to support an un-degassed lower mantle and a degassed upper mantle.

Opponents of the two-layer view of the mantle have pointed out that the statistical differences between MORB and OIB $^{3}He/^{4}He$ ratios are not as robust as has previously been thought (Meibom et al., 2003). Furthermore, a detailed geochemical study by Class and Goldstein (2005) has shown that the least degassed mantle melts, with respect to their $^{3}He/^{4}He$ ratios, are still strongly depleted in other trace elements – casting further doubt on the idea of a primitive undegassed mantle reservoir. Class and Goldstein (2005) propose that the observed difference in $^{3}He/^{4}He$ ratios between MORB and OIB is the product of inefficient degassing of mantle melts, rather than the existence of a chemically isolated, lower mantle source for OIB.

3.1.5.2.3 The global heat budget

Just as geochemical mass balance arguments have been used to argue for a two layer mantle, so the global heat budget can be used in the same way. The total heat production of the Earth is about 41 TW (1 TW = 10^{12} W). About 5 TW can be accounted for in the heat production of the upper continental crust leaving 36 TW as the heat production of the mantle. This is the composite of heat produced by radioactive decay plus heat produced by cooling since the formation of the Earth, for both the upper and lower mantle. Of this 31 TW emerges through the sea floor and 5 TW through the continental lithosphere. Using the measured heat-producing element concentration of the upper mantle it is possible to calculate the heat production of the upper mantle as 1 TW. Further, from the mass of the upper mantle it is possible to estimate the heat production of the upper mantle from cooling as 3 TW. These calculations show that of the total mantle heat production of 36 TW, only 4 TW can be ascribed to the upper mantle and the remaining 32 TW must come from the lower mantle (Davies, 1998) – see Fig. 3.19.

Geochemists have used these observations to demonstrate that the upper and lower mantle have very different heat producing element concentrations and that the lower mantle has retained a higher level of heat producing elements than the upper mantle – which has lost its heat producing elements to the continental crust. The lack of mixing between these two layers indicates, the argument goes, that the mantle is made up of two separate layers through which there is virtually no mass flow.

Geophysicists, however, have turned the geochemists' global heat budget argument on its head, and used it to argue for whole mantle convection, rather than, as the geochemists would argue, for a layered mantle. Their argument

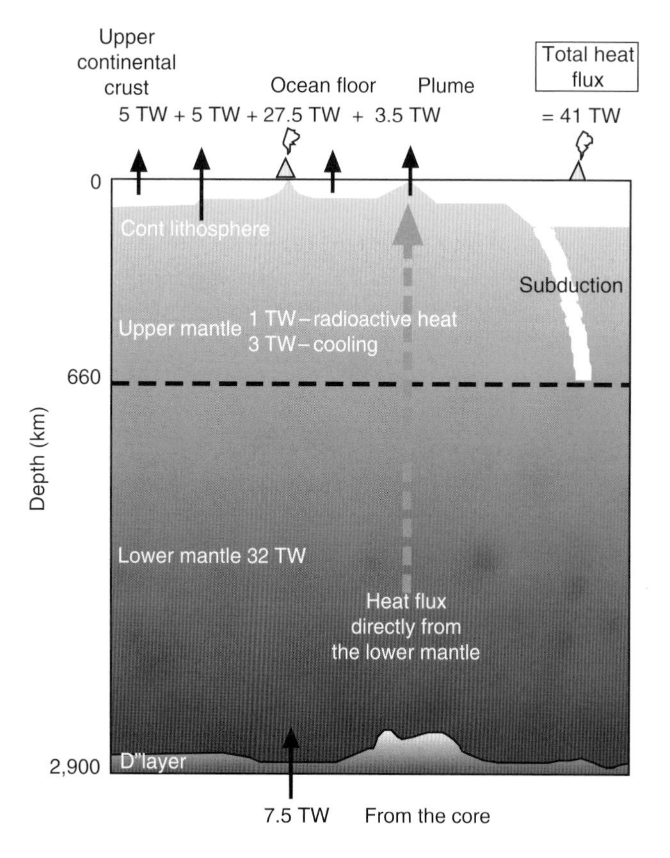

FIGURE 3.19 Estimates of mantle heat fluxes, showing the difference in heat production in the upper and lower mantle. Heat flux values are in TW (10^{12} W), from Davies (1998) and Yukutake (2000).

hinges on the manner in which heat is transferred from the lower to the upper mantle (Davies, 1998). If heat is transferred from the lower to the upper mantle by conduction, a hot thermal boundary layer will form which will be buoyant. This will give rise to a topographic expression which will be similar in volume to, but distinct from the ocean ridge system. It is argued that no such topographic expression is seen on the Earth's surface, and for this reason heat must be transported by advection, that is, by mass flow. This means that there is a mass flow, from the lower to upper mantle which is responsible for the 32 TW of heat coming from below the 660 km discontinuity. This mass flow implies whole mantle convection.

3.1.5.2.4 Summary

The traditional arguments from geochemistry for chemically distinctive deep and upper mantle reservoirs separated by the 660 km discontinuity are no longer sustainable. Rare gas arguments based on Ar and He isotopes are being shown to be much less robust than was previously supposed. Similarly arguments based on the distribution of heat producing elements are now open to alternative interpretations.

3.1.5.3 Mantle plumes

Mantle plumes are thought to be the mechanism whereby extensive volumes of basaltic melt are delivered rapidly to the Earth's surface. They were first invoked to explain the "hotspot" volcanism of the linear, time-progressive Emperor-Hawaii ocean-island chain. Subsequently they have been used to explain the origin of other ocean islands such as Iceland, the origin of large igneous provinces (LIPs flood basalts), and volcanic activity in Archaean greenstone belts. After 30 and more years of acceptance, mantle plumes have recently come under close scrutiny and at the present time there is intense discussion between "the plume lobby" and those who have moved "beyond the plume hypothesis," as the two groups have become known (see the recent review by Foulger et al., 2005). In a "two-layer mantle world" plumes represent an effective way of sampling the otherwise isolated lower mantle, although plumes can also neatly fit into a whole-mantle convection model.

A conventional view of a mantle plume is that it originates in the thin D″ layer of the lower mantle, at the core–mantle boundary, and represents a narrow (100 km in diameter) vertical column of hot mantle (the plume "stem") which traverses the entire thickness of the mantle. Where the hot mantle column impinges on rigid lithosphere the hot mantle spreads beneath the lithosphere into a mushroom-shaped "plume head" several hundred kilometers across. Within the plume head there is extensive melting of the mantle which results in voluminous volcanism. Basalts are the principal product of melting, but picrites – a highly magnesian "basalt" representing a higher temperature melt of the mantle – are also commonly found. This extensive volcanism can manifest itself as a large igneous province, comprising flood basalts if erupted on continental crust, or as an oceanic plateau if erupted on oceanic crust. During the lifetime of a plume a plate may drift over the hot column of mantle, and a linear chain of basaltic volcanoes develops, showing time progression. The linear time progression of such volcanic chains suggests that their source is fixed relative to plate movements and is located beneath the level of mantle convection associated with plate tectonics.

Unlike ocean-ridge and arc volcanism, there has never been a unifying theory for hotspot volcanism, and mantle plumes appear to behave independently of the process of plate tectonics. Davies (1993) has summarized the role of plate tectonics as that of cooling the mantle, whereas plumes cool the core. Some authors (e.g. Condie, 1998) have argued that mantle plumes have developed episodically throughout geological time and may be responsible for the major growth stages of the continents.

Since the first mantle plumes were identified in the 1970s an increasing number of mantle plumes have been recognized. For this reason Courtillot et al. (2003) sought to highlight those modern plumes which are of truly deep mantle origin. They identified between seven and ten mantle plumes which may originate at the core–mantle boundary. They showed that a further 20 plumes originate from the bottom of the mantle transition zone and another 20 in the upper mantle. These observations indicate that not every hot spot is underlain by a mantle plume and have opened up a huge debate about the precise nature of mantle plumes.

3.1.5.3.1. Advocates of the plume hypothesis

Advocates of the plume hypothesis point to the Hawaiian-Emperor volcanic chain, to Iceland and to the formation of large igneous provinces as evidences for *discrete, focused, persistent, and deeply sourced convective mantle upwellings* (Saunders, 2003). The Hawaiian-Emperor volcanic chain has sustained a high average basalt production rate for over 80 Ma, an enhanced oceanic topography, a deep seated velocity layer in the mantle underlying the volcanically active areas, magmatism that ceases as the islands are carried away from the topographic anomaly, and a clear time progression along the volcanic chain. Iceland, similarly, represents an area of abnormally thickened ocean crust, beneath which there is a layer with a low seismic velocity representing hotter mantle. In this case the hot mantle anomaly is cylindrical in form and extends to a depth of at least 440 km. LIPs also represent very large volumes of basalt erupted over a short time interval and many are connected via a chain of extinct volcanoes to a currently active volcanic center – the presumed site of the plume at the present time (Saunders, 2003). Geochemical arguments based upon trace elements and isotope ratios demonstrate that Hawaiian and Icelandic basalts are not the same as those derived from a midocean-ridge basalt source. More recent arguments have proposed a lower mantle/core geochemical signature in plume basalts (see Section 3.2.3.4).

3.1.5.3.2 Arguments against the plume hypothesis

One of the principal difficulties with the plume hypothesis is that it has been used by a very large number of geoscientists to explain a wide range of phenomena. This of course does not invalidate the idea, but it does give the hypothesis a poor image. More serious is the criticism that the plume hypothesis "cannot be tested, has many variants, exceptions, rationalizations, ad-hoc adjustments and failed predictions" (Anderson, 2003). The principal scientific arguments against it are:

- Mantle tomography has not been able to image narrow vertical structures traversing the whole mantle that would represent plume stems (although see Montelli et al., 2004). Such features are recorded only from the upper mantle.
- The high temperatures required by mantle plumes are not supported by the expected positive heat flow anomalies.
- The temperature contrast between plumes and "normal mantle" is not as extreme as that claimed by the advocates of mantle plumes, because lateral temperature fluctuations in the upper mantle are greater than has previously been acknowledged.
- Physical models of the mantle suggest plumes are impossible because of the high pressures at depth which suppresses the buoyancy of the material and prevents the formation of narrow plumes.
- The geochemical arguments used in support of the plume hypothesis are ambiguous and do not *require* the deep and shallow sources proposed by plume theorists.

(Sheth, 1999; Anderson, 2000; MacKenzie, 2001; Foulger and Natland, 2003; Meibom et al., 2003).

Of course in order to satisfactorily refute a scientific theory it is important to provide a better explanation of the facts. In the case of mantle plumes it is suggested that plate tectonic processes can explain all the features ascribed to mantle plumes. In the case of hot spots this may be achieved by combining a source of melt in the upper mantle with either a propagating fracture zone or with continental margin-edge driven convection in the shallow mantle (Foulger & Natland, 2003).

3.1.5.3.3 Conclusion

One of the issues which both the plume debate and the layered versus whole mantle convection debate highlight is the origin of chemical heterogeneities in the Earth's mantle. Do we, as has traditionally been argued, have upper and lower mantle layers which are compositionally distinct, or are there other, less tidy models, which can equally well accommodate the geochemical data? Further, when and how have such heterogeneities been created? We will return to these issues in the final section of this chapter.

3.1.6 The modern mantle as a geochemical system

One of the most important contemporary questions about the Earth's mantle is to enquire how exactly the modern mantle works as a geochemical system. In this section we shall summarize some of the ideas considered in earlier sections and present them in the form of a mantle model (Fig. 3.20).

3.1.6.1 Inputs – fluxes into the mantle

The principal mechanism by which new material is added to the mantle is subduction. The results of seismic tomography have shown that the mass flux into the mantle at subduction zones, at least in part, penetrates below the 660 km discontinuity into the lower mantle, and there is a view that some of this material may reach the D″ layer at the core–mantle boundary (see Fig. 3.2). The largest mass flux into the mantle is oceanic lithosphere with its associated oceanic crust, although this is crust which has been processed through the "subduction factory" and is therefore strongly depleted in water-soluble elements (Niu & O'Hara, 2003). A smaller amount of crustal material, in the form of sediment is also subducted. In addition, there is a significant flux of water into the mantle through subduction zones, although the depth to which it can penetrate is debated (see Chapter 4, Section 4.1.5)

An additional, unquantified contribution is also made by the delamination of the lower crust and the subcontinental lithosphere (Chapter 4, Section 4.5.2.1).

3.1.6.2 Outputs – fluxes out of the mantle

The main outputs from the Earth's mantle are basaltic melts (see Section 3.1.5). Basaltic and

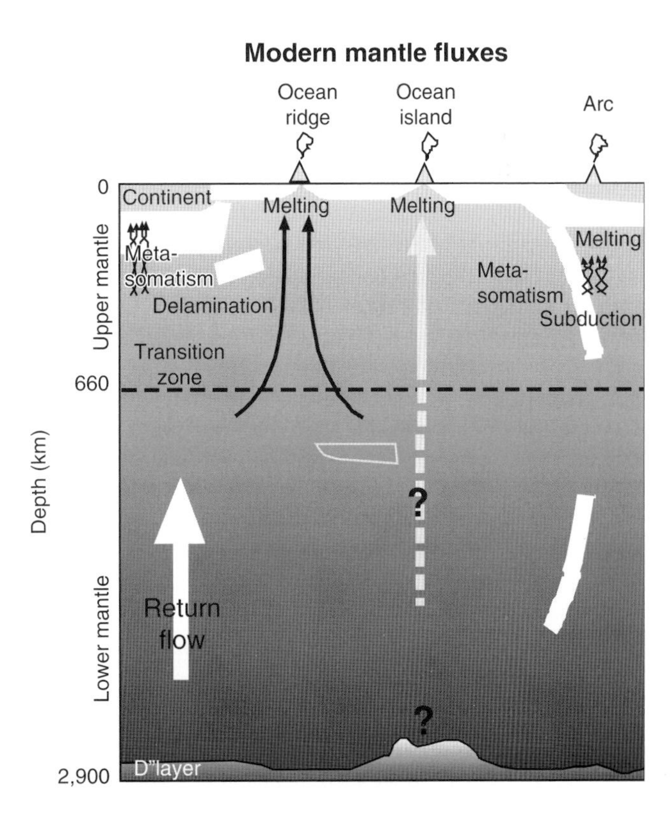

FIGURE 3.20 Cartoon showing the main fluxes in and out of the modern mantle.

related melts are produced at ocean ridges, at the site of ocean islands, in arcs and beneath continents. The subtle variations in chemistry between these different basalt types are the combined effect of differing mantle sources and the different melting processes which take place within the mantle.

As part of the longer term evolution of the Earth, the reprocessing of oceanic crust in the "subduction factory" leads to the production of continental crust. Thus continental crust is also an output of the upper mantle, but as the product of a two-stage process.

Associated with the flux of melt to the Earth's surface is the flux of volatiles, dissolved in basaltic melt, reflecting the mantle contribution to the oceans and atmosphere (see Chapter 5).

3.1.6.3 *Fluxes within the mantle*

Not all melt produced within the mantle reaches the Earth's surface. Instead it may migrate and freeze in a different part of the mantle from where it was produced. This is the process of mantle metasomatism, likely to be an important process in both the subcontinental lithosphere and in the mantle wedge above an arc. In the case of the subcontinental lithosphere, metasomatism, is an important mechanism for modifying older subcontinental mantle, and its effects may be detected in subsequent melts of subcontinental mantle. Melt and fluid migration above a subduction zone are also important in initiating mantle melting in the mantle wedge.

The evidence for fluxes between the lower and upper mantle is more contentious. Whilst there is clear evidence of the transfer of material from upper to lower mantle through the sinking of subducted slabs, the ultimate destination of this material is uncertain. More controversial, as the current debate on mantle plumes is demonstrating, is the upward flux, from a lower to upper mantle reservoir.

3.1.6.4 *Mass fluxes into and out of the mantle*

The total mass flux of magma to the Earth's surface has been estimated by Crisp (1984) to be between 26 and 34 km^3/a. This comprises a mass flux at ocean ridges of between 20 and 26 km^3/a, at arcs between 5 and 7 km^3/a and in ocean islands, km^3/a. Using a density for basalt of 2.9 g/cm^3 this equates to a total flux of 7.5–9.9 $\times$ 10^{13} kg/a. Schubert et al. (2001) calculated the flux at ocean ridges by using a seafloor creation rate of 2.8 km^2/a and assumed a processing depth of 60 km. This yields the mass of mantle processed annually beneath ridges as 5.4 $\times$ 10^{14} kg/a, which for between 5 and 10% partial melting indicates a basaltic flux of 2.7–5.4 $\times$ 10^{13} kg/a. Schubert et al. (2001) also estimate that the total volume of mantle processed in mantle "hot spots" is 85 km^3/a, which for 5–10% melting translates into a basaltic flux of 1.4–2.7 $\times$ 10^{13} kg/a. These estimates indicate that the total *mass flux out of* the mantle is about 10^{14} kg/a.

Using a subduction rate of about 3 km^2/a and a slab thickness of 100 km Davies (1998) estimated that the mass flow into the mantle through subduction was about 10^{15} kg/a. Hence the *mass flux into* the mantle is about 10^{15} kg/a. The mismatch between this value and the mass flux out of the mantle arises because we are equating two different fluxes. One is the flux of ocean crust (flux out) the other is the flux of ocean crust plus oceanic lithosphere (flux in).

Davies (1998) proposed a model for the mantle in which he distinguished between active and passive mass flow (Fig. 3.21). Active flow is where there is a driving force such as the negative buoyancy forces of deep subduction. The principle of the conservation of mass requires that active mass flow is matched by a return flow. This is passive mass flow. Support for the concept of passive return flow comes from the heat flow arguments described in Section 3.1.5.2. In addition it has been proposed that the Darwin Rise topographic anomaly in the Pacific ocean and the African superplume – a large-scale, thermal upwelling, 200 km thick and 7,000 km long, which rises 1,200 km from the core–mantle boundary (Ni & Helmberger, 2003) – could represent regions of deep warm mantle upwelling. These two structures represent a combined *mantle* mass flow of 7 $\times$ 10^{14} kg/a.

Given these calculated fluxes it is possible to estimate the time it takes for the entire

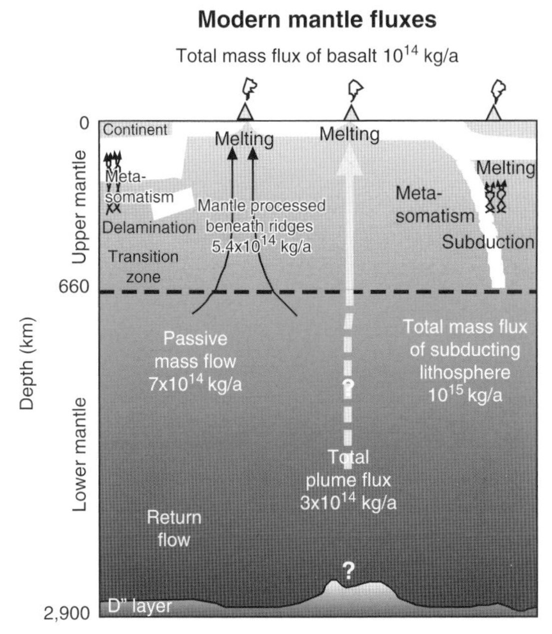

FIGURE 3.21 Mantle fluxes from Davies (1998) and Schubert et al. (2001).

upper mantle to be replaced. Assuming an upper mantle mass of 1.3×10^{27} g and mass flow of ca. 10^{18} g/a, Davies (1998) calculated an upper mantle replacement time of between 0.7 and 1.0 Ga. If whole mantle convection is a more appropriate model, and the mass of the whole mantle is taken as 4.3×10^{27} g, then the replacement time is between 2.2 and 3.2 Ga. Schubert et al. (2001) calculate longer replacement times of 1.94 Ga for the upper mantle and 7.52 Ga for the whole mantle. Both estimates indicate that a significant portion of the Earth's mantle has been processed during the lifetime of the Earth. Such an understanding has to be an important component in mantle models.

3.2 The Earth's earliest mantle

Our knowledge of the modern mantle leaves the clear impression that it records a history. The methods of geophysics and geochemistry reveal both subtle and more obvious traces of the mantle's prehistory as we probe its present form. Now we turn to examine the Earth's earliest mantle, by using, as much as is possible, samples of the Archaean mantle preserved in the Earth's oldest rocks. Our goal here is to establish, as best as we can, the way the Earth's mantle worked when the Earth was a young planet.

3.2.1 Samples of the Archaean mantle

Unlike the modern mantle, direct samples of the Archaean mantle are rare and geophysical measurements impossible. This means that our prime evidence comes from the study of Archaean mafic and ultramafic magmas.

3.2.1.1 Archaean basalts

Basaltic rocks are common in Archaean terrains and are particularly abundant in Archaean greenstone belts. However, they are frequently altered and metamorphosed, and only the best preserved samples should be used to obtain information about the composition of the Archaean mantle. Recent research has shown that Archaean greenstone belts formed in a number of different tectonic environments (see Chapter 1, Section 1.3.1). This means that greenstone belt basalts potentially offer an insight into a variety of mantle environments in the early Earth.

Arndt et al. (1997) compared the compositions of Archaean tholeiites with those of modern basalts from ocean-ridge, ocean-island, and arc environments (Fig. 3.22). They showed that Archaean tholeiites have higher SiO_2 and FeO, and lower incompatible trace element concentrations (although enriched in Rb and Ba relative to modern MORB), compared to their modern equivalents. Many of these geochemical features of Archaean basalts match those of modern arc basalts, but there are also important differences. This led Arndt et al. (1997) to the important conclusion that "the major- and trace-element characteristics of Archaean basalts are matched by no common type of modern basalt." The uniqueness of Archaean tholeiitic basalts requires that they are either from a source different from that of modern tholeiites, or are the product of a different melting process. Arndt et al. (1997) interpreted the low incompatible trace element concentrations and high Si, Fe, Ni,

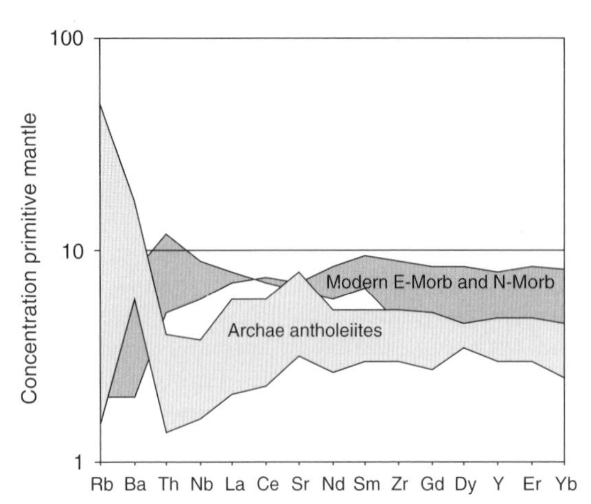

FIGURE 3.22 The trace element composition of Archaean tholeiites compared to modern enriched-MORB and normal-MORB, normalized to the primitive mantle (after Arndt et al., 1997). See Text Box 2.2.

and Cr to signify a high degree of partial melting. From what we know of melting processes in the modern mantle this would indicate that Archaean tholeiites are produced by hotter, deeper mantle melting – a thicker melt column – than are modern MORB.

3.2.1.2 Komatiites

Komatiites are ultramafic magmas which are almost exclusively confined to the Archaean. They are common in many greenstone belts where they are interbedded with tholeiitic basalts. They were discovered in the 1970s, in the Barberton Greenstone Belt in South Africa, outcropping in the Komati river, and it is from this locality that they derived their name. The relatively recent recognition of a new class of igneous rock caused something of a stir within the Earth Science community at the time, and komatiites quickly became recognized as an important window into the Earth's early mantle. Komatiites, however, have had more than their share of controversy, as will be seen from the discussion below.

Typically komatiites are ultramafic rocks with volcanic features such as glassy flow tops, flow top breccias, amygdales, and pillows. They form lava flows varying in thickness from a few centimeters to over 100 m. They are highly magnesian with more than 18 wt% MgO and in their primary state they contain olivine and pyroxenes in a glassy matrix. The best preserved examples display "spinifex texture" – slender, elongate blades of olivine – which are indicative of very rapid cooling. Very commonly komatiites are associated with komatiitic basalts – less magnesian ultramafic lavas, with between 12 and 18 wt% MgO – dominated by clinopyroxene and sometimes showing spinifex texture.

The very high MgO-content of komatiite melts has been taken to indicate that they represent extremely high-temperature melts of the mantle. The exact temperature of melting is reflected in the MgO-content of the parental melt. However, herein lies a difficulty, for determining the original MgO-content of a komatiitic melt is not straightforward (Fig. 3.23). There are two reasons why this is so. First, many komatiite flows are highly differentiated, leading to major compositional differences between chilled flow margins and domains which are cumulus-enriched. In fact the unwitting analysis of olivine-rich komatiite cumulates has, in the past, led to an overestimate of the MgO-content. Second, komatiites are almost always chemically altered as a result of submarine or subaerial weathering after eruption and subsequent metamorphism. By screening out differentiated and altered samples, Arndt (1994) proposed a maximum melt composition of about 30% MgO. Such melts would have had an eruption temperature of about 1,600°C and, as will be shown later, probably segregated from a source several hundred degrees hotter than the ambient mantle.

3.2.1.2.1 Contaminated komatiites

One of the consequences of such high eruption temperatures is that komatiite lavas had very low viscosities. Flow rates would have been very high producing extensive, turbulent lava flows, which experienced very rapid cooling (Huppert et al., 1984). This observation led to a major revolution in thinking about komatiites, for a feature of superhot, turbulent lava flows is that they are thermally corrosive

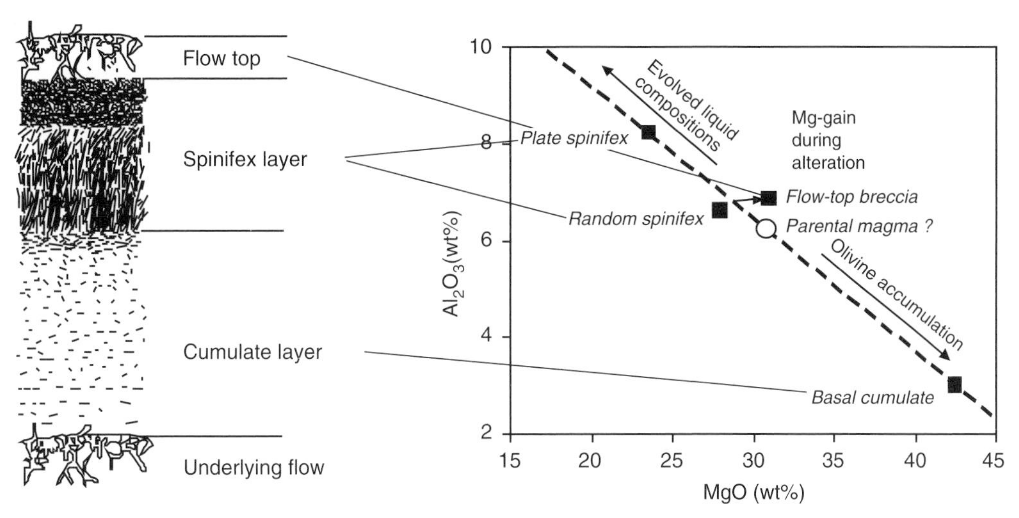

FIGURE 3.23 Compositional variation within a komatiite flow. The stratigraphic section on the left shows an idealized komatiite flow comprising a chilled flow top, a spinifex textured section, and a cumulate base. Typically such flows are a few meters thick. The MgO–Al_2O_3 plot on the right shows the compositional variation within the different layers of the flow as a consequence of olivine fractionation. A small amount of the compositional variation can also be attributed to Mg-gain during the alteration of the flow top. The circle shows the probable composition of the parental magma, although this composition may not be precisely represented by any rock composition within the flow (after Arndt, 1994).

and have the capacity to assimilate a significant proportion of the rocks through which and onto which they are emplaced. This prediction was verified by the description of thermal erosion channels at the base of komatiite flows in the Kambalda greenstone belt in western Australia (Groves et al., 1986).

The recognition that komatiites were most probably chemically contaminated sent the world of komatiite studies into confusion for a time, for whilst previously komatiites had been held as the premier window into the Archaean mantle, it now appeared that this window was rather dirty. Fortunately, the very high percentage of contamination predicted (40%) has not been realized (Foster et al., 1995). At the present time trace element and isotopic studies are used to discriminate between contaminated and uncontaminated komatiites (Arndt, 1994) and identify the extent of contamination. So the pendulum has swung back and currently komatiites are regarded as an important source of information about the composition of the Archaean mantle.

3.2.1.2.2 Komatiite types

Komatiites have been subdivided into a number of different chemical types. The most frequently used classification is that of Arndt (1994) and recognises Al-depleted komatiites (or Barberton-type komatiites, from the greenstone belt where they were first recognized) and Al-undepleted types (or Munro-type komatiites, named after their type locality in the Abitibi greenstone belt, Canada). To these can be added a class of Ti-depleted komatiites, found in the Abitibi Greenstone Belt and similar in composition to the Cretaceous komatiites from Gorgona Island (Sproule et al., 2002), and a further group of extremely Ti-depleted-Si-enriched komatiites from the Kaapvaal Craton (Wilson, 2003). There is a strong relationship between the level of Al-depletion in a komatiite and its REE content. On a Al_2O_3/TiO_2 versus $[Gd/Yb]_n$ diagram there is an inverse correlation between the Al_2O_3/TiO_2 and $[Gd/Yb]_n$ ratios (Fig. 3.24). Al-depleted komatiites have relatively high REE concentrations and $[Gd/Yb]_n > 1.0$, Al-undepleted komatiites have lower REE concentrations

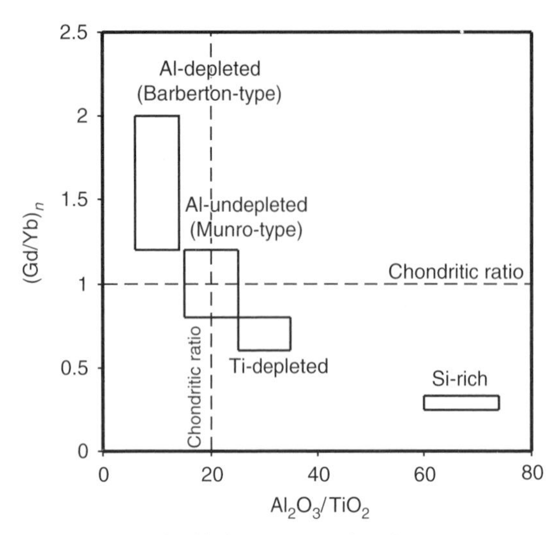

FIGURE 3.24 $(Gd/Yb)_n$ versus Al_2O_3/TiO_2 ratio plot showing the differences between the different chemical types of komatiite. These element pairs are selected, in part because they are thought to be unaffected by later chemical alteration of the komatiite. The $(Gd/Yb)_n$ ratio is a measure of the degree of fractionation of the REEs. The subscript "n" indicates that the values are chondrite normalized.

and chondritic Al_2O_3/TiO_2 and $[Gd/Yb]_n$ ratios. The extremely Ti-depleted- Si-enriched komatiites have very low REE concentrations and are strongly depleted in light REE with $[Gd/Yb]_n < 1.0$.

The significance of the different komatiite types has become clear from a combination of experimental and trace element studies. Al-depleted komatiites are generally thought to have formed in the presence of an aluminous phase, such as garnet at depths of 200–400 km or majorite-garnet at depths of 400–660 km. Al-undepleted komatiites formed in the absence of an aluminous phase either in the shallow mantle, where garnet is consumed during melting, or at great depth, below the mantle transition zone (>660 km) where perovskites become the major mantle phases. Xie and Kerrich (1994) have argued that the phase changes which take place with depth in the mantle can be fingerprinted geochemically and can be used to identify the depth of melting of the different types of komatiite. They show that the ratio of HFSEs – Nb, Zr and Hf – to the REE can be used as a discriminant between olivine (a shallow mantle phase), majorite (from the mantle transition zone), and perovskite (a lower mantle phase) fractionation. They propose, on the basis of different trace element signatures found in the komatiites of the Abitibi Greenstone Belt, Canada, that Al-undepleted komatiites formed by a high degree of partial melting of the shallow mantle (<300 km), whereas Al-depleted komatiites formed in equilibrium with majorite garnet in the depth range 400–700 km. It should be noted that experimental studies of similar compositions record slightly lower pressures (Herzberg, 1992). More extreme however is the finding by Xie and Kerrich (1994) that a second group of Al-undepleted komatiites, also from Abitibi, have a trace element signature indicative of perovskite fractionation, implying a depth of melting in excess of 700 km.

3.2.1.2.3 *Melting conditions*

The present consensus is that komatiites are high-temperature melts, the most magnesian of which erupted at temperatures close to 1,600°C. These high temperatures require melting deep in the mantle in the depth range 300–700 km. They also represent a relatively high fraction of melting. Herzberg (1992) suggested on the basis of experimental studies that Munro-type komatiites represent 35–40% melting whereas Barberton type may represent as much as 50% melting. The only serious grounds on which these conclusions might be challenged is if komatiites were the product of wet mantle melting and this leads to the most recent controversy to surround komatiites.

3.2.1.2.4 *Were komatiites wet?*

Researchers working on the komatiites of the Barberton Greenstone belt have recently argued that komatiites are the product of wet mantle melting. Their evidence comes from both field observation and from experiment. Field observations of vesicles in komatiite lavas and pyroclastic komatiites lead to the inference that the lavas were degassing water when they erupted. In addition, experimental studies of pyroxenes in Barberton komatiites (Parman et al., 1997, 2003) show that they crystallized from a melt containing a minimum of 6 wt% water. Further support for the wet melting model comes from the work of Collerson and Kamber (1999) who showed that the Nb/Th and Nb/U ratios of uncontaminated komatiites and komatiitic basalts were progressively fractionated during the Archaean. They have suggested that the principal control on this fractionation is via hydrous (subduction zone) processes.

The counter argument, summarized by Arndt (2003) and Arndt et al. (1998), stresses the inability of low viscosity komatiitic melts to contain dissolved volatiles. Furthermore, most komatiites are chemically depleted implying that they are the melts of a source which was previously chemically fractionated. If this was the case, then any volatiles in the source would be expected to have been removed in the first melt and would not be present to enter a komatiitic melt.

One of the most appealing aspects of a hydrous mantle origin for komatitiites is that it solves the problem of high komatiite melt temperatures and reduces them to what would be considered "normal" for the Earth's mantle. For example, Inoue et al. (2000) found that in a hydrous melting experiment, Al-depleted, Barberton-type komatiites can be produced at relatively low pressures (8 GPa, i.e. 200 km) and low temperatures (1,300–1,500°C) compared with the ca. 1,900°C calculated for dry melting (Section 3.2.2.2). Thus, a wet melting model for komatiites challenges the view that the Archaean mantle was particularly hot. This is important, for until recently a "hot" Archaean mantle was a fundamental assumption about the Archaean Earth. At present the debate is polarized between a "dry-komatiites-are-produced-in-mantle-plumes" group and a growing group of scientists arguing for "wet-komatiites-of-subduction-zone-origin."

3.2.1.3 Boninites

Boninites are lavas with a high MgO (10–22 wt%), high SiO_2 (52–58 wt%), mg nos. greater than 60 (mg number, or mg#, is the atomic fraction of Mg relative to total (Mg + Fe(II)) in the rock) and low levels of incompatible elements. Geologically recent boninites are found in arc environments. They are thought to have formed through a two stage process. Their mantle source is depleted through one or more melting events and then the refractory harzburgite residue is enriched and remelted by the fluxing of fluids from a subducting slab. Melting takes place in anomalously hot supra-subduction zone conditions at high temperatures (ca. 1,300°C) and at shallow depth (<10 kb) (Crawford et al., 1989). Geochemically, modern boninites are characterized by U-shaped REE patterns which reflect the two processes involved in their generation. LREE enrichment from the subduction zone fluids is imprinted on an already light- and middle-REE depleted mantle, depleted through an earlier melt extraction event.

The recognition of boninitic or boninite-like magmas in late Archaean (Kerrich et al., 1988; Smithies, 2002) and early Archaean greenstone belts (Polat et al., 2002), albeit as rather rare melts, has been used by some authors to suggest that mantle processes similar to those involved in the genesis of modern boninites were operating at that time. In addition, it has been proposed that some komatiitic basalts may also be Archaean equivalents of modern boninites (Parman et al., 2001), and it is possible that the high-Si komatiites described by Wilson (2003) formed in a similar manner.

Because they strongly implicate "modern-type" subduction processes, claims of Archaean boninites must be examined carefully. Smithies et al. (2004) in a recent review evaluated the evidence and propose two groups of Archaean "boninites." They show that there are Archaean boninites which are very similar in composition to modern boninites. The oldest of these are found in the 3.12 Ga Whundo Group, in the Pilbara of western Australia, where they form part of a greater than 10 km thick sequence of mafic, andesitic, and rhyolitic lavas. Similar lavas are known from the 2.8 Ga Superior Province in Canada. These are thought to have formed by modern-type subduction processes and imply subduction at 3.12 Ga. A second group of boninite-like lavas is less enriched in light REE. These include lavas from the Abitibi Greenstone belt in Canada and the 3.8 Ga Isua greenstone belt in west Greenland. These boninite-like lavas are thought to have formed by second-stage melting, in the same way as true boninites, but are the product of an "inefficient subduction enrichment process, during some earlier style of Archaean subduction" (Smithies et al., 2004), hence their lower concentrations of fluid-mobile elements.

3.2.1.4 *Fragments of the Archaean mantle: do Archaean ophiolites exist?*

Recognizing fragments of the Archaean mantle is fraught with problems for it is difficult to distinguish mantle rocks from other ultramafic rocks such as komatiites and ultramafic intrusions especially when they have been metamorphosed and deformed. It has been suggested that Archaean mantle rocks are preserved in Archaean ophiolites in a manner analogous to modern environments (Section 3.1.1.4). However, the existence of Archaean ophiolites is itself hotly debated and the evidence for their existence is equivocal. Yet, they are a subject of great importance, for, if they do exist, not only would they provide actual samples of the Archaean mantle, but they would also give firm evidence for the operation of plate tectonics early in Earth history.

Originally all Archaean greenstone belts were thought to be ophiolites, obducted slices of Archaean ocean floor, but now it is generally agreed that this is not normally the case (Bickle et al., 1994). Most Archaean "ophiolites" have been "identified" on the basis of their similar stratigraphy to the classical ophiolite igneous stratigraphy (See Section 3.1.1.4). Such claims have been made for the lower Kam Group in the Yellowknife Greenstone Belt, Canada (Helmstedt et al., 1986), the Jamestown Complex in the Barberton Greenstone Belt, South Africa (De Wit et al., 1987), the Wind River Greenstone Belt in Wyoming (Harper, 1985), and from the Yilgarn Block in western Australia (Fripp & Jones, 1997). An intense debate on the origin of the Belingwe Greenstone belt in Zimbabwe centered on whether or not there was a thrust plane within ultramafic rocks near the base of the succession (Kusky & Kidd, 1992) and hence whether the greenstone belt represented a thrust slice tectonically emplaced upon older rocks. The consensus now is that this succession is not an ophiolitic fragment (Blenkinsop et al., 1993).

Of relevance here are those Archaean ophiolites which may contain an ultramafic mantle section. One possible example is the Dongwanzi ophiolite in China (Kusky et al., 2001) from which harzburgite tectonites up to 70 m thick, dunites, pyroxenites and wehrlites have been described. These rocks also contain podiform chromitites with nodular textures, thought to be unique to mantle chromitite occurrences (Kusky et al., 2001; Kusky, 2004). This ultramafic assemblage is interpreted as depleted mantle from which a basaltic melt has been extracted.

One of the first claims for fragments of Archaean mantle was that of Collerson et al. (1991) who reported 3.8 Ga spinel-bearing metaperidotites and pyroxenites with mantle-like chemistry, interleaved with felsic gneisses in northern Labrador. Similar claims have also been made for ultramafic xenoliths found in Archaean high-grade felsic gneisses from the greater than 3,800 Ma Itsaq Gneiss Complex in west Greenland. Friend et al. (2002) described dunites and harzburgites from the Itsaq Gneiss Complex and argued on geochemical grounds that these rocks represent the "best characterized sample of the early Archaean mantle." Very similar dunites and harzburgites from the same region of the Itsaq Gneiss Complex were described by Rollinson et al. (2002), and yet these samples are chromite-bearing and display magmatic layering indicating a magma chamber origin. These conflicting interpretations illustrate the difficulty in unequivocally assigning a mantle origin to ultramafic xenoliths in gneiss terrains.

3.2.1.5 *Fragments of the Archaean mantle – xenoliths from the subcontinental lithosphere*

As geochronological determinations on xenoliths from the subcontinental lithosphere become more robust, a consistent observation is emerging – that the subcontinental lithosphere beneath Archaean continental crust is very ancient (see Section 3.1.3.2). Recent studies using Os-isotopes have shown that the subcontinental mantle beneath Archaean cratons has been chemically isolated from the convecting mantle since the late Archaean (Pearson, 1999). These are perhaps our most certain samples of the Archaean mantle. However, it is known that the Archaean subcontinental lithosphere was extremely chemically depleted (Fig. 3.12) and represents only a small fraction of the mantle as a whole. Furthermore many of these samples have been altered through interaction with later mantle melts. Thus, whilst kimberlite-borne xenoliths from the subcontinental mantle may be the most abundant samples of the Archaean mantle, they are not particularly useful in building Archaean mantle models.

One particular class of Archaean mantle xenoliths has received special attention. These are eclogites recovered from the subcontinental lithosphere. Richardson et al. (2001) showed that some Archaean eclogite xenoliths have the trace element and Os-isotopic characteristics of a basaltic protolith and an isotopic history which shows a significant time gap between basalt generation and eclogite crystallization. These properties are typical of subducted ocean floor and indicate that the subcontinental lithosphere beneath the Kaapvaal Craton contains fragments of 2.9 Ga subducted ocean floor.

3.2.1.6 The composition of the Archaean mantle

Calculating the composition of the Archaean mantle is a task which has occupied geochemists for some decades. It is not a trivial task because the nature of the Archaean mantle is that it was constantly changing in composition. However, estimates of the primitive mantle composition, also known as the composition of the BSE – the mantle as it was after the core has been extracted but before the continents were formed – are given at various points in this chapter, as follows: estimates of the major element composition of the primitive mantle are given in Table 3.1; the composition of the Archaean subcontinental lithosphere in Table 3.4; the trace element composition of the primitive mantle in Table 3.2; the initial compositions and the temporal evolution of the Nd-, Hf-, Pb- and Os-isotopic systems are given in Section 3.2.3. Finally, the volatile content of the Archaean mantle is discussed in Chapter 5, Section 5.2).

3.2.2 Thermal constraints on the nature of the Archaean mantle

Quantifying the thermal evolution of the Archaean mantle is a subject of great importance. For mantle temperatures strongly influence the viscosity of the mantle, which in turn controls the pattern of mantle convection and tectonic style. However, this is a difficult subject because there are many uncertainties. Interpreting the thermal history of the Earth requires a detailed knowledge of heat transport mechanisms and of heat source distributions. In addition thermal models are strongly dependent on the extent to which the Earth's mantle is or has been layered. All of these variables are only partially known, rendering any discussion of the thermal evolution of the mantle in the Archaean subject to large uncertainties.

One of the starting points in calculating the thermal history of the early Earth is the observed mismatch between the present-day heat production rate and the rate of heat loss. Today the total thermal flux for the Earth is about 41 TW (Davies, 1998). When this is compared with the present-day radioactive heat production of 20 TW it is clear that the Earth has another source of heat in addition to its radioactive heat. This heat source is thought to be the residue of "old heat" dating from the formation and differentiation of the Earth.

This observation alone suggests that the Earth's mantle was once much hotter than it is now and has cooled over time. What will also become apparent from the discussion below is that the Earth probably experienced its most dramatic cooling *during* the Archaean, although when exactly this cooling took place is much debated. Published estimates of secular cooling rates vary greatly. Values of between 30°K/Ga and 100°K/Ga have been proposed by Jackson and Pollack (1984), and Christensen (1985). Below we consider the evidence for a hotter

Archaean mantle drawn from a knowledge of the Earth's radioactive heat content and mantle potential temperatures calculated for komatiites and basalts.

3.2.2.1 Evidence from radioactive heating
The radiogenic heat production of the Earth's mantle has declined over the lifetime of the Earth, although the exact magnitude of this decline depends upon the concentrations of the heat producing elements K, U and Th in the mantle. The ratios K/U and Th/U are well known for the mantle and so models are dependent only on a knowledge of the U content of the mantle. This is thought to vary between 20 and 26 ppb (Table 3.2). Some of this uncertainty may be due to the progressive fractionation of U from the mantle into the continental crust over time. Recent estimates of the K/U ratio of the BSE from Helffrich and Wood (2001) combined with the heat production calculations of Pollack (1997) indicate that mantle heat production was slightly more than twice its present value at 2.5 Ga and about 4 times the present value at 4.0 Ga. This is illustrated in Richter's (1988) secular cooling curve shown in Fig. 3.25.

3.2.2.2 Evidence from komatiites
It has been shown experimentally that the MgO-content of komatiitic liquids is directly related to their eruption temperature. However, precisely determining the MgO-content of komatiitic liquids is not simple (see Section 3.2.1.2). Nisbet et al. (1993) discuss the most robust estimates for magnesian komatiites and conclude that there is strong evidence for lavas with 25.6 wt% MgO and more limited evidence for liquids up to 29 wt% MgO. They calculate that a liquid with a MgO-content of 25.6 wt% had an eruption temperature of about 1,520°C, whereas a liquid with an MgO-content of 29 wt% had an eruption temperature of about 1,580°C. Combining these results with pressure–temperature models of mantle melting Nisbet et al. (1993) calculated that komatiites with 25.6 wt% MgO would have melted at a temperature of 1,800°C at a depth of 12 GPa. They therefore indicate a mantle potential temperature of 1,800°C. Al-depleted komatiites with a MgO-content of 29 wt% MgO and an eruption temperature of about 1,580°C would have melted in the majorite stability field at 1,900°C at a depth of 18 GPa and represent a mantle potential temperature of 1,900°C.

Until recently most models for komatiite genesis would have interpreted these very high mantle potential temperatures as evidence for a plume origin and indicative of temperatures substantially higher than the ambient mantle.

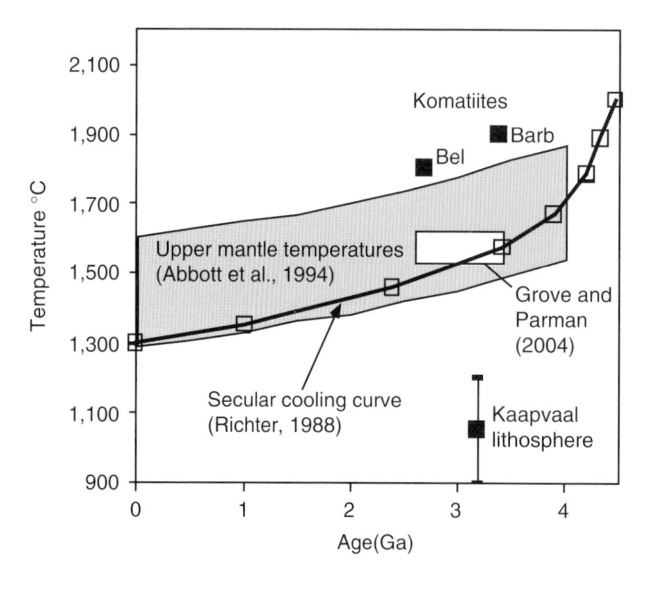

FIGURE 3.25 The secular cooling curve for the potential temperature of the Earth's mantle (from Richter, 1988) and calculated upper mantle potential temperatures based on basalt chemistry (Abbott et al., 1994 – shaded field). Also shown are the potential temperatures for komatiites from Barberton Greenstone Belt (Barb) and the Belingwe Greenstone Belt (Bel), estimated komatiite temperatures from Grove and Parman (2004) and the calculated temperature of the subcontinental lithosphere from beneath the Kaapvaal Craton (Richardson et al., 1984).

However, the new ideas on a wet mantle origin for komatiites challenge this view. Calculations by Grove and Parman (2004) based on a wet melting model indicate mantle temperatures between 1,500 and 1,600°C (Grove & Parman, 2004), rather than the 1,800–1,900°C suggested by Nisbet et al. (1993). The implications of a wet, cool origin for komatiites are that the mantle must have cooled very early in Earth history, and for much of the Archaean was only slightly hotter (ca. 100°C) than at present.

3.2.2.3 Evidence from basalt chemistry

Abbott et al. (1994) calculated the liquidus temperatures of basalts from ophiolite suites and Archaean greenstone belts from geochemical data for the last 3.8 Ga. From these results they calculated potential temperatures for their mantle sources. Their calculations show that over geological time basalts show a wide range of melting temperatures (over 300°C), the lower limit of which is similar to the secular cooling curve of Richter (1988) (Fig. 3.25). Two important conclusions can be drawn from this study. First, that Archaean basalts are no different from more recent basalts inasmuch as they display the same range of mantle potential temperatures. This indicates that, in many respects, the Archaean mantle behaved much as does the modern mantle. Second, average upper mantle potential temperatures have declined since the late Archaean by an estimated 137 ± 8°C (based on the temperature range in Fig. 3.25), or by 187 ± 42°C (based on mean temperatures).

3.2.2.4 Evidence from subcontinental mantle xenoliths

One of the important discoveries about the subcontinental lithosphere to come from xenolith and diamond inclusion studies is that beneath many Archaean cratons the subcontinental mantle is extremely ancient. The results of thermobarometry on these mantle xenoliths have been used to construct paleogeotherms for the subcontinental mantle (Fig. 3.4) which show that Archaean subcontinental lithosphere is significantly cooler than more recent lithospheric mantle. The early results of Richardson et al. (1984) on diamond inclusions from beneath Kaapvaal Craton in South Africa indicated temperatures of 900–1,200°C at depths of 150–200 km, for the time of diamond genesis in the Archaean (Fig. 3.25). This observation shows that the subcontinental lithospheric mantle was significantly cooler than both the calculated secular cooling curve for the mantle and the mantle from which komatiites was generated. For example, Nisbet et al. (1993) calculate that komatiites generated at a depth of about 150 km would have a potential temperature of about 1,600°C, compared with a temperature of about 1,000°C for the subcontinental lithospheric mantle at a similar depth. These observations are not surprising, for the insulating effect of a cool mantle "keel" is necessary to preserve ancient continental crust. In addition, however, this result demonstrates that there were strong lateral temperature variations within the Archaean mantle.

3.2.2.5 Melt production in the Archaean mantle

If, as many suppose, the Archaean mantle had a higher potential temperature than the modern mantle, it is important to examine the implications of this for melt production during the early history of the Earth. The relationship between mantle potential temperature and melt thickness during adiabatic melting was outlined in Section 3.1.4.3 and may be briefly summarized by stating that as mantle potential temperature increases so will the melt production, as expressed in the depth of the melt column and the melt thickness. This is illustrated in Fig. 3.26, which shows how deeper, higher-temperature melting should lead to the formation of a thicker oceanic crust.

3.2.2.6 The cooling history of the mantle

Figure 3.25 shows the secular cooling curve for the mantle from Richter (1988), the calculated potential temperatures for Archaean komatiites and basalts, and an estimated temperature for the Archaean subcontinental mantle. Two important conclusions follow. First, it is clear that the Archaean mantle had both hot and cool regions. Potential temperatures calculated from dry komatiite melting temperatures imply an anomalously hot mantle source,

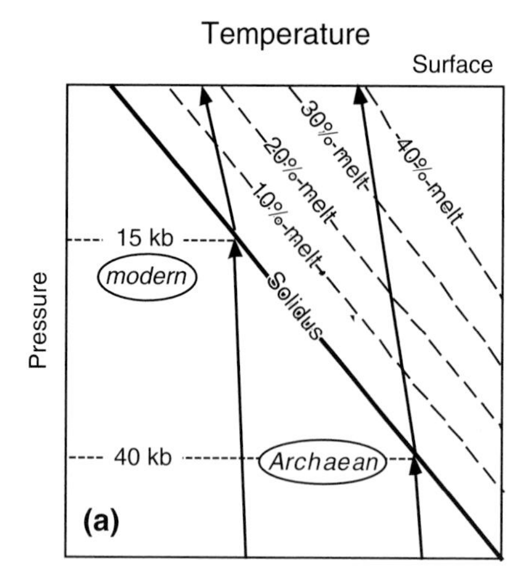

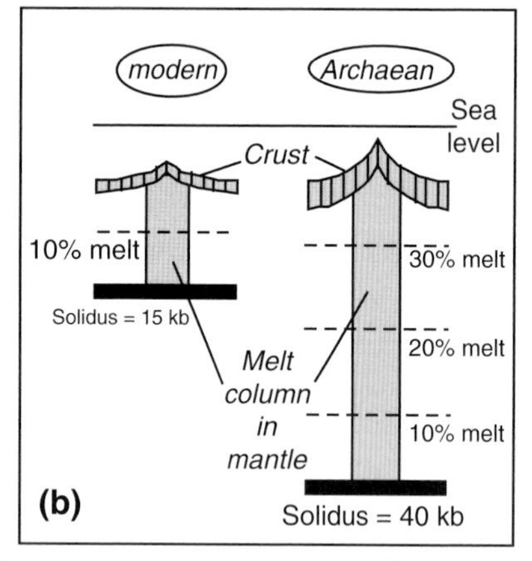

FIGURE 3.26 Pressure temperature diagram and oceanic crust and mantle cross section to illustrate how different mantle potential temperatures affect the melting of mantle peridotite during adiabatic decompression. (a) Typically in the modern mantle the adiabat crosses the mantle solidus at shallow depth (15 kb) and undergoes between 10 and 15% melting. In a hotter Archaean mantle the adiabat crosses the mantle solidus at greater depth and at higher temperature and undergoes a much higher degree of melting (up to 40%). These melts erupted at a higher temperature than modern shallower mantle melts. (b) Melting of the modern mantle at relatively shallow depth, with up to 15% melting, gives rise to an ocean crust about 7 km thick. Melting of the Archaean mantle with a higher potential temperature and a deeper melt column would give rise to a much thicker oceanic crust (after Grove & Parman, 2004).

whereas the temperature of the Kaapvaal subcontinental lithosphere implies that there were also regions of cool mantle. Even allowing for the wet melting of komatiites, mantle temperatures are still both higher and lower than the secular cooling curve of Richter (1988).

Second, whilst there is no agreement on the detail, it is evident that much of the Archaean mantle was hotter than the modern mantle. This has implications for mantle viscosity. The viscosity of the Earth's mantle shows a strong dependence on temperature to the extent that an upper mantle 250°C hotter than the present day mantle would be 60 times less viscous (Davies, 2006). A lower viscosity will influence patterns of mantle convection and the nature of Archaean tectonics, and would have influenced the size plates in the precursor to modern plate tectonics.

Richter's cooling curve suggests a secular cooling rate of about 100°C/Ga. Mantle potential temperatures from basalts and from wet komatiites suggest a similar value, although komatiites, if they are dry melts, imply a higher cooling rate. For many years models of the Archaean mantle have been based upon high, and in some cases, very high mantle temperatures. Now we are less certain, for some of the evidence suggests that, although the Archaean mantle was hotter than the present mantle, it was not very much hotter. If this turns out to be true than we will have to revise our models of Archaean tectonics.

3.2.3 Mantle evolution through time – radiogenic isotope constraints

Radiogenic isotopes have proved a most powerful tool for understanding mantle processes. By studying mantle peridotites and mantle-derived melts from a variety of different geological time periods it is possible to define evolutionary curves for the different isotopic systems within the mantle. These curves, when plotted on isotope ratio versus time diagrams, can be used to characterize the chemical evolution of the mantle over time. Deviations from the chondritic trend are used to identify chemical fractionation events in the mantle during Earth history. Of particular importance is the memory of now-extinct isotope systems in Archaean samples, which record differentiation events within the very early Earth, and which have subsequently been destroyed in the convecting mantle.

In this section mantle evolution curves are presented for neodymium (Nd), hafnium (Hf), lead (Pb) and osmium (Os) isotopes. A summary of these isotopic systems is given in Text Box 3.2. Earlier studies based upon the study of Sr isotopes in the mantle (e.g. Bell et al., 1982) are now known to be unreliable because of its high geochemical mobility (Goldstein, 1988). The significance of the mantle evolution curves described here is that they demonstrate that the mantle does not operate as an isolated system but that it has evolved in its composition over time, in response to core formation, crust extraction, and the recycling of crustal material.

3.2.3.1 The neodymium isotopic evolution of the mantle

Nd and Sm (samarium) are members of the REE group. Both behaved as refractory, lithophile elements during Earth accretion, and so their mantle ratio is expected to be chondritic. This has given rise to the concept of a CHondritic Uniform Reservoir (CHUR), the now accepted reference frame for Nd-isotopic evolution. Most commonly the Nd-isotopic evolution of the mantle is defined using the isotope ^{143}Nd, the product of ^{147}Sm decay (with the half-life 106 Ga). There is, however, another rarer isotope of Nd, which is particularly useful in exploring the earliest evolution of the mantle. ^{142}Nd is the product of ^{146}Sm decay, a short-lived isotope with a half-life of only 103 Ma, produced only in the first few hundred million years of Earth history.

3.2.3.1.1 ^{143}Nd evolution

The initial ^{143}Nd/^{144}Nd ratio of the bulk silicate Earth is obtained from the measurement of initial isotope ratios in primitive chondritic meteorites. Over time, therefore, the Earth should evolve along an evolution line, defined by the Sm/Nd ratio of chondrites. However, during mantle melting both Nd and the parent element Sm are incompatible (i.e. they are readily incorporated into the melt phase during mantle melting) and are removed from the Earth's mantle in basaltic and ultramafic melts, leaving the mantle depleted in Nd and Sm. Although both Nd and Sm prefer to enter the melt phase rather than remain in the mantle residue during melting, Nd has a stronger preference to enter the melt than does Sm. For this reason melt-depleted mantle becomes more depleted in Nd than Sm and so has an elevated Sm/Nd ratio relative to chondritic meteorites. This understanding of mantle chemistry allows us to make sense of the observation that many geologically recent mantle-derived rocks have higher than chondritic ^{143}Nd/^{144}Nd-isotope ratios. These melts must have come from a source region that had already experienced some melt extraction, a portion of the mantle known as the "depleted" mantle.

Deviations in measured ^{143}Nd/^{144}Nd-isotope ratios from the chondritic evolution curve are expressed in parts per thousand deviation using the ε_{Nd} notation. The evolution of this parameter with geological time has been documented by many authors (see summary in Rollinson, 1993). A recent compilation is given by Nagler and Kramers (1998) in Fig. 3.27. This curve is based upon a carefully screened database of mantle melts, whose age is well constrained and whose isotope ratios show no evidence of crustal contamination or disturbance during metamorphism.

The Nagler and Kramers (1998) ε_{Nd} mantle evolution curve shows two important features. First, that in the early Archaean the average ε_{Nd} value is almost constant and about 1.0. Thereafter, from about 3.0 Ga to the present, ε_{Nd} values steadily increase to a present-day value of +10. Nagler and Kramers (1998) suggest that the average ε_{Nd} value of about 1.0 for the early Archaean has no geological meaning but is in part an analytical artifact and in part the result of the uncertainty in ε_{Nd} measurements on chondrites.

In contrast, the steady increase in ε_{Nd} values in the Earth's mantle with time after 3.0 Ga does have a geological meaning and implies the progressive extraction of the light rare Earth elements from the Earth's mantle. The most popular explanation of this fractionation

TEXT BOX 3.2. Isotopic systems relevant to the long-term evolution of the Earth's mantle

Parent isotope	Daughter isotope	Isotope ratio measured	Decay constant	Reference Composition used for terrestrial evolution	Isotopic notation used
Neodymium isotopes					
^{147}Sm	^{143}Nd	^{143}Nd/^{144}Nd	6.54×10^{-12} yr^{-1} (half-life $= 1.06 \times 10^{11}$ years)	Present-day Chondritic ^{143}Nd/^{144}Nd ratio = 0.512638; ^{147}Sm/^{144}Nd = 0.1966	$\varepsilon_{Nd} = 10000 \times$ the deviation from the chondritic ^{143}Nd/^{144}Nd ratio at the time of interest
^{146}Sm	^{142}Nd	^{142}Nd/^{144}Nd	6.74×10^{-9} yr^{-1} (short lived isotope with a half life of 1.03×10^{8} years)	Several standards are in use. The present-day ^{142}Nd/^{144}Nd ratio in the Ames standard is 1.141838; solar system initial ^{146}Sm/^{144}Nd = 0.008	$100 \times \varepsilon^{142}$Nd – the deviation from the ^{142}Nd/^{144}Nd ratio of a given reference standard at the time of interest, expressed as ppm
Hafnium isotopes					
^{176}Lu	^{176}Hf	^{176}Hf/^{177}Hf	1.865×10^{-11} yr^{-1} or 1.983×10^{-11} yr^{-1}	Present-day Chondritic ^{176}Hf/^{177}Hf ratio = 0.282772; ^{176}Lu/^{177}Hf = 0.0332	$\varepsilon_{Hf} = 10,000 \times$ the deviation from the chondritic ^{176}Hf/^{177}Hf ratio at the time of interest
Lead isotopes					
^{238}U	^{206}Pb	^{206}Pb/^{204}Pb	1.55×10^{-10} yr^{-1}	Initial ratio of ^{206}Pb/^{204}Pb in the mineral troilite in the Canyon Diablo meteorite = 9.307	μ = the ^{238}U/^{204}Pb ratio measured in a mantle or crustal source region
^{235}U	^{207}Pb	^{207}Pb/^{204}Pb	0.985×10^{-9} yr^{-1}	Initial ratio of ^{207}Pb/^{204}Pb in Canyon Diablo troilite = 10.294	—
^{232}Th	^{208}Pb	^{208}Pb/^{204}Pb	0.049×10^{-9} yr^{-1}	Initial ratio of ^{208}Pb/^{204}Pb in Canyon Diablo troilite = 29.474	κ = the ^{232}Th/^{238}U ratio measured in a mantle or crustal source region
Osmium isotopes					
^{187}Re	^{187}Os	^{187}Os/^{188}Os	1.666×10^{-11} yr^{-1}	Present-day Chondritic ^{187}Os/^{188}Os ratio = 0.1270; ^{187}Re/^{188}Os = 0.40186	γ^{187}Os = the percentage deviation from the chondritic ^{187}Os/^{186}Os ratio at the time of interest
^{190}Pt	^{186}Re	^{186}Os/^{188}Os	1.54×10^{-12} yr^{-1}	Bulk silicate Earth value = 0.119834	ε^{186}Os = $10000 \times$ the deviation from the bulk silicate Earth value for ^{186}Os/^{188}Os at the time of interest

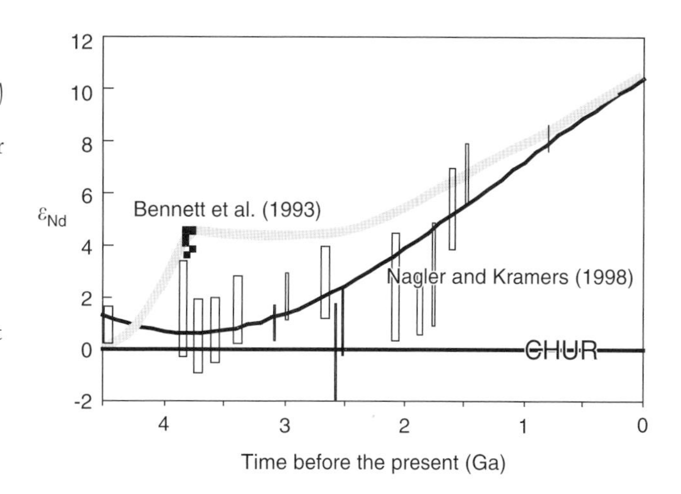

FIGURE 3.27 Mantle ε_{Nd} versus time plot showing mantle evolution according to the Nagler and Kramers (1998) curve (black line) relative to the chondritic uniform reservoir (CHUR). Selected error boxes from the Nagler and Kramers (1998) dataset are also shown. The curve is defined by the polynomial $\varepsilon_{Nd}(T)_{sample} = 0.164T^3 - 0.566T^2 - 2.79T + 10.4$. Data points from Bennett et al. (1993) for the Amitsoq gneisses (squares) showing high ε_{Nd} are also plotted, together with the mantle evolution curve proposed by Bennett et al. (1993) on the basis of these data (grey curve).

event is the extraction from the mantle of a felsic, light REE enriched continental crust. The ε_{Nd}-time evolution curve for the Earth's mantle presented in Fig. 3.27 is independently supported by forward modeling calculations (Nagler & Kramers, 1998). These calculations are constrained by the isotopic compositions of crust and mantle reservoirs and strongly support the notion of continental crust extraction as the primary cause of mantle depletion.

Alternative views of early Archaean mantle evolution require that mantle depletion started as early as ca. 4.5 Ga (see compilation in Rollinson, 1993). These models imply significant mantle Sm–Nd fractionation in the very early Archaean and have major implications for the differentiation of the early Earth. One such study is that of Bennett et al. (1993) who measured very high ε_{Nd} values (+3.5 to +4.5) in 3.81 Ga Amitsoq gneiss samples. Collerson et al. (1991) also calculated an isochron ε_{Nd} value of +3.0 for 3.8 Ga-old peridotites from northern Labrador. The extreme deviation from CHUR early in Earth history (Fig. 3.27) was interpreted by Bennett et al. (1993) as evidence for an extreme and very early fractionation of the Earth's mantle relative to CHUR. Such an event implies the formation of extensive continental crust prior to 3.8 Ga, for which there is no independent geological evidence. This apparent paradox and the claim for very early extensive mantle differentiation led to a detailed reexamination of the Bennett et al. (1993) dataset by Moorbath et al. (1997) and Vervoort et al. (1996). They showed that Nd-isotopes in the Amitsoq gneiss samples used by Bennett et al. (1993) were mobilized during later metamorphism and do not faithfully record the isotopic compositions at 3.8 Ga, the time of formation.

3.2.3.1.2 ^{142}Nd evolution

^{142}Nd is produced by the decay of ^{146}Sm–a short-lived isotope which became extinct within the first 400–500 Ma of Earth history. For this reason ^{142}Nd has the potential to identify very early events in the Earth's mantle. Positive or negative deviations of ε-^{142}Nd from the bulk Earth (chondritic) value would indicate the presence of a REE fractionated reservoir during the first 100–200 Ma of Earth history (see Text Box 2.3).

With this in mind, a number of research groups have searched for ^{142}Nd anomalies in very early basalts and sediments with variable success. In part this is due to the very low abundances of ^{142}Nd in natural samples. The first study was by Harper and Jacobsen (1992) who reported elevated ε-^{142}Nd values in sediments from the 3.8 Ga Isua greenstone belt, although this finding was later disputed (McCulloch & Bennett, 1993). More recently, two research groups, using improved analytical methods, have reported the presence of a ^{142}Nd anomaly in rocks from the Isua greenstone belt (Boyet et al., 2003, Caro et al., 2003, 2006).

These results require that the mantle source of the 3.8 Ga basalts and related sediments experienced Sm/Nd fractionation within 50–200 million of years of planetary accretion and that this fractionation was preserved from that time until at least 3.8 Ga. This finding was strikingly confirmed by Boyet and Carlson in 2005, who redetermined the ε^{142}Nd value of chondritic meteorites. These new ultrahigh precision measurements showed that the chondritic value is 20 ppm lower than that for terrestrial rocks, implying an ubiquitous ^{142}Nd anomaly in the Earth's mantle.

The recognition of a ^{142}Nd anomaly within the mantle implies that the Earth experienced a major, very early differentiation event. The study by Boyet and Carlson (2005) showed that lunar basalts also have elevated ^{142}Nd/^{144}Nd ratios relative to primitive chondrites, implying that the Moon was formed from an Earth that had already experienced major differentiation. This means that the early differentiation of the Earth took place within 30 Ma of the formation of the solar system. Whilst the precise nature of this differentiation event is not known, a favored model is the formation of an Fe- and trace element-enriched basaltic crust, perhaps as an initial crust to a magma ocean. It is postulated that this crust is now isolated from the convecting mantle and is located deep within the lower mantle.

The implications of a very early differentiation event in the Earth's mantle have far-reaching consequences for geochemical models for the silicate Earth. It means, for instance, that if a major trace element-enriched component was removed from the mantle very early in Earth history, the mantle has had a composition similar to the modern MORB source for most of Earth history.

3.2.3.2 The hafnium isotopic evolution of the mantle

Hf-isotopes behave in a very similar way to Nd-isotopes during most magmatic processes in the Earth's mantle. The elements Lu (lutetium) and Hf are refractory and lithophile during planetary accretion, and like Sm and Nd, their mantle ratio is expected to be chondritic. Oceanic basalts show strong correlations between their Nd-and Hf-isotope ratios (Vervoort & Blichert-Toft, 1999) reflecting the geochemical similarity between the two isotope systems and their coupled isotope evolution in the mantle (Fig. 3.28). Hf-isotopes, however, have a particular advantage over Nd-isotopes, making them especially useful in unraveling events in the very early history of the mantle. During mantle melting the fractionation of Lu relative to Hf is more extreme than for the Sm–Nd system, resulting in a stronger suprachondritic signal for ε_{Hf} than for ε_{Nd} in the same melting event. This is seen in isotope correlations for modern oceanic basalts where ε_{Hf} is up to 1.5 times greater than ε_{Nd} (Vervoort & Blichert-Toft, 1999).

Experimental studies show that the mantle mineral most likely to discriminate between Lu and Hf is the mineral garnet. Thus mantle melts showing Lu/Hf fractionation

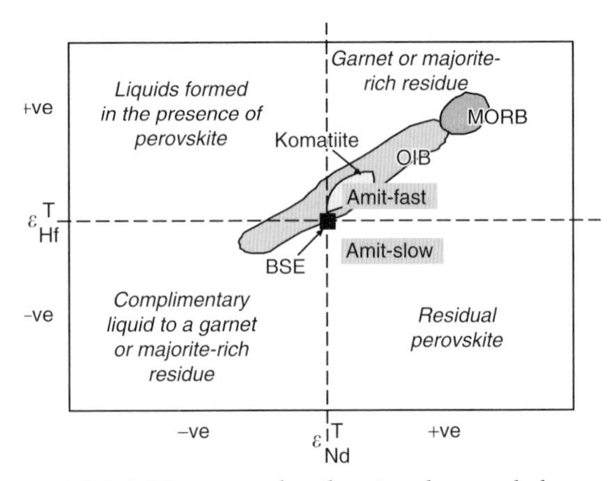

FIGURE 3.28 ε_{Hf}–ε_{Nd} plot showing the coupled evolution of the two isotope systems (after Blichert-Toft & Albarede, 1997). Depicted is the mantle array (MORB – midocean ridge basalt, OIB – ocean island basalt and the field of Archaean komatiites from Blichert-Toft & Arndt, 1999). The bulk silicate Earth value (BSE) is from Blichert-Toft and Albarede (1997) and lies to the low side of the mantle array. The diagram is divided into quadrants showing the effects of mantle mineralogy on ε_{Hf} and ε_{Nd}. The different interpretations of the ε_{Hf}–ε_{Nd} data for the Amitsoq gneisses (Vervoort & Blichert-Toft, 1999) according to the fast and slow ^{176}Lu decay schemes are shown (Amit-fast, Amit-slow).

as evidenced by a positive ε_{Hf} are thought to indicate derivation from a mantle source from which a basaltic melt has been previously extracted, leaving a garnet enriched residue (Blichert-Toft & Albarede, 1997). An alternative mechanism for Lu/Hf fractionation is by the mineral perovskite. In this case, however, Hf is more readily accommodated in the perovskite structure than is Lu, and so the resultant melts will have positive ε_{Hf} and residues after melting negative ε_{Hf} (Blichert-Toft & Albarede, 1997). Since perovskite is a deep mantle mineral, negative ε_{Hf} implies a deep mantle origin for such melts (Fig. 3.28).

There are however some problems with using Hf-isotopes to trace the evolution of the mantle, which is why, despite their obvious superiority to Nd isotopes, they have not been more widely used. First, they present formidable analytical difficulties, although this problem is now being resolved with the advent of new techniques (Vervoort & Blichert-Toft, 1999). Second, there is uncertainty about the extent to which Hf is mobile during metamorphism (Blichert-Toft & Frei, 2001). Most important, however, is the uncertainty in the half-life for ^{176}Lu-decay. Currently there are several different decay constants in use, the extremes of which are represented by the slow decay constant ($\lambda^{176}Lu = 1.865 \times 10^{-11}$) of Soderlund et al. (2004) and a fast decay constant ($\lambda^{176}Lu = 1.983 \times 10^{-11}$) advocated by Bizzarro et al. (2003). Until recently it was thought that the differences were related to the materials studied with the slow decay constant derived from terrestrial samples and the fast on meteorites. However, recently Amelin (2005) showed that the slow decay constant is also applicable to meteorites. At present this matter is unresolved. Switching between the two decay constants strongly influences calculated ε_{Hf}-values. For example, rocks formed at 3.6 Ga show a difference of 5 ε_{Hf}-units between the two decay schemes. This huge discrepancy has profound geological implications. A final complication is that there are also differences in the chondritic reference values in use (ca. Blichert-Toft & Albarede, 1997; Salters & White, 1998; Blichert-Toft & Arndt, 1999). This adds an additional uncertainty of ± 1.0 ε_{Hf}-units to calculated values.

3.2.3.2.1 A mantle curve for Hf-isotopes

The Hf-isotope mantle evolution curve calculated by Vervoort and Blichert-Toft (1999) is given in Fig. 3.29a. This curve is in two sections. From 4.5 Ga until ca. 3.0 Ga the mantle appears to have experienced an early, extreme fractionation, whereas after ca. 3.0 Ga the curve conforms to the expected time integrated Lu/Hf fractionation for the mantle. However, since the mantle evolution curve is based upon calculated ε_{Hf}-values which will vary according to the decay constant used, there is need for a reexamination of the mantle Hf-evolution curve in the light of the controversy over decay constants. For this reason the $\varepsilon_{Hf}{}^{T}$-data presented in Figures 3.29b, care calculated according to the fast (Bizzarro et al., 2003) and slow (Soderlund et al., 2004) decay schemes, respectively.

The results of the fast decay scheme (Fig. 3.29b) are similar to the original curve by Vervoort and Blichert-Toft (1999) and imply a very early (pre-3.5 Ga) mantle differentiation event, followed by less extreme mantle depletion. The early depletion event was most likely caused by the production of a very early mafic crust. Thus the ultimate source of the parental magmas of the Jack-Hills zircons, the Amitsoq gneisses, and some Barberton komatiites, was mantle that had already been depleted in the early Archaean. Bizzarro et. al. (2003) proposed that this early differentiation of the mantle took place prior to 4.33 Ga. This result is consistent with evidence from ^{142}Nd for a very early mantle melting event, as outlined in the previous section of this chapter.

If, however, the mantle evolution curve is calculated based upon the slow decay scheme of Soderlund et al. (2004) a very different interpretation emerges. In this case the ε-Hf evolution of the mantle is a single stage process from about 3.5–4.0 Ga until the present (Fig. 3.29c). However, using this decay constant many early Archaean samples plot with negative $\varepsilon_{Hf}{}^{T}$ in the "enriched source" region of the diagram. A possible interpretation of this result is that a significant amount of

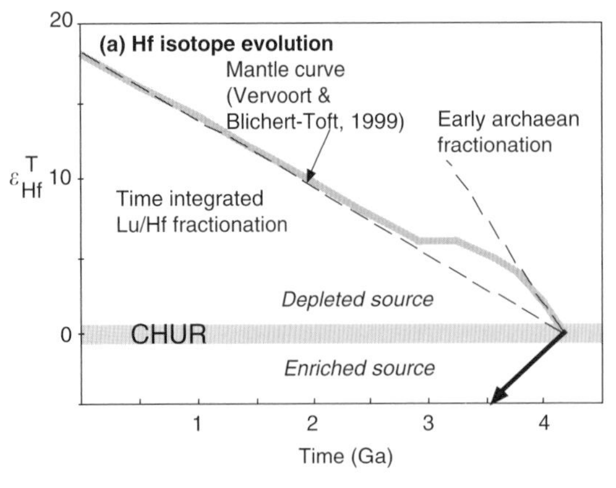

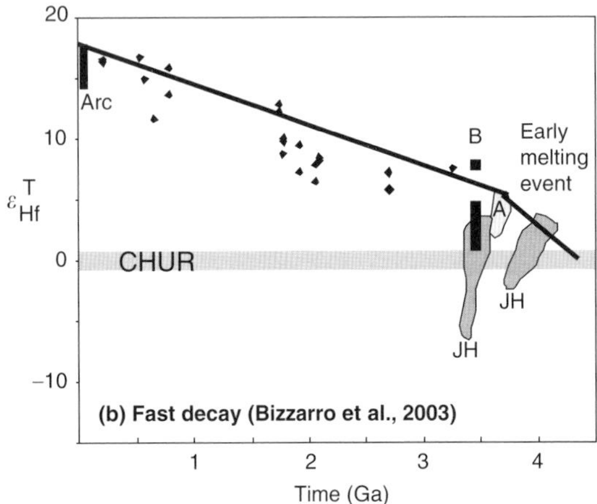

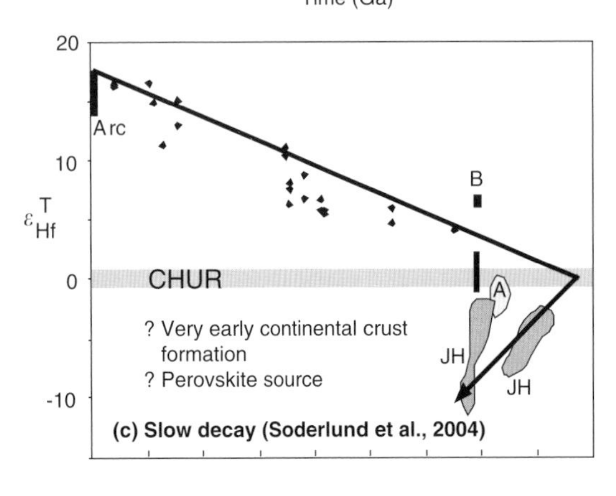

FIGURE 3.29 ε_{Hf}–time diagrams for the evolution of the Earth's mantle. (a) The mantle evolution curve of Vervoort and Blichert-Toft (1999) showing pre-3.0 Ga fractionation followed by a less extreme fractionation event. The mantle curve indicates that the mantle was depleted in Lu relative to Hf. Negative ε_{Hf}-values would imply derivation from an enriched source such as enriched mantle or ancient continental crust. (b) A proposed ε_{Hf}-mantle evolution curve based upon ε_{Hf}-values calculated using the fast decay sheme of Bizzarro et al. (2003) – $\lambda = 1.983 \times 10^{-11}$. This dataset supports the mantle evolution curve proposed by Vervoort and Blichert-Toft (1999) shown in (a). The data are taken from Blichert-Toft and Arndt (1999), Vervoort and Blichert-Toft (1999), Amelin et al. (1999), and Amelin et al. (2000). Symbols: diamonds, most radiogenic samples; Arc, modern arc samples; A, zircons from Amitsoq Gneisses, west Greenland; B [black squares], komatiites and basalts from Barberton Greenstone Belt, South Africa; JH, zircons from the Jack Hills, Australia. (c) A proposed ε_{Hf}-mantle evolution curve based upon ε_{Hf}-values calculated using the slow decay scheme of Soderlund et al. (2004) – $\lambda = 1.865 \times 10^{-11}$. This dataset does not support the very early differentiation of the mantle but implies the very early existence of continental crust. Data sources as in (b). ε_{Hf}^{T} values were calculated using the chondritic values of Blichert-Toft and Albarede (1997) – $^{176}Hf/^{177}Hf_{present\text{-}day} = 0.282772$, $^{176}Lu/^{177}Hf = 0.0332$. A consensus is now emerging in favour of the slow decay scheme, consistent with interpretation (c) in this figure.

continental crust had formed prior to 3.5 Ga. Alternatively the negative ε_{Hf}-values indicate a mafic precursor which was derived from a perovskite-rich lower-mantle source (see Fig. 3.28) by the partial melting of the deep mantle. The conclusion from the slow decay scheme, that there was a significant amount of very early continental crust, is not consistent with the geological evidence (see Chapter 5, Section 5.2.1.2).

A further factor which relates to the Hf-isotopic evolution of the mantle arises from the choice of the current chondritic reference value (Blichert-Toft & Albarede, 1997). On a $\varepsilon_{Nd}{}^{T}$–$\varepsilon_{Hf}{}^{T}$ plot (Fig. 3.28) the chondritic value (BSE) is displaced toward the low side of the correlated terrestrial Nd–Hf-isotope array. There are two important implications of this observation. First, mass balance constraints require that there is another terrestrial reservoir (see e.g. Bizzarro et al., 2002), which is complementary to the present terrestrial array. This may be the buried ancient proto-crust, as inferred from ^{142}Nd isotope measurements. Second, and contrary to the inference of some workers (e.g. Hart, 1988), no present-day basalts have isotopic compositions representative of a primitive mantle source (Blichert-Toft & Albarede, 1997).

Blichert-Toft and Arndt (1999) proposed an alternative different set of reference values for the BSE. If these were to be adopted, then it is necessary that either the Earth accreted at a different time from the normally accepted 4.57 Ga or from a bulk composition different from that of CI chondrites. This latter option was explored by Patchett et al. (2004) who found significant variation in the Hf-isotopic composition of chondritic meteorites, indicating that the future resolution of terrestrial Hf-isotope systematics is likely to be found in an appropriate choice of meteoritic parent for the Earth.

3.2.3.3 The lead isotopic evolution of the mantle

The U (uranium)–Th (thorium)–Pb (lead) isotopic system represents three independent decay schemes and is a powerful but complex tool with which to unravel the history of the Earth's mantle (Text box 3.2). During planetary accretion U and Th are refractory, lithophile elements and will reside in the mantle. Pb on the other hand is a volatile and chalcophile/siderophile element and may in part, be stored in the core. Initial U and Th concentrations are derived from chondritic meteorites, and initial Pb isotope compositions are taken from the iron-sulfide troilite phase in the Canyon Diablo meteorite. The initial bulk Earth U/Th ratio was 4.0 ± 0.2 (Rocholl & Jochum, 1993).

During mantle melting U, Th, and Pb are incompatible (i.e. preferentially enter the melt phase during partial melting rather than remain in the refractory residue), with U the most incompatible. Typically U/Th ratios do not change during mantle melting whereas U/Pb ratios do, leading to elevated U/Pb ratios in the upper continental crust relative to the mantle. At the Earth's surface U is easily mobilized in an oxidizing environment and so has the potential to be recycled back into the mantle. Thus elevated U/Th and U/Pb ratios in the mantle may be the product of recycled U.

3.2.3.3.1 The U/Pb ratio of the mantle

A plot of initial Pb-isotope compositions for crustal and mantle rocks on a ^{206}Pb/^{204}Pb versus ^{207}Pb/^{204}Pb diagram shows a "banana-shaped" field (Fig. 3.30; Kramers & Tolstikhin, 1997). A particular feature of this plot is the wide variation in ^{207}Pb/^{204}Pb ratio at constant ^{206}Pb/^{204}Pb in the lower (older) part of the curve.

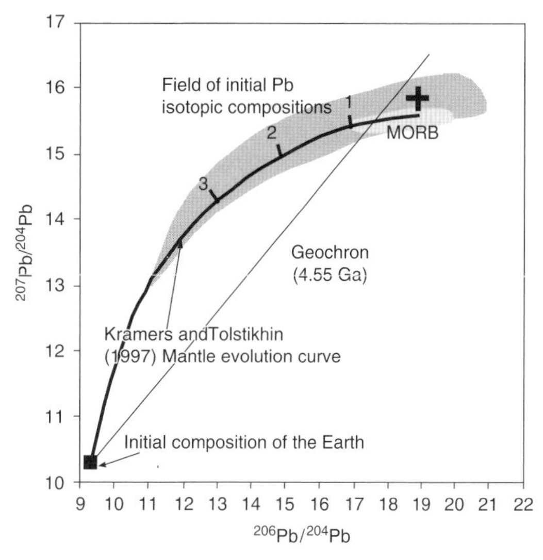

FIGURE 3.30 The Kramers and Tolstikhin (1997) Pb-isotopic evolution curve for the Earth's mantle, given an initial μ(^{238}U/^{204}Pb ratio) of 11.2, relative to the field of initial Pb-isotope compositions, the MORB field, the 4.55 Ga Geochron and the primordial Pb-isotopic composition. The numbered marks on the mantle evolution curve indicate Ga before the present. The cross marks the composition of the average continental crust.

Some authors have suggested this requires a major U–Pb fractionation event during the early Archaean or Hadean (Vervoort et al., 1994; Kamber et al., 2003).

3.2.3.3.2 The future Pb-paradox

A particular feature of the Pb-isotope evolution curve for the mantle it that it extends into the future relative to the 4.55 Ga Geochron line. (The Geochron defines the locus of present-day Pb-isotope ratios for the Earth, given the initial primordial compositions and an age of the Earth of 4.55 Ga.) The same finding would also be true for a 4.57 Geochron. This problem, known as the "Pb-paradox," or the "future Pb-paradox," is a mass balance problem and relates to the mass balance of the different terrestrial Pb-isotope reservoirs. The Pb-isotope composition of the bulk Earth should plot on this line. However, the two reservoirs likely to dominate the Pb-budget of the silicate Earth, the MORB reservoir and the continental crust, plot to the right of the Geochron (Fig. 3.30). When this is considered in the light of the mass balance of the silicate Earth, two conclusions follow. First, the MORB-source mantle and the continental crust cannot be complementary reservoirs, as has been argued by some workers (see Chapter 4, Section 4.5.1.2). Second, there must be at least one more terrestrial reservoir, with a low U/Pb ratio, which plots to the left of the Geochron line. Current solutions to the future Pb-paradox include (1) a low U/Pb reservoir in the lower continental crust (Kramers & Tolstikhin, 1997), (2) the partitioning of Pb into the core, with the effect that the Bulk Earth composition no longer lies on the Geochron (Elliott et al., 1999) and (3) low U/Pb in subducted oceanic crust (Murphy et al., 2002).

A further implication of the mantle evolution curve shown in Fig. 3.30, is that mantle compositions which plot to the right of the geochron imply time integrated U/Pb ratios in the MORB-mantle higher than those in the BSE. This means that there must have been an increase in the U/Pb ratio of the mantle over time (White, 1993). This is the opposite of what might be expected, since it is widely accepted that the continental crust, with a high U/Pb ratio, was extracted from the mantle. A resolution to this dilemma can be found by concluding that the upper mantle has not had its high U/Pb for very long. This leads to a dynamic view of the mantle in which the upper mantle is replenished with fluxes of high U/Pb whilst at the same time losing U to the continental crust. The mechanisms through which this is achieved have been long debated.

3.2.3.3.3 HIMU mantle

There is a further mantle source, sampled by some OIBs, which is characterized by high $^{238}U/^{204}Pb$ ratios. This ratio is known as the μ-value (Greek letter mu), hence the term HIMU given to this mantle reservoir (Zindler & Hart, 1986). One model for the origin of this HIMU source is that it is ocean crust which was altered during hydrothermal activity at a midocean ridge, subducted, and then chemically isolated within the mantle. In this case the presence of HIMU mantle is evidence for the deep recycling and long-term storage of subducted oceanic crust. An alternative model for the origin of HIMU mantle is that it is the product of mixing between depleted mantle and melts of metasomatized mantle (Kamber & Collerson, 1999).

3.2.3.3.4 The Th/U ratio of the mantle

Further insight in understanding the Pb-isotopic evolution of the mantle comes from the study of the isotope ^{208}Pb, the product of ^{232}Th decay. Galer and O'Nions (1985) showed that the time integrated U/Th ratio of the mantle, calculated using Pb-isotopes, is about 3.8. However, this ratio does not agree with present-day measured elemental ratios, for in the modern MORB-source the U/Th ratio is about 2.6. This paradox has been dubbed the kappa conundrum after κ, the Greek symbol for the $^{232}Th/^{238}U$ ratio.

The solution to the kappa conundrum has to be that the mantle is an open system with respect to U and Th and that its Th/U ratio has changed with time, reducing from the bulk Earth value of about 4.0 to the present-day value of about 2.5 and with a time-integrated value of 3.8. There are a number of different

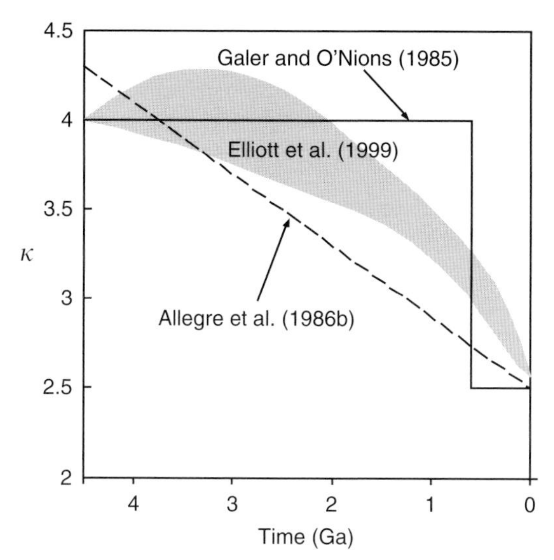

FIGURE 3.31 Evolution curves for the Th/U evolution of the Earth's mantle. The plot shows the change in κ ($^{232}Th/^{238}U$) values with geological time. Three models are shown, the stepwise model of Galer and O'Nions (1985) (solid line), the secular evolution model of Allegre et al. (1986b) (dashed line) and solutions to the U-recycling model of Elliott et al. (1999) (shaded field).

models to explain this change in ratio and these are illustrated in Fig. 3.31.

Galer and O'Nions (1985) were the first to recognize the κ-conundrum. They proposed a stepwise model whereby, through much of Earth history, the mantle had a nearly constant κ with values equivalent to that of the BSE and only within the last 600 Ma was the ratio reduced to its present value. They suggested that the mantle operated as an open system in which there was an exchange of U and Th between the upper and lower mantle. In this two-layer mantle model, the removal of U and Th from the upper mantle to the crust by partial melting is counterbalanced by the flux of high-κ lower mantle into the upper mantle (Fig. 3.31).

Allegre et al. (1986b) proposed a linear, secular decrease in Th/U with time from an initially high bulk Earth value of 4.3 to the present-day value of 2.6. They linked this decrease to the extraction of the continental crust from the mantle. In an alternative approach, Elliott et al. (1999), following Zartman and Haines (1988) suggested a model which is a hybrid of the previous two. In this model the Th/U of the mantle remained close to the bulk Earth value during the Archaean and then declined to its present value. The model of Elliott et al. (1999) is based upon two important assumptions. First, that the Th/U ratio of the mantle declines because of the extraction of the continental crust from the mantle. Second, that in the post-Archaean period, with increasing oxidizing conditions (see Chapter 5, Section 5.3.1), crustal U was recycled back into the mantle, thereby accelerating the decrease in Th/U ratio (Collerson & Kamber, 1999; Xie & Tackley, 2004). The calculations of Elliott et al. (1999) shown in Fig. 3.31 successfully satisfy the constraints of a bulk Earth Th/U ratio of 4.0, a modern MORB ratio of 2.5 and a time-integrated U/Th ratio for the MORB source of 3.8 and fit with measured mantle Th/U ratios in kimberlitic zircons since 2500 Ga (Zartman & Richardson, 2005). The model also predicts κ-values greater than Bulk Earth in the early Archaean (Fig. 3.31), in agreement with measurements on early Archaean rocks (Frei & Rosing, 2001). In this model the decrease in mantle Th/U began about 2.0 Ga ago.

3.2.3.4 The Osmium isotopic evolution of the mantle

Os is one of the PGEs and the isotope ^{187}Os is the product of ^{187}Re decay, with a half-life of 41.6 Ga. For recent review of this isotopic system see Carlson (2005). Both Re and Os behaved as refractory and siderophile elements during the accretion of the Earth, and so their principal abundances are expected to be in the Earth's core. Hence, the best estimates of the initial $^{187}Os/^{188}Os$ composition of the solar system are derived from iron meteorites (Smoliar et al., 1996). Nevertheless, the $^{187}Os/^{188}Os$ *ratio* of the silicate Earth is thought to be chondritic. The chondritic reference values most commonly used are those of Shirey and Walker (1998) (see Fig. 3.32), although the appropriateness of these values is debated, for there are differences of up to 2.0 γ_{Os}-units between the present-day isotope

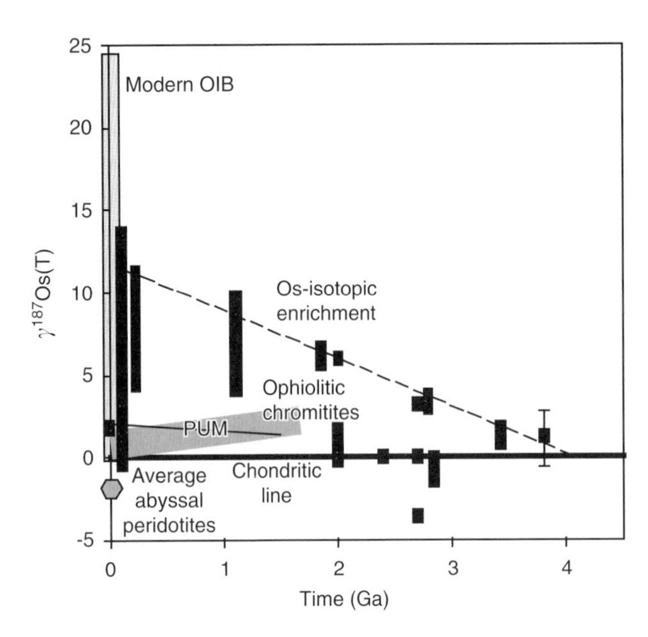

FIGURE 3.32 γ^{187}Os(T) plot versus time for mantle-derived samples. This plot shows that some mantle-derived samples have been derived from a chondritic mantle source, whereas others are from an enriched source. A single sample lies below the chondritic line at 2.7 Ga, and may have been derived from Re-depleted oceanic lithosphere. [γ^{187}Os(T) is the percentage deviation from the chondritic reference value at the time of separation from the mantle.] Present-day chondritic reference values from Shirey and Walker (1998) – ^{187}Re/^{188}Os = 0.40186, ^{187}Os/^{188}Os = 0.1270, λ ^{187}Re = 1.666×10^{-11} yr^{-1}. Data sources (black boxes and squares) are mostly from the compilation in Walker and Nisbet (2002) with additions from Frei and Jensen (2003) and Bennett et al. (2002). Ophiolitic chromitites are the grey band sloping to 0 γ-Os, whereas Primitive Upper Mantle (PUM) is the thin black line with a negative slope.

ratios of the different chondrite groups (Walker et al., 1999). γ_{Os} is the percentage deviation from the chondritic reference value at the time of separation from the mantle.

During mantle melting Re is mildly incompatible (tends to enter the melt phase) whereas Os is strongly compatible (i.e. remains in the unmelted residue). This leads to high Re/Os ratios in oceanic crust and relatively low Re/Os ratios in the depleted mantle. In contrast, in felsic melts Re and Os are both incompatible, and so their concentrations are very low in the continental crust making this a very different isotope system from the Nd-, Hf- and Pb-systems. These geochemical properties mean that the Os-isotopic evolution of the upper mantle will reflect the extraction of basaltic melts but is unlikely to have any memory of the formation of the continental crust. Mantle melts which have supra-chondritic Os-isotopic ratios are thought to be derived by the melting of ancient, high Re/Os, basaltic crust, which must have been recycled back into the mantle and stored there, making Os-isotopes an important "tracer" of recycled ancient oceanic crust.

3.2.3.4.1 Estimates of the Os-isotopic composition of the upper mantle

For many isotope systems the composition of the upper mantle can be estimated from the narrow compositional range of midocean ridge basalts. In the case of Os-isotopes however, MORB samples have a wide range of ^{187}Os/^{188}Os ratios and identifying a representative depleted upper mantle composition is less straightforward. The great variability in MORB Os-isotope ratios can be caused either by seawater contamination or by the rapid radiogenic ingrowth of ^{187}Os in melts which have a high initial Re-content. For this reason a number of other methods have been used to estimate the composition of the upper mantle and these are outlined below.

3.2.3.4.2 Ophiolitic chromites

The mineral chromite tends to have high Os and low Re concentrations and is resistant to chemical change after crystallization. It is therefore a robust phase and because of its low Re-content does not require a large age correction in ancient samples. Massive chromitites are commonly found in ophiolites and so provide a reliable record of the Os-isotope composition of the upper mantle. Walker et al. (2002) found that the Os-isotope composition of chromitites from ophiolites formed over the last 1.2 Ga define a linear trend, with a slight positive slope and with a present-day intercept at 0.12809 (γ_{Os} = 0.9) (Fig. 3.32). The close similarity between this value and chondritic values supports the view that the convecting upper mantle has a chondritic Os-isotopic ratio.

3.2.3.4.3 Xenoliths from the subcontinental lithosphere and the composition of the primitive upper mantle

Many mantle xenoliths, derived from the subcontinental lithosphere show a correlation between an indicator of melt depletion such as Al_2O_3 and their $^{187}Os/^{188}Os$ isotope ratio. Rather surprisingly, samples of Proterozoic subcontinental mantle from many different continents plot on trends which intersect at a common point implying that they were derived from a compositionally homogeneous source that had not experienced melt-depletion. This undepleted source has been designated the PUM (Meisel et al., 2001). The present-day composition of the PUM is estimated to be 0.1296 (γ_{Os} = +2.0) and this value has increased slightly over geological time (Fig. 3.32). The significance of this result is unclear, for the subcontinental lithosphere is a comparatively small mantle reservoir with a complex history of melt extraction and migration and unlikely to be representative of the upper mantle as a whole.

3.2.3.4.4 Abyssal peridotites

Os-isotopic measurements on abyssal peridotites – the residues of recent midocean ridge basalt melting – show a wide range of present-day compositions ($^{187}Os/^{188}Os$ = 0.120–0.129, γ_{Os} = −5.5 to +1.6) with an average $^{187}Os/^{188}Os$ ratio of 0.1247 (Snow & Reisberg, 1995). Meibom et al. (2002), working on peridotite-derived platinum-group-element-rich alloys, described a similar heterogeneity and an almost identical average $^{187}Os/^{188}Os$ ratio of 0.1245. They argued that this chemical heterogeneity reflects a real variability in the mantle which they suggest is the product of random mixing between a radiogenic and a nonradiogenic end-member. They propose that this result supports the Statistical Upper Mantle Assemblage (SUMA) model of the mantle advocated by Meibom and Anderson (2003) – see Section 3.3.2.2 – also known as the "plum pudding" model of the mantle.

The range of values measured on mantle peridotites overlaps with other estimates for the composition of the convecting upper mantle such as PUM and ophiolitic chromites. Nevertheless, the mean γ_{Os} value of ca. −2.0 differs from the estimate based on ophiolitic chromitites by 3.0 γ_{Os} units and from PUM by about 4.0 γ_{Os} units (Fig. 3.32). This result does not sit easily with a simple chondritic model for the upper mantle.

3.2.3.4.5 Enriched Os-isotopic reservoirs

The data in Fig. 3.32 show that a combination of samples of different ages, mostly derived from plume-related magmas, have strongly supra-chondritic Os-isotopes ratios. This features is seen in many, but not all, Archaean komatiitic lavas and in modern ocean-island basalts. Evidence for this isotopic enrichment has been strengthened recently by studies of a second isotope of osmium – ^{186}Os, the product of the radioactive decay of ^{190}Pt. Combined $^{186}Os/^{188}Os$, and $^{187}Os/^{188}Os$ measurements are now used to study the fractionation of the highly siderophile elements pairs Pt–Os and Re–Os in the Earth. Some modern plume related magmas are simultaneously enriched in $^{186}Os/^{188}Os$, and $^{187}Os/^{188}Os$ and are derived from a source with supra-chondritic Pt/Os and Re/Os ratios (Brandon et al., 2003). This source is likely to be ancient, simply because the long half-lives of the parent isotopes need a long period of time to develop the appropriate isotope ratios. Two different explanations for the origin of such an enriched source have been proposed. One idea is that the Pt/Os and Re/Os fractionation took place within the Earth's core and that mantle samples with elevated Os-isotope ratios may have had an extremely deep origin and interacted with the Earth's outer core. However, Schersten et al. (2004) have shown that such a view is inconsistent with new tungsten isotope studies.

Alternatively, enriched Os-isotope signatures might be acquired from ancient subducted oceanic crust and associated pelagic and ferromanganese materials, now residing in the mantle (Baker & Jensen, 2004).

3.2.3.4.6 Early Archaean data

The earliest terrestrial $^{187}Os/^{188}Os$ data come from the early Archaean (3.81 Ga) chromitites of the Itsaq Gneisses in west Greenland (Bennett et al., 2002; Rollinson et al., 2002). These samples show a range of compositions but with a mean value which is slightly suprachondritic (Frei & Jensen, 2003). This could indicate that the enriched Os-isotope reservoir discussed above was in existence even in the very early Archaean and indicates the very early recycling of basaltic crust.

3.2.3.4.7 The Os-isotopic evolution of the mantle

It is clear that the Earth's mantle has at least two Os-isotopic reservoirs – a plume-related isotopically enriched reservoir and a chondritic upper mantle reservoir. Both have long histories (Fig. 3.32). The variations in composition within the upper mantle reservoir reflect Re-depletion and enrichment related to melt extraction. The isotopically enriched plume reservoir represents chemically isolated, rhenium-enriched, recycled oceanic lithosphere. There is some evidence to suggest that this enriched reservoir may have been in existence since the early Archaean (Walker & Nisbet, 2002) and was the source of some Archaean komatiites and the 3.81 Ga Itsaq Gneiss chromitites. If this is true, then basaltic crust was being created and recycled even before 4.0 Ga. Estimates of the present size of this high Re/Os basaltic reservoir vary from 5% to >10% of the whole mantle (Bennett et al., 2002; Walker et al., 2002).

3.2.3.5 The isotopic evolution of the Earth's mantle through time

Each of the different radiogenic isotope systems discussed above provides its own insight into the history of the Earth's mantle. Here these different threads are brought together to provide an overview of mantle evolution through the history of the Earth.

As an aside, it is also worth noting that a close look at the isotopic evolution of the Earth's mantle challenges the chondritic model for the Earth. This is particularly clear for the Lu–Hf and Re–Os isotope systems, where there is uncertainty about which meteorite type is the most appropriate starting point for the isotopic evolution of the Earth. Recent studies of chondritic Sm–Nd ratios by Boyet and Carlson (2005) further support the nonchondritic view of the bulk Earth.

A study of the radiogenic isotope memory of the Earth's mantle clearly shows that the mantle is neither an independent part of the Earth system nor has it been for a long time. But rather, it records a history of the extraction and recycling of both basaltic and continental crust. This raises two very important questions. First, is our isotopic record of the mantle representative of the whole mantle or only the upper mantle? Second, is there any primitive, undifferentiated mantle still preserved within the large mass of the differentiated Earth's mantle?

Below, four important stages in the history of the Earth's mantle are briefly described.

3.2.3.5.1 I – The pre-4.5 Ga differentiation of the mantle

Studies of the short-lived isotope ^{142}Nd in primitive chondritic meteorites have recently shown that the Earth's mantle does not have a chondritic ^{142}Nd–isotope ratio (Boyet & Carlson, 2005). The implication of this finding is that the Earth experienced a major differentiation event in which an Fe-rich, trace element-enriched basaltic crust formed on a magma ocean. This crust has subsequently been removed and isolated from the convecting mantle and may now be represented by the D″ layer (Tolstikhin & Hofmann, 2005). Similar ^{142}Nd results on lunar samples suggest that this differentiation took place before the formation of the Moon, that is, within 30 Ma of the formation of the solar system (Boyet & Carlson, 2005).

3.2.3.5.2 II– Early Archaean mantle differentiation related to the extraction of basaltic crust

There is an increasing body of evidence to show that there was a very early differentiation event in the mantle. Hints of this process come from the ^{143}Nd-, Pb- and Hf-isotope studies outlined above. Os-isotopes indicate that this differentiation event led to the formation of an early basaltic crust, and Walker and Nisbet (2002) used the Os-isotopic composition of Archaean komatiites to estimate that there must have been about 5% of recycled basaltic crust in the mantle by 4.2 Ga. Recent ^{142}Nd evidence from Caro et al. (2005) also supports this view and shows that the mantle differentiated between 50 and 200 Ma after the formation of the solar system and that this differentiated reservoir was preserved until between 3.6 and 3.8 Ga when it was sampled by metabasalts and tonalitic orthogneisses, now preserved in west Greenland. The modeling of Caro et al. (2005) shows that, paradoxically, the early mantle seems to have been able to maintain an isolated, depleted reservoir and yet experienced a much shorter stirring time than the modern mantle, implying that convective processes in the Hadean and early Archaean mantle were different from those of the present day.

3.2.3.5.3 III – Mantle differentiation related to the extraction of the continental crust

The ^{143}Nd-isotope record of the Earth's mantle provides a record of the progressive extraction of the continental crust, from about 3.0 Ga to the present-day (Fig. 3.27). Continental crust is created from the mantle in two steps. First, basaltic crust is produced by the partial melting of the mantle, leaving a depleted mantle residue with a positive ε_{Nd} signature. Second the newly-formed oceanic crust is either melted or dehydrated during subduction to trigger continental crust formation.

The link between the creation of oceanic crust and the formation of continental crust is important and needs to be explored. On the one hand there is Os-isotope evidence to suggest the formation of basaltic crust as early as 4.2 Ga, and on the other hand basaltic crust is not "processed" on a large enough scale to form continental crust until 3.0 Ga. This seems to suggest that at 3.0 Ga the nature of mantle processes changed from "ancient" to modern (Fig. 3.33). This time interval may mark the inception of the subduction process, and even large-scale mantle convection. Prior to this time, it is possible that basaltic crust was returned to the mantle by gravitational processes but in a more localized manner.

3.2.3.5.4 IV – The beginning of recycling of crustal material into the mantle

A further stage in the history of mantle evolution is the return into the mantle of material previously fractionated into the continental crust. This stage is possible only after a significant volume of continental crust and continent-derived sediment have been formed and the subduction process established. The principal evidence for crustal recycling comes from solutions to the second Pb-paradox, the kappa conundrum. Crustal recycling explains

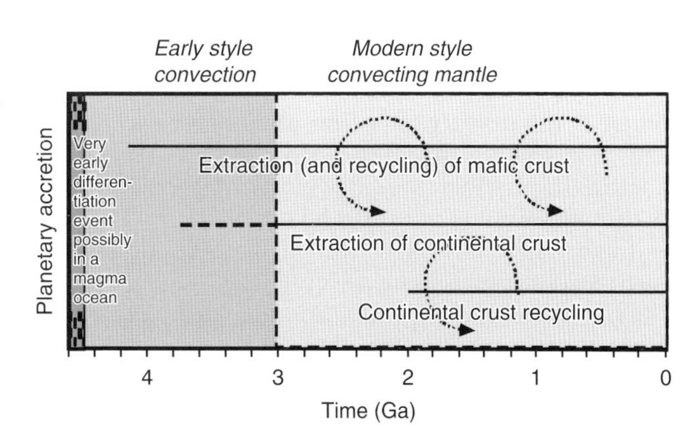

FIGURE 3.33 Time-line of mantle evolution showing the onset of mafic and continental crust extraction and recycling, based upon radiogenic isotope studies.

the disparity between Th/U ratios in the Bulk Earth and in the present-day convecting mantle and may also explain the secular increase in the U/Pb ratio observed in the MORB source. The model of Elliott et al. (1999) is based on the recycling of U into the mantle from about 2.0 Ga and links the increased mobility of U to the rise of atmospheric oxygen at this time (Collerson & Kamber, 1999).

3.3 Mantle models

Our view of the mantle is in a state of transition. Earlier in this chapter we explored the current debate about mantle convection, the evidence for a two-layer mantle and the evidence for a deep mantle source for mantle plumes (Section 3.1.5). It is now becoming clear that geophysicists have persuasive evidence of mantle-scale mass transfer, which challenges the two-layer view of the mantle prevalent amongst geochemists. This is forcing geochemists to reexamine their assumptions about the sources of basaltic magmas and areas of uncertainty previously ignored are now being explored in some detail.

Any successful model of the mantle therefore must satisfy the inferences from seismic tomography about deep slab subduction, be capable of producing the varied geochemical signatures of the different types of ocean basalt, produce enough heat to account for the global heat flux, and be dynamically consistent (Kellogg et al., 1999). At present there are two broad schools of thought about the structure of the mantle. They were neatly summarized by Carlson (1988) with the terms "layer cake-" and "plum pudding-" mantle models. Those who adhere to the layer cake model see the mantle as distinctly chemically stratified, whereas the devotees of plum puddings view the mantle as a pudding with "plums" of chemically distinct source regions randomly distributed throughout the whole mantle.

In the early history of the Earth there are further issues to be addressed. These include the likelihood of a magma ocean, the nature of early mantle convection, the timing of the onset of modern mantle convection and the timing of the onset of subduction. Thus, in this next section we first consider some contemporary models which seek to explain how the modern mantle works. Then, in the light of these insights we discuss models for the Archaean mantle.

3.3.1 Layer-cake models for the modern mantle

Until recent years the physical contrasts above and below the 660 km discontinuity were regarded as the critical evidence in support of two layered mantle convection (Fig. 3.17). Now, however, there is strong evidence from seismic tomography that subducted slabs penetrate through this discontinuity and descend, in some cases, deep into the lower mantle. Mineral physics has also demonstrated that the discontinuity is isochemical, and can be explained simply by phase transformations, and that the Earth's mantle is not chemically layered. Thus geochemists have had to reconsider their view that the mantle below the 660 km discontinuity is a geochemically and isotopically distinctive source for oceanic basalts and chemically different from the upper mantle. As a result of this thinking a number of new models for the mantle have emerged. Several of these models have preserved the concept of a chemically layered mantle but have shifted the upper-lower mantle boundary to a deeper level in the Earth.

3.3.1.1 The lava-lamp model

The lava-lamp model for the Earth's mantle (Albarede & van der Hilst, 1999; Kellogg et al., 1999) takes its name from that household curiosity of the 1960s based upon the fluid-dynamic properties of liquids with near identical densities. The model is based upon seismic observations of an irregular, deep boundary layer in the mantle at 1,600–2,000 km depth.

In this model the mantle is differentiated, into an upper layer (1,600–2,000 km thick), and a compositionally distinct lower layer, enriched in iron and heat producing elements. This lower layer, making up 20–30% of the mantle, is thought to have originated either during the early differentiation of the Earth or by the burial of subducted oceanic crust.

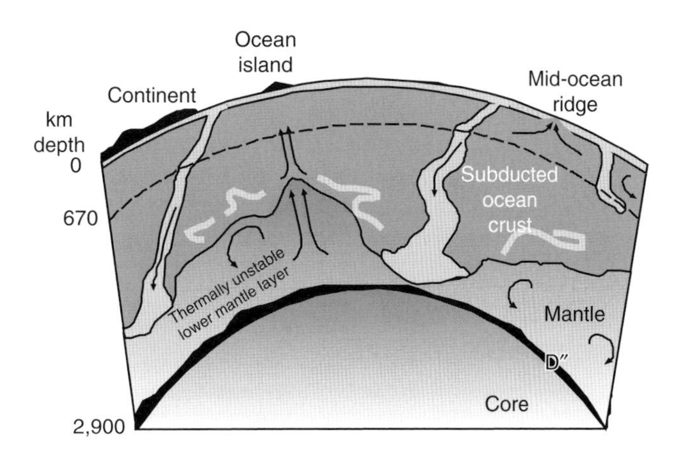

FIGURE 3.34 The lava-lamp model for the Earth's mantle (after Kellogg et al., 1999).

The layer is internally heated from its radioactive element content leading to a thermal expansion such that there is almost no density difference between the upper and deep mantle. This gives rise to a boundary between the upper mantle and deep mantle which is highly irregular. In places where dense lithospheric material descends the lower mantle layer is thin, and elsewhere, there are deep mantle upwellings which rise though the upper mantle, to produce "plume related" ocean island basalts (Fig. 3.34).

This model adequately incorporates the observations from seismic tomography of deep subduction, whilst keeping the geochemical distinctiveness of long-lived deep (OIB) and shallow (MORB) mantle geochemical reservoirs. Support for the model comes from recent tomographic maps showing large-scale chemical heterogeneities throughout the lower mantle (Trampert et al., 2004). However, the model is susceptible to the outcome of the current debate about the deep mantle origin of plumes.

3.3.1.2 The D″ model of Tolstikhin and Hofmann

In order to accommodate whole mantle convection into a layered mantle model Tolstikhin and Hofmann (2005) and Tolstikhin et al. (2006) have shifted the critical layering of the mantle to the base of the lower mantle, to the D″ layer. Again, this approach preserves both deep-mantle subduction and at least two geochemically distinct mantle reservoirs. They propose that the D″ layer contains about 20% of the inventory of the silicate Earth's heat producing elements and that it is enriched in iron, rare gases, and incompatible elements. Fluxes from the D″ layer are seen in plume magmas and, to a lesser extent, some D″ material may be entrained in the mantle by convective flow.

This model is the development of an idea originally proposed by Hofmann and White in 1982. In its original form this model was proposed as an alternative to the then prevalent view that OIB is from a primitive mantle source. In it mantle plumes were derived from subducted oceanic lithosphere, accumulated and stored at the core–mantle boundary.

In this present version of the model the D″ layer is thought to have originated very early in Earth history, as an early, incompatible element- and metal-rich basaltic crust, enriched during late accretion (4,540–4,000 Ma) with chondritic material. There is support from Nd- and Hf-isotopes for the existence of this very early differentiate of the mantle (see Sections 3.2.3.1 and 3.2.3.2). This crust, when subducted, had a bulk density which exceeded that of the mantle and numerical modeling experiments confirm that it would have stabilized at the core–mantle boundary (Davies, 2006).

3.3.1.3 The water filter model

Bercovici and Kurato (2003) have proposed a model for whole mantle convection which is

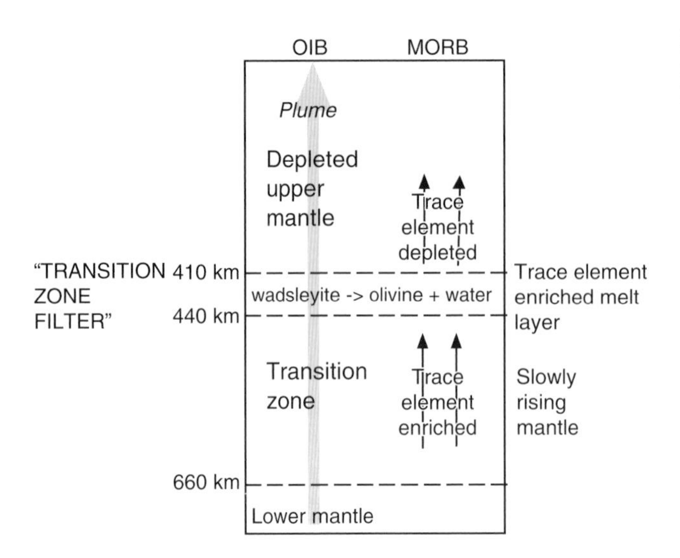

FIGURE 3.35 The water filter model for the Earth's mantle (after Bercovici and Kurato, 2003; Hofmann, 2003).

consistent with geochemical observations by suggesting that the mantle transition zone acts as a filter to mantle melts rising from deep within the Earth (Fig. 3.35). This model is predicated upon a wet mantle, and is based upon the observation that the mantle transition zone (between 410 and 660 km) is both a significant phase boundary in the mantle and is marked by a high water content (0.2–2.0 wt% water), due to the ability of the high-pressure polymorphs of Mg_2SiO_4, wadsleyite (β-Mg_2SiO_4) and ringwoodite (γ-Mg_2SiO_4) to store water in their lattice. Some support for a high water content in the mantle transition zone comes from recent conductivity studies which show a higher than normal water content in the transition zone beneath the Pacific (Huang et al., 2005) and evidence of melting in the mantle transition zone beneath the Andes (Booker et al., 2004).

The model also assumes an incompatible trace element-rich lower mantle and a depleted upper mantle. Bercovici and Kurato (2003) propose that when deep mantle material rises the mineral wadsleyite transforms to olivine (Fig. 3.1), releasing water. The liberated water triggers melting in the transition zone to produce a melt which is rich in water and incompatible trace elements. This melt layer, which is recycled in either a solid or molten state in the transition zone, acts as a chemical filter to all rising mantle material, ensuring that any upwelling mantle is chemically depleted. The exception to this process is a mantle plume whose higher temperatures suppress the solubility of water in wadsleyite allowing it to rise through the "filter" in an unmodified enriched state.

The water filter model implies that the source of midocean ridge basalts is the lower mantle and that it is depleted in the transition zone, as it rises. It adequately explains the geochemical differences between MORB and OIB but still requires deeply sourced mantle plumes. One problem that this model does not resolve however, is that of the isotopic difference between MORB and OIB (see Fig. 3.5b). For the removal of trace elements in the mantle transition zone by small fraction melting by the "water filter process" does not alter isotope ratios and will not therefore impart to MORB and OIB their isotopic distinctiveness (Hofmann, 2003).

3.3.2 A "plum pudding" model for the mantle

An alternative to the layer-cake model for the mantle is that it is like a plum pudding in which chemically distinct regions are scattered throughout the mantle-like plums in a pudding (Carlson, 1988). Previously Allegre and Turcott (1986) had used the alternative culinary analogy of "marble cake" to describe the same feature. The heterogeneities implied in this model occur on a variety of scales and represent high viscosity "blobs" in a lower viscosity matrix. It is the viscosity difference between "blobs" and the host mantle that helps to preserve the chemical differences, even in a stirred mantle. The density, shape, size, and depth in the mantle of these heterogeneities can vary, and they may take the form of elongate layers within the mantle as well as elliptical blobs.

Given the great antiquity of the Earth's mantle there is plenty of time available for mixing. In addition there is scope for large variations in temperature so that a range of mixing scenarios can be envisaged whereby different mantle domains remain variously incorporated into or distinct from one another. Some authors have argued that heterogeneities of this type will be destroyed by convective movements in the mantle, although Albarede (2003) has pointed out that the concept of heterogeneity is largely dependent upon the scale of sampling, which in the case of the mantle reflects the size of the region sampled during a particular melting event. In addition, complex patterns of heterogeneity do not necessarily indicate imperfect mantle mixing and a lack of convection but rather reflect the random redistribution of material by the convection process.

A "plum pudding" view of the mantle is an alternative marriage of both geophysical and geochemical data for the Earth's mantle. In this model, deep subduction through geological time has progessively added enriched material to the lower mantle, creating two or more chemically distinct reservoirs located randomly throughout the whole mantle.

3.3.2.1 The Helffrich and Wood whole mantle convection model

The principal observation behind the Helffrich and Wood (2001) model of the mantle is that the scattering of seismic waves in the lower mantle indicates that small heterogeneities are present, mostly about 4 km in size. As with all seismic methods the physical cause of the seismic scattering anomalies could theoretically be either thermal or compositional in origin. Helffrich and Wood (2001) argue that the anomalies are chemical, because diffusivity rates show that they are too small to retain their thermal identity for longer than about 200,000 years. They conclude that they are most probably fragments of subducted lithospheric slabs.

In this model, mantle heterogeneity is the product of subduction which, over geological time, has recycled back into the mantle "enriched" oceanic crust, together with a small amount of continental crust as sediment and depleted (sterile) oceanic lithosphere. During the process of subduction slabs break up and the enriched, oceanic crust component becomes preserved as geochemically distinct domains. That is, as the "plums" in the upper and lower mantle. Helffrich and Wood (2001) used a mass balance calculation based on the heat producing element content of the bulk silicate Earth and other incompatible trace elements to show that the whole mantle is made up of about: 9% recycled oceanic crust, 5% ancient recycled oceanic crust (dating from a very early differentiation event in the Earth's mantle), 0.3% recycled continental crust, and about 85% "sterile," highly depleted mantle. MORB and OIB are derived by the melting of depleted mantle which comprises a mixture of sterile mantle and recycled oceanic crust. The differences in MORB and OIB chemistry can be explained in the different contribution of recycled oceanic crust to the melting.

It was shown earlier (Section 3.1.6) that the present mass flow within the mantle is sufficient to account for the recycling of the entire mantle over the last 4.0 Ga. Thus Helffrich and Wood (2001) argue that more than 90% of the mantle has been recycled, that is, processed through a midocean ridge, so that virtually the whole mantle is a mix of chemically enriched "plums," within a matrix of sterile "pudding."

There is some geochemical support for this model as it has been long recognized that the isotopic character of the HIMU ocean island basalt source is consistent with an origin by oceanic crust subduction. There is also some evidence that such sources may be upto 2.0 Ga old (Salters & White, 1998). Further, a recent geochemical-diffusion study of ocean island basalt sources showed that bodies as small as a few meters wide might be preserved for as long as 1.0 Ga in the convective mantle (Kosigo et al., 2004). However, not all geochemical arguments support the idea of oceanic crust subduction as the source of OIB. The new Li-isotope evidence of Elliott et al. (2004) appears to show that OIB does not have a recycled MORB signature and Niu and O'Hara (2003) have argued in a similar way from trace elements and believe that OIB is derived from metasomatized oceanic lithosphere.

3.3.2.2 The SUMA mantle model

Further geochemical support for a "plum pudding" mantle comes from the work of Meibom et al. (2002) and Meibom and Anderson (2003). However, unlike Helffrich and Wood (2001) these authors apply their model only to the Earth's *upper* mantle. Meibom and Anderson (2003) argue on isotopic grounds that the geochemical differences between OIB and MORB have been overplayed and propose that the chemical heterogeneity of the mantle is actually a function of the sampling of the mantle during partial melting. Thus in this model separate MORB and OIB reservoirs are not necessary. The SUMA model views the mantle as heterogeneous on length-scales varying from 100 m to 100 km – the Statistical Upper Mantle Assemblage. The heterogeneities are created by recycling and plate tectonics. When this assemblage is melted it is averaged in a number of different ways, depending upon exactly what is melted and in what proportions, so that sampling upon melting and averaging takes place in different ways to produce OIB and MORB.

A central point in the argument of Meibom and Anderson (2003) is that a lower mantle source is not necessary as a source for OIB. Hence all the processes of partial melting and sampling of the mantle that they describe can take place in the upper mantle. However, there is no reason in principle why the Statistical Upper Mantle Assemblage model of the mantle cannot be applied to the whole mantle.

3.3.3 Archaean mantle models

The layer-cake versus plum-pudding debate about the nature of the mantle is unresolved and both sides have strong advocates. What is clear, however, is that the 660 km discontinuity in the mantle is not an impenetrable barrier to subduction. Whether or not compositional layering exists at a deeper level in the mantle is not yet established. Certainly, the plum pudding model has its strengths, particularly since it is linked to well-established dynamic processes, processes which have probably operated throughout much of Earth history.

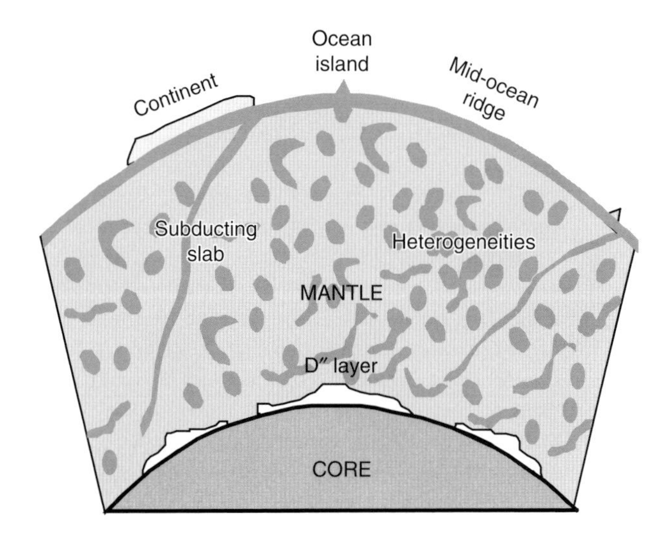

FIGURE 3.36 A "plum pudding" model for the mantle, in which subducted oceanic lithosphere becomes mixed by convective stirring into the mantle and so introduces chemical heterogeneities, rather like "plums in a pudding." In many cases the heterogeneities are too small to be observed seismically. (After Helffrich & Wood, 2001.)

Now however, using our understanding of modern mantle models, it is important to turn to mantle models for the early Earth, in order to gain an understanding of how the earliest mantle functioned. Here we not only need to make use of our knowledge of the modern mantle and the history it contains but also to include processes which took place early in Earth history, whose identity may be all but destroyed by modern mantle processes.

3.3.3.1 Layered mantle models for the early Earth

Modern mantle models of the "layer-cake type" require that the Earth's mantle became layered at some stage in its history. Thus far in this discussion the existence of this layering has been accepted although its origin has not been discussed. A number of mechanisms have been proposed through which a layered mantle might have formed during the very early stages of mantle evolution in the early Archaean and these are now discussed.

3.3.3.1.1 Layering as a result of a Hadean magma ocean

As the Earth accreted, high-energy impactors would have triggered internal melting and favored the formation of a magma ocean. Such a magma ocean could have differentiated by the processes of crystal fractionation or melt compaction to give a stratified mantle in which there was a layer comprising an accumulated crystal pile and a separate layer comprising the residual mantle.

3.3.3.1.2 A model based on olivine fractionation

One of the first indications that the mantle might have acquired some stratification in a magma ocean, developed during the 1980s from an understanding of komatiitic magmas. This model focused on the observation that at the pressures and temperatures of komatiite generation olivine is less dense than komatiitic magma and will float on a komatiitic magma layer (Nisbet & Walker, 1982). This led to a model for the mantle in which there was a 130 km thick olivine-rich refractory cap underlain by a stagnating and differentiating komatiitic magma ocean, about 90 km thick. In the last two decades models of komatiite generation have changed and our understanding of magma oceans has improved; nevertheless the concept of a layered mantle produced in a magma ocean is still very much alive.

3.3.3.1.3 Models based on perovskite fractionation

Our understanding of the mineralogy of the lower mantle has progressed in recent years and now models of mantle layering center around the fractionation of the principal lower mantle mineral – Mg-perovskite in a Hadean magma ocean. In this model a magma ocean was created during accretion and probably crystallized from the bottom up, to produce a perovskite-rich crystal pile in the lowermost mantle, a basaltic "crust" and a "residual mantle" component. In recent years the evidence for a Hadean magma ocean has increased and identifying a geochemical signal of perovskite fractionation has become the quest of recent experimental studies on lower mantle minerals. Tantalizing clues lie in the interpretation of nonchondritic element ratios in PUM rocks. Walther et al. (2004) used a mass balance approach to calculate whether there is sufficient chemical difference between a primitive peridotitic BSE and the composition of the present convecting mantle (based on undepleted peridotites) to allow for the existence of an additional reservoir made up of lower mantle minerals. Their calculations showed that an additional reservoir is permissible and that the composition of the present convecting mantle is consistent with the removal of 10–15% of the mineral assemblage 90% Mg-perovskite + 10% (Ca-perovskite + ferropericlase). In a related study Corgne et al. (2005) emphasize the importance of the minor phase Ca-perovskite in controlling the trace element mass balance in this process. They argue that only 8% fractionation is permissible without disturbing the trace element concentrations of the present convecting mantle. The resulting crystal pile would be enriched in U and Th and therefore be a significant reservoir of heat producing elements. It would also have high Sm/Nd, U/Pb, Sr/Rb and Lu/Hf, similar to the character of the mantle HIMU reservoir, identified from some ocean island basalts.

These calculations lead to us to conclude that there may well be a rheologically strong, chemically isolated layer made up of cumulus phases from the lower mantle. This lower mantle layer might be identified with the D″ core–mantle boundary layer, or the deep mantle Fe-rich layer identified by Kellogg et al. (1999) and described above (Section 3.3.1).

3.3.3.1.4 The subduction of Hadean basaltic crust

A number of models for the very early Earth propose the existence of a Hadean (ca. 4.5 Ga) basaltic crust. The evidence for such a model comes from the study of ^{142}Nd and is consistent with the observed extreme fractionation of ^{207}Pb (Kamber et al., 2003), and with Hf-isotope studies on zircons (Bizzarro et al., 2003). For this reason it has been proposed that an early, trace element enriched mafic crust, created perhaps as a lid to a Hadean magma ocean, was subducted and buried deep in the mantle, where it is thought to be still stored (Galer & Goldstein, 1988, 1991). Hence, this material is a candidate for a lower mantle layer and is another possible explanation for a layered mantle created very early in Earth history.

3.3.3.1.5 Layering due to the progressive in-mixing of volatiles into the mantle

It was shown earlier that the presence of volatiles in the Earth's mantle, particularly the presence of water, has a profound effect on mantle viscosity. On this basis McCulloch (1993) argued that the pattern of mantle convection would change over time as the water content of the mantle changed. The basis of this model is that in a hotter Archaean mantle the normal mode of subduction would be that of young warm oceanic crust. Such crust would not penetrate deep into the mantle but would melt during subduction losing its volatiles to the surrounding mantle, leading to a hydrated, low viscosity layer in the upper 200–400 km of the mantle.

In this view of the early Archaean mantle there would be an upper, low viscosity hydrous layer and a deeper, dry layer with no deep transport of volatiles. Underlying the hydrated layer would be a convecting mantle transition zone and beneath that, a convecting lower mantle. McCulloch (1993) proposed that over Earth history, as the planet cooled, the mode of subduction changed, eventually giving rise to the deeper penetration of subducted slabs and a more thorough rehydration of the mantle. In this model a hydrous mantle layer has progressively extended to deeper levels in the mantle over time.

3.3.3.2 Plume driven whole mantle convection

Campbell and Griffiths (1993) proposed a plume driven model for whole mantle convection which includes a major shift in mechanism during Earth history. The model is based upon the following premises: (1) plumes can tell us about the geochemical structure of the Earth's mantle; (2) plumes rise from the core–mantle boundary and (3) plume chemistry is recorded in komatiitic and picritic melts. On this basis Campbell and Griffiths (1993) report a shift in plume trace element chemistry over time, from trace-element depleted during the Archaean to enriched and OIB-like since the Proterozoic, which they attribute to a change in plume source chemistry over time. To account for this difference they propose a major change in mantle dynamics to explain the change in plume source compositions. They propose that in the early Archaean (pre-4.0 Ga) the Earth had a cool buoyant lithosphere overlying a cool, chemically depleted, unstable boundary layer. This unstable layer formed cold plumes which descended to the core-mantle boundary to form a deep layer of depleted mantle (Fig. 3.37). At this boundary the depleted mantle is heated, in contact with the core, and ascends in the form of a hot plume. With the advent of wide-scale subduction, during the late Archaean, there is a change in tectonic style and descending oceanic lithosphere reaches the core-mantle boundary providing an enriched source for mantle plumes (Fig. 3.37). This explains the existence of trace element enriched plume magmas in the geological record after 2.0 Ga.

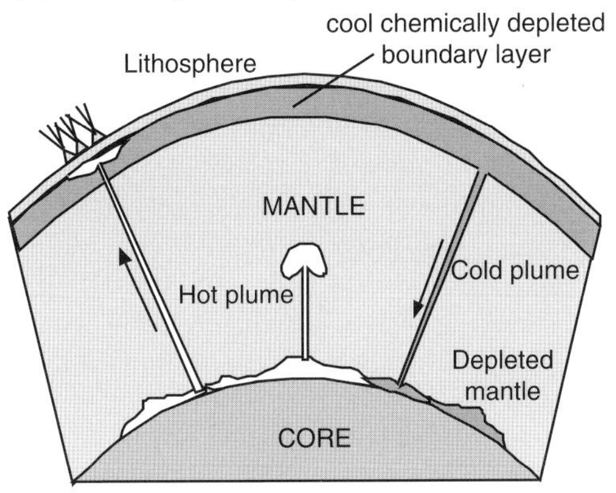

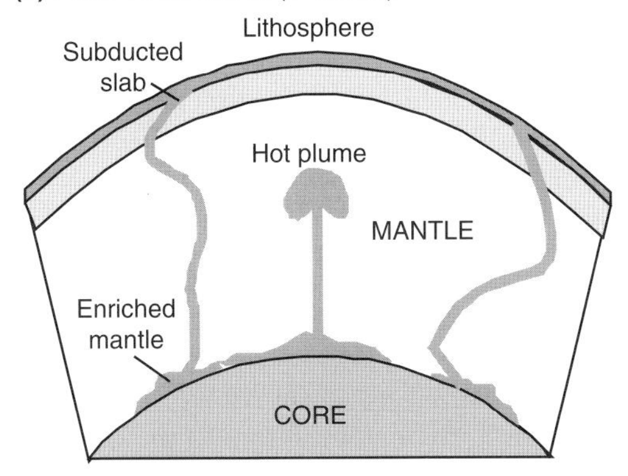

FIGURE 3.37 The Campbell–Griffiths (1993) plume-driven mantle convection model. (a) the pre-4.0 Ga Earth with descending, depleted cold plumes, and (b) the 2.0–0 Ga Earth with ascending enriched plumes.

3.3.3.3 Questions about the early mantle

3.3.3.3.1 When did "modern-style" mantle convection begin?

It is likely that the earliest events in the Earth's mantle were not the product of "normal" mantle convection but rather related to planetary processes such as fractionation within a magma ocean. In this way we can explain the very early differentiation of the Earth (pre-4.5 Ga), proposed on the basis of short-lived Nd-isotopes. Similarly, the extreme volatile element loss from the Earth might be explained in this way. These early processes are thought to have ceased within the first 100 Ma of Earth history (Yokochi & Marty, 2005).

Heterogeneities which formed after this time provide evidence for convective processes within the early mantle, although there may not have been the same pattern of convection as is found today. Examples of such heterogeneities are: basalts from Isua in west Greenland derived from a high Sm/Nd source, preserved in the mantle until at least ca. 3.8 Ga; felsic gneisses from Greenland and Labrador formed between 3.65 and 3.8 Ga whose Pb-isotope compositions record derivation from a mantle source which separated between 4.1 and 4.3 Ga (Kamber et al., 2003); basalts from the ca. 2.7 Ga Abitibi greenstone belt in Canada which contain Pb-isotopic evidence for mantle heterogeneities which must have originated within the first few hundred million years of Earth history (Vervoort et al., 1994).

Gangopadhyay and Walker (2003) argue from Os-isotope evidence that the mantle was homogenized by 2.7 Ga, indicating that "modern-style" mantle convection was in operation at that time.

3.3.3.3.2 When did subduction begin?

There is evidence from the study of mantle xenoliths for the cycling of basaltic oceanic crust back into the mantle during the late Archaean. For example, Os-isotope studies of eclogite from the subcontinental lithosphere beneath the Kaapvaal Craton show that they have the character of 2.9 Ga subducted oceanic crust (Richardson et al., 2001). Some Os-isotope data suggest that there was an even earlier recycled basaltic crust, formed at about 4.0 Ga (Section 3.2.3.4). There is also noble gas evidence that crustal materials were recycled into the mantle during the Archaean (see Chapter 5). The existence of boninites as old as 3.12 Ga from Pilbara, and 3.8 Ga boninite-like magmas from Isua, also strongly suggest the recycling of water into the early Archaean mantle. All these lines of evidence point to the initiation of subduction during the Archaean, and most probably in the early Archaean.

3.3.3.3.3 Has any primitive mantle been preserved in the modern mantle?

The concept of a primitive mantle reservoir, with the composition of the Bulk Silicate

Earth, has been prominent in models of the mantle for several decades (see Section 3.1.2). Such models are based upon a chondritic starting composition for the Earth, modified during core formation and perhaps in a magma ocean, and presuppose that there was a point in time when the mantle was totally homogenized. Whether the remnants of this primitive mantle can be identified today in modern basalts, or even in the early history of the Earth in Archaean basalts, is the subject of some debate. In favor are mantle melts with chondritic Os-isotope ratios (Fig. 3.32) and against are calculations which show that it is possible that the entire mantle has been processed through the subduction system (Section 3.1.6.4) during the history of the Earth.

Different authors have variously located primitive mantle in the present-day lower mantle beneath the 660 km discontinuity, in a deep layer beneath a 1,600 km discontinuity, in the D″ layer at the core–mantle boundary, or as "blobs" within the lower mantle (Becker et al., 1999). It is arguable whether any modern basalts have been derived from such a primitive reservoir, and it is possible that such a reservoir does not exist. In fact models of whole mantle convection, such as that of Helffrich and Wood (2001), in which the majority of the mantle has been processed through the subduction process, render the preservation of primitive mantle most unlikely.

However, Archaean basalts do have average compositions which are different from modern basalts (see Section 3.2.1.1) which could imply a difference in mantle source composition and raises the possibility that they sampled primitive mantle. This argument has been developed by Condie (2005) who used immobile incompatible trace element ratios to demonstrate a difference in source composition between modern and Archaean plume basalts. One of his principal findings is that Archaean plume basalts have Zr/Nb and Nb/Th ratios close to that of primitive mantle implying that a primitive mantle reservoir was present in the early mantle.

3.3.3.4 The Hadean mantle conundrum

Perhaps one of the best ways to summarize our current understanding of the earliest Archaean mantle is to follow Frei et al. (2004) in identifying "the Hadean mantle conundrum." The Hadean Mantle conundrum recognizes that the Hadean mantle is heterogeneous on a variety of length scales and shows features of isotopic enrichment (Os- and Pb-isotopes) and isotopic depletion (Nd- and Hf-isotopes). These results show that even during the Hadean "the upper mantle underwent a complicated time-integrated evolution that seems to have included processes such as melt extraction and the subsequent recycling of components back into previously depleted regions of the upper mantle" (Frei et al., 2004). Such processes sound surprisingly modern and suggest that a modern understanding of the mantle may "work" back into the earliest Archaean.

Our understanding of the exact nature of the heterogeneities in the Archaean mantle is, at present, still incomplete. On the one hand, there is good evidence for layering in the Hadean mantle, formed perhaps by crystal fractionation in a magma ocean and/or the subduction of an ancient basaltic "magma ocean lid." On the other hand, there is evidence for subduction going back to the early Archaean and maybe to the Hadean, which means that heterogeneities could also have been introduced into the mantle as subducted oceanic crust and lithosphere, resulting in an Archaean "plum pudding" type mantle. Discriminating between these two causes of Hadean mantle heterogeneity – subduction and primordial differentiation – is an important part of any future research agenda.

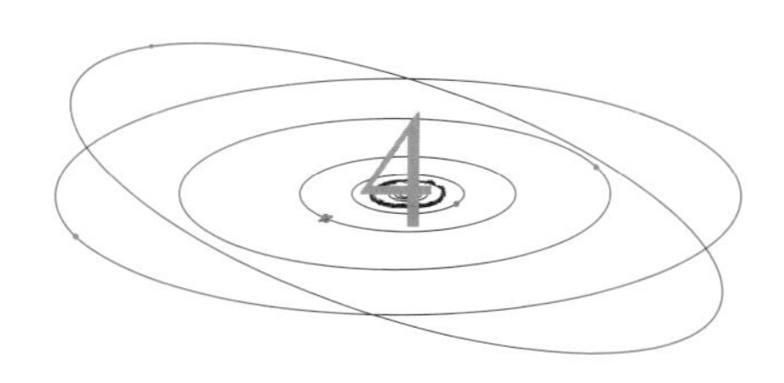

THE ORIGIN OF THE CONTINENTAL CRUST

The origin of the continental crust – the big picture

The Earth is the only planet amongst the rocky planets of the solar system to possess a well-developed, felsic continental crust. This crust was progressively extracted from the Earth's mantle over the last 4.0 billion years, and although it only represents 0.6% of the mass of the silicate Earth, it contains up 70% of the Earth's budget of highly incompatible elements. Over Earth history, the reservoir of depleted mantle, from which the crust was created, has grown in volume to mirror the growth of the continental crust.

The principal site of modern crustal growth is in arc environments in association with destructive plate margins. These are sites where new crust is created through partial melting in the mantle wedge, and where the principal flux from mantle to continental crust is basaltic. Destructive margins are also sites where crustal material is recycled back into the mantle as subducted sediment, and some current models indicate that at present mantle melting and recycling are in balance so that the present crustal growth rate is zero.

The most robust crustal growth models indicate that the continental crust grew progressively with time, perhaps episodically – reflecting periods of intense crustal growth at certain periods in Earth history. There was almost certainly a period of rapid growth of the continental crust during the late Archaean. Furthermore, there is now good evidence that the composition of continental crust-forming magmas was different at that time. In the Archaean the principal flux was magmas of the Si–Na-rich tonalite–trondhjemite–granodiorite (TTG) association. Experimental and geochemical studies show that these magmas most probably formed by the partial melting of hydrous basalt. The most efficient means of generating the large volumes of these magmas required to make new crust is through the partial melting of a wet basalt slab in a subduction zone. A modern analog is the relatively rare, high-silica adakite magmatic suite. These are slab melts which have interacted with the mantle wedge en route to the crust. The complement to TTG magma production is the formation of a refractory, melt depleted, eclogitic slab residue which is returned to the mantle.

The average, time-integrated composition of the continental crust provides an important constraint on crust evolution models. Our best estimate suggests that the bulk crust is andesitic in composition and is strongly enriched in incompatible elements. Given that in modern arcs the principal flux from mantle to crust is basaltic in composition, this presents a major paradox. Solutions to this paradox include a process of lower crust delamination in which ultramafic material is removed from the root of the continents and returned to the mantle. Alternatively, the composition of the continents has been modified by weathering and selective element recycling. However, there are difficulties with both these models, and instead it is suggested that the paradox is resolved through the changing composition of the continental crust with time. Thus the present andesitic composition is a time averaged composition between the present basaltic flux and a more felsic (TTG) flux during the Archaean.

Current models of Archaean crust generation imply that subduction is a relatively ancient process operating since at least about 3.5 Ga. An outstanding question is the relative contribution that plume magmatism has made to the generation of felsic continental crust.

The Earth's continental crust occupies 41.2% of the surface area but represents only 0.35% of the total mass of our planet (0.6% of the mass of the silicate Earth). Nevertheless, it is an important geochemical reservoir for many incompatible trace elements, and it contains up to 70% of the total Earth budget of certain elements (Rudnick & Fountain, 1995). In contrast to the relatively thin (7 km), and young basaltic crust which underlies the oceans (all less than 200 Ma), the Earth's continental crust is, in part, extremely ancient and geologically is very varied. The continental crust is composed of a very wide range of different rock-types and contains rocks varying in age from 4.0 Ga to the geologically recent, reflecting its long and complex history.

Until recently it was thought that the Earth was unique amongst the terrestrial planets in possessing a chemically evolved, granitic, continental crust. We now know that, although the Martian crust is predominantly basaltic, felsic rocks are present in the Syrtis Major caldera on Mars (Christensen et al., 2005). This new discovery is not unexpected and raises interesting parallels between the surface of Mars and the Earth's earliest crust.

It has long been recognized that the presence of liquid water on Earth is one of the principal differences between the Earth and the other terrestrial planets. More than 20 years ago Campbell and Taylor wrote a paper entitled *No water, No Granites – No Oceans, No Continents*, in which they proposed that water was *the* vital ingredient in the formation of a felsic continental crust (Campbell & Taylor, 1983). Hence, it may be no coincidence that the Earth, the only planet with abundant liquid water, is also the only planet with a substantial felsic continental crust.

The link between liquid surface-water and the presence of a stable continental crust becomes more apparent when looking at the history of Mars and Venus. The surface of Mars records evidence of only minor recent tectonic activity (Hartmann et al., 1999), and yet was once the home to violent outpourings of water. This can be seen in its deeply channeled surface. Venus on the other hand has been geologically active in its recent history, and has a surface temperature of 740 K so it cannot sustain liquid water. Since neither planet appears to have an extensive continental crust of the type found on Earth, this comparison shows that it is neither thermal energy alone, as on Venus, that gives rise to a continental crust nor is it the presence of water alone, as on Mars. It is a balance between the two.

The purpose of this chapter therefore, is to explore in detail how the Earth came to be so different from its planetary neighbors in possessing an extensive, evolved continental crust. In particular it will be argued that a major portion of the Earth's continental crust had formed by the end of the Archaean and that

> *Archaean crust-forming processes play a pivotal role in shaping the structure of the Earth's continental crust, . . . so that . . . we need to increasingly focus our attention on the distant past, as the key to understanding the present (Kemp & Hawkesworth, 2003)*

First, however, we briefly consider the processes of crust formation and crustal growth in the modern Earth.

4.1. Modern Crust Formation – Models and Mechanisms

Underlying all recent models for modern crust formation are a number of assumptions. First, it is widely agreed that the continental crust has been extracted from the Earth's mantle. Geochronological and geochemical evidences strongly indicate that the continental crust was not a product of the Earth accretion process but formed later. Second, density considerations suggest that, unlike the oceanic crust, once created, the continental crust is very difficult to destroy, certainly by large-scale recycling back into the mantle. An exception to this assumption is the recycling of the erosion products of the continental crust as sediments. Third, crust formation requires thermal energy and this is supplied in different ways, dependent on the model of crust formation. Finally, as already outlined, the presence of water is vital to the process of continental crust formation, as will be discussed in detail below (Section 4.1.2).

TEXT BOX 4.1 Glossary of terms related to continent generation

adakite	Volcanic rocks, formed in an arc environment; adakites represent a rock series which ranges in composition from andesite to rhyolite; typically they have SiO_2 > 56 wt%, Na_2O >3.5 wt%, <7.5 wt%, and low K_2O; see also *high magnesium andesite*
amphibolite	A metamorphic rock containing abundant plagioclase and hornblende ± garnet; has the composition of basalt and represents a hydrated basalt; subducted ocean floor basalt transforms to amphibolite at depth
amphibole–peridotite	Mantle peridotite which is hydrated and so contains amphibole in addition to, or instead of, clinopyroxene. Will melt at a lower temperature than "dry" mantle
andesite	The most abundant rock-type in a well developed volcanic arc. Andesites can be defined in terms of their silica and alkali contents (see Fig. 4.10).
basaltic andesite	An important rock-type in a volcanic arc. They are more siliceous than basalts but less so than andesites (see Fig. 4.10)
CIPW normative feldspar content	The composition of feldspar in a particular rock as calculated from the major element chemistry of the rock using a specific set of rules – the (CIPW) normative calculation. The feldspar content is expressed in terms of the weight % proportion of the Na-, K- and Ca-feldspars (Albite (Ab), Orthoclase (Or) and Anorthite (An))
high-alumina basalt	Basalt from an arc environment which typically has about 20 wt% Al_2O_3, rather than the ca. 15 wt% found in oceanic basalts
high-field strength element	Trace elements which have a small ionic radius but which carry a high ionic charge. These tend to be the immobile elements during fluid alteration
high-magnesium andesite	An andesite contains 2–3 wt% MgO; a high-magnesium andesite contains 4 wt% MgO, and in some cases up to about 9 wt% MgO. Some authors distinguish high *mg#*-andesites as those with a mg# > 0.45. There is some overlap with the term boninite (see Chapter 3, Section 3.2.1.3).
high-magnesium basalt	A primary midocean basalt has about 9 wt% MgO. A high-magnesium basalt has a higher concentration of MgO than this. Broadly synonymous with the term picrite. Represents a higher-temperature melt of the mantle than *MORB*.
initial isotope ratio	The ratio of a radiogenic daughter isotope to a reference nonradiogenic isotope present in a magmatic rock at the time of crystallization. It represents the isotope ratio present in the rock before radiogenic ingrowth increases the isotope ratio to its present-day measured value
komatiite	An ultramafic magma produced during the Archaean see (Chapter 3, Section 3.2.1.2).
low field strength element	Trace elements which have a large ionic radius but a low charge (1+ or 2+). These tend to be elements which are mobile during fluid alteration
mg# (magnesium number)	The atomic fraction of Mg relative to total (Mg + Fe(II)) in the rock
model Hf age	The age of sample calculated from the length of time it has been separated from the mantle. Calculations of this type assume a particular mantle growth curve
MORB	Mid Ocean Ridge Basalt
primary melt	A melt which has remained unmodified in composition since it formed in its source region. Hence, a melt which has not experienced either crystal fractionation or reaction with the matrix through which it has traveled
primitive arc basalt	A basalt formed by partial melting of the mantle wedge beneath an arc and which is effectively unmodified in composition and may closely approximate to a *primary melt*
SCLM	The Sub Continental Lithospheric Mantle – that part of the Earth's mantle which lies beneath and is coupled to the continental crust
TTG	Granitoid magmas in the compositional range Tonalite–Trondhjemite–Granodiorite. These are the magmas which have built the Archaean continental crust

4.1.1 Crustal growth at destructive plate boundaries

The principal site of geologically recent crustal growth is at active, destructive plate boundaries. The general model (MacDonald et al., 2000, integrating results from many sources) is illustrated in Fig. 4.1, in which the numbered circles refer to points 1 to 4 in the text. The model requires that:

1. Amphibolite, formed from hydrated ocean floor basalt, in the upper layer of a subducting slab begins to dehydrate at about 50–60 km depth, releasing fluids into the overlying mantle wedge.
2. The hydration of the mantle wedge results in the formation of amphibole in mantle peridotite. This amphibole–peridotite is dragged deeper into the mantle through viscous coupling to the subducting slab to a depth of about 110 km, where the amphibole breaks down and releases aqueous fluid into the overlying mantle, initiating partial melting in the mantle wedge.
3. The partial melt zone (melt and matrix) migrates upwards as a diapir into a hotter region of the wedge, thereby promoting further melting.
4. The diapir is dragged into the corner of the wedge by convective flow within the wedge, where the melt is focused.

During the 1980s and early 1990s there was a long debate about the composition of the primary melts parental to arc magmas. One group argued for a "wedge source" for arc magmas and believed that arc magmas are derived from high magnesian basalts derived from the mantle wedge (see e.g. Gust & Perfit, 1987). A second group argued for a "slab source" and that the parental melts to arc magmas are high-alumina basalts and basaltic andesites, produced by the melting of the subducted ocean crust in the down-going slab (Brophy & Marsh, 1986). Since we know that the average composition of the continental crust is andesitic (see Section 4.2.2.3), and that andesites are volumetrically the most important component of arcs, the slab source would seem to be the obvious choice. However, theoretical and experimental studies of slab melting showed otherwise, for thermal modeling

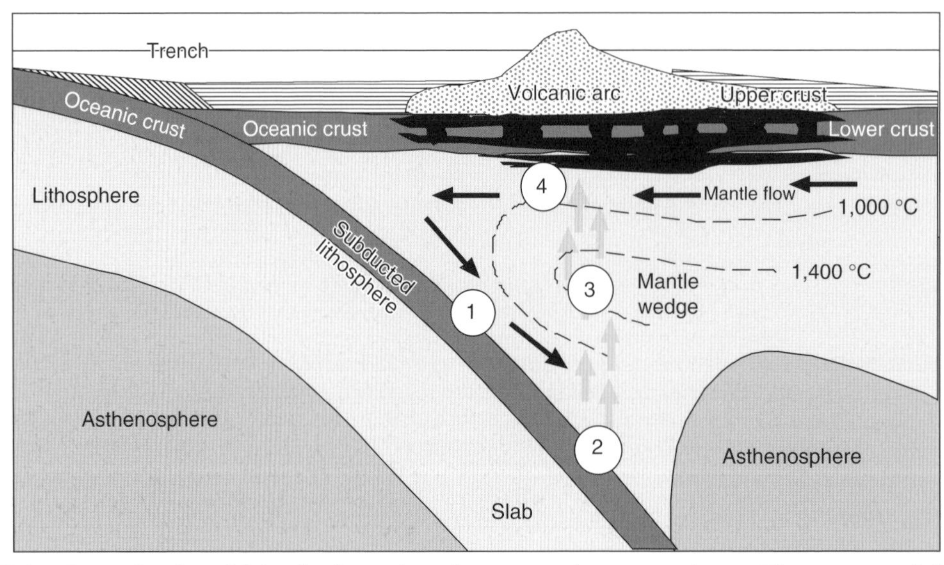

FIGURE 4.1 Generalized model for the formation of new crust in an oceanic arc. The young arc is built on oceanic crust. The arc is differentiated into upper crust, comprising andesitic lavas and sediments, and lower crust made up of former oceanic crust and mafic and ultramafic cumulates. The numbered circles show the process of fluid transfer from the subducted oceanic crust to mantle and melting within the mantle wedge.

calculations made it clear that slab melting rarely happens because there is not enough heat in a subduction zone to give rise to slab melting (Peacock et al., 1994). Further, if basalt melting does take place under slab conditions the melts that are produced are felsic (Rapp et al., 1991), not high-alumina basalt, as supposed. These results lead us to the conclusion that a "slab source" model is inappropriate as a general model for arc magma genesis. In contrast, experimental studies of a primitive arc basalt, in the presence of water, show that they originated in the mantle (Pichavant et al., 2002), strongly supporting the "wedge model" for arc magma genesis. This conclusion is very important for models of crustal evolution, for if the flux into an arc is basaltic, and yet the volumetrically dominant magmas are andesitic, then there should be significant volumes of mafic and ultramafic cumulates in the lower crust and upper lithospheric mantle of many arcs (Fig. 4.1).

4.1.2 Arc magma sources

The quest of recent research into arc magmatism has been to assess the relative contributions of the different components of arc magmatism. These include the mantle wedge, a fluid phase, subducted sediment, and the subducted slab. In each case these differing contributions can be identified by a geochemical fingerprint. A further potential magma source, present in some arcs, is continental crust, which has a geochemical signature similar to that of subducted sediments. In order to avoid this ambiguity, most definitive studies of arc magma sources have been made in oceanic arcs, where older continental crust is absent, although crust-derived sediment may be present.

4.1.2.1 The mantle wedge

The mantle wedge is the principal source of arc magmatism. Basalts produced from the melting of the mantle wedge are the main contribution to modern crustal growth in an arc environment. Seismic imaging provides an understanding of the thermal structure of the mantle wedge (Abers et al., 2006) and geochemistry provides a helpful insight into the details of the melting process. In this section the geochemical controls on the partial melting process are explored.

A comparison between trace element concentrations in primitive arc magmas and typical midocean ridge basalts (Fig. 4.2) shows that there is a decoupling between the high field strength and low field strength incompatible trace elements. HFSEs have flat trace element patterns, although with lower concentrations than are found in MORB, indicating that their source was more depleted than a MORB-source. This means that the mantle wedge must have been depleted by the removal of a basaltic melt, prior to the generation of arc magmas (Elliott et al., 1997). More generally,

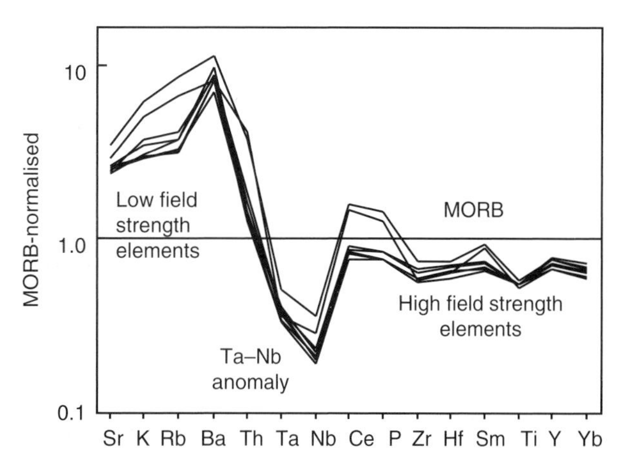

FIGURE 4.2 The trace element content of typical arc magmas, expressed relative to concentrations in midoceanic basalts showing three groups of elements – high concentrations of low field strength (soluble) elements, low concentrations of high field strength (insoluble elements) and with MORB-like elemental ratios, and a negative Ta–Nb anomaly. Data from the Marianas Arc, a "depleted" arc mantle source (Elliott et al., 1997). (See Text Box 2.2.)

trace element studies using Nb–Y relationships and Sm/La ratios indicate that arc mantle may be either depleted or enriched (Pearce & Parkinson, 1993; Plank, 2005). This variability is consistent with a model of magma generation in which there are multiple melting events during mantle flow in the wedge (Fig. 4.1).

Estimates of the extent of melting in the mantle wedge vary. Hochstaedter et al. (1996) estimated between 8 and 20% melting for the Kamchatka arc, similar to the range for MORB generation, although other authors argue that the process of hydrous melting will produce a higher melt fraction than is found in MORB. There is also a relationship between the amount of melting and the thickness of the arc crust. Since melting beneath an arc takes place as a consequence of amphibole breakdown at a more or less constant depth of ca. 100 km, the thickness of the melt column beneath the arc is limited by the thickness of the arc above. The evidence for this finding comes from a study by Plank and Langmuir (1988) who showed that in arc magmas the concentration of Na increases with increasing crustal thickness. They argued that in a thick arc, the higher concentration of an incompatible element such as Na in the melt implies a smaller amount of melt production and so a shorter mantle melt column. In contrast, beneath thin arc crust there will be a longer melt column, a greater amount of melt produced and so a lower concentration of incompatible elements.

A second prominent feature of arc magma chemistry is the marked Nb–Ta depletion, seen in trace element patterns (Fig. 4.2). This depletion is thought to indicate a phase within the unmelted mantle residue which has an affinity for Nb and Ta. Identifying this phase has been problematical. There are two principal candidates – a titaniferous phase such as ilmenite, sphene or rutile – in which Nb or Ta will substitute for Ti, or an amphibole. Increasingly, there is experimental evidence to indicate the importance of amphibole in the process of Nb–Ta depletion (Tiepolo et al., 2001).

4.1.2.2 *A fluid phase*

Evidence for the involvement of a fluid phase in arc magma genesis comes from the elevated concentrations of low field strength, incompatible trace elements (Sr, K, Rb, Ba and Pb) relative to MORB, the third characteristic of MORB-arc basalt comparisons (Fig. 4.2). The low field strength elements are those which are fluid-mobile and their enrichment in the source of arc magmas indicates that they have been transported there in a fluid phase. As discussed earlier, it is thought that this fluid phase is derived from the subducted slab and that fluid-mobile elements are leached from the slab and transferred into the mantle wedge. The precise mechanism of fluid release is complex and involves a series of dehydration reactions in which metamorphic minerals such as chlorite, lawsonite, zoisite, amphibole and chloritoid are progressively broken down to a depth of between 150 and 200 km (Schmidt & Poli, 1998).

Many trace element studies use the ratio of fluid-mobile to immobile elements such as Ba(mobile)/La(immobile) and Pb/Ce in arc magmas as a monitor of fluid transfer into the mantle wedge (see e.g. Ayers, 1998). The higher the Ba/La or Pb/Ce ratio, the greater the fluid flux. In addition, further evidence for the presence of a fluid phase in the source of arc magmas is their relatively high water content.

Important here are the observations that (i) the abundance of water in the mantle strongly influences the amount of melting that takes place, and (ii) the subarc mantle is in places relatively cool, and yet it still melts because where water is present the mantle solidus is lowered and melting occurs.

4.1.2.3 *Slab melts*

Although there is now general agreement that the prime contribution to arc magmatism is from a mantle source, there are special cases where a contribution from the subducted slab is also thought to be significant. There are two lines of reasoning, which make this probable. First, Peacock et al. (1994) showed from thermal modeling calculations that slab melting will take place when the slab is unusually warm, as in the case of the subduction of a spreading ridge, or very young oceanic crust. The second argument is based upon the composition of a rather rare form of high magnesian andesite known as adakite, found in some arc environments. Experimental studies show that adakites have the right compositions to be slab melts, indicating that slab melting can not only take place but also does take place in some rare instances.

Adakites are normally a minor component of modern arc magmatism. Sometimes they form a discrete component, but more normally they contribute to a particular magmatic suite though mixing, along with the other components discussed in this section (Stern and Kilian, 1996). Early in Earth history, however, slab melting may have made a much more important contribution to crustal growth than it does today (Martin, 1986). We will address this topic later in the chapter (Section 4.4.2.1).

4.2.2.4 Subducted sediment

There are convincing arguments, from isotope and trace element geochemistry, that arc magmas also contain a component derived from the melting of subducted sediments. One of the first indications that this is the case comes from the study of the cosmogenic isotope ^{10}Be in some arc magmas (Brown et al., 1982). ^{10}Be is only produced in the upper atmosphere from the action of cosmic rays on oxygen and nitrogen and is transferred to Earth in rain and snow. ^{10}Be also has a short half-life so it has effectively disappeared after about 10 Ma. Thus the presence of ^{10}Be in arc magmas from the Aleutians (Brown et al., 1982) is a strong indication that arc magmas contain a component which originated at the Earth's surface.

Further evidence in support of a sedimentary component in arc magmas comes from the study of Sr, Nd and Pb isotopes. On a $^{143}Nd/^{144}Nd$ versus $^{87}Sr/^{86}Sr$ mixing diagram some arc magmas define a compositional array between the MORB source and a component enriched in Sr and unradiogenic Nd, which looks very much like average Atlantic sediment (Fig. 4.3). Oceanic sediment has a high Nd/Sr ratio, and so a small amount of subducted sediment can have a significant effect on the bulk $^{143}Nd/^{144}Nd$ of the resultant arc magma (Davidson, 1983; Ellam & Hawkesworth, 1988a). Seawater on the other hand is enriched in Sr relative to Nd and so seawater mixing can be identified from a vector of constant $^{143}Nd/^{144}Nd$ (Fig. 4.3).

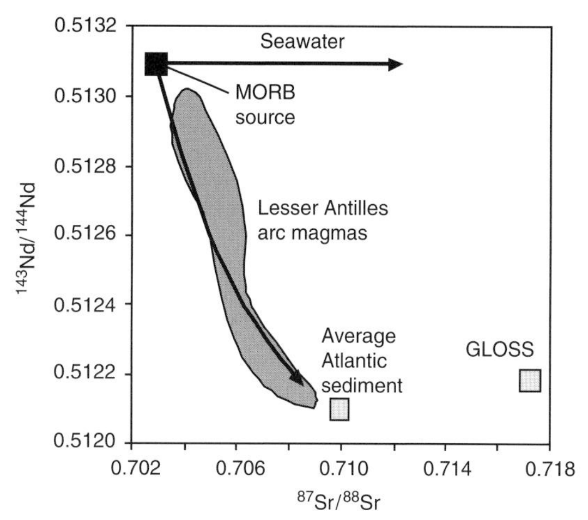

FIGURE 4.3 A Nd–Sr isotope ratio plot showing the compositional range of arc magmas from the Lesser Antilles. They lie between the composition of MORB-source and average Atlantic sediment, implying that they contain a component of subducted sediment (after Davidson, 1983). Mixing with seawater would not significantly change the Nd-isotope ratio. GLOSS is the average composition of GLObal Subducting Sediment from Plank and Langmuir (1998).

A recent trace element study by Plank (2005) provides further support for a sedimentary component in arc magmas. Plank showed that in volcanic arcs where there is a high sediment flux, there is an almost 1:1 correlation between the Th/La ratio in the magma and that of the subducting sediment (Fig. 4.4). This relationship suggests that subducted sediment strongly controls the Th/La ratio in these particular arc magmas.

4.1.3 Crustal growth through intraplate magmatism

Although the dominant site of contemporary crustal growth is at convergent plate boundaries, there are two other tectonic settings in which crustal growth can also take place. The first is through the formation of Large Igneous Provinces, and the second is at passive continental margins.

LIPs are sites of extensive basaltic volcanism, which originate in a manner which is separate from the processes of sea-floor spreading.

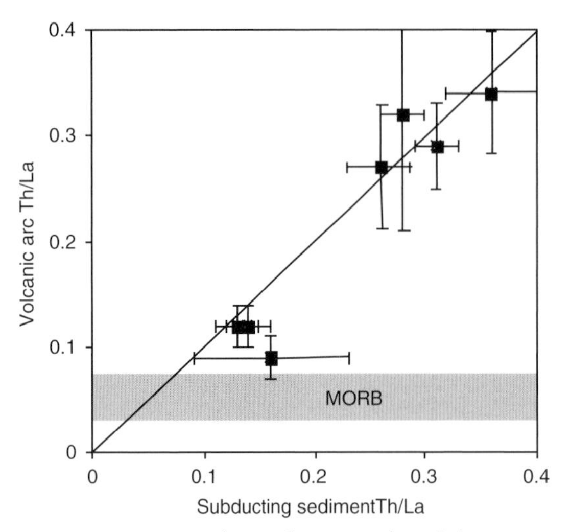

FIGURE 4.4 Correlation between the Th/La ratio of subducting sediment in high sediment flux arcs and Th/La in volcanic arc magmas (after Plank, 2005).

They include continental flood basalts of the type found in the Deccan plateau in India or the Karoo province of southern Africa, and, where the volcanism is within oceanic crust, oceanic plateaus. Most Mesozoic and Cenozoic LIPs had an original areal extent of more than a million km^2 and represent magma volumes of up to 4 million km^3. The Ontong Java oceanic plateau has a lava volume of 6 million km^3 and when extrusive and intrusive components are aggregated a total volume of 44.4 million km^3 (Ernst et al., 2005).

Most LIPs are emplaced very rapidly with the bulk of the magmatism occurring within a few million years. This combination of huge volumes of melt emplaced over a short period of time requires a very particular set of processes, requiring very high melt production rates in the mantle. Most authors invoke hotter than average mantle beneath LIPs within which extensive decompression melting takes place. Whether or not these hot spots are mantle plumes is the subject of some debate although many models of LIP magmatism are connected to the emplacement of mantle plumes (Ernst et al., 2005).

How LIPs are incorporated into the continental crust is the subject of discussion, most of which has centered around oceanic plateaus, for, once created, these are not easily destroyed either by erosion or, because of their great size, by subduction. There are two basic models. In one, oceanic plateaus are accreted onto existing continental crust through collision, imbrication, and obduction (Saunders et al., 1996). Alternatively, or in addition, partial melting may take place within the thick basaltic section of the plateau giving rise to felsic rocks, thus converting oceanic plateau mafic rocks into "true" felsic crust. A possible example of this mechanism is the Aruba batholith in the Caribbean oceanic plateau (White et al., 1999). Mechanisms of these types have added support to the "mantle plume" model of continental growth (Abbott & Mooney, 1995; Stein & Goldstein, 1996). This model is currently gaining momentum but is not yet the consensus view (see Section 4.4.2.4). Calculations by Barth et al. (2000) based on the crustal La/Nb ratio suggest that this model of crustal growth accounts for, at most, between 5% and 20% of the total crustal mass over geological time.

4.1.4 Crustal growth at rifted margins

A further site of crustal growth has recently been recognized from seismic reflection studies of volcanic rifted continental margins. Studies have shown that 90% of all passive continental margins around the world are volcanic rifted margins (Menzies et al., 2002). Continental margins of this type are characterized by thick sequences of subaerially erupted flood basalts, with volcanic sequences up to 7 km, associated with rifting and uplift. Related intrusive rocks are present in the underlying middle and lower crust. Seismic studies show that the seaward, submarine segment of the continental margin is a transition zone with true oceanic crust, in which there are strongly developed seismic reflectors indicating alternating layers of volcanic and sedimentary rocks. There is also an area of the lower crust marked by high seismic velocities, indicating abundant igneous rocks.

It is now recognized that volcanic rifted margins are closely related to the LIPs discussed in the section above, and that they represent

LIPs associated with the process of continental break-up. What is not yet clear is whether the volcanism associated with continent separation is driving the break-up process or whether the separation process is focusing the volcanism. Nevertheless, the extensive volumes of magma concentrated along volcanic rifted margins cannot be ignored in models of crustal growth.

In each of the three tectonic settings discussed above – convergent margins, within-plate (oceanic and continental crust) and at rifted margins – the dominant flux from the mantle to the continental crust is basaltic. This however poses a major problem, for the continental crust is not basaltic in composition, rather, as will be shown below (Table 4.2), the average bulk composition of the continental crust is andesitic. This paradox is one of the major problems to be addressed in determining the origin of the continental crust.

4.1.5 Crust–mantle fluxes

The most important site of present-day crustal growth is in an arc environment, either through the creation of a new arc or through arc accretion. In this section we examine present-day fluxes between the mantle and the crust in an arc environment, in an attempt to begin to quantify the process of crustal growth.

In a study of 17 different island arcs Reymer and Schubert (1984) estimated that the present-day, worldwide, magmatic crustal addition rate is 1.65 km_3/yr. This is the principal present-day flux from the mantle to the crust. In contrast, the return of material to the mantle is much more complicated and involves a number of different components (Table 4.1).

TABLE 4.1 Recent estimates of present-day fluxes into and out of the continental crust and mantle during subduction.

Flux	Volume	Mass	Reference
Flux into the crust (from mantle)			
Magmatic Continental growth rate			
Total	1.65 km^3/yr		Reymer and Schubert (1986)
Net crustal growth rate			
Phanerozoic	1.0 km^3/yr		Reymer and Schubert (1984, 1986)
Present	0.0 km^3/yr		Plank and Langmuir (1998)
Average over 4.0 Ga	1.94 km^3/yr	Total mass – 7.76 $\times 10^9$ km^3	Reymer and Schubert (1986)
Flux back into the mantle			
Rate of subduction of oceanic crust			
Total	20 km^3/ yr		Crisp (1984)
Rate of sediment subduction			
Global average	0.73 km^3/ yr	1.8×10^{15} g/yr	Plank and Langmuir (1998)
Average (from the literature)	0.5–0.7 km^3/yr		Plank and Langmuir (1998)
Tectonic erosion and subduction of sediment			
Long-term average	0.90 km^3/ yr		von Huerne and Scholl (1991)
Sedimentary pore water			
		0.86×10^{15} g/yr	Plank and Langmuir (1998)
Structurally bound water			
In slab (some released through arc magmatism)		$0.7–1.1 \times 10^9$ g/m^2	Schmidt and Poli (1998)
Subducted CO_2			
Total slab		2.2×10^{14} g/yr	Peacock (1990)

These include oceanic crust, sediment, and fluids – principally water.

Present-day spreading rates allow us to estimate the volume of oceanic crust, which is subducted annually. Estimates are between 18 and 20 km^3/ yr, the differences being in part due to different assumptions about the average thickness of the oceanic crust. If this rate of subduction has been constant over geological time, then the Earth's mantle should contain 5% recycled oceanic crust (Helffrich & Wood, 2001), although if this were concentrated in the upper part of the mantle then, of course, the proportion would be considerably greater.

Estimates of the volume of subducted sediment are subject to greater uncertainty, for there are two separate components to this process. On the one hand, sediment is subducted directly from the ocean floor, along with the slab. In addition, however, there is sediment which initially becomes part of a subduction accretion complex built onto the ocean floor but which is later subducted by a process known as "tectonic erosion." Plank and Langmuir (1998) calculated the global mean sediment flux to the mantle from plate convergence rates, trench lengths, and the thickness of the sedimentary column in a large number of different trenches. They also took into account the uncertainties which arise through the "loss" of sediment to accretion complexes in the overlying plate. Estimates of the annual volume of subducted sediment vary between 0.38 and 0.75 km^3/yr, although Plank and Langmuir's (1998) average of 0.5–0.7 km^3 is probably the most robust (Table 4.1). To the subducting component should also be added the tectonic erosion element of 0.9 km^3/yr (von Huerne & Scholl, 1991).

Plank and Langmuir (1998) have pointed out that the sum of the two sedimentary subduction components is almost the same as the magmatic growth rate for the continents, indicating that at the present time the continents are in a steady state. However, the mass balance is not simple, for it has already been noted that the flux from the mantle to the continents is basaltic, whereas the flux from the continents into the mantle is closer to the composition of average continental crust and is andesitic (Plank & Langmuir, 1998). Helffrich and Wood (2001) estimated that, given a constant rate of subduction for 4.0 Ga, the mantle should contain about 0.3% recycled sedimentary material, although we only know recent sediment recycling rates with any certainty.

Water is also part of the subducting sedimentary column and is present as pore water in sediments and as structurally bound water in metabasalts and gabbros. Estimates of the amount of structurally bound water vary from about 6 wt% at 20–30 km depth reducing to 2 wt% at 60 km and 1 wt% at 200 km (Poli & Schmidt, 1995). Schmidt and Poli (1998) estimated that the water input into a subduction zone from oceanic lithosphere is $0.71–1.1 \times 10^9$ g/m^2, although between 18 and 37% of this water will be released in the generation of arc magmas.

Figure 4.5 summarizes in a qualitative manner the principal inputs and outputs into and out of the continental crust at the present time. In differing proportions these are also the principal elements of continental growth and recycling. The inputs, representing a flux from the mantle to the crust, are magmatic contributions in the subduction, intracontinental, and continental margin environments. The main quantifiable output is erosion-related sediment return to the mantle through subduction. Less quantifiable, and of uncertain significance, is the delamination of the lower continental crust (see Section 4.5.2). Also relatively unknown is the role of intracrustal differentiation and the transfer of material from the lower to upper continental crust.

It is these basic parameters which we will continue to explore as we now turn to the problem of the origin of the continental crust.

4.2 First order constraints on the origin of the continental crust

There are two principal observations which we can use to constrain models for the origin of the Earth's continental crust. The first is a knowledge of the age structure of the crust and the second is its average composition. Both of

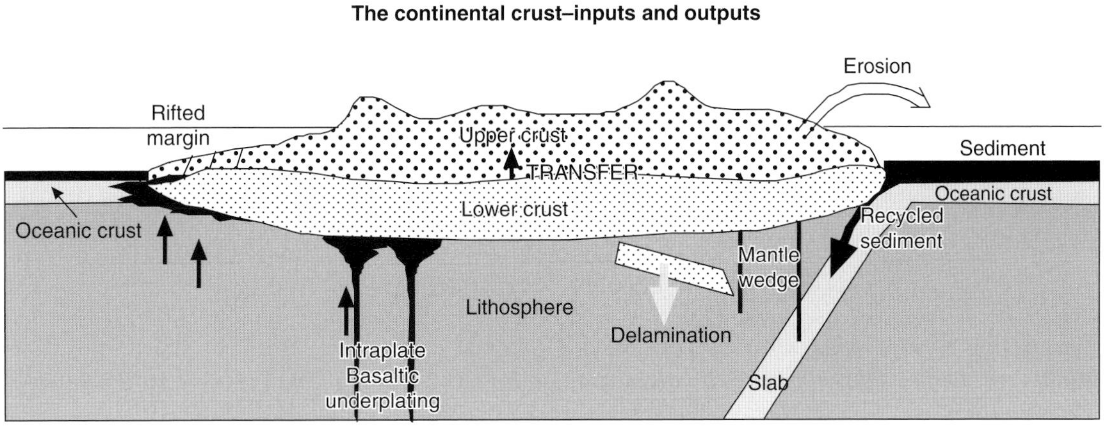

FIGURE 4.5 A model for the fluxes into (black arrows) and out of (white arrows) the continental crust.

these parameters, however, are hard to determine and inevitably therefore are the subject of some debate.

4.2.1 Crustal growth through time

Determining an appropriate growth curve for the Earth's continental crust has been the subject of considerable controversy for the last 30 years and more. In 1969 Hurley and Rand proposed, on the basis of the age distribution of crustal rocks, that the continents had grown progressively with time. During subsequent decades more data on the age distribution of crustal rocks have become available and the model has been refined. Recent compilations of crustal age patterns still confirm the progressive growth of the continental crust but suggest that it has grown in a nonlinear fashion over time (see Condie, 1998). In particular it appears that perhaps as much as 50% of the continental crust formed before the end of the Archaean at 2.5 Ga ago. If these observations are correct then two important questions arise. First, what was the cause of the episodicity in continental growth, and second why was there particularly rapid crustal growth early in Earth history?

An entirely different view of crustal growth, also current for more than 30 years, was espoused by Armstrong (1968) who took seriously the observation that continental crust can be destroyed, recycled, and recreated through sediment subduction. Armstrong (1968) and more recent advocates of this model (Sylvester et al., 1997) suggest that the entire present-day volume of the continental crust formed very early in the history of the Earth but has subsequently been recycled. In this model, the observed age distribution seen in crustal growth curves simply reflects the subtractions and additions to the crust over time. In this "no-growth," or "steady state" model for crustal evolution, the process of crustal recycling must have been of major importance throughout geological time.

4.2.1.1 Constraining crustal growth models

The proponents of the two different crustal growth models have supported their arguments with a number of rather different lines of evidence, drawn from both geochemistry and geophysics. In detail the arguments are intricate but a summary of the major claims and counter claims is given below.

4.2.1.1.1 The age distribution of crustal rocks

An estimate of the relative volumes of continental crust of different ages is an obvious way to assess crustal growth models. However, making such an estimate is not straightforward for two reasons. First, it cannot be assumed that the age of the continents is the same at depth as it is at the surface (Corfu, 1987). Second, the reworking of older crust will make the observed crustal ages younger than they really are. This latter problem seriously distorted some early attempts to assess crustal growth through time, but this has been circumvented in more recent studies through the use of robust isotopic methods. Some of these are outlined below.

4.2.1.1.1.1 U–Pb zircon ages The U–Pb age of zircon grains records the time of crystallization of the host igneous rock. Thus a global compilation of zircon ages for crustal rocks should provide an accurate assessment of the rate of crustal growth over geological time. Global compilations of zircon ages from juvenile crust were made by Condie (1998, 2000) and Rino et al. (2004). Both studies show similar growth curves (see Fig. 4.8 for Condie's growth curve) which support the progressive crustal growth model, although the age distributions show distinct peaks at 2.7, 1.9, and 1.2 Ga, implying some episodicity in crustal growth.

4.2.1.1.1.2 Zircon ages combined with Nd-model ages A particular difficulty in determining the pattern of crustal growth through time is knowing whether or not a particular suite of granitoids has formed directly from the Earth's mantle and is therefore a juvenile addition to the crust or whether it is reworked older crustal rocks. This problem can be resolved by determining the Nd-isotopic composition at the time of crystallization (determined from the zircon age) as illustrated in Fig. 4.6. The logic here is that if we know the time that the crust was extracted from the mantle (from the U–Pb zircon age) then the Nd-isotope ratio calculated from the measured value for the sample, at that time, can be compared to that of the depleted mantle at the same time. Deviations from the depleted mantle value may imply mixing with older crust, and this proportion can be quantified (see Fig. 4.6).

Patchett and Arndt (1986) used this approach to calculate the proportion of newly-formed crust in a large province of 1.7–1.9 Ga Proterozoic rocks from Europe, Greenland, and North America and demonstrated that a large volume of new crust was created at that time. Similarly, Patchett and Samson (2003) showed that a very large volume of the North–American-European continent was juvenile and grew significantly during the short time interval 1.7–1.9 Ga. They calculated that the growth rate of this region was equivalent to that of the present-day, global, arc growth rate, implying that if there was crustal growth

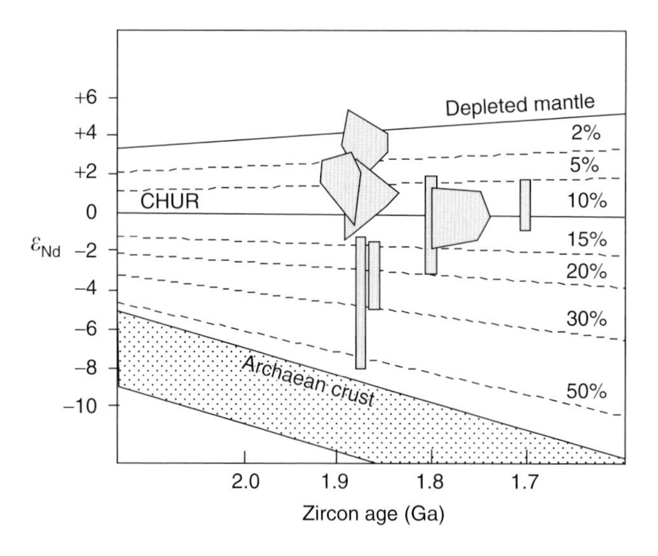

FIGURE 4.6 A plot of ε_{Nd} versus time for samples of Proterozoic crust from North America and Europe (grey boxes). The age of the samples was fixed from their U–Pb zircon age. The Nd-isotope compositions are interpreted as mixing between a depleted mantle source, indicative of juvenile crust, and recycled Archaean crust. The % curves indicate the proportion of reworked Archaean crust incorporated into Proterozoic crust and hence the proportion of truly juvenile crust formed (data from Patchett & Arndt, 1986) after Rollinson (2006).

anywhere else on the Earth in this time, then the period 1.7–1.9 Ga represented a time of very rapid crustal growth during Earth history.

More recently Nd-model ages have been used to identify a large area in Central Asia, the Altaid Tectonic Collage, which was the site of voluminous juvenile crust generation during the Phanerozoic. Studies in this remote, relatively unknown region have estimated that up to 50% of the crust is juvenile and of Paleozoic age (Patchett & Samson, 2003; Jahn, 2004).

4.2.1.1.1.3 Model Hf ages in sedimentary zircons A further "robust" geochronological method is the use of model Hf ages in zircons, for model Hf ages in zircon are a measure of the time at which the granitoid parent to the zircon was extracted from the mantle. Stevenson and Patchett (1990) used this method to investigate the steady state crustal growth model in the light of the apparent absence of large volumes of early felsic crust. They argued that if there were large volumes of felsic crust in existence early in Earth history, then evidence for their existence will be preserved in clastic sediments and that these rocks will contain zircons which will record Hf-isotope model ages older than the depositional age of the sediments. If, on the other hand there was little older felsic crust, then the Hf model ages of the zircons would be similar to the age of sediment deposition. They examined clastic sediments from Archaean and Proterozoic basins and found that in most late Archaean sedimentary sequences (2.6–3.0 Ga) zircon Hf model ages are within 10% of the depositional age, implying a small volume of preexisting felsic crust. Proterozoic sediments in contrast showed a 400 Ma disparity between the zircon Hf model age and their depositional age, signifying a large volume of felsic crust. These results imply that a substantial volume of felsic crust was created in the time interval 3.0 to 2.5 Ga.

Taken together, the age data for crustal rocks, for which there is now a very extensive database, strongly support a progressive model for the growth of the continental crust. Geochronological studies and the distribution pattern of juvenile crust provide very little support for an extensive, very early granitic crust. However, other geochemical and geophysical arguments have also been used to determine the pattern of crustal growth over time. These are discussed in the following sections.

4.2.1.1.2 The trace element evolution of the mantle over time – a proxy for continental growth?

Given that the continental crust was originally extracted from the Earth's mantle, it is to be expected that the mantle will preserve some geochemical record of crust extraction. This is the basis for the trace element and isotopic arguments developed in this and the following section.

The incompatible trace elements Nb and U are enriched in the continental crust relative to the primitive mantle but are also strongly fractionated during the crust-forming process. The primitive mantle is thought to have had a Nb/U ratio of about 30, whereas the modern mantle has a ratio of about 47. This difference is attributed to the extraction of the continental crust (Nb/U ~ 10) from the primitive mantle (Fig. 4.7). The large difference in Nb/U ratios between the primitive mantle and the modern mantle was used by Hofmann et al. (1986) and Sylvester et al. (1997) to monitor the growth of the continental crust over geological time. They argued that if the modern mantle Nb/U ratio could be tracked back in time via 'modern looking' Nb/U ratios in ancient mantle melts, then, on the grounds of mass balance, a complimentary continental crust must have existed. Geochemical studies of basalts as old as 3.5 Ga (Green et al., 2000) show that they have Nb/U ratios equivalent to that of modern OIBs and MORBs (Fig. 4.7).

This result could imply that there was an extensive volume of continental crust in existence at 3.5 Ga, strongly supporting the "no-growth" crustal growth model of Armstrong (1968). However, there are problems with this interpretation. First, because Nb exchange between the crust and mantle is not a simple two-component system (see Section 4.5.1.2.3) and second, because U is mobile in an oxidizing

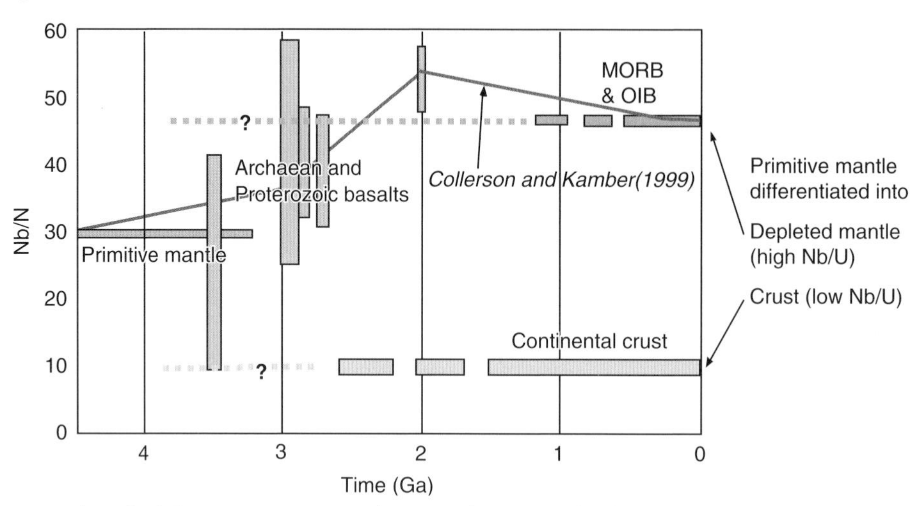

FIGURE 4.7 A plot of Nb/U ratio over time, showing the range of values for primitive mantle, the continental crust and the depleted mantle (represented by MORB and OIB and various Archaean and proterozoic basalts). Also shown is the curve for Nb/U versus age from Collerson and Kamber (1999).

environment and so has become increasingly mobile during Earth history as the atmosphere has become more oxygenic (Collerson & Kamber, 1999). Further, as U has become more mobile, it has been more readily recycled back into the mantle. Collerson and Kamber (1999) therefore proposed that the Nb/U ratio of the mantle has changed over time, initially increasing, accompanying crust extraction until about 2.0 Ga, after which it began to decrease as U began to be recycled back into the mantle (Fig. 4.7). Thus the Nb/U ratio cannot be used to argue for "no-growth" models of the continental crust.

4.2.1.1.3 The Nd-isotopic evolution of the mantle as a proxy for continental growth

During mantle melting Nd and Sm are both partitioned into the melt phase but also fractionated relative to each other. The net effect of this process is that over geological time the mantle has become depleted in the element Nd but enriched in the isotope ratio relative to its initial chondritic composition (Fig. 3.27). It is commonly accepted that the missing Nd is now located in the continental crust and that the record of $^{143}Nd/^{144}Nd$ isotope evolution in the mantle is a useful proxy for the growth of the continents.

A number of models have been proposed which link crustal evolution with the mantle Nd-isotope evolution curve, the most realistic of which is probably the transport-balance model of Nagler and Kramers (1998). This model is based upon their empirically derived Nd-isotope mantle evolution curve and assumes that the upper mantle melts to form basaltic oceanic crust, which is then reprocessed to form continental crust. An important aspect of the model is that it also includes crustal recycling, such that as the volume of continental crust grows with time, proportionately some crust is recycled back into the mantle through erosion and subduction.

Their best-fit models are shown in Fig. 4.8 and support a progressive growth model for the continental crust, with about 50% crust formed by the end of the Archaean. Their calculations show that there was little crustal recycling before the mid-Proterozoic.

4.2.1.1.4 Continental freeboard arguments

The continental freeboard is the position of sea-level relative to the continental masses. It is thought that the continental freeboard has remained approximately constant, within 1 km of the present level, for at least 2,000 Ma. If this observation is true, then it suggests that the

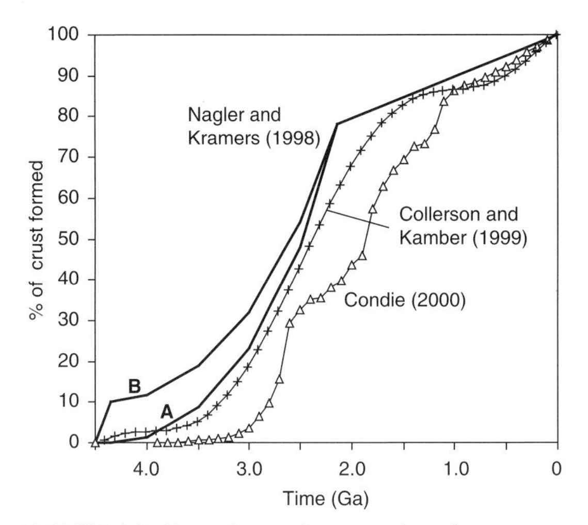

FIGURE 4.8 Crustal growth curves based upon transport-balance modeling (Nagler & Kramers, 1998), U-Pb zircon ages from juvenile crust (Condie, 2000) and Nb/U ratios for the depleted mantle (Collerson & Kamber, 1999). The Nagler and Kramers model assumes either 0% crust at 4.4 Ga (A) or 10% crust at 4.4 Ga (B), and their calculations are consistent with Pb isotope modeling and Hf-isotope studies.

continental volume has remained essentially constant, relative to the oceans, throughout the last 2,000 Ma, thus supporting a "no-growth" scenario for the past 2.0 Ga. However, a major assumption of the freeboard argument is that the volume for the ocean basins has also remained constant. This was questioned by Reymer and Schubert (1984), who pointed out that the volume of the ocean basins should expand with time because of the secular decline in terrestrial heat flow. If this is the case then continental *growth* is necessary to maintain an approximately constant freeboard, reversing the original argument for no growth in favor of progressive continental growth.

4.2.1.2 Crustal growth curves

Crustal growth curves based upon Nd-isotopic modeling, the Nb/U ratio in the mantle, and U–Pb zircon ages of juvenile crust are shown in Fig. 4.8.

4.2.1.2.1 The steady state (no-growth) model

On balance none of the evidence presented above supports a steady state model for crustal growth. Although extremely old zircons are now known, signifying the existence of felsic rocks at 4.4–4.5 Ga (Harrison et al., 2005), on balance the age distribution of crustal rocks still weighs against this idea. Furthermore, the steady state model is not supported by isotopic studies of the depleted mantle. Some years ago it was thought that mantle Nb/U ratios provided evidence for an extensive early continental crust, but this has now been shown to be subject to major uncertainties and also must be rejected. Continuing investigations, using new isotopic techniques continue to find no support for the steady state model. For example, a recent, novel study of the isotope ^{40}Ca, the decay product of the short-lived isotope ^{40}K, failed to find evidence of significant volumes of Hadean felsic crust (Kreissig & Elliott, 2005).

4.2.1.2.2 The progressive growth model

Instead, much of the geological data support a progressive growth model for the continental crust. This is consistent with the great rarity of very old felsic rocks and very old zircons and also with the trace element and isotopic memory of the mantle. In fact it is very difficult to find any evidence to support the existence of large volumes of felsic rocks present in the very early Earth.

4.2.1.2.3 The "episodic" progressive growth model

A variant of the progressive crustal growth model is the episodic crustal growth model most recently advocated by Condie (1998, 2000). Condie has argued that the age distribution of juvenile crustal rocks, as determined from U–Pb zircon ages, is distinctly episodic, with peaks at 2.7, 1.9 and 1.2 Ga. There is some support for this model from Nd model ages and there may be evidence for an additional peak in the Phanerozoic (as discussed above). Condie (1998, 2000) has suggested that the peaks in crustal growth represent mantle plume activity triggered by avalanching of subducted material into the mantle. If this model is correct we should expect to find a significant plume signature in the geochemistry of the continental crust.

The major difficulty with the episodic distribution of crustal ages is understanding what the patterns mean. Are they sampling the result of a preservation bias and do not fully represent crustal growth as it really was? Alternatively, do they represent a single mechanism of crustal growth (say subduction related) with the peaks in the growth curve representing accelerated periods of crustal growth at a certain point in time (Kemp & Hawkesworth, 2003)? A third possibility is that there are two different mechanisms of crustal growth, one of which is the "baseline mechanism" and which may be subduction related and another which is episodic and maybe, as proposed by Condie (1998, 2000), plume related.

4.2.1.2.4 Which growth curve?

Three different crustal growth curves are shown in Fig. 4.8. The Nagler and Kramers (1998) curves are based upon the Nd-isotopic composition of the mantle. The Collerson and Kamber (1999) curve is based upon the Nb/U ratio of the mantle and the Condie (2000) curve is based upon U–Pb zircon ages in juvenile crust. All show slightly different rates of crustal growth, particularly in the late Archaean and early Proterozoic.

The relative usefulness of these curves can be assessed in the light of a new study by Hawkesworth and Kemp (2006), Kemp et al. (2006) who used a combination of oxygen and Hf isotopes in zircon to assess the time at which their parent granitoids were derived from the mantle. In this case they used oxygen isotopes in zircon to identify juvenile, mantle derived grains, and zircon Hf-isotope model ages to establish the time at which they were extracted from the mantle. In a case study from the granitoids of the Lachlan fold belt in Australia, they showed a mismatch between the U–Pb crystallization age of the zircons and the time of their derivation from the mantle of up to 1.5 Ga. This result implies that the crystallization age of some zircons is not the same as the time of the derivation of their parent granitoid from the mantle, which may have been much earlier. Some zircon crystallization ages therefore represent the process of crust differentiation, not crust generation. This finding therefore challenges the long held assumption that zircon U–Pb crystallization ages are the same as the time of granitoid crust formation. The implications of this study are that crustal growth curves based upon U–Pb zircon ages, such as those of Condie (2000) and Rino et al. (2004), should be considered as minima and probably underread the true crust formation age. For this reason the curves of Collerson and Kamber (1999) or Nagler and Kramers (1998) are to be preferred.

4.2.1.3 Crustal growth rates

Many authors have sought to calculate growth rates for particular segments of continental crust, for particular points in geological time. The benchmarks are the Phanerozoic magmatic crustal growth rate of 1.65 km^3/yr, and net crustal growth rate of 1.0 km^3/yr, of Reymer and Schubert (1984), based on their study of Phanerozoic arcs (Table 4.1). Where crustal growth rates exceed the Phanerozoic growth rate, processes of crustal growth other than that of arc accretion, such as plume models, are sometimes invoked (Stein & Goldstein, 1996).

Estimates of the present-day, net crustal growth rate vary from about 1.0 km^3/yr (Reymer & Schubert, 1984) to zero (Plank & Langmuir, 1998), although the time-integrated growth rate for the entire continental crust since 4.0 Ga is about 2.0 km^3/ yr (Table 4.1), much greater than present-day estimates, implying that the continental crust has grown very rapidly at some time or times in the past. This observation adds some credibility to a feature which is apparent on all the growth curves, and particularly apparent on the two preferred curves, that a major proportion of the continental crust appears to have formed between about 3.0 Ga and 2.0 Ga (Fig. 4.8). This signifies that there were major episodes of crustal growth early in the history of the Earth. What this means is important for us to discover.

4.2.2 The average composition of the continental crust today

Making an estimate of the average composition of the Earth's continental crust is not a trivial task for it requires a knowledge of the chemical composition of each of the different crustal layers in additions to a knowledge of the relative proportions of these layers

4.2.2.1 Estimating the composition of the upper continental crust

Perhaps the most straightforward task in determining the average composition of the Earth's continental crust is estimating the average composition of the upper crust, for there are a number of natural materials which can be used as proxies. There are two popular approaches. Taylor and McLennan (1985) showed that the average REE pattern of post-Archaean shales shows remarkable uniformity and suggested that marine shales can be used to provide the average insoluble element composition of the upper crust. Loess, fine grained glacial sediments, have also been used as a proxy for average upper crust. Here the logic is that glacial erosion mechanically grinds and blends large areas of bedrock and that these compositions are not subjected to the fractionation processes which take place during chemical weathering. They may therefore more faithfully record crustal compositions than shales, although Gallet et al. (1998) have shown that loess cannot totally escape the effects of chemical weathering.

4.2.2.2 Estimating the composition of the lower crust

More difficult is the task of estimating the composition of the *lower* continental crust. There are three facets to this problem. First, is that it requires the successful integration of geophysical and geochemical observations. Second, is that it requires a judgment about the relative importance of information present in the rare outcrops of lower crustal rocks versus the more common, but fragmentary, xenoliths of lower crustal material. Third, is the thorny issue of whether or not the manner in which the lower crust is formed today is different from early in Earth history.

Deep seismic reflection experiments in the continental crust show that P-wave velocities increase with depth and that in many cases there is a lower crustal layer several kilometers thick which has a P-wave velocity greater than 6.9 km/sec (Durrheim & Mooney, 1991; Rudnick & Fountain, 1995). Many studies have equated this high velocity layer with basaltic compositions and thus infer that the lower continental crust is basaltic (see for example Zandt & Ammon, 1995). This view is also supported by xenolith studies, for the majority of lower crustal xenoliths are mafic (Rudnick & Taylor, 1987).

In contrast, deep crustal rocks found in metamorphic belts, granulites, are much more varied in composition and include both felsic and mafic varieties (Rollinson & Blenkinsop, 1995). This is consistent with a recent study in the North China Craton which showed that there the lower crust can be divided into two layers – an upper lower crust with velocities between 6.7 and 6.8 km/sec and a lowermost crust, with velocities between 7.0 and 7.2 km/sec (Gao et al., 1998), which is interpreted as a mixture of 20–40% mafic and 80–60% felsic rocks. Niu and James (2002) report a study of seismic velocities in the lower crust of the Kaapvaal Craton in South Africa where they found that the base of this 34–40 km thick, and mid-Archaean, crust is made up of felsic to intermediate rocks underlain by a very sharp Moho. It is possible therefore that some regions of lower continental crust are felsic.

4.2.2.3 Models

A particularly influential model for the average composition of the continental crust is the andesite model of Taylor and McLennan (1981, 1985). This model is based upon the observation that modern continental growth primarily takes place at a convergent margin. The model has two principal assumptions. First, that the continental crust could be divided into geochemically distinct upper and lower portions which are present in the proportions of 1 to 2. Second, that the bulk composition of the continental crust is andesitic. Given these premises and an estimated average composition for the upper continental crust, it was possible for Taylor and McLennan to calculate the composition of the lower continental crust, which they found to be basaltic.

The andesite model attracted a great deal of criticism, for there seemed to be a mismatch between Taylor and McLennan's (1981) calculated lower crustal composition and observed lower crustal compositions (Weaver & Tarney, 1982). There was also criticism of their choice of average Australian shale for their upper crustal average and over their choice of crustal thickness in their mass balance model (Christensen & Mooney, 1995). This has led to a number of revisions of the original model.

The two most serious drawbacks with the andesite model are that there is no empirical estimate of lower crustal compositions and there is no serious attempt to quantify the relative proportions of upper and lower crust. So, in a significant leap forward, Rudnick and Fountain (1995) combined their joint expertise in geochemistry and geophysics respectively, to solve these two problems. They used seismic reflection measurements and heat flow data from the lower crust to estimate its broad composition, and then refined their results using compositional data from xenolith suites. They calculated the relative proportions of upper and lower crust by using a weighted average derived from crustal sections taken from several different tectonic settings, where the proportions of upper to lower crust is known either from observation or seismic evidence (Fig. 4.9).

A refinement of this model (Rudnick & Gao, 2003) includes an estimate of the composition and volume of the middle continental crust, ignored in the earlier model of Rudnick and Fountain (1995) and a revised estimate for the composition of the upper continental crust. In this revised model upper, middle, and lower continental crust are in the proportions 31.7%, 29.6%, and 38.8%. The composition is given in Table 4.2 and, to date, it is the best estimate we have for an average composition of the whole crust.

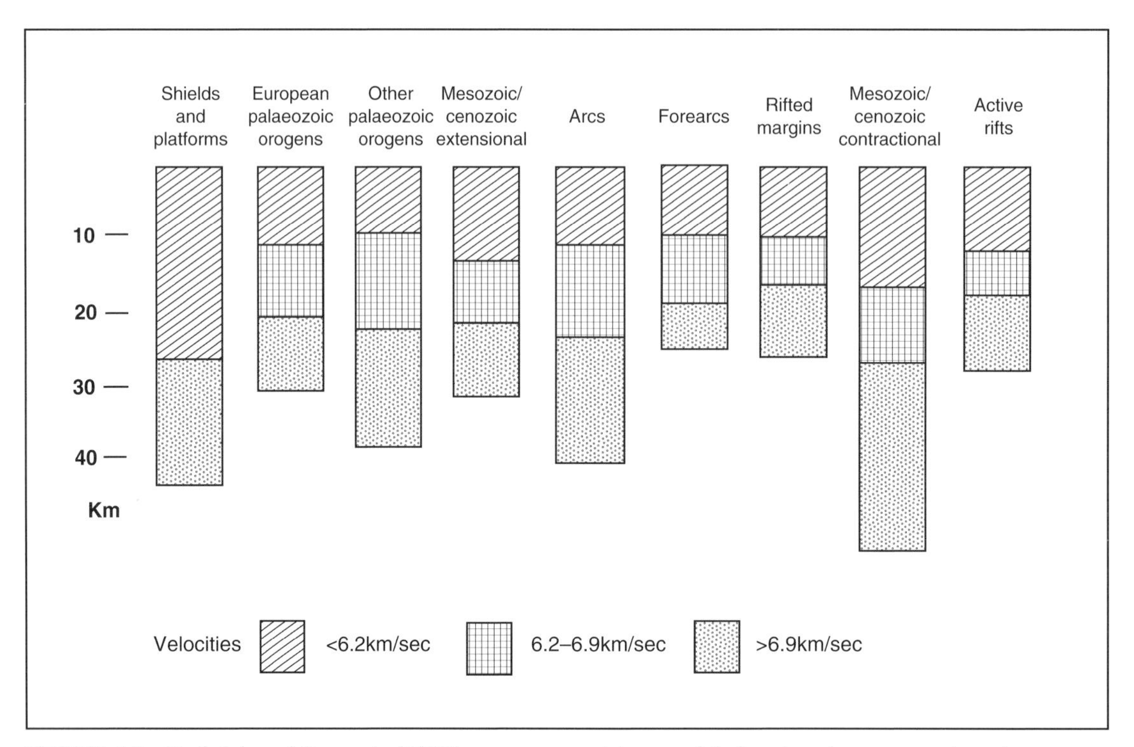

FIGURE 4.9 Rudnick and Fountain (1995) crust composition model showing the type sections for different tectonic provinces used in the model.

TABLE 4.2 Average compositions of the continental crust and Archaean TTGs. 1. Rudnick and Gao (2003), 2. Rudnick and Fountain (1995), Nb and Ta from Barth et al. (2000), 3. Weaver and Tarney (1984), 4. Taylor and McLennan (1985, 1995) and McLennan et al. (2006), 5. Condie (1997), 6. Wedepohl (1995), 7. Condie (2005) – early Archaean, 8. Condie (2005) – late Archaean, 9. Martin et al. (2005) – late Archaean, 10. Smithies (2000), 11. Kamber et al. (2002) – early Archaean.

	Continental crust						Archaean TTG				
	1	2	3	4	5	6	7	8	9	10	11
SiO_2	60.6	59.10	63.20	57.30	59.70	61.61	70.40	68.30	68.36	70.20	68.58
TiO_2	0.7	0.70	0.60	0.90	0.68	0.67	0.31	0.42	0.38	0.33	0.54
Al_2O_3	15.9	15.80	16.10	15.90	15.70	15.04	15.20	15.50	15.52	15.74	14.25
FeO(tot)	6.7	6.60	4.90	9.10	6.50	5.56	2.48	3.04	2.91	2.84	4.24
MnO	0.1	0.11	0.08	0.18	0.09	0.09	0.06	0.07	0.05	0.04	0.06
MgO	4.7	4.40	2.80	5.30	4.30	3.65	0.96	1.39	1.36	1.09	1.03
CaO	6.4	6.40	4.70	7.40	6.00	5.39	2.74	3.26	3.23	3.17	3.41
Na_2O	3.1	3.20	4.20	3.10	3.10	3.18	4.71	4.51	4.70	4.87	4.12
K_2O	1.8	1.90	2.10	1.33	1.80	2.58	2.22	2.20	2.00	1.88	2.52
P_2O_5	0.1	0.20	0.19		0.11	0.17	0.10	0.14	0.15	0.12	0.10
	100.1	98.41	98.87	100.51	97.98	97.94	99.18	98.83	98.66	100.28	98.85
Mg#	55.6	54.3	50.5	50.9	54.1	53.9	40.8	44.9	45.4	40.6	30.3
Cs	2	2.6	–	1.5	–	3.4	–	–	–	–	2.1
Rb	49	58.0	61.0	37.0	53.0	78.0	76	67	67	50	61.78
Ba	456	390.0	707.0	250.0	429.0	584.0	500	769	847	746	295.89
Th	5.6	5.6	5.7	4.2	5.5	8.5	4.1	8.1	-	6	1.67
U	1.3	1.4	1.3	1.1	1.4	1.7	1.2	1.5	-	1	0.28
K	14,942	15,772	17,432	10,999	14,942	21,399	18,428	18,262	16,602	15,606	20,927
Nb	8	8.0	13.0	11.0	8.0	19.0	6.1	6.2	7	5.4	3.69
Ta	0.7	0.7	–	–	–	1.1	0.41	0.84	–		0.34
La	20	18.0	28.0	16.0	18.0	30.0	22	36	30.8	29.8	20.71
Ce	43	42.0	57.0	33.0	42.0	60.0	40	65	58.5	51.6	42.51
Pb	11	12.6	15.0	8.0	13.0	14.8	–	–	–	–	14.8
Pr	4.9	5.0	–	3.9	–	6.7	–	–	–	–	5.6
Sr	320	325	503	260	299	333	362	515	541	495	272
Nd	20	20.0	23.0	16.0	–	27.0	16	25	23.2	19.9	21.63
Sm	3.9	3.9	4.1	3.5	4.0	5.3	2.9	4.2	3.5	2.7	4.23
Zr	132	123.0	210.0	100.0	118.0	203.0	152	154	154	149	169.25
Hf	3.7	3.7	4.7	3.0	3.4	4.9	3.8	4.7	–		4.58
Eu	1.1	1.2	1.1	1.1	1.2	1.3	0.82	1.07	0.9	0.91	1.18
Ti	4,197	4,197	3,597	5,396	4,077	4,010	1,858	2,518	2,278	1,978	3,261
Y	19	20.0	14.0	20.0	21.0	24.0	8.5	9.1	11	6.8	13.07
Ho	0.77	0.8	–	0.8	–	0.8	–	–	–		0.5
Yb	1.9	2.0	1.5	2.2	1.9	2.0	0.82	0.71	0.63	0.46	1.17
Cr	135	119.0					45	35	50		32.97
Ni	59	51					17	22	21		19.62

Interestingly, the differences in major element chemistry between the different estimates of average crustal composition are not great (Rudnick, 1995). Most models have between 57 and 64 wt% SiO_2, and between 0.50 and 0.55 molar Mg/(Mg+Fe), indicating bulk compositions in the andesite range (Fig. 4.10).

More instructive for models of crustal evolution are the trace element concentrations. Average crust is marked by a moderate enrichment of incompatible trace elements and the strong enrichment of the highly incompatible elements such as Cs, Rb, Ba, Th, U, K relative to the Earth's primitive mantle (Fig. 4.11). The elements U, Th, and K are the Earth's principal

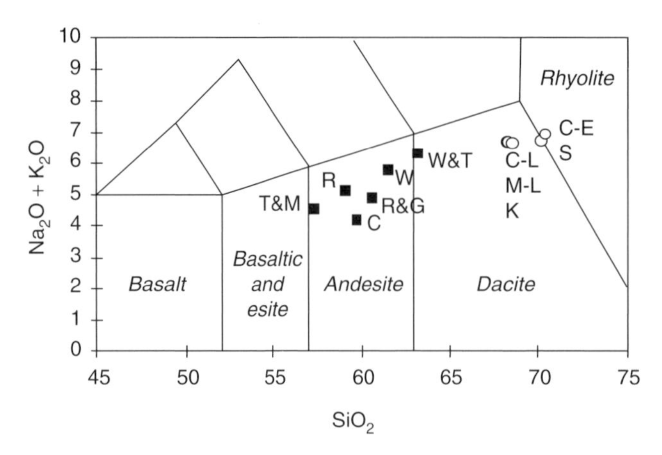

FIGURE 4.10 Total alkalis versus silica classification for the average crustal compositions listed in Table 4.2. Solid squares, average continental crust after Taylor and McLennan (1985) [T&M], Condie (1997) [C], Rudnick and Fountain (1995) [R], Rudnick and Gao (2003) [R&G], Weaver and Tarney (1984) [W&T], Wedepohl (1995) [W]; circles, average TTG compositions after Martin et al. (2005) [M-L (late Archaean)], Condie (2005) [C-E (early Archaean), C-L (late Archaean)], Smithies (2000) [S], Kamber et al. (2002) [K].

heat producing elements, and so their enrichment in the continental crust has implications for the Earth's thermal budget. Relative to the smooth decrease in incompatible elements from Cs to Y on Fig. 4.11 there is a strong positive anomaly for Pb and marked negative anomalies for the elements Ta, Nb, and Ti.

These then are the raw data that must be accommodated in any model for the origin of the continental crust. One way in which they can be used is illustrated by Rudnick (1995) who showed that the crustal La/Nb ratio (1.5) was significantly different from that for modern arc magmas (ca. 3.0), implying perhaps that there was another component present. One possibility is that this component is derived from plume magmatism, and given a plume magma La/Nb ratio of ca 0.8 it is possible to calculate that about 10% of the continental crust is plume derived. However the more recent upward revision of the crustal La/Nb ratio to 2.5 (Plank & Langmuir, 1998; Rudnick & Gao, 2003) has meant that this distinction is now much less secure.

It is important to remember that average compositions are also *time integrated* compositions. This means that, if the composition of the continental crust has changed with time, then this variation is lost in the crustal average, masking what could be an important component of crustal evolution. This is the subject of the next section of this chapter.

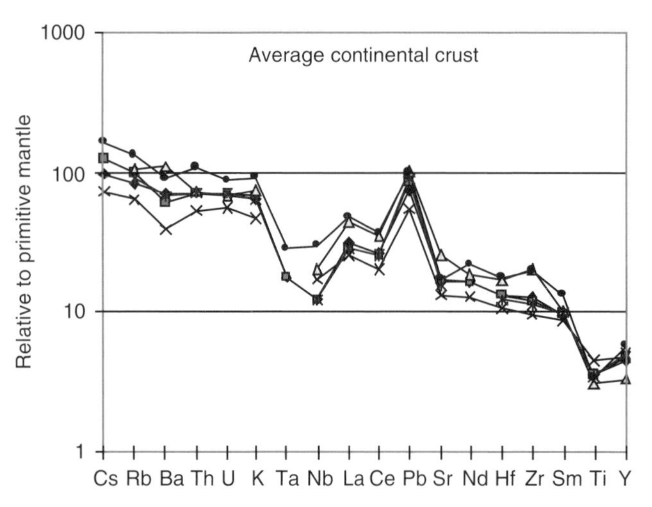

FIGURE 4.11 Trace element concentrations in the average crustal compositions listed in Table 4.1, normalized to the composition of the primitive mantle (after McDonough & Sun, 1995). (See Text Box 2.2.)

4.3 The secular evolution of the Earth's continental crust

A major premise which lies behind much of this book is the belief that Earth has changed, and Earth processes have changed during Earth history. In fact it has already been demonstrated that the chemical composition of the suboceanic, depleted mantle and the subcontinental mantle has changed over time (see Chapter 3). A particularly widely held view in this respect is the belief that the mantle was hotter early in Earth history than today. If this is true, then, as will be demonstrated in Section 4.4, this factor will strongly influence the process of continent generation. In addition factors such as the change in the oxygen concentration of the atmosphere will also exert an influence on element partitioning between continental crust and mantle – see for example Section 4.2.1.1. In this section therefore, we examine the evidence from geophysics and from geochemistry for secular change in the structure and composition of the continental crust and the change of process that this may imply.

4.3.1 Geophysical evidence for the secular evolution of the continental crust

Geophysical studies of the continental crust have raised the possibility that there are differences in both crustal thickness and heat production between Archaean continental crust and juvenile crust that has formed more recently.

4.3.1.1 A change in crustal thickness over time

It was shown earlier that there is some discussion over the composition of the lower continental crust and whether it is felsic or basaltic. One possible solution is that lower continental crustal compositions have changed with time. Durrheim and Mooney (1991, 1994) studied seismic sections through Archaean and Proterozoic cratons and suggested that Archaean continental crust is between 27 and 40 km thick and has a thin mafic layer at its base, making up between 5 and 10% of the total crustal thickness. In contrast Proterozoic crust is between 40 and 55 km thick and has a lower mafic layer which makes up between 20 and 30% of the crustal thickness. However, more recent reviews have argued against this finding, suggesting that observed differences in crustal thickness are more likely the product of differences in tectonic setting (Wever, 1992; Christensen & Mooney, 1995; Rudnick & Fountain, 1995; Zandt & Ammon, 1995). These authors argue that, for example, an average crustal section in an active rift is different from an average crustal section of an ancient shield or platform (Fig. 4.9). For this reason they find that the most important differences in crustal structure arise as a consequence of tectonic setting, rather than as a product of changing crustal thickness over time.

4.3.1.2 Changes in crustal heat flow with time

A related debate focused on heat flow data from different regions of the continental crust. Nyblade and Pollack (1993) showed that average heat flow measurements in Archaean cratons are lower than those for Proterozoic cratons. This observation has, however, been interpreted in two quite different ways. On the one hand, it has been argued that cratons of different age have different bulk compositions. and so have different concentrations of heat producing elements (U, Th and K), hence different levels of heat production. Alternatively, the observed differences in heat flow do not derive from the crust but reflect different lithospheric thicknesses between Proterozoic and Archaean cratons reflecting different mantle heat flow contributions (Rudnick et al., 1998; Nyblade, 1999).

More recently the results of Nyblade and Pollack (1993) have been challenged by Jaupart and Mareschal (1999) who argue that "average" heat flow values for Archaean and Proterozoic cratons are meaningless because the variation within each group of values is so large. Instead, these authors emphasize the importance of the differences in crustal heat flow values *within* Archaean cratons and *within* Proterozoic cratons and suggest that the important observation is that there are different types of crust in both Archaean and Proterozoic cratons.

Despite the failure of these geophysical arguments to unequivocally demonstrate a difference in bulk crustal composition with time, there are a number of geochemical arguments which do support a difference. These are reviewed below.

4.3.2 Geochemical evidence for the secular evolution of the continental crust

The dominant rocks of the Earth's early continental crust are the felsic magmas of the tonalite–trondhjemite–granodiorite suite, or the TTG suite, as it is known (Rollinson, 2006). Compared to average crust TTGs are more felsic, more sodic and have a lower mg-number, and are clearly different from average "andesitic" crust (Table 4.2, Fig. 4.10). This implies that the magmatic flux across the Moho has changed with time (Kemp & Hawkesworth, 2003).

4.3.2.1 Secular variation in the trace element composition of Archaean granitoids

4.3.2.1.1 REE in Archaean granitoids

In a study which has subsequently had a profound influence on our thinking about Archaean crustal evolution, Martin (1986) showed that the REE chemistry of granitoids has changed with time. He showed that granitoids can be divided into two distinct groups according to their age (Fig. 4.12). These differences can be summarized on a plot of $(La/Yb)_n$ versus Yb_n – the chondrite normalized ratio of light to heavy members of the rare Earth element group, that is, the slope of the REE series, plotted against the normalized concentration of the heavy REE Yb. On such a plot granitoids which are older than 2.5 Ga define a field of high $(La/Yb)_n$ ratios and low Yb_n concentrations, whereas granitoids younger than 2.5 Ga plot with lower $(La/Yb)_n$ ratios and higher Yb_n concentrations.

The most interesting aspect of this geochemical difference between Archaean and post-Archaean granitoids, is why this should be so. Martin (1986) concluded that the geochemical differences reflect different mechanisms of formation, one involving garnet in the source and the other not involving. This important difference can be used to begin to define involving the melting process which gave rise to TTG magmas and set limits on the composition of the material which melted. This is a topic to which we will return in the next section of this chapter.

4.3.2.1.2 Rb-Sr in Archaean granitoids

Exploring in some detail the way in which Rb/Sr ratios may change during the process of magma genesis related to the formation of juvenile crust, Ellam and Hawkesworth (1988b) found that there was a difference in geochemistry which, again could be linked to age. They found that Archaean volcanic and plutonic rocks from Zimbabwe show a positive

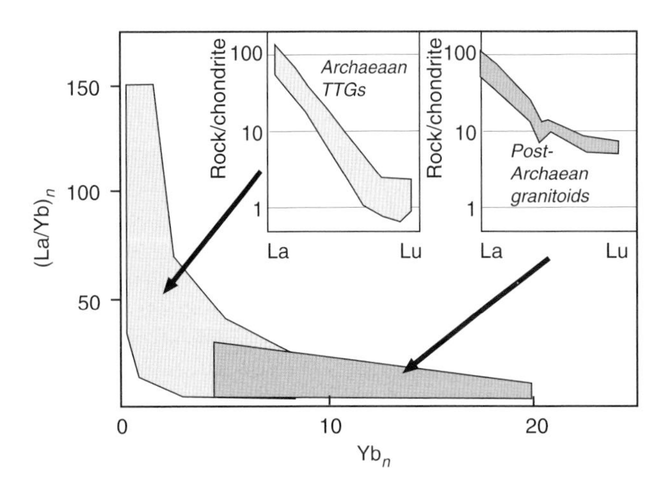

FIGURE 4.12 Chondrite-normalized plot of La/Yb versus Yb showing the field of Archaean granitoids (high $(La/Yb)_n$, low Yb_n) and the field for post-Archaean granitoids (low $(La/Yb)_n$, high Yb_n). The inset diagrams show typical REE patterns for Archaean TTGs and post-Archaean granitoids (data from Martin, 1994, after Rollinson, 2006).

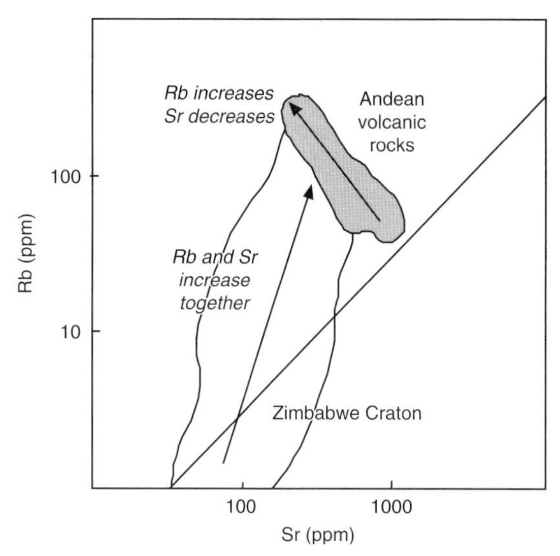

FIGURE 4.13 Rb–Sr concentrations in Archaean felsic igneous rocks from Zimbabwe, and post-Archaean felsic igneous rocks from the Central Andes (after Ellam & Hawkesworth, 1988b).

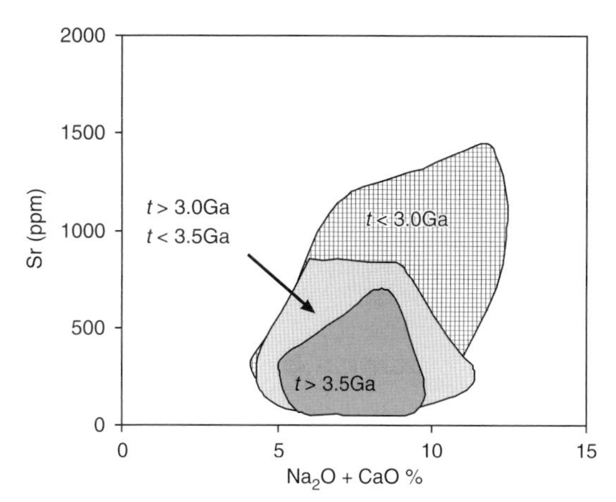

FIGURE 4.14 Compositions of Archaean TTG as a function of age, Sr content and Na_2O + CaO content showing increasing Sr and Na_2O + CaO with decreasing age during the Archaean (after Martin & Moyen, 2002).

correlation between Rb and Sr, whereas geologically recent andesites and dacites from the Andes show a negative correlation (Fig. 4.13). They explained the difference between the two magma series as being due to the influence of plagioclase in the melting process, since Sr preferentially partitions into plagioclase relative to Rb in crustal magmatic systems. What this might mean is that in the Archaean, crustal rocks were produced from a source from which plagioclase was absent, whereas geologically recent crustal rocks are produced in the presence of plagioclase, either in the source or during subsequent fractionation of the melt.

4.3.2.1.3 Secular change in granitoid chemistry during the Archaean

Since some of the most extreme changes that have taken place within Earth history have taken place within the Archaean, there is a case for supposing that secular change might also be visible *within* the Archaean. This is the thesis of Martin and Moyen (2002) who found that chemical composition of Archaean TTGs evolved over time from 4.0 Ga to 2.5 Ga. They show that during the Archaean TTGs become progressively enriched in the elements Mg, Ni, Cr, Sr, and (Na_2O+CaO) (Fig. 4.14). They interpret these differences in terms of the progressive importance of the mantle and the decreasing importance of plagioclase in the source region of TTGs, which probably reflects an increase in the depth of melting of the TTG source region during the Archaean.

4.3.2.2 Trace elements in Archaean sediments

It has already been shown that some sedimentary rocks, notably shales, can provide a very useful average composition for the upper continental crust. An obvious extension of this argument is to compare shales from the modern with those from early in Earth history to see if they record a secular change. Much of this work has been done by Taylor and McLennan, summarized in their book of 1985 and in subsequent papers (see e.g. McLennan et al., 2005).

Chemical differences are indeed evident between average shale composition. Post-Archaean sediments are characterized by higher levels of incompatible elements, higher K/Na, Th/Sc, Th/La, Th, and U than Archaean sediments (Fig. 4.15). A comparison of rare Earth

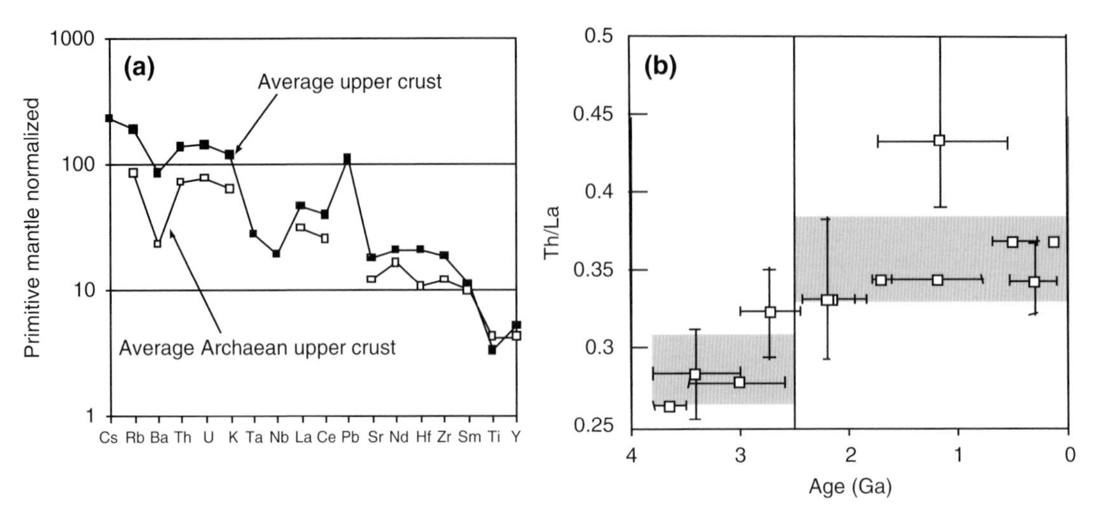

FIGURE 4.15 (a) Primitive mantle-normalized compositions for the present day upper continental crust and average Archaean upper crust, from McLennan et al. (2005). The comparison show the relative enrichment in incompatible elements in present-day upper continental crust (see Text Box 2.2). (b) The secular change in Th/La ratio in the upper continental crust is from the compilation by Plank (2005). The grey bands are the Archaean and post-Archaean averages from Taylor and McLennan (1985).

element patterns shows that post-Archaean sediments are characterized by a negative Eu anomaly, relatively flat heavy REE and steeper light REE than Archaean sediments.

The meaning of a secular change in sediment composition has been the subject of much debate. As mentioned above, the conventional understanding is that it reflects a change in the average composition of the Earth's upper continental crust. However, what a change in the composition of the upper continental crustal compositions might mean is more problematical. Is the difference a reflection of different mechanisms of crust formation, or does it indicate different amounts of intracrustal fractionation over geological time? Furthermore, to what extent does this change also reflect different weathering and erosion regimes during Earth history, driven in part by changes in atmospheric composition?

4.4 Crustal growth during the Archaean

The dominant rock-types of the Archaean continental crust are those granitoids of the tonalite–trondhjemite–granodiorite (TTG) magmatic suite. Chemically they are defined by their CIPW normative feldspar content, as plotted on an O'Connor granitoid classification diagram, where parental melt compositions plot close to the tonalite–trondhjemite–granodiorite triple point (Fig. 4.16). Compositionally TTGs are different from average crust, in that they are enriched in Na_2O and SiO_2 and depleted in CaO, FeO, MgO and TiO_2 (Table 4.2, Fig. 4.16).

4.4.1 Constraints on the origin of Archaean TTGs

Many Archaean TTG suites have mantle-like Sr and Nd initial isotope ratios indicating that they were originally derived from the mantle. In this section we use the constraints that can be obtained from their geochemistry, from recent experimental studies on basaltic melting, and from thermal modeling to set limits on their origin. Determining the detail of this process is complex, for many of the key variables are not known with any degree of certainty. Not least is the problem of evaluating the relative importance of the source composition and of the melting process on TTG melt compositions. This was illustrated recently in a debate on the origin of TTGs which has

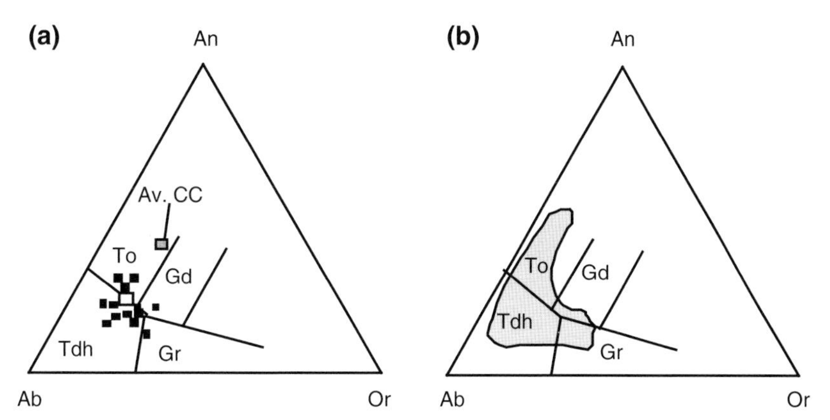

FIGURE 4.16 O'Connor Ab–An–Or diagram showing the fields of Tonalites (To), Trondhjemites (Tdh), Granodiorites (Gd) and Granites (Gr); (a) typical TTG compositions; the average TTG of Martin (1994) is shown as the white square; the average continental crust of Rudnick and Gao (2003) is the grey square; (b) typical melt compositions produced during the dehydration melting of amphibolites (after Martin, 1994 and Rollinson, 2006).

foundered on this very difficulty, one group emphasizing the importance of source chemistry, the other the importance of process (Foley et al., 2002; Rapp et al., 2003).

4.4.1.1 Experimental studies

A large number of experimental studies have demonstrated that when basalt is melted under hydrous conditions the resultant melts closely resemble TTGs in composition (Fig. 4.16b) – see reviews by Wyllie and Wolf (1993); Wyllie et al. (1997) and Rapp (1997). It is the detail of these experiments, however, which is useful in constraining the process of TTG genesis. Crucial to the partial melting process is the presence of water, for water lowers the basalt solidus into a temperature range where melting can more readily take place. The availability of water, therefore, is a major constraint on models of TTG genesis. Other key controls are the precise composition of the parent basalt, a factor which is not yet fully known, and the degree of partial melting, thought to be normally between 10 and 20%. These different parameters govern the mineralogy of the residue in equilibrium with a TTG melt.

The critical reactions were summarized by Wyllie et al. (1997) and are shown in Fig. 4.17. The point at which melting begins is defined by the solidus. This is S-shaped so that at pressures below about 1 GPa melting does not commence until about 900°C, whereas at higher

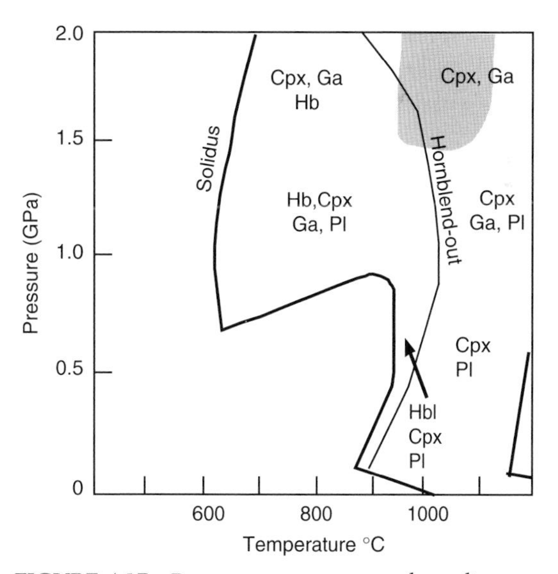

FIGURE 4.17 Pressure–temperature phase diagram showing the amphibolite solidus and the residual minerals in equilibrium with TTG melts at different pressures and temperatures (after Wyllie et al., 1997; Rollinson, 2006). The shaded field is the preferred melting region of Moyen and Stevens (2005) based upon their review of the experimental evidence.

pressures it begins at between 600 and 700°C. The hornblende-out (the point at which all the mineral hornblende has been consumed in the melting reaction), garnet-in, and plagioclase-out mineral stability curves subdivide the melting region into six assemblages of residual minerals coexisting with a TTG melt. These vary from amphibolitic, that is, they include hornblende to eclogitic (are rich in garnet and clinopyroxene).

It is the identity of these mineral assemblages which is important to a full understanding of the process of TTG genesis. For example, a knowledge of the residual phases coexisting with TTG melt makes it possible, through geochemical modeling, to predict the trace element composition of TTG melts formed during basalt melting and compare them with those actually found. The inverse process allows the measured trace element concentrations in TTGs to be used to make semiquantitative estimates of the pressure and temperature conditions of melting (see e.g. Rollinson, 1996). This approach allows models of TTG genesis to be tested. For example, it can be seen from Fig. 4.17 that some TTG melts are in equilibrium with an eclogitic, that is a garnet + clinopyroxene residue. Such an association was demonstrated by Rollinson (1997) from the Archaean West African Craton, where crustal TTGs and eclogite xenoliths from 150 km beneath that craton, now in kimberlites, coexist and have compositions suggesting geochemical complementarity.

An area of some uncertainty in basalt–TTG dehydration melting experiments is the choice of an appropriate basaltic starting composition. This is extremely important in the modeling of TTG trace element compositions. In an empirical study, Rollinson and Fowler (1987) showed from geochemical modeling, that incompatible element rich meta-tholeiites form the most appropriate source for TTGs, a point which has been recently repeated by Foley et al. (2002) and Kemp and Hawkesworth (2003). This suggests that TTGs are not the product of the melting of a depleted MORB-like basalt, nor a typical Archaean tholeiite (Chapter 3, Section 3.2.1.1), but a basalt which is enriched in incompatible elements. Such an observation could point toward a plume basalt contribution in the genesis of the continental crust.

4.4.1.2 Geochemistry

When compared with geologically recent granitoids Archaean TTGs are enriched in LREE and Sr, and depleted in heavy REE, Sc, and Y. As already noted (Fig. 4.12) they have steep REE patterns with concave heavy REE, and the negative Eu anomaly frequently found in Phanerozoic granites is absent. On a primitive mantle-normalized trace element diagram they have a marked negative Nb–Ta anomaly (relative to La) and when compared with average crust have lower Nb/Ta and high Zr/Sm ratios (Fig. 4.18). They also have a negative Ti anomaly (relative to Sm) and low Ni, Cr, and V concentrations (Martin, 1994).

These geochemical characteristics are consistent with the derivation of TTGs from a hydrated basaltic source, as outlined in the previous section, provided that the partial melting takes place in the garnet stability field (thus accounting for the low heavy REE, Y, Sc) and in the absence of plagioclase (explaining the high Sr, no Eu anomaly). The low Nb/Ta, high Zr/Sm ratios and the concave shape of the REE pattern are attributed to the presence

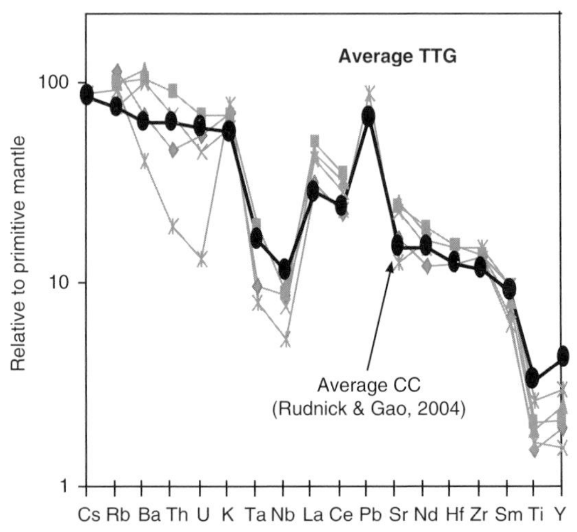

FIGURE 4.18 Plot of trace elements in average TTG (grey lines with symbols) (from Table 4.2) compared with trace elements in the average crust (heavy line with black symbols) (values of Rudnick & Gao, 2003). (See Text Box 2.2.)

of amphibole in the source, indicating that the basaltic source was hydrated (Foley et al., 2002).

These geochemical arguments imply that TTGs have been derived from the mantle in two stages. First the mantle melted to form a basalt and then that basalt became hydrated and remelted to form a TTG magma. However, the fact that TTGs have mantle-like initial Sr and Nd ratios means that the basaltic TTG precursor had a short crustal residence time (i.e. there was a short time interval between the creation and destruction of the basaltic crust), the normal case for oceanic crust.

Recently there has been some debate, based upon trace element partitioning between TTGs and their residual phases, over the depth of basalt melting during TTG genesis. Using similar arguments Foley et al. (2002) have argued for relatively shallow melting in amphibolite facies, whereas Rapp et al. (2003) prefer deeper, eclogite facies melting. A recent reassessment of the experimental data for basalt melting by Moyen and Stevens (2005) has shown, from trace element distributions, that the likely P–T range for the production of TTG melts are >15 kbar and 900–1,100°C (Fig. 4.18), a narrower range than previously proposed, but within the eclogite facies as proposed by Rapp et al. (2003).

4.4.1.3 Thermal modeling

A logical test of the hypothesis that TTG melts are produced by the dehydration melting of wet basalt is to explore the conditions under which such melting might take place. In a series of thermal modeling experiments Peacock et al. (1994) found that the most important control on the partial melting of a subducting basaltic slab was the age of the slab. They found that hot, young oceanic lithosphere melts more readily than cooler, older oceanic lithosphere. Their calculations showed that for melting to take place, the oceanic lithosphere must be less than 2 Ma old if subduction is fast and less than 5 Ma old if subduction is slow, implying that a slab can melt only under rather special conditions (Fig. 4.19). It has been argued that a hotter Archaean mantle makes such a process more probable early in Earth history.

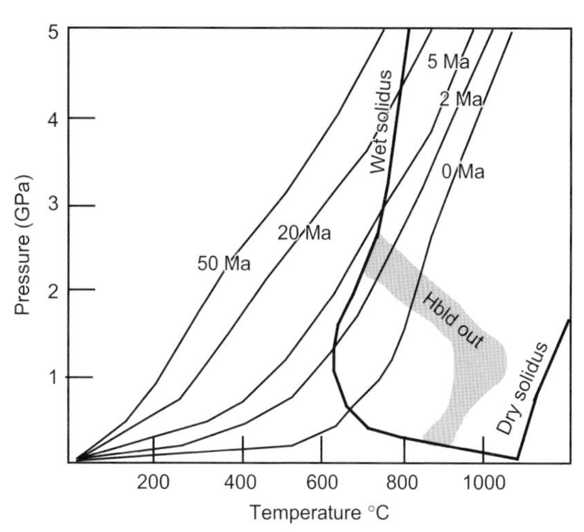

FIGURE 4.19 Pressure–temperature phase diagram showing the results of the thermal modeling experiment of Peacock et al. (1994) for the subduction of ocean crust at a rate of 1 cm/yr. The curves labeled 0 Ma to 50 Ma show the P–T path for subducted ocean floor 0–50 Ma old. Also shown are the wet and dry solidi for basalt melting and the hornblende-out curve. It can be seen that only the P–T trajectories of young ocean floor intersect the wet basalt solidus (after Rollinson, 2006).

However, even if melting does take place within a subducting slab some authors have suggested that this does not necessarily mean that the melt can be easily extracted from its source. This problem was investigated in a thermal modeling experiment by Jackson et al. (2005) who show that large volumes of TTG melt can segregate from a basaltic parent over geologically realistic timescales. They also showed that the composition of the melt is as much a function of the timescale of melting (rapid or slow) and the precise manner in which the melt segregates and migrates, as it is pressure, temperature, and starting composition. Hence, a variety of melts can be produced from an identical source.

4.4.2 Models

The evidence from experimental studies, geochemical and thermal modeling outlined above, converges to support the view that the most probable mechanism for TTG genesis is the partial melting of hydrated basalt. What is not agreed, however, is the tectonic setting in which this partial melting took place. This is the subject of considerable discussion, and of great importance, for a proper knowledge of the mechanism whereby TTG magmas are produced will provide important insights into one of the major tectonic processes in the early Earth system.

4.4.2.1 *The slab melting model*

For many years it has been argued that TTG genesis took place in a subduction zone through the partial melting of a subducting basaltic slab. This is an attractive model because it is efficient, it provides a mechanism for water to be introduced into the system, through the hydration of ocean floor basalts, and allows the generation of large volumes of TTG melts in a uniformitarian manner. However, this approach makes two assumptions. First, that plate tectonics was operating during the Archaean, and second that slab melting, a rather rare process in the modern, was the main mechanism of TTG genesis. The principal argument in support of this model is that heat production was greater during the Archaean and the Earth's mantle was hotter, and under these conditions slab melting would have been possible.

4.4.2.1.1 *The adakite analogy*

A particularly powerful argument in support of TTG melt production by slab melting is the adakite analogy. Adakites are rather rare, modern, TTG-like magmas. They take their name from an island in the Aleutian chain where they were first documented (Kay, 1978). Adakites are geochemically different from normal arc magmas and show a large number of geochemical similarities with Archaean TTGs (Fig. 4.20). They are produced in subduction settings where there is unusually high heat flow, such as sites of ridge subduction. A close examination of this process by Thorkelson and Breitsprecher (2005) shows that melting most readily takes place in tectonic settings where the slab has broken during subduction and that melting takes place along the thinned edge of the slab. This tectonic environment is known as a slab window. Thus the evidence from the tectonic setting of modern adakites strongly indicates that they are slab melts (see Section 4.1.2).

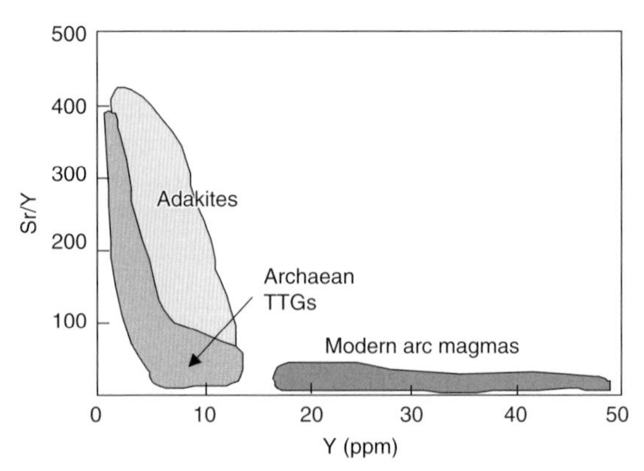

FIGURE 4.20 Plot of Sr/Y versus Y showing the fields for adakites, Archaean TTGs and modern arc magmas (after Drummond & Defant, 1990 and Rollinson, 2006).

The geochemical similarity between Archaean TTGs and adakites strongly suggests that TTGs are also slab melts (Martin, 1999).

4.4.2.2 *Composite models – multiple sources during subduction melting*

A variant of the slab melting model for Archaean TTG generation draws on the observation that modern subduction zone magmas are the product of mixing and derived from multiple sources (Section 4.1.2). Rollinson (2006) pointed out the inadequacy of a simple slab melting model by showing that (1) many TTGs contain a mantle component, (2) some contain an older crustal component, either from recycled sediment or ancient crust (Berger & Rollinson, 1997; Henry et al., 1998; Halla, 2005), and (3) that some TTGs have a contribution from a dehydrated slab (Rollinson & Tarney, 2005). These observations combined with our understanding of the process of modern arc magma genesis indicate that TTGs might be a mixture of at least three different components – a slab melt, a melt from the mantle wedge, and melted subducted sediment. This is in addition to an older crustal component. Martin and Moyen (2002) outlined a range of possible scenarios in which this mixing might take place. Their model is based upon slab melting, in which the main variable is the angle of dip of the slab. Where the angle of dip is low, the slab will melt at shallow depth and there will be minimal interaction with the mantle wedge. At a higher angle of dip, slab melting will take place at a deeper level, and some interaction between the slab melt and the mantle wedge is more probable.

4.4.2.3 Underplating and the melting of basaltic lower crust

TTGs can also be formed by the partial melting of basaltic lower crust underplated at a continental margin. In this model the basic premise of basalt melting is retained, but the tectonic setting is different and so is an important alternative to crust generation models which require subduction. This is the explanation given for geologically recent TTGs formed in one of the granitoid batholiths of the Andean continental margin (Atherton & Petford, 1993), where there is no evidence for subduction contemporaneous with TTG genesis. The model might be extended into the Archaean to envisage basalt melting beneath arc-like crust or at the base of a thick pile of oceanic basalts.

This model was given added impetus by Smithies (2000) and Kamber et al. (2002) who showed that there are significant chemical differences between Archaean TTGs and adakites, weakening the adakite–TTG-slab melting analogy. Smithies (2000) showed that adakites are less siliceous and have a higher mg# than TTGs, and Kamber et al. (2002) showed that they have different fluid-mobile element (e.g. Sr, K, Rb, Ba) characteristics. Given that the mantle is highly magnesian, the difference in mg# between TTG and adakite is thought to reflect the different extent to which the magmas have "seen" the mantle *en route* to the surface. Adakites, therefore, with a high mg# must have interacted with the mantle wedge on their way to the surface, whereas TTGs for some reason have not. If both melts were produced at the same depth, which is what the experimental evidence suggests, then TTGs must have found their way to the surface through lithologies less magnesian than the peridotites of the mantle wedge. On this basis it is proposed that TTGs are not slab melts but rather are melts of basaltic lower crust, in a thick basaltic lava pile. One environment today where very thick basaltic lava sequences form is in an oceanic plateau.

Additional support for a lower crustal melting model comes from thermal modeling experiments, in which the thermal effects of multiple basaltic intrusions injected into basaltic lower crust are calculated (Jackson et al., 2005). There are three important results from these experiments, First, that significant melting can take place in the lower crust of a basaltic lava sequences, second that the melt can be extracted from its source, and finally that the melt is TTG in composition.

4.4.2.4 Plume/oceanic plateau models

This model is a variant of the basalt lower crust melting model described above, but in this case basalt melting takes place at the base of thick basaltic crust. Crust of this type may form through imbrication and stacking of normal-thickness oceanic crust or may form as initially thick crust in an oceanic plateau as the product of mantle plume-related magmatism (Chapter 3, Section 3.1.5).

An advocate of this model is Albarede (1998) who argues that the episodicity of continental growth is linked to activity in the mantle, which strongly suggests growth through mantle plume activity. Similar arguments have also been advanced by Condie (2000) – see Section 4.2.1.2. A much cited example of this form of rapid crustal growth is the Birimian Shield in west Africa. Here, a huge area of juvenile crust formed within 50 Ma, starting with extensive oceanic plateau basalts at 2.1 Ga and rapidly followed by the generation of large volumes of granitoids (Abouchami et al., 1990 Boher, et al. 1992).

In this model of crust generation TTG genesis may take place through partial melting at the base of a thick basaltic lava pile, such as is found in an oceanic plateau. In this case the sinking of a mafic eclogite crustal root into the underlying mantle may be the trigger for melting (Zegers & van Keken, 2001). A possible modern analog of this process is the tonalite of the Aruba batholith, part of the Caribbean oceanic plateau (White et al., 1999).

4.4.2.5 Discussion

Over the past few years the tectonic setting of Archaean TTG generation has been hotly debated. Most recently the focus of the debate has been between the melting of basalt in either the lower crust or in a subducting slab. In summary, because the adakite analog for TTGs was thought to be incorrect a mafic lower crustal source for TTGs had been proposed. However, this conflict has now been resolved as a result of a new analysis of adakitic lava chemistry. A recent comprehensive study has shown that there are two major groups of adakites. High-silica adakites ($SiO_2 > 60\%$) have the character of slab melts which have interacted with the mantle wedge, whereas low-silica adakites are thought to be melts of metasomatized mantle (Martin et al., 2005). This finding provides a solution to the adakite–TTG analogue problem, for now there is greater clarity over what is meant by the term adakite. Most important is the realization that there is compositional similarity between middle and late Archaean TTGs (<3.3 Ga) and modern high-silica adakites, strongly supporting the earlier view that TTGs are indeed slab melts (Martin et al., 2005).

Further evidence confirming slab melting as an important process, at least after the early Archaean, come from the revised estimates of the probable TTG melting regime. Moyen and Stevens (2005) propose that melting took place between 900 and 1,100°C at >15 kbar (ca. 50 km) depth, implying a relatively cool geothermal gradient of about 20°C/km. They argue that such geothermal gradients are most likely found in a subduction zone and that the likely depth of melting (>50 km) is more likely attained in a subduction zone rather than in deeply buried basaltic crust.

What is not yet resolved is the origin of the early Archaean, low-mg# TTGs. These rocks have melt compositions which have not interacted with the mantle wedge. This might mean, as discussed above, that they were produced in a thick basaltic lava pile. Alternatively, they were derived by slab melting but the melt did not interact with the mantle because the melting took place at shallow depth, because of the low angle dip of the slab. This secular evolution in TTG compositions linked it would seem, to progressive slab-melt mantle interaction over geological time, lends itself to a tectonic model in which the angle of dip of the subducting slab increased with time (Martin et al., 2005). In summary it would seem that there was some change in the process of crust generation which took place *during* the Archaean.

The presence of water in the melting environment is also important for TTG genesis and it is difficult to see how enough water could be provided to the base of thick basalt crust to sustain extensive melting. This is a major difficulty for the underplating model. In contrast, a subduction setting provides a continuous supply of ocean water through metamorphic dehydration reactions (Kamber et al., 2002), again making it a more probable tectonic environment for melting.

A central element in the debate over the origin of TTGs is the competing contributions of an intraplate, plume-related process and subduction-related processes. For a long time the prevailing view has been that a subduction-related process is the most probable. However, there are two lines of evidence, which keep reappearing, which suggest that a plume component cannot be ruled out. Kemp and Hawkesworth (2003) point out that the basaltic source for TTG magmatism was enriched in Th relative to MORB, but had a primitive mantle-like Th/La ratio, implying a source which is enriched, but not with a subduction or crustal component. In addition there is evidence from the apparent episodicity of the crustal growth curve, which may also imply a mantle plume contribution to crustal growth. Thus there is an important future agenda to identify and to quantify the relative contributions of the subduction and plume components during Archaean crust generation. For this we shall need to start by collecting more precise trace element data for Archaean TTGs.

4.5 CRUST–MANTLE INTERACTIONS – RESERVOIRS AND FLUXES

It is generally agreed that the ultimate source of the Earth's continental crust is the mantle. Thus the growth history of the continental crust through time is also the history of its separation from the upper mantle. This section briefly reviews current ideas on how the crust and mantle reservoirs might be linked and how these linkages may have changed through time. It then considers the evidence for ancient and modern fluxes from the upper mantle to the continental crust.

4.5.1 Reservoirs

There is good trace element and isotopic evidence to support the view that the continental crust and the depleted upper mantle are complementary geochemical reservoirs relative to the composition of the Earth's primitive mantle. The trace element enriched nature of the continental crust relative to the primitive mantle was illustrated in Fig. 4.11, and, it is argued, the depleted mantle is the complementary reservoir. At first sight the trace element enriched character of the Earth's continental crust looks like a 1% melt of the upper mantle (O'Nions & McKenzie, 1988). However, that is not to say that this is the precise mechanism of its extraction, for McKenzie (1989) showed that it is not possible to extract such a small melt fraction from the mantle. Thus as already discussed, the extraction of the continental crust from the mantle must be a multistage process.

In this section we first examine a basic model which outlines the key crust–mantle relationships, and then, in the light of additional data we explore more complex models of crust–mantle interaction.

4.5.1.1 Crust–mantle reservoirs – the basic model

The simplest way of thinking about the relationship between the continental crust and mantle is to use a three reservoir box model, in which an initial primitive mantle composition (reservoir 1) is progressively differentiated through time into a depleted mantle reservoir (reservoir 2) and the continental crust (reservoir 3). This simple approach provides a good explanation of the Nd- and Sr-isotope compositions of the mantle and crust. On a Nd–Sr isotope plot, crustal compositions are complementary to the mantle isotopic array (Fig. 4.21a) relative to the composition of the primitive mantle. Expressing the Nd-isotope data in a slightly different manner, an ε_{Nd} versus time

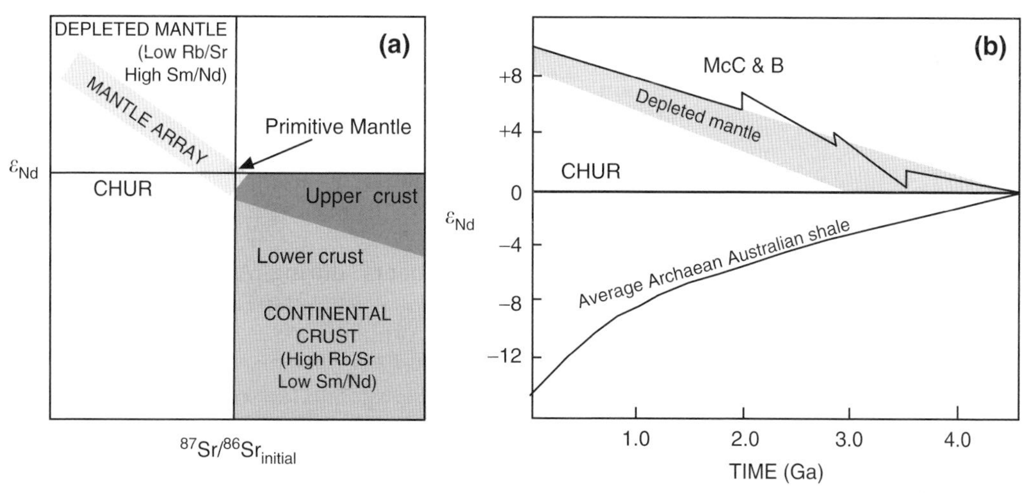

FIGURE 4.21 (a) ε_{Nd} versus initial Sr plot showing the complementary compositions of the depleted mantle and the continental crust relative to that of the Primitive Mantle; (b) ε_{Nd} versus time diagram for the depleted mantle (grey shaded area; also shown is the stepwise curve of McCulloch and Bennett, 1994 – McC&B) and average Archaean Australian shale – a proxy for the composition of the continental crust, showing the complementary evolution of each with time, relative to the chondritic reservoir (CHUR) (after Rollinson, 1993).

diagram also shows the complementary nature of the continental crust and depleted mantle. Over time, as the continental crust was extracted, the mantle has become progressively depleted in Sm and Nd, but at the same time Sm and Nd were fractionated relative to each other leading to an *increase* in ε_{Nd}, relative to CHUR. The complementary process is seen in the continental crust, illustrated in this case by a proxy – the shale average, showing that over time the continental crust becomes more *depleted* in ε_{Nd} (Fig. 4.21b).

4.5.1.2 How many mantle reservoirs are required to make the continental crust?

As the volume and quality of isotopic and trace element data for the continental crust and mantle have improved in recent years, it has become increasingly clear that the qualitative, three reservoir model outlined above is inadequate and cannot fully explain all the geochemical features of the crust–mantle system. As more trace element data are considered it is apparent that additional reservoirs need to be considered in addition to the primitive mantle, the depleted mantle, and the continental crust.

4.5.1.2.1 A Contribution from the Lower Mantle

A central question for models of crust extraction from the mantle is whether or not the continental crust has been extracted from the whole mantle or just from the "depleted" upper mantle. The debate between geophysicists and geochemists over whether there is whole mantle convection, or whether the mantle convects as two independent layers was outlined in Chapter 3, Section 3.1.5. Hofmann et al. (1986) argued that the mass balance of trace element ratios for Nb/U and Ce/Pb in the continental crust and upper mantle require that the continental crust be extracted from a volume of mantle that equates to about 50% of the whole mantle – a volume much larger than that of the modern depleted mantle as defined by the 660 km discontinuity. More recently Helffrich and Wood (2001) have confirmed this estimate on the basis of calculations using K, U, and Th concentrations.

Stein and Hofmann (1994) sought to resolve this problem by proposing that, over geological time, there have been periodic influxes of lower mantle material into the upper mantle to "recharge it" with the necessary trace elements to make the continental crust. These periodic influxes were thought to have taken the form of mantle plumes originating in the lower mantle, and some workers see this model as support for a plume model of continental growth.

Calculations of this type can, however, be turned on their head and used to argue that since the volume of mantle required to make the continental crust is inconsistent with the volume of the depleted mantle the mantle cannot be chemically layered. In that case we arrive at a model for the mantle in which there is a depleted portion, from which the continental crust has been extracted, and an undepleted portion, but the two are not confined to discrete layers. In addition trace element constraints outlined below will show that other reservoirs are also present.

4.5.1.2.2 Buried ancient basaltic crust

There is good evidence from Nd-, Hf- and Pb-isotopes that our planet experienced a very early differentiation of the mantle. Each of these isotope systems preserves a record of an extreme mantle depletion event very early in Earth history – now known to have taken place within 30 Ma of the formation of the solar system (Boyet & Carlson, 2005 – see Chapter 3, Section 3.2.3.1). The nature of this depletion event, however, is not well understood and could have resulted from the extraction of either felsic crust, or enriched mafic crust (Galer & Goldstein, 1988, 1991). Evidence of the existence of very early felsic continental crust is lacking (see Section 4.2.1), leaving LILE enriched mafic crust as the likely complement to the depleted mantle reservoir (Boyet & Carlson, 2005). It is likely that this early mafic crust was subducted and buried deep in the mantle, where it is thought to be still stored and isolated from the convecting mantle (see Chapter 3, Section 3.2.3.1).

Some of the initial evidence for very early mantle depletion was based upon Nd-isotopes (e.g. Collerson et al., 1991; Bennett et al., 1993; Bowring & Housh, 1995) and was challenged by Moorbath et al. (1997), who argued that the isotopic evidence for it was unreliable and the product of later, open-system behavior. Nevertheless, despite the specifics of some localities, there is a growing body of evidence to support an early mantle differentiation and depletion event, and this has to be explained.

Particularly powerful in this respect are Hf (hafnium)-isotopes, for when these data are expressed as ε_{Hf}, they are more sensitive than their geochemical twin ε_{Nd} to the processes of mantle differentiation. Bizzarro et al. (2000) demonstrated this for carbonatite melts and showed that some carbonatites have a Hf-isotope signature which indicates that they are derived from an ancient, enriched mantle reservoir, with an entirely different character from that of depleted mantle. This reservoir has subchondritic Lu/Hf and suprachondritic Sm/Nd and is thought to represent an ancient mafic component. The existence of such an unradiogenic-Hf reservoir would also explain the mismatch between the Nd–Hf composition of modern depleted mantle and the BSE (Fig. 3.28). Bizzarro et al. (2000) suggest that it could account for up to 10–15% of the total mass of the silicate Earth. Support for such a model has been equivocal.

McCulloch and Bennett (1994) pointed out that the existence of deeply buried mafic crust in the mantle is inconsistent with the very high temperatures of the early Archaean mantle and that any deeply buried crust would be rapidly homogenized back into the mantle. However, Bizzarro et al. (2000) dispute this and point out that, with respect to Hf-isotopes, the early to middle Archaean mantle was poorly mixed and much more heterogeneous than the modern mantle so that a mafic component could maintain its distinctiveness.

4.5.1.2.3 Eclogite slabs in the mantle

A different four reservoir model was proposed by McDonough (1991) and Barth et al. (2000) who showed that, whilst the depleted mantle and the continental crust are complementary reservoirs for many trace elements, this is not the case for the elemental ratios Nb/La, Nb/Ta and Ti/Zr. Average continental crust has subchondritic Nb/Ta and Nb/La ratios, as does MORB – and by inference, the depleted mantle. Hence, with respect to the elements La, Nb, and Ta, the continental crust and the depleted mantle cannot be complementary silicate reservoirs. This means that there has to be another "unidentified" silicate reservoir enriched in Nb and Ta to balance the composition of the silicate Earth.

McDonough (1991) suggested that this reservoir enriched in Nb and Ta was eclogite, the refractory residue of a sinking slab, which became isolated in the deep mantle. In a more recent study Rudnick et al. (2000) showed that Nb and Ta are enriched in the accessory phase rutile in eclogitic xenoliths preserved in kimberlites, from regions of Archaean crust. These eclogites have supra-chondritic Nb/La, Nb/Ta, and Ti/Zr ratios indicating that they are the enriched missing reservoir (Fig. 4.22). Hf-isotope studies on eclogitic xenoliths show that they are very heterogeneous, implying that subducted ancient ocean crust is a very heterogeneous reservoir (Jacob et al., 2005).

Mass balance calculations suggest that this eclogitic reservoir is of considerable size. Kamber and Collerson (2000) estimated the mass of deeply subducted ocean crust over 4.3 Ga to be about 1.4×10^{26} g, that is, about 3% of the mass of the silicate Earth, greater than the mass of the continental crust, and equal to about 20% of the mass of oceanic crust subducted over time. Where this reservoir is located is not precisely known. However, Kamber and Collerson (2000) argued that the Nb and Ta contents of MORB worldwide are very well correlated, implying that, for these elements, the depleted mantle is a well-mixed reservoir, and unlikely therefore to contain high Nb/Ta slabs. For this reason they propose that the eclogitic slab reservoir is located in the lower mantle.

4.5.1.2.4 The progressive growth of the depleted mantle with time

In an attempt to solve the paradox of very early mantle depletion in the absence of early felsic crust, McCulloch and Bennett (1994) proposed a model for crustal growth in which the

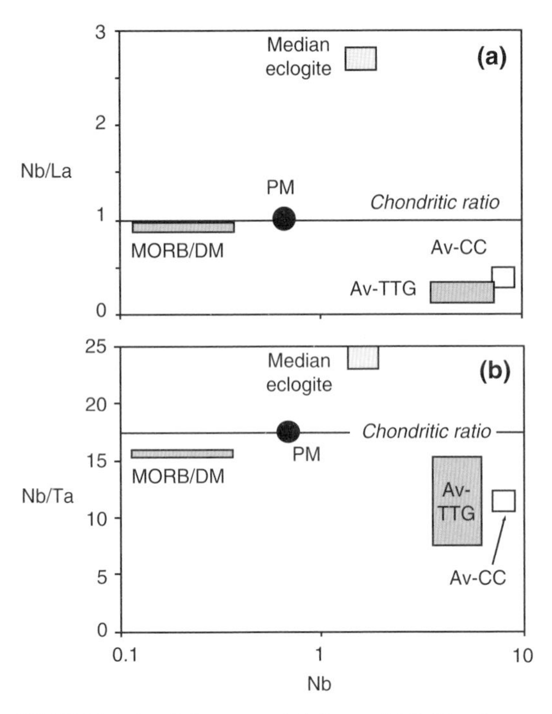

FIGURE 4.22 Nb versus Nb/La and Nb/Ta plots showing the compositions of the primitive mantle (PM), the depleted mantle (MORB/DM), the continental crust (Av-CC), average TTG (Av-TTG) and the median eclogite composition (after Barth et al., 2000; Rudnick et al., 2000). The relationships illustrate that with respect to Nb–Ta–La the continental crust and depleted mantle cannot be complementary reservoirs derived from the primitive mantle and require an additional reservoir with a higher Nb/Ta than that of the primitive mantle. Eclogites appear to represent this missing reservoir. A similar relationship seems to hold for Archaean TTGs.

depleted upper mantle reservoir has grown in volume with time, in parallel with the growth of the continental crust. In this way a small volume of highly depleted mantle could have existed very early in Earth history as the complement to a very small volume of felsic crust. Their model is constrained by isotopic data and is based upon an episodic model of crustal growth. The growth of the continental crust and the complementary depleted upper mantle reservoir is modeled in a stepwise manner, such that between 4.5 and 3.6 Ga the depleted mantle reservoir represents only 10% of the volume of the whole mantle, growing to 40% at 1.8 Ga – the last major continental-crust forming episode in their model (Fig. 4.21b).

It should be noted, however, that a model in which the depleted mantle makes up 40% of the volume of the whole mantle faces the same difficulties as those identified by Hofmann et al. (1986) and Stein and Hofmann (1994) (Section 4.5.1.2.1 above), for such a volume extends below the 660 km discontinuity and raises questions about a one- or two-layer mantle. Furthermore, the progressive growth model of McCulloch and Bennett (1994) also does not adequately accommodate the presence of an early enriched mantle reservoir, described above (Section 4.5.1.2.2).

4.5.1.2.5 The role of the subcontinental lithosphere in the growth of the continental crust

A potentially significant reservoir, and one which a number of authors have suggested is important in the context of continent formation, is the subcontinental lithospheric mantle (SCLM). Kramers (1987, 1988), suggested that the TTG magmas of the Archaean crust formed in an open-system magma layer in the early Earth, the cumulates from which are now preserved as the SCLM. More recently Abbott et al. (2000) proposed a model of continental growth founded upon the premise that the continental crust was extracted from the SCLM.

The subcontinental lithosphere is made up of Mg-rich peridotite, depleted in the elements Fe, Ca, and Al, which has equilibrated at relatively low temperatures (850–1,100°C) and forms a thick mantle "keel" several hundred kilometers thick beneath the continental crust. The presence of this keel is thought to account for the long-term preservation of the continental crust. In detail, the mass of the subcontinental lithosphere is ill-constrained and is at most about 2% of the mass of the mantle (McDonough, 1991). It is compositionally heterogeneous and typically has a multistage history. For further details the reader is referred to Chapter 3, Section 3.1.3.2.

Os-isotope studies have confirmed that there is a close link between the continental crust and its underlying SCLM, inasmuch as there is a relationship between the time of crust formation and the age of the SCLM (see e.g. Reisberg & Lorand, 1995; Pearson, 1999). What is not well understood however, is whether or not the connection between the SCLM and the continental crust has anything to do with the process of crust formation. Our understanding of modern crust formation, in an arc environment, would suggest that there is a link between the creation of new continental crust and the process of basalt extraction from the mantle wedge, and so it is plausible that subarc SCLM is former, melt-depleted mantle wedge. How this relates to crust formation in the Archaean is less clear, not least because there are compositional differences between modern and Archaean SCLM (the Archaean is higher in SiO_2 and lower in CaO, Al_2O_3, see Fig. 3.12). As indicated above, some authors interpret the temporal link between the SCLM and continental crust as genetic and would regard them as complementary reservoirs, whereas others, such as Nagler et al. (1997) do not.

In order to elucidate this problem further Fig. 4.23 shows the relationship between some of the main crust and mantle reservoirs on

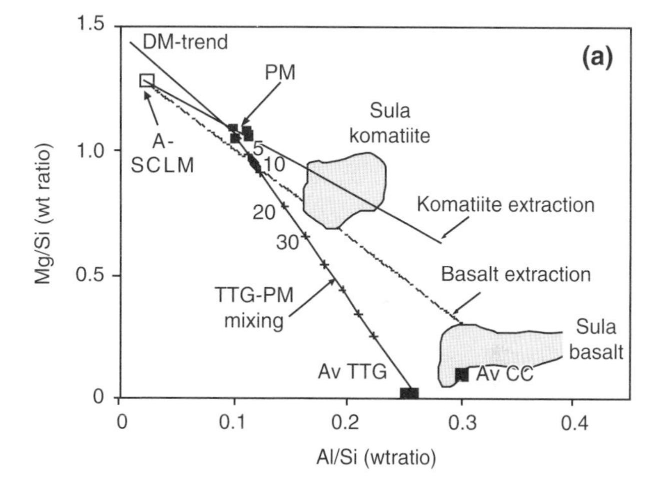

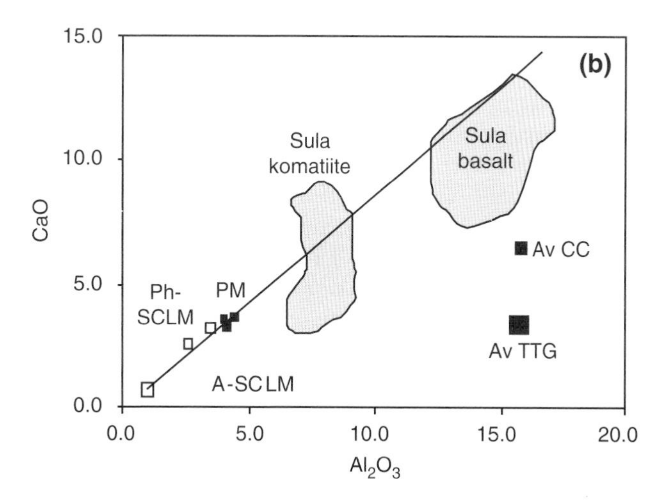

FIGURE 4.23 Geochemical relationships between some the main crust and mantle reservoirs.
(a) Mg/Si versus Al/Si wt ratio diagram (after Jagoutz et al., 1979) showing the Si-enriched nature of the Archaean SCLM (A-SCLM) relative to modern depleted mantle (DM-trend), and possible models for the origin of the A-SCLM. One model is that of komatiite extraction. This is illustrated with the complementary relationship between Munro-type komatiites from the Sula Mountains greenstone belt, Sierra Leone (Rollinson, 1999) and A-SCLM, relative to the primitive mantle (PM). It also shows that Archaean SCLM did not attain its present composition by the extraction of average crust (Av CC) or average TTG (Av TTG). An alternative model for the origin of the SCLM is that it is derived from PM mixed with a small fraction of TTG melt (the mixing line is shown with crosses marked 5–30% TTG mixing). According to this model A-SCLM could be formed by the removal of Archaean basalt from PM plus 5–10% TTG melt (black oval shape) by basalt extraction (dotted line).
(b) CaO–Al_2O_3 plot illustrating the komatiite melt extraction model for the Archaean SCLM and further illustrating that average crust (Av CC) and average TTG (Av TTG) are not complementary to Archaean SCLM. Ph-SCLM is the composition of the Phanerozoic SCLM.

Mg/Si and Al/Si ratio and CaO versus Al_2O_3 plots. These diagrams explore potential complementary relationships between the reservoirs assuming primitive mantle (PM) as a common starting composition. It should be noted that Archaean SCLM is enriched in Si relative to the modern depleted mantle and so lies below the present day depleted mantle (DM) trend. In contrast Proterozoic and Phanerozoic SCLM lie on the DM-trend.

It can be seen from these diagrams that the complement to Archaean SCLM (A-SCLM) is Archaean komatiite (of the undepleted, Munro-type, as exemplified by the Sula Mountains lavas) and not Archaean basalt and not Archaean TTG. These relations show clearly that Archaean TTG crust cannot be the complement to the Archaean SCLM and that Archaean TTG and SCLM are not related as melt and residue.

A connection between komatiite melt extraction and the Archaean SCLM is not a new idea, and was proposed in 1989 by Boyd and again, more recently by Abbott et al. (2000) and Parman et al. (2004). However, there are two problems with this analysis. First, the relationships illustrated in Fig. 4.23 suggest that Archaean SCLM has been depleted by 40–50% komatiite extraction. Given the large volume of the Archaean SCLM this represents a very large volume of komatiitic melt, which is not known in the geological record. Second, these ideas do not sit comfortably with the conventional wisdom on komatiite genesis (hot and deep) for the SCLM is relatively shallow and cool.

Given that there are difficulties with the komatiite model for the origin of the SCLM, an alternative model is proposed which starts with the premise of crust formation through slab-melting. A consequence of this process is that the mantle wedge becomes impregnated with a low percentage of TTG melt. If this enriched primitive mantle is chosen as the starting composition for the crust–SCLM system, then the removal of about 30% basaltic melt will leave a residue with the composition of the Archaean SCLM. In this model, the removal of a more modest fraction of basalt, rather than a large fraction of komatiite, is more consistent with what is observed in Archaean greenstone belts.

These different models illustrate that the relationship between the Archaean SCLM and the continental crust is something of an enigma. On the one hand there is a very clear temporal relationship between the two, indicating that they are part of a common process. On the other hand they do not represent a complementary melt and restite pair. Here it is proposed that the link is indirect. It is suggested, on the basis of the geochemical relationships shown in Fig. 4.23, that the SCLM formed in a subduction environment in which new Archaean crust was being created though slab melting. As a consequence of this process the mantle wedge was enriched with a small fraction of trapped felsic melt. The subsequent melting of the mantle wedge and the extraction of a basaltic melt can explain the formation of a highly depleted subcontinental lithosphere.

4.5.1.3 Crust–mantle reservoirs – a summary

Table 4.3 below summarizes the different reservoirs thought to be present in the modern silicate Earth. However, of these, only the depleted mantle and buried eclogitic slabs have compositions which make them directly complementary to the composition of the continental crust. It is suggested here that Archaean SCLM is not closely related in a geochemical sense to Archaean felsic crust but is the product of basalt extraction, basalt which now is emplaced within the Archaean crust. In contrast, Phanerozoic SCLM may be the restite-complement of the basaltic precursor to modern crust. The proposal that there was an early-formed (pre 3.7 Ga), enriched, basaltic crust (Section 4.5.1.2.2), if confirmed, has important implications for the balance between the major Earth reservoirs, not least because the primitive mantle and the bulk silicate Earth can no longer be regarded as compositionally identical.

These relationships indicate that about 53% of the mantle was involved in the formation of the continental crust and remaining 46% was not. A subject of continuing discussion

TABLE 4.3 The relationship between the different reservoirs in the modern silicate Earth.

Reservoir	% BSE (by mass)	Relationship to the continental crust	Comments
Reservoirs in the modern silicate Earth			
Continental crust	0.6%		Grown progressively over time
Depleted mantle	50%	Complementary compositions	Grown progressively over time
Ancient oceanic crust – early-formed enriched basalt	10–15%	Unrelated – formed before the formation of the continental crust	Located in the deep mantle
Oceanic crust Eclogitic slabs	3%	Complementary minor reservoir	Located in the deep mantle – increased over time – and in the SCLM
Subcontinental lithospheric mantle (SCLM)	2%	Not known – but not compositionally complementary	Grown progressively over time
Delaminated continental crust and SCLM	1.3%		Assume delaminated blocks comprise 50% crust, 50% SCLM. Located in the upper mantle on buoyancy grounds
Primitive mantle	28–33%	Not related – the initial (parental) composition	% mass determined by difference

is the spatial relationship within the Earth of the different mantle reservoirs identified above.

4.5.2 Fluxes

4.5.2.1 The paradox of modern mantle fluxes

One of the most profound problems in understanding the origin of the Earth's continental crust is the discrepancy between the composition of the continental crust, which is andesitic, and the composition of the present-day flux across the Moho, which is basaltic. Two types of solution have been proposed. On the one hand it has been argued that the composition of the continental crust has been modified after it formed in order to adjust its composition from a basaltic protolith to andesite. Alternatively, the balance of fluxes across the Moho has changed over time so that modern processes provide an incomplete explanation for the time-averaged composition of the continental crust. These two models will be examined in turn.

4.5.2.2 Processes which can modify the composition of the continental crust

In order to transform a basaltic protolith into crust with an andesitic composition it is necessary to remove a mafic or ultramafic component from the continental crust. This can be done in one of two ways – either by a magmatic-tectonic process (delamination) or through the process of weathering (Albarede, 1998).

4.5.2.2.1 Lower crustal delamination

Lower crustal delamination, or crustal foundering as it is sometimes called, describes a theoretical model whereby lower continental crust might become detached from the rest of the continental crust and sink into the mantle. This model provides a mechanism whereby the lower continental crust, together with its lithospheric keel, is recycled back into the mantle. The delamination of mafic or ultramafic lower crust is thought to take place when thickened basaltic lower crust or lower crustal cumulates are converted to eclogite, and so undergo a density inversion, such that they become detached from the base of the continental crust and sink into the mantle (England & Houseman, 1989; Kay & Kay, 1993). This is most likely where mantle temperatures are relatively high, and so is restricted to arc environments, volcanic rifted margins or continental regions undergoing extension (Jull & Kelemen, 2001).

Plank (2005) has advanced a general geochemical argument for lower crustal foundering, based on crustal Th/La ratios. She argues that the continental crust has an unusually high Th/La ratio (0.28–0.31), a ratio which is significantly higher than that found in oceanic basalts (<0.2) or in arc magmas (0.2). This high Th/La ratio can be shown to be an intrinsic feature of the continental crust and not inherited through sediment recycling. This means that the crust's high Th/La must be created within itself. Plank (2005) proposes a process of magmatic fractionation within the crust – a hypothesis which is in accord with high Th/La values in the upper continental crust (0.36) and lower Th/La in the lower crust (0.15). In this model basaltic magmas are added to the crust and crystallize leaving low Th/La cumulates in the lower crust and slightly higher Th/La melts in the upper crust. Later partial melting further fractionates high Th/La into the upper crust. In order to maintain this high Th/La ratio the high density lower crust, with a low Th/La ratio, has to be removed by foundering into the mantle, to leave the bulk crust enriched in Th relative to La.

Perhaps the best documented case for lower crustal delamination comes from the eastern block of the North China Craton. This is a region where the continental crust is thin, and has a predominantly felsic lower crust, leading Gao et al. (1998) to propose that its mafic root has been removed. In support of this view Gao et al. (2004) cite Jurassic adakites from this region with very unusual geochemical properties. The adakitic composition of the lavas indicates, as discussed earlier in this chapter (Section 4.4.2.1), that they are melts of a mafic source with an eclogitic mineralogy. However, they also carry a geochemical signature of reaction with mantle peridotite. In addition, they contain xenocrystic Archaean zircons and yet have no geochemical features indicative of interaction with Archaean felsic crust. Gao et al. (2004) suggest that the best way to integrate all these observations is to interpret these adakites as melts, produced during the Jurassic, of Archaean mafic crust, the source of the zircons, which has foundered into the convecting mantle, and interacted with mantle peridotite. The Archaean mafic crust is interpreted as mafic lower crust from the North China Craton which foundered into the mantle during the Jurassic.

Further insights into the process of lower crustal delamination in North China come from the study of Os-isotopes in kimberlitic xenoliths, derived from the subcontinental lithospheric mantle, from this same region. These xenoliths show a marked absence of samples of Archaean age (Wu et al., 2003), which is unusual, since Archaean crust is normally underlain by a thick Archaean SCLM. This observation suggests that the original Archaean sub continental lithosphere has also been removed from beneath the North China Craton, along with the mafic lower crust, and suggests the coupled foundering of both lower crust and SCLM into the convecting upper mantle.

Similar observations have been made in the younger crust of the Sierra Nevada batholith in the USA, where Ducea and Saleeby (1998) proposed that the very thin granite crust (30–40 km), underlain by peridotitic upper mantle, can be explained by the delamination of a thick eclogitic root. Boyd et al. (2004) support this view with evidence from a seismic tomographic study which shows the descent of a two layer slab comprising an eclogitic upper part and a peridotitic lower part, into the mantle. This may be the only place on Earth where dense material is currently being removed from the continental crust into the mantle.

It is clear, from the examples cited above, that lower crustal delamination takes place. What is less certain is the relative importance of this process, for the number of convincing examples of lower crustal delamination is small. Kramers and Tolstikhin (1997) argued on the basis of their Pb-isotope forward transport model that lower crustal delamination is a minor process and not important in modifying crustal compositions. In contrast, Plank (2005) argued that between 40 and 50% lower crustal loss is required over geological time, to achieve the present-day Th/La ratio of the bulk crust. Whilst this volume equates to only 0.3% of the mantle mass, and less than 10% of the mass of subducted slabs over geological time, it is still a significant proportion of the continental crust.

4.5.2.2.2 Weathering and Mg-recycling

At its simplest, the transformation of a basaltic protolith to andesitic crust requires the removal of Mg and an enrichment in Si. This describes the process of chemical weathering in igneous rocks and has been advocated by Albarede (1998) as the means whereby the composition of the continental crust can be modified. The chemical weathering of crustal rocks results in the removal of MgO, Na_2O and CaO in solution and the retention of SiO_2, Fe_2O_3, TiO_2, and K_2O in the residue (Rudnick, 1995). Of the dissolved species, CaO and Na_2O end up, for the most part, in sediments, whereas MgO becomes fixed in the oceanic crust through hydrothermal exchange at oceanic spreading centers. Ultimately, therefore, some of the MgO in the continental crust is returned to the mantle through the subduction of Mg-enriched oceanic crust and some Ca may be returned as subducted limestone.

Trace element ratios can also be used to support the modification of crustal compositions by weathering. Kemp and Hawkesworth (2003) showed that on a Nb/La versus Sr/Nd plot average crustal compositions lie away from the expected intraplate-arc mixing line and are displaced to lower Sr/Nd ratios. They conclude that the Sr/Nd ratio of the bulk crust has been modified, such that Sr has been recycled back into the mantle. The mechanism for this removal is not unique, however, and might include the removal of plagioclase-rich cumulates, as well as the removal of Sr to the oceanic crust by weathering and erosion.

The principal issue is whether or not the mechanism of weathering and recycling can modify the composition of a basaltic protolith enough to produce an average crust composition which is andesitic. There are two lines of reasoning which suggest that it may not and that crust modification by weathering and recycling is not a major process. First, average crust does not closely resemble the composition of a sediment – it is metaluminous – suggesting that its composition is not dominated by a weathered component (Rudnick, 1995). In addition, Rudnick and Gao (2004) suggest that MgO sequestration to the oceanic crust may not be as significant as has previously been thought.

4.5.2.2.3 The secular evolution of the continental crust

It has already been shown that the composition of the Earth's continental crust and the Earth's mantle have evolved chemically over time (see Section 4.3 and Chapter 3, Section 3.2.3). Hence, as the continental crust has grown, so has its composition changed, as is apparent from the differences in the REE content and Rb/Sr ratio of granitoids and the Th/La ratio of sediments (Section 4.3.2). These chemical differences could indicate that the mechanism of crust formation has also changed with time. Further support for this hypothesis comes from Plank's 2005 study of crustal Th/La ratios, discussed above. Plank argued that the present-day high Th/La ratio (0.28–0.31) of the continental crust is the product of internal crustal fractionation. However, Archaean continental crust has a much lower Th/La ratio (0.18) than modern continental crust, and does not require intracrustal differentiation, and so may have formed in a different manner.

Here we explore the idea that the discrepancy between the average composition of the continental crust, which is andesitic, and the modern flux from the mantle to the continental crust, which is basaltic, can be explained in terms of a change in the composition of the crust-mantle flux over time. The hypothesis adopted here is that Archaean crust had a TTG composition, formed from a TTG melt, and was not fractionated into lower basaltic and upper felsic components. Modern crust on the other hand has a basaltic bulk composition but has been modified to andesitic through the fractionation and the removal of a mafic lower crustal component (Rudnick & Taylor, 1987). Evidence for the absence of a mafic lower crust in the Archaean comes from Archaean lower crust preserved as granulite terrains, such as the Lewisian (Rollinson & Tarney, 2005), the Limpopo Belt (Berger & Rollinson, 1997), and the lower crust of the Kaapvaal Craton

(Niu & James, 2002), all of which tend to be TTG in composition. As already outlined, Plank (2004) has shown that the low Th/La ratio of Archaean felsic crust (0.18) does not require an origin by fractionation and delamination but is instead a product of (two-stage) mantle differentiation.

This hypothesis can be further tested with a mass balance calculation in which it is assumed that the composition of the Earth's continental crust was TTG until the end of the Archaean at 2.5 Ga. Crustal growth curves (Fig. 4.8) show that between 43 and 54% of the continental crust had formed by this time (the Condie growth curve is considered to be a minimum). Knowing that 43–54% of the bulk crustal composition is TTG means that it is possible to calculate the average composition of the post-Archaean flux from mantle to crust, using the average TTG compositions presented in Table 4.2.

The results of these calculations are shown in Fig. 4.24. Calculated compositions range between 49 and 55 wt% SiO_2, 7.2–9.1 wt% MgO, 0.95–1.7 wt% K_2O and mg# in the range 57–63. These compositions are close to those of primitive arc picrites described by Leat et al. (2002) and the high-mg# andesites described by Kelemen et al. (2003). Trace element compositions are more difficult to constrain, largely because there are few good trace element data averages for TTGs and because the concentrations of some elements changed *during* the Archaean. However, trace element plots on a mantle normalized diagram (not shown) are enriched in the LILEs and have negative anomalies for Nb and Sr and a positive Pb anomaly, showing an arc-like character. Hence, this mass balance approach is consistent with the view that the relative fluxes from mantle to crust have changed with time – from the Archaean, when the flux was felsic, to more recent times when it is basaltic, and that the basaltic contribution is arc-like in character.

4.5.3 Input–output model for Archaean crust

It has been argued in this chapter that the continental crust has grown over geological time and that a major phase of its growth was during the Archaean. Models for Archaean crust generation start with the premise that the mantle potential temperature was higher in the Archaean. On this basis oceanic crust would be thicker, the angle of subduction would be shallower and slab melting more common. In addition, if slab melting was a common processes then there would also be abundant eclogitic restite, the residue of slab melting, returned to the mantle.

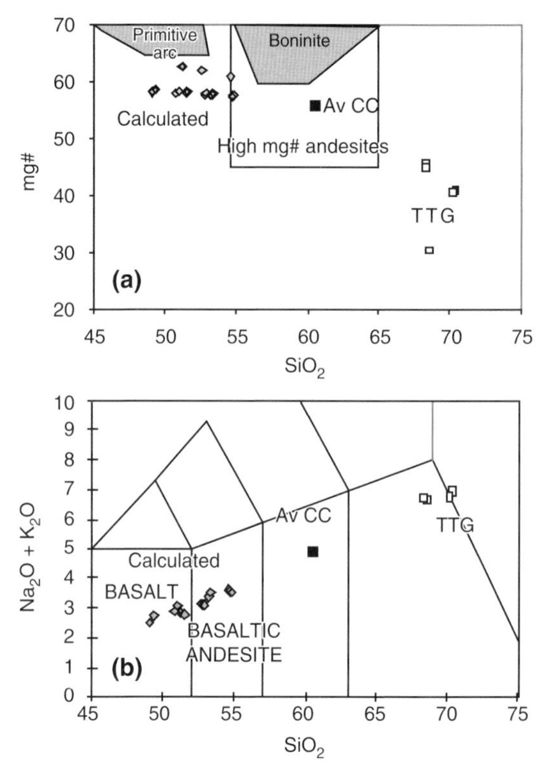

FIGURE 4.24 Calculated compositions of the flux from mantle to crust since the Archaean expressed as (a) mg# versus SiO_2 and (b) total alkalis versus SiO_2. The plots show the results of a mass balance calculation in which it is assumed that between 43 and 54% of the continental crust is TTG in composition and formed before 2.5 Ga. The calculations are based upon five different average TTG compositions (TTG), and the bulk crustal composition of Rudnick and Gao (2003) (Av CC). All the data are presented in Table 4.2. In (a) the composition of primitive arc picrites and boninites is taken from Leat et al. (2002) and high mg# andesites from Kelemen et al. (2003).

However, as was shown in Chapter 3, the evidence of higher mantle potential temperatures in the Archaean is the subject of much debate. On the one hand, there is no doubt that the decay of radioactive elements in the mantle during early in Earth history made for a hot early mantle. What is less certain is whether or not this period extended into the mid- and late-Archaean, for the previously convincing evidence from komatiites is now looking more equivocal.

What we do know from geochemistry is that:

- Archaean felsic crust looks like slab melt (Section 4.4.2)
- This melt interacted and probably equilibrated with the mantle during its ascent (Kelemen, 1995)
- The mafic source of this melt was trace element-enriched relative to modern MORB (perhaps PM rather than DM)
- Archaean crust generation is coupled to the formation of the SCLM but not through the extraction of TTG melt from SCLM (Section 4.4.2.5).

What is unknown is the contribution to Archaean crust generation from sediment recycling, and the role, if any, which intracrustal fractionation played.

4.5.3.1 *How much sediment recycling?*

The transport balance model of Kramers and Tolstikhin (1997) for the U–Th–Pb isotope system is particularly sensitive to the extent to which crustal material can be recycled into the mantle and offers some constraints on the mechanism of crustal recycling. They concluded that the rate of recycling of crustal material back into the Earth's mantle has increased with time, particularly since 2.0 Ga. We know from Pb-isotope studies that some sediment recycling took place during the Archaean (Halla, 2005), but it would seem as though the proportion was small.

4.5.3.2 *Arc versus plume contributions*

One of the important debates explored earlier in this chapter (Section 4.4.2.5) concerns the relative contributions of intraplate (plume) and arc processes to the formation of the continental crust. Here again we are faced with an apparent contradiction, because on the one hand crustal growth curves indicate that there were periods of rapid crustal growth during Earth history, which are best explained by a plume contribution to the crust. On the other hand, the average composition of the continental crust looks very much like arc crust, leaving little room for a plume contribution.

One of the more robust measures of plume versus arc contributions to the continental crust is the La/Nb ratio, for intraplate magmas typically have La/Nb < 1.0, whereas arc magmas have La/Nb > 1.0. This ratio was used by Rudnick (1995) to estimate that between 10 and 35% of the continents formed from an intraplate component. More recently, using revised crustal Nb concentrations, Barth et al. (2000) calculated that between 5 and 20% of the continental crust is of intraplate origin. However, Plank (2005) urges some caution in the acceptance of these proportions, for the impact of sediment recycling could overinflate the arc contribution to crustal growth.

More pertinent here is the extent to which plume magmas might have contributed to the formation of *Archaean* continental crust. Condie (2000) has suggested that, given that the mantle potential temperature was higher in the Archaean, there was a greater plume contribution to Archaean continental crust than to the modern. In addition it was noted earlier that Archaean TTGs appear to have been derived from an enriched mafic parent, possibly a plume source, although of course we are less certain about the geochemical composition of plume magmas during the Archaean.

4.5.3.3 *Was there intracrustal fractionation in the Archaean?*

Much of the discussion of intracrustal fractionation in the Archaean crust has centered on the generation of granites and the creation of lower crustal granulite restites. Rollinson and Tarney (2005) have recently disputed this hypothesis and argue that lower crustal granulites have compositions consistent with primary felsic crust. There is little evidence to date for the removal of a mafic root from Archaean felsic crust, *during the Archaean*, in

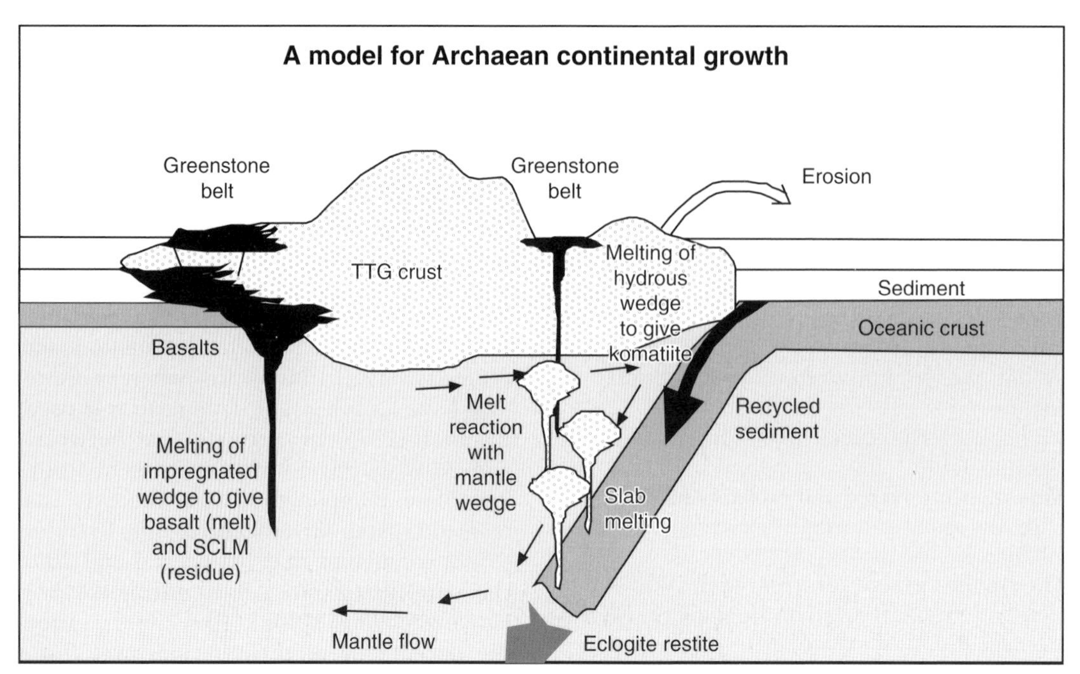

FIGURE 4.25 A cartoon model indicating the different probable contributions to Archaean continental growth.

a manner analogous to delamination models for more recent continental crust.

4.5.3.4 Origin of the Archaean SCLM

The Archaean SCLM is very depleted in Ca and Al, relative to the primitive mantle and has a composition which lies to the high-silica side of the depleted mantle trend. This was the basis for the argument presented earlier, that it cannot be a restite after TTG extraction but rather might be the residue after komatiite extraction. However, as argued above, there are problems with this model, for komatiite melting is thought to be hot and deep, and this is not the character of the SCLM. In addition, the komatiite extraction model would require larger volumes of komatiite magma than are typically found in Archaean terrains.

Thus it may be that a different process is responsible for the origin of the Archaean SCLM. If, as suggested above, some TTG melt from slab melting remains in the mantle wedge, shifting the composition of the mantle wedge to a more Si- and Al-rich composition, the subsequent melting of this enriched mantle could yield a basaltic melt and leave a highly depleted mantle with a composition similar to the Archaean SCLM (Fig. 4.23). The tectonic setting for such a process is a subduction environment in which there is slab melting, that is, the appropriate setting for Archaean TTG crust formation. The advantage of this model is that it provides an integrated solution to the SCLM problem in that there is an explanation for the temporal link between SCLM and Archaean crust, whilst the lack of a direct genetic relationship is maintained.

4.5.3.5 Summary

A summary of the different possible contributions to Archaean crustal growth is illustrated in Fig. 4.25.

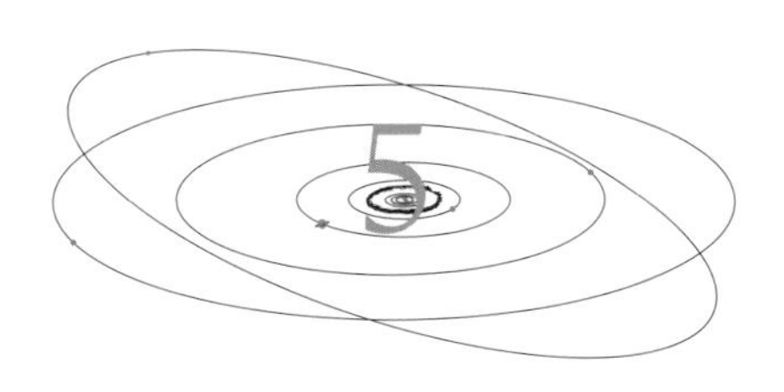

THE ORIGIN OF THE EARTH'S ATMOSPHERE AND OCEANS

The origin of the Earth's atmosphere and oceans – the big picture

The volatiles which now make up the Earth's atmosphere and oceans were acquired very early in the history of the planet, and their origin is most easily explained through the standard model of planetary accretion. During early accretion the Earth gravitationally captured gases from the solar nebula. These were subsequently lost during the later stages of accretion but are preserved in the mantle source of OIBs. Later accretion took place in a solar-gas free environment in which carbonaceous chondrite grains and planetesimals were irradiated by the young sun to form the rare gas phase Ne–B, now preserved in the upper mantle. At this stage the Earth also acquired its major volatiles – water, nitrogen, carbon, and sulfur – from its carbonaceous chondrite building materials. There is now evidence to show that a liquid water ocean was present on Earth at a very early stage in its history, perhaps as early as the late stages of accretion.

The concentration of volatiles in the outer Earth is significantly lower than that found in carbonaceous chondrites indicating that the Earth has lost a major part of its volatiles acquired during accretion. It is likely that some of this volatile loss took place through impact degassing. This atmospheric loss probably took place in the presence of a liquid water ocean and so led to enhanced volatile loss. Subsequent Earth history records that the remaining volatiles were recycled from the Earth's surface back into the mantle through subduction and then returned again to the Earth's surface reservoirs.

In the modern Earth, the largest reservoirs of water, carbon, nitrogen, and sulfur are the deep mantle (water) and the core (carbon, nitrogen, and sulfur). Water is recycled into the deep Earth through subduction, although the majority of subducted water is released in the shallow mantle. In contrast carbon is not easily subducted, and so carbon recycling is restricted to between the surface reservoirs and the shallow mantle. Nitrogen cycling is even shallower and is principally between the surface reservoirs – surface sediment, the oceans, and the atmosphere. The modern sulfur cycle is between the modern surface reservoirs (sediments and the oceans) and the mantle. However, this cycle is sensitive to redox processes and so behaved in a very different manner in the anoxic Archaean, such that the surface sulfur reservoirs have grown in size with time.

The modern Earth's atmosphere is a disequilibrium atmosphere and very different from that of its planetary neighbors both in pressure and in composition and is thought to have evolved in its composition over time. A number of lines of geological evidence show that the partial pressure of oxygen in the Earth's atmosphere has increased over time and that it was particularly low ($<10^{-5}$ the modern level) in the Archaean. Oxygen levels increased dramatically at about 2.3 Ga, although at present there is no consensus as to why that happened. There is also geological evidence for high levels of CO_2 in the Archaean atmosphere, although climate modeling indicates that there was not enough CO_2 to keep the Earth from being frozen. Since there is no evidence for an extensive Archaean snowball Earth, an additional greenhouse gas – methane – has been invoked as an important constituent of the Archaean atmosphere.

It is likely that the progressive oxidation of the atmosphere also influenced the oceans. Certainly the chemistry of the Archaean oceans was different from that of modern oceans and was richer in Ca, Mg, K, Fe, Ba, Ce, and Si and lower in S, P, Mo, Re, U, and Os. These differences, in part, reflect the different redox conditions of the Archaean Earth but also reflect a different oceanic biology and a different balance from today of the hydrothermal and weathering inputs into the Archaean ocean. Iron-isotopes studies of pyrite in black shales indicate a shift in the oxidation state of the oceans between about 2.3 and 1.8 Ga similar to the oxidation of the atmosphere.

The Earth is unique amongst the terrestrial planets in possessing a hydrous ocean and a CO_2-poor, oxygenic atmosphere. In this chapter we investigate how the Earth acquired its volatiles and how the compositions of the atmosphere and oceans have evolved through Earth history. In so doing we shall find ourselves seeking to understand early-Earth interactions between the Earth's surface reservoirs and the solid Earth.

When measured against the mass of the whole Earth the oceans (0.02% of the Earth's mass) and atmosphere (one millionth of the Earth's Mass) appear insignificant. However, despite their small mass, the Earth's volatile reservoirs play a disproportionate role in the Earth system. This is illustrated in four very important interactions within the Earth system.

- First, volatiles exert an important control on the physical properties of the mantle. For example, the presence of water reduces the strength of olivine aggregates and seriously alters the viscosity of the mantle. Experimental studies show that at 300 MPa, in the presence of water, the viscosity of olivine aggregates deformed in the dislocation creep regime is reduced by up to a factor of 140. Thus a wet mantle is a low viscosity mantle. Conversely a mantle that is dried out by partial melting will be stiffer and more refractory, as is the case for the lithospheric "lid" to modern oceanic mantle. Thus, if it is possible to estimate the volatile content of the mantle both now and in the Archaean, it will be possible to set some physical constraints on models of mantle evolution over time.
- Second, volatiles have played an essential role in controlling the evolution of life on Earth. The principal insight of the Gaia hypothesis discussed in Chapter 1 is that there is a relationship between the biosphere and the other Earth reservoirs. In particular volatiles have played a vital role in permitting the right environment to foster the evolution of life on Earth. This will be the subject of the next chapter.
- Third, the Earth would not have a continental crust if it had not had a hydrosphere. Many years ago Campbell and Taylor (1983) argued that the presence of water on Earth is the key to the origin of its continental crust as captured in the pithy title of their paper *No Water, No Granites – No Oceans, No Continents*. This view is still widely held.
- Finally, the Earth owes its cool, stable climate to the dynamic interactions between the atmosphere, oceans, and the Earth's surface, for as long as perhaps 4.0 Ga.

Here the Earth's volatiles are divided into two groups. There are the **major volatiles** – H, C, N, S, Cl, Br, and I, which together with oxygen dominate the Earth's atmosphere and oceans and are the main topic of this chapter. In addition there are the **noble gases, or rare gases** – He, Ne, Ar, Kr, and Xe, which form a very small proportion of the Earth's volatile budget, but play a central role in our understanding. The noble gases are blessed with a plethora of isotopes, which provide vital clues to processes operating during Earth accretion and provide information about the ultimate origin of the Earth's volatile budget and the mechanisms by which they were incorporated into the mantle.

First, however, we shall consider the distribution of volatiles in the different Earth reservoirs, and then in the second section of this chapter we will explore how the Earth acquired its present inventory of volatiles. In the last two sections we examine the geological evidence for the composition of the atmosphere and oceans in the early Earth and chart their changing composition through geological time.

5.1 THE VOLATILE BUDGET OF THE MODERN EARTH

Of the major volatile elements described above, water, a variety of carbon compounds, nitrogen and sulfur are the volatile compounds which dominate in the modern Earth. In this section we review the modern-Earth geochemical cycles for water, carbon, nitrogen, and sulfur and look in some detail at volatile mass balances between the Earth's surface reservoirs and the deep Earth. Then, having established how the modern Earth works we seek to determine how these geochemical cycles might have operated in the early Earth.

5.1.1 Water

5.1.1.1 The water budget of the modern Earth
The water budget of the modern Earth might be loosely divided into the outer Earth reservoirs and the mantle. In the outer Earth reservoirs there is about 1.4×10^{21} kg of water currently stored at the Earth's surface as the oceans, ice, freshwater, and groundwater. This makes up 0.0002 of the Earth's mass and is equivalent to 1.515×10^{23} g of hydrogen (Lecuyer et al., 1998). Assessing the water content of the mantle is more difficult and depends mostly upon experimental studies of the solubility of water in mantle minerals and melts. These concentrations are low, for most upper mantle minerals are nominally anhydrous, but the mantle is large, and at depth dense hydrous magnesian silicates may play an important role in storing water.

Experimental studies tell us that the water content of upper mantle minerals is in the range from a few tens to a few hundreds of ppm and takes the form of hydrogen located along structural defects in the mineral phases. At present the best estimates of the water content of the upper mantle are made through the measurement of the volatile content of basaltic magmas. This too is a difficult enterprise because most basalts lose their gas component on eruption through degassing. However, Saal et al. (2002) used pristine melt inclusions in olivine to overcome this problem and reported gas contents of primitive midocean ridge magmas and their source (see Table 5.1). Their water contents for the MORB source are 142 ± 85 ppm in good agreement with a previous estimate of 125 ± 75 ppm of Hirth and Kohlstedt (1996). Hauri (2002) measured the volatile content of ocean island basalts from Hawaii and found that water contents are much higher than in midocean ridge basalts (3,600–6,000 ppm) suggesting that the mantle is heterogeneous with respect to its volatile content (Table 5.1). From these estimates, and assuming that the mantle is a mixture of MORB source and OIB source in a ratio of 9 : 1, and that the mantle has the mass 4.3×10^{24} kg,

TABLE 5.1 The volatile content of ocean island and midocean ridge basalts, and the calculated volatile content of the mantle source of midocean ridge basalts (data from Hauri, 2002; Saal et al., 2002; Lecuyer et al., 1998).

	Ocean island basalt (Hauri, 2002) (ppm)	Midocean ridge basalt (Saal et al., 2002) (ppm)	Mantle source for ocean ridge basalt (Lecuyer et al., 1998) (ppm)
H_2O	3,600–6,000	370–1,220	142 ± 85
CO_2	not measured	44–244	72 ± 19
Cl	30–300	1–21	1 ± 0.5
F	380–540	50–135	16 ± 3
S	400–1600	495–1,024	146 ± 35
δD	−80 ± 10	−40 to −80	~−80

TABLE 5.2 Estimates of the fraction of water and mass of water in the different mantle reservoirs. Also shown are estimates of water balance in the modern mantle.

Reservoir	Fraction of water %	Mass of water	Reference
Water budget			
Surface water	100	1.4×10^{21} kg	Lecuyer et al. (1998)
Sedimentary rocks	0.11	0.23×10^{21} kg	Lecuyer et al. (1998)
Upper mantle	0.02–0.03	$0.2–0.3 \times 10^{21}$ kg	Data from Saal et al. (2002)
Mantle transition zone	0.5–1.0	$2.6–5.2 \times 10^{21}$ kg	Ohtani (2005)
Lower mantle	0.15–0.2	$3.4–4.5 \times 10^{21}$ kg	Ohtani (2005)
Total Mantle	**0.16–0.26**	**$7.0–11.0 \times 10^{21}$ kg**	
Water balance			
Total water returned to mantle (subduction)	1.83×10^{12} kg/yr		Jarrard (2003)
Total outgassing (arc + ridge)	2.0×10^{11} kg/yr		Peacock (1990)

the mass of water in the upper mantle can be estimated to be about $0.3–0.4 \times 10^{21}$ kg (Table 5.2).

Our knowledge of the water content of the deeper mantle comes from high-pressure experimental studies on the phases wadsleyite (β-Mg_2SiO_4) and ringwoodite (γ-Mg_2SiO_4) found in the mantle transition zone. These minerals have the capacity to store as much as 3 wt% H_2O (Bolfan-Casanova et al., 2000), indicating that the transition is water-rich. Ohtani (2005) too has estimated that the mantle transition zone may store up to 2 to 3 ocean masses of water. Additional support for the hydrated nature of the mantle transition zone comes from seismic studies by Van der Meljde et al. (2003) who report a widening of the 410-km discontinuity, indicative of the presence of water within mantle minerals, and Huang et al. (2005) who report similar results from an electrical conductivity study.

At even deeper levels of the mantle, solubility studies show that the phases Mg-perovskite, Ca-perovskite and magnesiowustite contain tens to thousands of ppm of H_2O (Bolfan-Casanova et al., 2000; Murakami et al., 2002) leading to an estimate of between 0.15 and 0.2 wt% water in the lower mantle. This equates to between 3.4 and 4.5×10^{21} kg of water, that is about 2.5–3 times the present ocean mass (Ohtani, 2005).

These observations show that

- The total water content of the Earth (excluding the core) is equivalent to between 7 and 8 oceans and, the silicate Earth contains between 0.16 and 0.26% water.
- The mantle is layered with respect to its water content, such that the upper mantle has a low water content (ca. 0.02 wt%), the transition zone has between 0.5 and 1.0 wt% water and the lower mantle an estimated 0.15–0.20 wt % water.

The low water content of the upper mantle may be principally the product of degassing, whereas the transition zone may operate as a "water trap," as identified by Bercovici and Kurato (2003) in their "water-filter" model of the mantle discussed in Chapter 3 (Section 3.3.1.2).

5.1.1.2 Water cycling on the modern Earth

Calculated fluxes of volcanic outgassing (2.0×10^{11} kg/yr – Peacock, 1990) compared with the total amount of subducted water (1.83×10^{12} kg/yr – Jarrard, 2003) indicate that there is a net flux into the mantle (Table 5.2). A similar, net influx into the mantle was calculated independently on the basis of geophysical modeling, by Franck and Bounamam (2001). However, Jarrard (2003) has estimated that a half to two-thirds of the subducted water escapes at subarc depths, triggering partial melting.

How water is transported into the mantle and the depths to which it can be carried have been investigated in high-pressure experiments. In brief, these show that whilst hydrous phases in basalt and subducted sediment can transport water down to the top of the transition zone at about 410 km, hydrous phases in mantle peridotite are capable of transporting water even deeper, certainly into the lower mantle and maybe into the core.

In basalts, lawsonite is the only hydrous phase which is stable in the deep mantle and this phase dehydrates at the depth of the mantle transition zone. Similarly in subducted pelagic sediments there is a series of hydrated aluminum silicates and oxides (including the phase Egg – $AlSiO_3.OH$) which also remain stable down to transition zone depths (Williams & Hemley, 2001).

In peridotites high-pressure hydrated magnesian silicate phases, named imaginatively after the letters of the alphabet, are capable of transporting water within a subducting slab deep into the lower mantle. Notable are the superhydrous phase B and phase D. This latter phase is stable to pressures of 50 GPa and can transport water deep into the lower mantle. If water penetrates even deeper than this it could be partitioned into the core as an iron hydride through reaction with native Fe metal in the lowermost mantle.

These experimental studies show that water in the upper part of a subducting slab, in basalt and sediment, is either returned to the surface, or is stored in the upper mantle, and/or will be trapped in the mantle transition zone. The importance of the water storage capacity of the mantle transition zone has only recently been recognized and has important implications for models of mantle evolution. However, in contrast to the upper part of the slab, some of the water in the lower, peridotitic part of a subducting slab is transported into and may be stored in the lower mantle (Fig. 5.1).

5.1.1.3 Water cycling in the Archaean

More difficult to predict is how water was distributed in the Archaean mantle. It will be shown later (see Section 5.2) that the Earth was initially volatile-rich, when it accreted and it subsequently lost water and other volatiles. Indicative here is the comparison between the water content of carbonaceous chondrites, the likely primitive material of Earth accretion (up to ca. 10 wt%), and the estimated water content of the present-day silicate Earth and hydrosphere (0.19–0.24 wt%, see Table 5.2).

Kawamoto et al. (1996) suggested that in the Archaean there would have been a water-rich transition zone, as is the case today. This may well be correct, given the Earth had a "wet" start. A further consideration is the extent to which water was subducted in the Archaean. The growth of the continental crust is probably

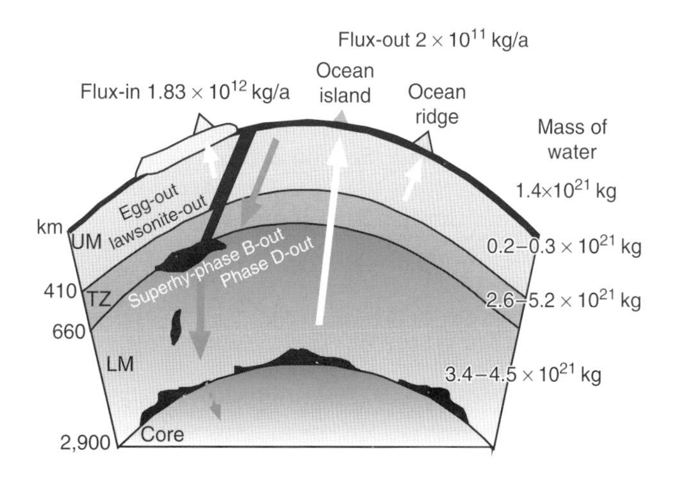

FIGURE 5.1 Model of water cycling in the Earth (after Ohtani, 2005). UM, upper mantle; TZ, mantle transition zone; LM, lower mantle. The grey arrows show water-in to the mantle, the white arrows shown water-out of the mantle. The approximate lower stability limit of hydrous phases in the mantle is shown as phase-out.

a useful monitor for the shallow subduction of water, but it may be that, given that the Earth's mantle was probably hotter in the Archaean, the deep penetration of water may well have been prevented (Bjornerud & Austrheim, 2004).

Intriguing in this regard are the current arguments over whether komatiites are the product of wet or dry mantle melting (Chapter 3, Section 3.2.1.2). Proponents of the wet-komatiite argument point to experiments which show that komatiites from the 3.5 Ga Barberton greenstone belt require at least 3 wt% magmatic water (Parman et al., 1997) to produce their observed mineralogy. In addition melt inclusion studies on the 2.7 Ga komatiites of the Belingwe greenstone belt show that the parental komatiite contained between 0.8 and 0.9 wt% water and that the mantle source should have contained ca. 0.5 wt% water (Shimizu et al., 2001) – significantly higher than in the modern MORB source, but perhaps closer to the wetter, OIB source.

If the advocates of wet komatiitic melts are correct, then the debate centers around the source of the water. Was this water acquired at the time of accretion, and komatiite magmatism was a late degassing event, or was this the recycling of surface water in a manner analogous to that in a modern subduction zone? Support for this latter argument comes from the association between komatiites and boninites in Archaean greenstone belts (Grove & Parman, 2004), for in the modern, boninites are known to be produced by hydrous melting of the mantle in a subduction zone environment. Although, it seems from more recent studies that only some Archaean boninites are true subduction boninites (Smithies et al., 2004 and see Chapter 3, Section 3.2.1.3). A further option, which has not been considered in detail, is that komatiites come from the hydrous mantle transition zone.

5.1.2 Carbon

5.1.2.1 The carbon budget of the modern Earth

The Earth contains significant quantities of carbon, principally as CO_2, in its atmosphere and oceans, in the crust, as carbonate-rich and organic-carbon (kerogen-bearing) sediments, in the mantle and probably also in the core. Abundance estimates of carbon in the surface reservoirs are relatively well known (see Fig. 5.2), but our knowledge of the amount of carbon in the mantle and core is less secure.

Making a correct estimate of the carbon content of the mantle is difficult. Total carbon concentrations in mantle xenoliths are between 1 and 10,000 ppm, with most samples containing less than 100 ppm (Deines, 2002). This is similar to early estimates of carbon in the MORB source which are around 400 ppm (Javoy & Pineau, 1991). However, more recent estimates of carbon in the MORB source are much lower. Saal et al. (2002) found only 20 ppm C in pristine melt inclusions in MORB olivine. A further difficulty with mantle carbon is that the precise location of the carbon is not well known. The solubility experiments of Keppler et al. (2003) on upper mantle silicate minerals show that they have very low carbon concentrations, in the range 0.1 and 1 ppm at pressures up to 3.5 Gpa. These values are much lower than expected and suggest that carbon is not located in a silicate phase but instead is as a separate carbon-rich phase. Holloway (1998) suggested that this carbon phase was graphite, although other workers have argued for carbonate (magnesite), a view which has some experimental support (Watson et al., 1990; Biellmann et al., 1993).

A first-order minimum estimate of the mass of carbon in the mantle can be made by assuming that the MORB-source is typical of the mantle as whole, and that it contains 20 ppm C, in keeping with the new results on low carbon in the MORB source and the low content of carbon in mantle minerals. In this case the mantle contains 8.6×10^{22} g (86×10^{6} Gt) of carbon. This estimate is smaller than, but of the same order of magnitude as, that of Zhang and Zindler (1993) who used $C/^{3}He$ ratios to estimate the carbon content of the degassed mantle reservoir as 22×10^{22} g (1.8×10^{22} mol).

Tingle (1998) estimated that the mass of carbon acquired by the Earth during accretion was between 10^{24} and 10^{25} g (10^{9}–10^{10} Gt), 0.02–0.2 wt% C. This is two to three orders of

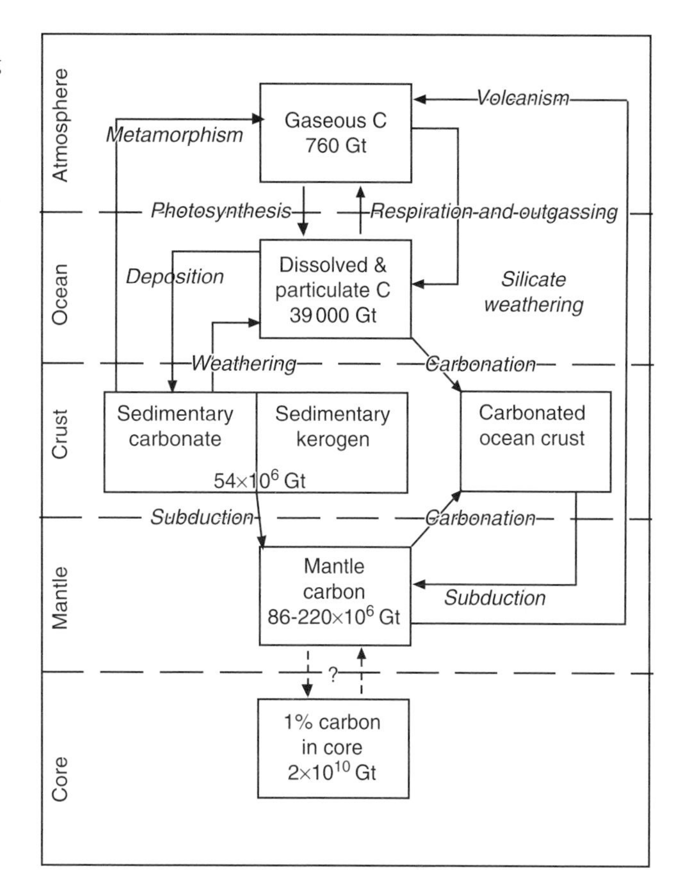

FIGURE 5.2 Schematic illustration of the modern "deep Earth" carbon cycle showing the main Earth carbon reservoirs and the pathways between them. The important fluxes are listed in Table 5.3. The mass of carbon in each reservoir is given in giga-tonnes of carbon (1 Gt = 10^{15}g). Figure adapted from Killops and Killops (2005).

TABLE 5.3 Estimates of present-day fluxes of carbon into and out of the mantle.

	Mols/yr	Grams of carbon/yr	Reference
Carbon fluxes into and out of the mantle			
Fluxes-out			
Mid ocean ridges	$2.3 \pm 0.6 \times 10^{12}$	2.76×10^{13}	Zhang and Zindler (1993)
	$0.8–1.6 \times 10^{12}$	$0.96–1.92 \times 10^{12}$	Holloway (1998)
	$2.2 \pm 0.9 \times 10^{12}$	2.64×10^{13}	Marty and Tolstikhin (1998)
	$9.3 \pm 2.8 \times 10^{11}$	1.12×10^{13}	Saal et al. (2002)
Arcs	2.5×10^{12}	3.00×10^{13}	Marty and Tolstikhin (1998)
Plumes	3×10^{12}	3.60×10^{13}	Marty and Tolstikhin (1998)
Sedimentary basins	$>8.3 \times 10^{7}$	9.96×10^{8}	Sherwood Lollar et al. (1997)
Total	Av. 6, range $4–10 \times 10^{12}$	7.20×10^{13}	Marty and Tolstikhin (1998)
Fluxes-in			
Veined MORB	4×10^{12}	4.80×10^{13}	Staudigel et al. (1989)
	3.4×10^{12}	4.08×10^{13}	Zahnle and Sleep (2002)
Subducted carbonate	2.3×10^{13} (max)	2.76×10^{14}	Zhang and Zindler (1993)
	1.4×10^{12}	1.68×10^{13}	Zahnle and Sleep (2002)
	2.16×10^{12}	2.59×10^{13}	Plank and Langmuir (1998)
Total slab	5.0×10^{12}	6.00×10^{13}	Peacock (1990)
Total	5.4×10^{12} (range $6.3–4.8 \times 10^{12}$)	6.48×10^{13}	

magnitude greater than the total carbon stored at the Earth's surface and in the mantle (Fig. 5.2), and implies either, massive carbon loss during the early history of the planet, or the presence of a large carbon reservoir in the Earth's core (Wood, 1993).

5.1.2.2 The global carbon cycle

Carbon is lost from the mantle through the volcanic outgassing of CO_2 and is reintroduced through the subduction of sedimentary organic carbon, carbonate-rich sediments, and carbonated oceanic crust. Estimates presented in Table 5.3 suggest that the present-day fluxes of CO_2 into the mantle broadly match the amount released, implying that at the present time the mantle is at a steady state with respect to carbon. An implication of this observation is that either all subducted carbon is returned to the Earth's surface or that the mass of carbon buried, through subduction, in the lower mantle is matched by the degassing of deep mantle carbon.

All of this raises the question of whether carbon, like water, can be returned to the lower mantle. High-pressure experimental studies on carbonated eclogites and peridotites indicate that carbonate-rich melts are released during the decarbonation and deep melting of the slab before it reaches the depths of the mantle transition zone. Only small amounts of carbon escape this process and these are subducted into the lower mantle (Hammouda, 2003; Dasgupta et al., 2004). The only evidence we have to support this idea comes from deep mantle diamonds which have carbon isotope values heavier than the mantle mean ($\delta^{13}C = -4.5$ instead of $\delta^{13}C = -5.5$, McCammon et al., 2004). Since marine carbonate tends to be enriched in the heavier carbon isotope, it is possible that these diamonds were formed through the reduction of subducted sediment which included a carbonate component.

5.1.2.3 Carbon isotope evolution over time

It is likely that Earth acquired its carbon during the early stages of accretion (Tingle, 1998). This carbon was probably in the form of hydrocarbons, not as CH_4, CO, and CO_2, since these common carbon compounds do not form ice at the temperatures of the inner solar system (Hunten, 1993). The isotopic composition of this initial carbon is not easily identified, for carbon was not homogenized in the solar nebular as indicated by the $\delta^{13}C$ values for carbonaceous chondrites which are between -25 and 0‰. Even more extreme is the isotopic value of individual organic constituents within these meteorites which can vary by as much as 75‰ (Lecluse et al., 1998; Sephton et al., 2003). More consistent are the iron meteorites which have $\delta^{13}C$ values of around -5‰ (see Text Box 5.1).

However, during the differentiation of the Earth carbon was isotopically homogenized to $\delta^{13}C = -5$‰, as indicated by isotopic measurements on midocean ridge basalts, carbonatites, and diamonds. The only exception to this are a small number of diamonds which have $\delta^{13}C$ values in the range -22 to -26‰. It has been suggested that these represent subducted biogenic carbon (Deines, 2002), although experimental studies have now shown that these low values are more likely to be the product of isotopic fractionations in the deep mantle during diamond crystallization (Maruoka et al., 2004).

Peridotitic diamonds of Archaean age have a similar carbon isotope value to that of the modern mantle mean, suggesting that the carbon isotope composition of the mantle has been relatively unchanged since the Archaean. If iron meteorites are representative of the isotopic composition of the early bulk Earth, then it is possible that the carbon isotope composition of the mantle might be unchanged since the homogenization of the Earth. However, matters may not be so simple, for there are some initial indications that carbon isotopes may have different histories in the upper and lower mantle. For example, diamond data indicate that there has been insufficient carbonate carbon penetration into the deep mantle to modify the Earth's carbon isotope ratio since its earliest differentiation. And yet the shallow mantle may be different, for Wilson and Spencer (2001) report a secular change in the $\delta^{13}C$ signature of carbonatites. Carbonatite carbon is carbon which is retained in the upper mantle and is released from subducting slabs

TEXT BOX 5.1 Stable isotope geochemistry

Most elements consist of more than one stable isotope. For example, in naturally occurring oxygen the isotope ^{16}O makes up 99.76% of all oxygen atoms, ^{17}O, 0.04% and ^{18}O, 0.20% (that is the stable isotopes occur in the ratio of 2500 : 1 : 5). Under particular circumstances the proportions of these isotopes can be altered and the resulting isotope ratio can be used to "fingerprint" particular geochemical processes. Traditionally geochemists have studied the stable isotopes of carbon, nitrogen, oxygen, and sulfur to unravel geochemical process in the solid Earth, the hydrosphere and atmosphere. More recently, with advances in mass spectrometry, a wealth of new elements have been added to this portfolio, in particular some of the transition metals. In this book we shall consider the geochemistry of the isotopes of carbon, nitrogen, oxygen, sulfur, and iron, and make passing reference to few more elements along the way.

Most common is the process of **mass-dependent fractionation**, in which the stable isotope ratio is altered as the consequence of physical processes differentially affecting atoms or molecules of different mass. Isotopes are fractionated relative to one another according to thermodynamic, kinetic, and diffusion processes. A simple example is the way in which oxygen isotopes in water molecules are fractionated during the process of evaporation. Water molecules containing the lower mass isotope ^{16}O are more likely to become water vapor than those containing the higher mass isotope ^{18}O. Hence the water vapor is enriched in isotope ^{16}O and the liquid water is enriched in isotope ^{18}O.

MDF follows specific rules based on the relative atomic mass differences between the fractionating isotopes, and as a result produces highly correlated relationships between isotope ratios for that element (Chapter 2, Section 2.3.3.2; Chapater 5, Section 5.3.1.3). MDF is expressed as a deviation from the appropriate standard as a "lower-case delta-value" in parts per thousand, for example, $\delta^{18}O$ ‰, "del oxygen eighteen" – which is computed from the expression

$$\delta^{18}O(‰) = [(^{18}O/^{16}O_{sample} - {}^{18}O/^{16}O_{standard})/(^{18}O/^{16}O_{standard})] * 1{,}000$$

Mass-independent fractionation, is less common and represents the processes whereby fractionations take place in a different manner from the simple proportionality predicted on the basis of atomic mass differences. This is particularly important in atmospheric chemistry, for it has been shown in laboratory experiments that, for example, the photochemical oxidation of sulfur in the atmosphere can lead to mass-independent fractionations (MIF) of the isotopes of sulfur. MIF may also be important in cosmochemistry and may explain isotopic variations in the solar nebula as recorded by different meteorite groups.

MIF is recorded as a deviation from a MDF line and is expressed using the convention "upper-case delta-value" in parts per thousand (e.g. $\Delta^{33}S$‰).

Carbon isotopes

The two naturally occurring stable isotopes of carbon are ^{12}C (98.89%) and ^{13}C (1.11%). The reference standard for carbon isotope measurements is a standard belemnite known as PDB. Carbon isotope ratios are expressed as $\delta^{13}C$ ‰. Carbon isotopes are important in meteorites, the mantle, the atmosphere, and in understanding biological processes. The mean mantle $\delta^{13}C$ is -5.5‰. Many biological reaction pathways discriminate against heavy ^{13}C in preference to the lighter ^{12}C. This leads to strongly negative $\delta^{13}C$ ratios in biological materials and can provide a geochemical fingerprint for biogenic processes (Chapter 6, Section 6.2.1).

Oxygen isotopes

The three isotopes of oxygen have already been described and are in order of abundance ^{16}O, ^{18}O, and ^{17}O. The reference standard for isotopic measurements is standard mean ocean water (SMOW). Isotope ratios are expressed most commonly as $\delta^{18}O$ ‰, but sometimes there is also interest in the ratio $\delta^{17}O$ ‰. Oxygen isotope measurements of different meteorite groups have been a particularly fruitful area of study (Chapter 2, Section 2.3.3.2). Results are presented on *x–y* plots of $\delta^{17}O$ versus $\delta^{18}O$ and allow us to classify different meteorite groups. These studies also reveal that there was significant oxygen isotope heterogeneity in the early solar system, at least in part due to mass-independent fractionation. Oxygen isotope ratios in surface water tend to be higher than those of the Earth's mantle, thus providing a fingerprint for surface water interaction with magmatic rocks. Zircons with elevated oxygen isotope ratios have been interpreted as having "seen" water from the Earth's surface.

Nitrogen isotopes

There are two naturally occurring isotopes of nitrogen ^{14}N (99.63%) and ^{15}N (0.37%). Isotope ratios are expressed as $\delta^{15}N$ ‰ and measurements are made relative to the nitrogen isotopic composition of the atmosphere. The present-day mantle has a value of about $\delta^{15}N = -5$ ‰. Nitrogen isotopes are fractionated during biological processes and a range of different fractionations has been identified in the

TEXT BOX 5.1 *(Cont'd)*

oxidation and reduction of nitrogen (Chapter 6, Section 6.2.3).

Sulfur isotopes

There are four naturally occurring isotopes of sulfur ^{32}S (94.9%), ^{33}S (0.76%), ^{34}S (4.29%) and ^{36}S (0.02%). Isotope ratios are expressed as either $\delta^{34}S$‰ or more rarely as $\delta^{33}S$ ‰ and measurements are made relative to an iron meteorite sulfide standard known as the Canon Diablo troilite.

Sulfur isotope fractionation is important in the study of atmospheric gases, sulfur in the oceans (Section 5.4.4.1) and in living organisms. Microbes use ^{32}S in preference to ^{34}S during microbial sulfate reduction providing a sulfur isotope fingerprint for a specific biological process (Chapter 6, Section 6.2.2). In addition, evidence for MIF of sulfur isotopes has provided important clues in the timing of the rise of oxygen in the Earth's early atmosphere (Section 5.3.1.3).

Iron isotopes

There are four naturally occurring isotopes of iron ^{54}Fe (5.85%), ^{56}Fe (91.75%), ^{57}Fe (2.12%) and ^{58}Fe (0.28%). Isotopic measurements are most commonly reported as $\delta^{56}Fe$ ‰. There is the potential for iron isotopes to be useful in biological fingerprinting, but at present the results are equivocal because fractionations of a similar magnitude are reported as a result of nonbiogenic processes (see Chapter 6, Section 6.2.4).

before they penetrate the mantle transition zone. This secular change is thought to reflect the increasing importance of sedimentary carbonate recycling over geological time.

5.1.3 Nitrogen

Nitrogen, unlike water and carbon, resides mostly in the outer Earth reservoirs and appears to have done so over most of Earth history. Zhang and Zindler (1993) have suggested that nitrogen was rapidly degassed very shortly after Earth accretion and reached present atmospheric levels after only about 200 Ma. It is thought that nitrogen levels have remained almost constant since this time, although there are currently no direct data on surface atmospheric pressures in the early Earth and on the contribution of nitrogen.

5.1.3.1 The modern nitrogen cycle

The principal flux of nitrogen in the modern Earth is *within* the surface reservoirs. Nitrogen is removed from the atmosphere by the processes of bacterial fixation as NH_3, and returned through the process of denitrification. Thus the surface nitrogen cycle consists of a wide range of biologically mediated redox reactions. The total flux of nitrogen in nitrogen fixation and denitrification is ca. 3.0×10^{13} mol N/yr (in Mather et al., 2004) indicating that the atmosphere is biologically replenished about every 10 Ma. In contrast the present volcanic flux of 2×10^{10} mol N/yr (Fischer et al., 2002), requires a length of time greater than the age of the Earth to replenish the atmosphere. Clearly therefore, surface processes, and, in particular, biogenic processes dominate compositional controls on the modern atmosphere (Fig. 5.3).

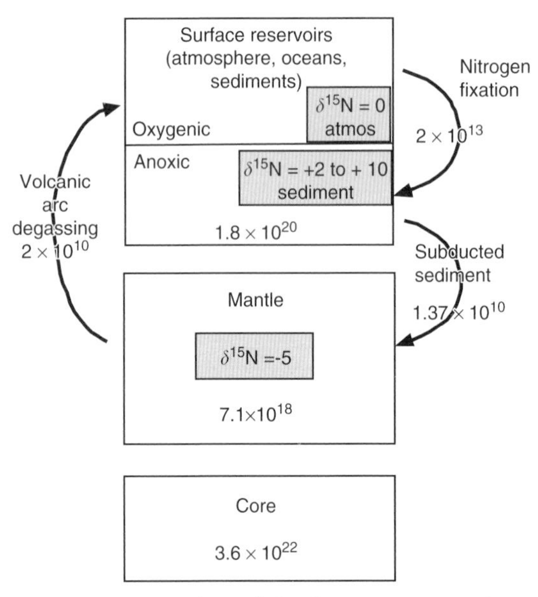

FIGURE 5.3 Outline of the deep nitrogen cycle in the modern Earth. The size of the reservoirs is taken mainly from Miyazaki et al. (2004) and is in mols. The nitrogen isotope values are from Marty and Dauphas (2003) and the fluxes are from Fischer et al. (2002) and are in mol/yr.

Miyazaki et al. (2004) have estimated from solubility experiments that the mass of nitrogen in the main Earth reservoirs is several orders of magnitude greater than that of the mantle (Fig. 5.3). However, because nitrogen has a siderophile character, the Earth's core is probably its largest nitrogen reservoir (Fig. 5.3).

The study of nitrogen isotopes, expressed as $\delta^{15}N$ (see Text Box 5.1), provides additional clues to understanding both the shallow and deep nitrogen cycles. The atmosphere provides the standard for nitrogen isotope measurements and so is, by definition, $\delta^{15}N = 0$. At the Earth's surface biogenic processes such as denitrification may lead to an enrichment of $\delta^{15}N$, by 6–7‰ so that modern sediments tend to be enriched in heavy nitrogen and are in the range +2 to +10 (Marty & Dauphas, 2003). This process is enhanced by an additional 2–3‰ during the metamorphism of sedimentary rocks. Consistent values in MORB, mantle xenoliths and diamonds suggest that present-day mantle has a value around $\delta^{15}N = -5$ (Figs. 5.4b, c).

These comparisons present us with something of a puzzle, for given that the Earth's atmosphere is likely to have been outgassed from the mantle, it is not easy to understand why the two reservoirs have such different nitrogen isotope signatures. There is also a large difference in the $N_2/^{36}Ar$ ratios of the mantle ($>10^6$) and atmosphere (ca. 10^4), which is also problematic, for nitrogen and argon are chemically similar and so should have behaved in a similar way during the degassing of the atmosphere.

Much debated is whether or not nitrogen is cycled into the mantle by subduction. Marty and Dauphas (2003) have argued in favor of such a process. They show that whilst MORB has a $\delta^{15}N = -5$‰, OIBs sources have values between +3 and +4 (Fig. 5.4c). They interpret the elevated values in OIB as evidence for the recycling of heavy, sedimentary carbon, with $\delta^{15}N =$ ca. +10‰, into the mantle through subduction. Some of the OIB sources they describe are as old as 2.0 Ga, indicating that the sedimentary recycling of nitrogen is an ancient process. These authors go as far as suggesting that most of nitrogen in the mantle is

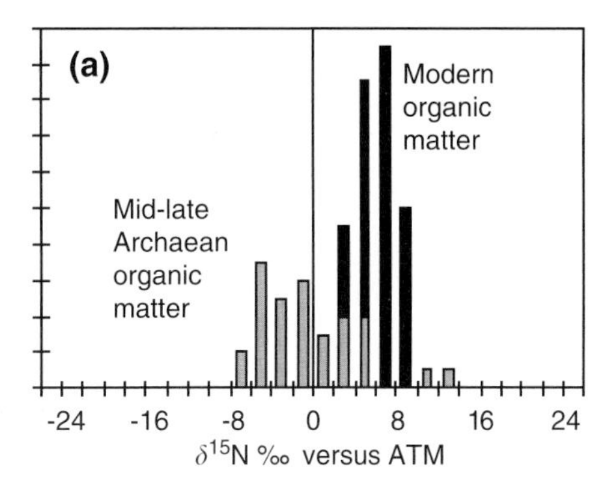

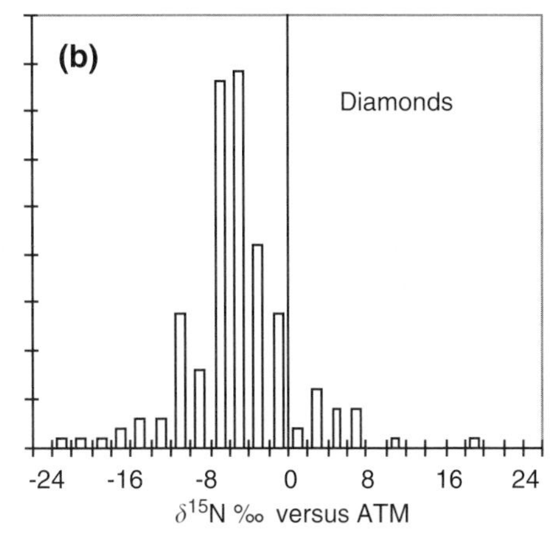

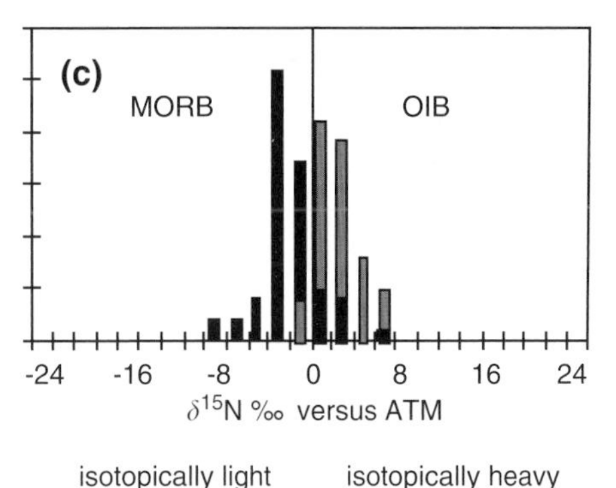

FIGURE 5.4 Variations in nitrogen isotopic composition in (a) modern organic matter in marine sediments (black) and organic matter in mid–late Archaean (2.9–3.4 Ga) cherts (grey); (b) peridotitic and eclogitic diamonds up to 3 Ga old; (c) mid ocean ridge basalts (MORB, black) and ocean-island basalts (OIB, grey); some of the OIB are derived from a source 2 Ga old. Note that all values are referenced to that of the present-day atmosphere. From the compilation by Marty and Dauphas (2003).

recycled and the present isotopic signature of the mantle reflects this process. Certainly this is plausible given the greater mass of nitrogen in the Earth's surface reservoirs, and, if true, is a powerful demonstration of the way in which biogenic processes have profoundly influenced the geochemical evolution of the solid Earth.

However, the ideas of Marty and Dauphas (2003) do not seem to represent a consensus and their model has been criticized by Cartigny and Adler (2003). Other authors have indicated that nitrogen recycling is of relatively small importance in the modern Earth (Zhang & Zindler, 1993; Lecuyer et al., 2000). Fischer et al. (2002) report nitrogen fluxes into and out of the Central American subduction system and find that they are balanced, implying that the subduction system acts as a barrier to the recycling of sedimentary nitrogen into the mantle. Scaling these results they find that the global flux of nitrogen at volcanic arcs is 2×10^{10} mol/yr, whereas the global mass of subducted nitrogen is 1.37×10^{10} mol/yr, indicating that subducted nitrogen is returned to the atmosphere in volcanic arcs. They conclude that the only nitrogen that could be returned to the mantle is that which is locked into subducted oceanic crust and lithosphere, and this is thought to be relatively unimportant.

5.1.3.2 Nitrogen in the Archaean mantle

The initial nitrogen isotopic composition of the Earth is not well known. However, nitrogen isotope measurements on 2.9–3.3 Ga diamonds from the subcontinental lithosphere have a mean $\delta^{15}N$ value of -5, and a similar C/N ratio to that of the modern mantle (Fig. 5.4), suggesting that there has been very little change since about 3.0 Ga (Marty & Dauphas, 2003).

In contrast, the nature of the surface nitrogen cycle in the Archaean is the subject of debate. Central to this discussion is when the nitrogen biogeochemical cycle began. For example, Mather et al. (2004) have suggested that in the Archaean nitrogen fixation was driven by nonbiological processes and that fixation rates were lower, probably of the order of 10^{11}. However, their model depends upon some assumptions about the extent to which N_2 is oxidized in a CO_2-rich atmosphere and the effectiveness of lightning in the process of nitrification. Beaumont and Robert (1999) argued for a difference in the N-isotopic signature of modern and Archaean sediments, implying that these differences reflect biogenic and nonbiogenic processes, respectively (see Fig. 5.4a). Again these authors are arguing that the nitrogen biogeochemical cycle had not begun in the Archaean. However, Pinti and Hashizume (2001) showed that the N-isotope values in Archaean sediments are very similar to those in modern hydrothermal vents where biological processes control the nitrogen cycle. In this case the nitrogen cycle might be very ancient but is expressed in Archaean sediments as a result of fractionations associated with biogenic hydrothermal processes, rather than the fractionations associated with the biologically driven redox reactions of the modern Earth.

It has already been noted that there is an unexpected difference in $N/^{36}Ar$ ratio between the mantle and atmosphere. Miyazaki et al. (2004) have suggested that this is very ancient and took place during core formation. They argue that during the formation of the core nitrogen was preferentially partitioned into the core relative to argon. This gave rise to huge differences in $N/^{36}Ar$ ratio between the core ($>10^9$) and the other Earth reservoirs (atmosphere $= 10^4$). Inefficiencies in the core formation process have led to the high $N/^{36}Ar$ ratios now found in the mantle (10^6) and explain the difference between mantle and atmosphere. This means that the Earth acquired its nitrogenous atmosphere very early and that this has been maintained since the earliest Archaean.

5.1.4 Sulfur

Sulfur occurs mostly in the solid Earth reservoirs and in oceans and is only a minor component of the atmosphere. Volcanic, sulfur-bearing gases emitted into the atmosphere have only a short residence time and are quickly transferred into the oceans. Because sulfur occurs in both the oxidized form – chiefly as the sulfate mineral gypsum, and in a reduced state – principally as the sulfide mineral pyrite, the element sulfur has an important part to play in balancing the Earth's redox budget. An example of this is the way in which the burial of biogenic sulfur in modern sediments contributes to the present-day oxygen flux (Section 5.3.1.6).

5.1.4.1 The modern sulfur cycle

Sulfur is removed from the continents by weathering, as sulfate, and is transferred to the oceans. In the oceans, sulfate may be either bacterially reduced to sulfide and precipitated in sediment, or removed as sulfate through evaporation in enclosed sedimentary basins. Sulfur is also added to the oceans from the mantle. There are three routes. One is via volcanism and is in the form of sulfur gases (SO_2 and H_2S), a second is through the removal of sulfur from the basaltic ocean floor through high-temperature hydrothermal activity, and the third through the low-temperature weathering of oceanic crust. Of these, the hydrothermal contribution is likely to be the greatest.

Sulfur is also returned to the mantle. This may follow one of two routes. The sulfur that is mobilized in the oceanic crust during high-temperature hydrothermal activity is a mix of ocean crust sulfide and seawater sulfate. Some of the seawater sulfate is converted into sulfide and fixed in the ocean crust and subsequently subducted into the mantle. In addition sulfur is removed from the oceans to form pyritized sediments and returned to the mantle during sediment subduction. Estimates of the fluxes are given in Canfield (2004) and are shown in Fig. 5.5.

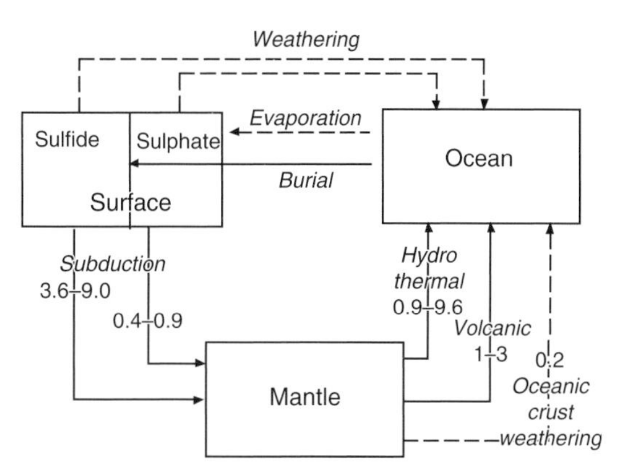

FIGURE 5.5 The modern sulfur cycle showing the fluxes in 10^{11} mol/yr (after Canfield, 2004). The parts of the cycle shown as dashed lines did not operate in the Archaean.

5.1.4.2 The Archaean sulfur cycle

There are two reasons why it is likely that the Archaean sulfur cycle might have been very different from that which is observed in the modern. First, as will be discussed later in this chapter, the Archaean atmosphere had very low levels of oxygen, so that there was no oxidative weathering of sulfide in crustal rocks at this time. This means that no sulfate was delivered to the Archaean oceans through weathering and consequently the Archaean ocean was probably very low in sulfate. Second, the removal of sulfate from the ocean is bacterially mediated. The operation of this process during the Archaean depends upon when sulfur reducing bacteria appeared during the Archaean. Both of these topics are discussed in some detail in this chapter (Section 5.4.3.1) and the next chapter (Section 6.4.3).

Hence the Archaean sulfur cycle (Fig. 5.5) would comprise inputs into the atmosphere and oceans from volcanic gases and into the oceans from hydrothermal activity but not river-borne sulfate. In addition, in the anoxic oceans, the oxidative alteration of the ocean floor would not take place. Thus the surface sulfur reservoir would have been small and most sulfur recycled back into the mantle as sulfide minerals. The sulfate part of the sulfur cycle is unlikely to have been fully operational until the late Proterozoic (Canfield, 2004).

5.2 THE ORIGIN OF THE EARTH'S ATMOSPHERE AND OCEANS

It is most likely that the volatiles which now make the Earth's atmosphere and oceans were acquired very early in the history of the planet. For this reason we need to return to the accretionary history of the Earth discussed earlier, in Chapter 2. Particularly useful here are the noble gases (Ne, Ar, Kr, Xe and He), for their abundances and isotopic ratios in the solar nebula are relatively well known, and so their evolution can be tracked relatively easily from the beginning of the solar system to the present. Also important are the carbonaceous chondrite meteorites. These are the most volatile rich of all meteorites and were probably the principal carriers of the Earth's volatile elements.

5.2.1 Possible sources of the Earth's volatiles

Understanding the origin of the Earth's volatiles is best done within the context of the standard model of planetary accretion (Chapter 2, Section 2.3.4.1). The terrestrial planets accreted from a swarm of planetesimals and planetary embryos, which had in turn formed from the solar nebula. Hence the Earth's volatiles are likely to have been derived from their planetesimal precursors and their chondritic "building blocks." Morbidelli et al. (2000) suggested that the most likely sources of the Earth's water were the outer asteroid belt, the giant planet regions, and the Kuiper belt, and they propose that most of the Earth's water was carried by a few planetary embryos from the outer asteroid belt, and that these were accreted in the final stages of Earth formation. Gases of the solar nebula were probably not an important source of volatiles, since the extreme solar wind associated with the T-Tauri phase of stellar evolution is likely to have blown the solar gas away. A further important aspect of the Earth's accretion is that it evolved within the "liquid water habitable zone" of the solar system (Kasting & Catling, 2003) and consequently was able to become home to life (see Chapter 6).

5.2.1.1 Constraints on the origin of the Earth's volatiles from volatile element abundance patterns

Kramers (2003) calculated major and minor (noble gas) volatile element abundance patterns in the Outer Earth Reservoir (the atmosphere, hydrosphere, oceanic and continental crust, and recycled components in MORB-source mantle). These are presented, normalized to solar abundances, together with data for chondrites in Fig. 5.6. The following observations can be made:

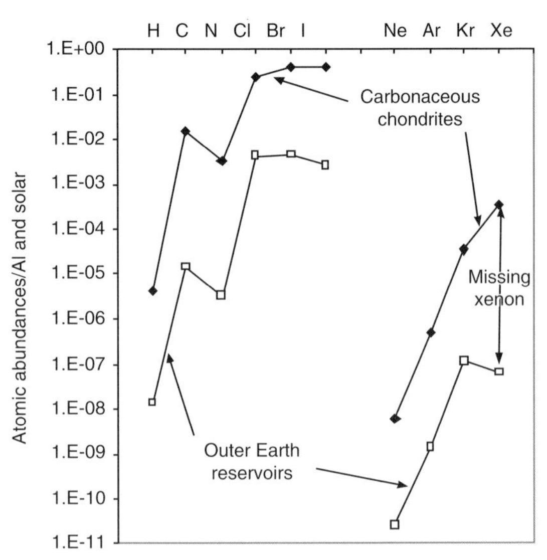

FIGURE 5.6 Major volatile element and noble gas abundances in the outer Earth reservoir of Kramers (2003) and in carbonaceous chondrites relative to Al and solar abundances. The data show that apart from xenon the Earth and chondritic meteorites have similar element distribution patterns and that both are strongly depleted in the noble gases and in H, C, and N relative to solar abundances.

1. The volatile abundances in the Earth are quite different from elemental abundances in the solar environment.
2. The relative volatile abundances in the Earth are similar to those found in chondritic meteorites. Ozima and Igarashi (2000) drew the same conclusion from their study of the isotopic compositions of the noble gases in the Earth's atmosphere (assuming that they approximate to the whole Earth noble gas inventory), and Lecuyer et al. (1998) found that the Earth's D/H ratio is the same as that in carbonaceous chondrites.
3. Although the relative volatile abundances are similar between Earth and chondrites, the absolute abundances are much lower in the outer Earth.
4. The element distributions in the Outer Earth Reservoir and chondrites are fractionationed relative to solar concentrations.
5. The concentration of xenon in the Outer Earth Reservoir is depleted relative to what might be expected from the chondritic abundance pattern. This has been dubbed "the Earth's missing xenon" and is a feature also

seen in Venus and Mars (Tolstikhin & O'Nions, 1994).

5.2.1.2 Evidence from the noble gases

Ozima and Igarashi (2000) found that noble gas ratios in the upper mantle are similar to those in the modern atmosphere. This means that the noble gases in the mantle and the atmosphere experienced the same fractionation event, an event which must have taken place before the formation of the Earth, in which case the fractionation may have taken place during chondrite formation or be an even older process within the solar nebula.

Firm evidence of a solar signature in the Earth's volatile budget comes from neon-isotope studies. This can be seen in the $^{20}Ne/^{22}Ne$ isotope ratio of mantle samples (MORB and OIB), which is greater than the value for air and close to the value for the solar nebula, as determined from the analysis of solar wind (Fig. 5.7). In detail neon-isotope systematics provide a wealth of information about the origin of the Earth's volatiles and the manner in which they were incorporated into the mantle. Elevated $^{21}Ne/^{22}Ne$ isotope ratios are controlled by terrestrial nucleogenic processes in which ^{4}He atoms, produced by the radioactive decay of U and Th, collide with ^{18}O in mantle minerals. In contrast, those $^{20}Ne/^{22}Ne$ values which are elevated above the atmospheric ratio of 9.8 are thought to contain a primordial component of solar neon ($^{20}Ne/^{22}Ne$ = 13.8). Both midocean ridge basalts and ocean island basalts display elevated $^{20}Ne/^{22}Ne$ isotope ratios (Fig. 5.7) indicating they have a solar component. A further member of the neon-isotope system is the component Ne–B with a relatively uniform $^{20}Ne/^{22}Ne$ ratio of 12.5. Ne–B has been identified in meteorites and represents an event in the early history of the solar system, during which meteorites and small planetesimals were irradiated by solar atoms and ions by a young sun, after the removal of solar nebular gases (Fig. 5.7).

A recent study by Ballentine et al. (2005) has established an upper limit for the $^{20}Ne/^{22}Ne$ isotope ratio in the upper mantle. Using the noble gas composition of gases exsolved from basalts derived from the upper mantle, they found that the maximum $^{20}Ne/^{22}Ne$ isotope ratio is between 12.2 and 12.5. They conclude that this ratio cannot be the product of mixing or fractionation

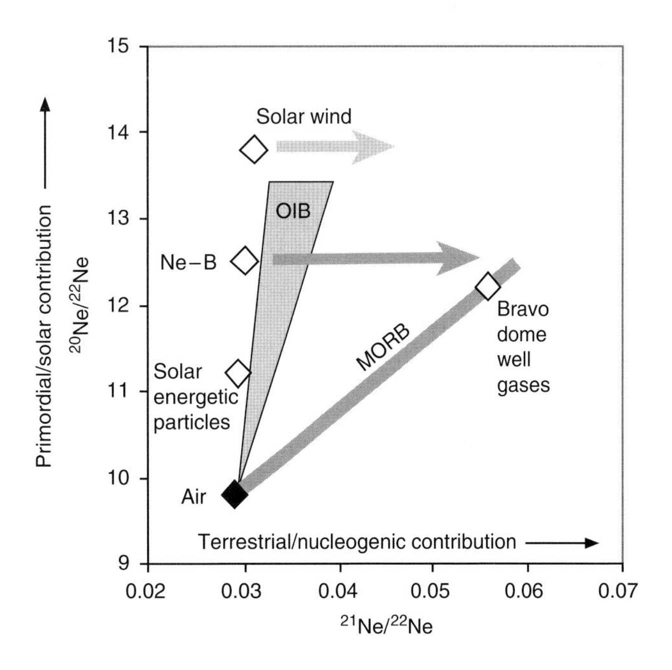

FIGURE 5.7 The neon-isotope systematics of the Earth's mantle showing the elevated (solar) $^{20}Ne/^{22}Ne$ isotope ratios in MORB and OIB. The Ne–B component is the product of a consistent mixing process involving solar wind (a proxy for the solar nebula) and solar energetic particles. Well gases from Bravo dome in New Mexico are the outgassing of upper mantle-derived basalts. The maximum value calculated by Ballentine et al. (2005) shows that they are identical in composition to the Ne–B component (after Ballentine et al., 2005; Graham, 2005).

processes but is a signal of a Ne–B component in the upper mantle. Ballentine et al.'s (2005) result for the upper mantle is significantly lower than that for OIBs, which have $^{20}Ne/^{22}Ne$ isotope ratios greater than 13, implying they contain a primordial, solar noble gas component.

The current interpretation of these observations is that there is a deep mantle source for ocean island basalts which formed during the earliest stage of planetary formation by directly trapping the gases of the original solar nebula. In contrast, the outer part of our planet, the convecting upper mantle, is made up of planetesimals which had been heavily irradiated by solar ions, during a later phase of accretion when the nebular gases had been swept away and the planetary debris could be directly irradiated by the young sun. The deep neon reservoir is thought to be the D″ layer of the mantle and the two reservoirs appear to have remained isolated from each other for the whole of Earth history.

5.2.1.3 The cometary "late accretion" model

Chyba (1987, 1990) argued that it was possible that a significant portion of the Earth's oceans were of cometary origin, delivered by comets of the Oort cloud between 4.5 and 3.5 Ga. His calculations, based on the lunar impacting record and an estimate of the proportion of impactors that would be dominantly water-bearing, suggest that between 0.2 to 0.7 ocean masses of water could have accreted to the Earth by this process (Chyba, 1990). However, there are some serious doubts about this model. Mass balance arguments presented by Dauphas and Marty (2002) indicate that the mass of comets present during late accretion was very small relative to the total mass of impacting bodies. A second reason, given by Chyba (1990) himself, pointed out that the addition of cometary water to Earth is not necessarily cumulative, for existing liquid water can potentially be removed by the addition of a later impactor. Third, Kramers (2003) presented geochemical arguments to show that the D/H ratio of all comets so far measured (the Oort Cloud comets) is twice that of the Outer Earth Reservoir. In addition the C/H and N/H ratios of comets are similar to those of carbonaceous chondrites and four times greater than that of the Outer Earth Reservoir and would give rise to higher C/H and N/H ratios than are observed. A final problem with this model is that the late delivery of volatiles through comets leaves little scope for later volatile loss – an important feature of the Earth's volatile abundance pattern (Section 5.2.2).

5.2.1.4 The source of the Earth's volatiles

In detail there were two sources for the Earth's volatiles. Neon isotopes show that neon at least, and probably all the noble gas inventory, was derived from a solar nebular gas. The second source was chondritic, and most of the CO_2 and H_2O were derived through the accretion of chondritic material. However, this is not the full story, for the Earth also experienced volatile loss – the topic of the next section.

5.2.2 Constraints on volatile loss

After the accretionary event in which the Earth acquired its volatiles, other processes took place which caused it to lose them. There are two lines of evidence which tell us about the early Earth's loss of volatiles. The first comes from a comparison between the volatile concentrations in the outer Earth and those of carbonaceous chondrite meteorites (the most primitive and most volatile-rich of all the meteorite groups). It is clear from Fig. 5.6 that the Outer Earth Reservoir has two to three orders of magnitude less volatiles than carbonaceous chondrites. In addition it is evident that the lighter major elements are more depleted than the heavy ones.

The second piece of evidence comes from the study of xenon concentrations and xenon isotopes. Figure 5.6 shows that xenon is more depleted in the Outer Earth Reservoir than it should be, compared to the other noble gases. This feature can be seen in both Xe abundances and in xenon isotope ratios. For example, ^{129}Xe and ^{136}Xe are radiogenic isotopes produced by the decay of ^{129}I and ^{136}Pu, respectively. Their initial concentrations can be fairly well constrained so that the amount of expected ^{129}Xe and ^{136}Xe can be calculated. Calculations of this type show that the Earth has lost

99% of its ^{129}Xe and between 60 and 90% of its ^{136}Xe (Kramers, 2003). Since Xe is a heavy gas, this means that at some stage in its history the Earth must have experienced the massive loss of its whole atmosphere. Alternatively, the strongly fractionated distribution of the volatile elements requires an even greater loss of the lighter gases. Calculations of this type require that between 97 and 99% of the original atmosphere has been lost to space.

5.2.2.1 *The timing of volatile loss*

The short-lived radiogenic isotopes of Xe also provide information on the time of volatile loss from the Earth. Kramers (2003) showed that I–Xe systematics place a minimum time limit on volatile loss of 90 Ma, and combined $^{129}Xe(I)$–$^{136}Xe(Pu)$ systematics indicate that Xe loss could have occurred until 100–200 Ma after the formation of the solar system. These results are consistent with extensive volatile loss during Earth accretion but imply that xenon-loss continued long after the early-Earth differentiation (Yokochi & Marty, 2005).

5.2.2.2 *Mechanisms of volatile loss*

Models for the loss of volatiles during the Earth's accretion history include **hydrodynamic loss** – a special case of the thermally driven atmospheric escape that occurs when the average thermal energy of hydrogen atoms or molecules in the upper atmosphere is comparable to the gravitational binding energy. When this happens the atmosphere attains an outward velocity at high altitudes comparable to the speed of sound so that a variety of atmospheric gases can escape together. A second mechanism of atmosphere loss is through the process of **impacting**. A giant impact of the size which created the Moon could lead to loss of the bulk atmosphere (Porcelli et al., 2001). A third mechanism of volatile loss is related to the **magmatic outgassing** of a magma ocean and the exchange of gases from magma to atmosphere that can take place during the lifetime of a magma ocean. In this model the differential solubilities of volatiles can explain their different concentrations in the outer Earth reservoir.

5.2.2.3 *Enhanced volatile loss in the presence of an early liquid water ocean*

A significant consideration in the discussion of volatile loss during impacting is whether or not the Earth had a liquid ocean at the time of impacting, for the "impact erosion" of an existing liquid water ocean, by a large impactor exerts a significant control on the loss of volatiles from the Earth.

Genda and Abe (2005) have recently argued that there would have been an ocean on the proto-Earth during the giant impacting stage. They used a radiative–convective equilibrium model for a H_2O–CO_2 atmosphere and calculated that protoplanets in an Earth orbit would have a liquid water ocean whereas protoplanets in the orbit of Venus would not. They showed that when atmospheric loss is modeled in the presence of a liquid water ocean it is greater than atmospheric loss on a dry planet. The reason for this is that during a giant impact the ocean vaporizes in the area of impact, but also vaporizes elsewhere due to the shock waves which travel through the ground. Thus the presence of a liquid ocean on Earth was likely during the giant impact phase of accretion and this would have led to enhanced volatile loss.

There is geochemical support for this hypothesis from Kramers' (2003) chondrite normalized distribution pattern of the terrestrial volatiles (Fig. 5.8). These data show a chondritic ratio for C and N, and an overabundance of H and Cl relative to C and N (Fig. 5.8). This is unexpected since hydrogen with its low atomic mass should be depleted more than the heavier carbon or nitrogen. A similar relationship is seen when volatiles are plotted as their likely molecular species. Again, H as H_2O and Cl as HCl do not lie on the trend of increasing abundance with atomic mass. This observation can neither be explained by a cometary origin of the Earth's water nor by a simple bulk volatile loss or even a fractionated volatile loss. A possible explanation for the increase abundance in Cl and H is that these elements were preferentially retained in a liquid water ocean. Thus whilst other volatiles were fractionated during impacting and the related hydrodynamic

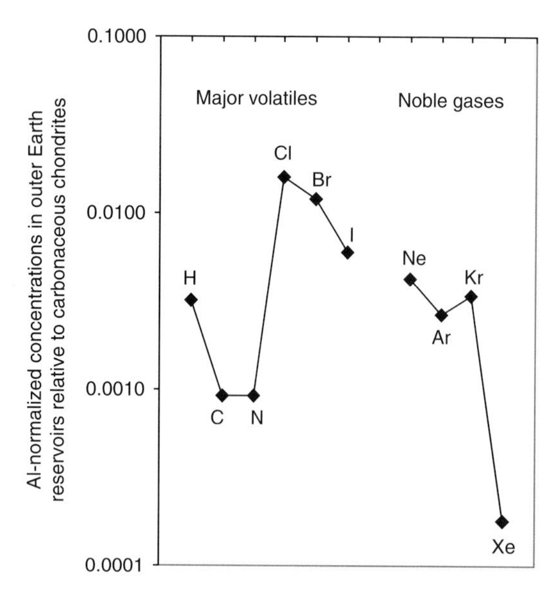

FIGURE 5.8 The Al-normalized concentrations of the major volatile elements (H, C, N, Cl, Br and I) and the noble gases (Ne, Ar, Kr, Xe) in the outer Earth reservoirs relative to Al-normalized concentrations in carbonaceous chondrites (after Kramers, 2003).

escape, H and Cl were retained in a liquid water ocean.

5.2.2.4 The Earth's missing xenon

Given that xenon is the heaviest of the noble gases, its low abundance in the Outer Earth Reservoir relative to chondrites (Figs. 5.7 and 5.8) is puzzling. Tolstikhin and O'Nions (1994) argued that xenon depletion is the opposite of what is expected during hydrodynamic escape given the large mass of these atoms. Instead they proposed that xenon loss has to be explained by fractional degassing. More recently Pepin (2000) has proposed that Xe was fractionated on the early Earth by hydrodynamic escape, perhaps because xenon alone amongst the rare gases has the ability to become ionized and escape in its ionized form. This view is consistent with the protracted period of Xe-loss calculated from Xe-isotopes.

5.2.3 A possible sequence of events for volatile gain and loss on the early Earth

Evidence for solar neon in the deep Earth and Ne–B in the upper mantle indicates that the Earth's volatiles were acquired in at least two different stages during accretion (Ballentine et al., 2005). In addition the low Earth abundances of volatiles relative to chondrites indicate that there were subsequent volatile-loss events.

A possible sequence of events is as follows:

1. During early accretion the Earth gravitationally captured gases from the solar nebula. These would have provided the proto-Earth with a dense primordial atmosphere and would have induced deep melting to produce a magma ocean. Exchange between the atmosphere and the magma ocean led to the dissolution of solar gases into the accreting planet (Porcelli et al., 2001) and their convection into the deeper part of the planet. It is possible that this early stage of accretion was reducing in which case carbon may have dissolved in metallic melts and subsequently removed to the core. In addition H_2O may have reacted with metallic melts giving rise to iron hydride which partitioned into the core and/or H_2 which would be lost by hydrodynamic escape (Kramers, 2003). During this early stage of volatile accretion it is likely that the noble gases made up a significant part of the volatile budget.
2. Later accretion took place in a solar-gas free environment in which carbonaceous chondrite grains and planetesimals were irradiated by the young sun to form Ne–B. At this stage the Earth acquired a large volume of volatiles from its carbonaceous chondrite building materials. These volatiles were already fractionated relative to solar concentrations as can be seen from the distribution pattern of carbonaceous chondrites (Fig. 5.6). This stage of accretion was probably more oxidizing (Kramers, 2003).
3. Several episodes of volatile loss then followed in which the overall concentrations of all volatiles were reduced by one or two orders of magnitude. It is likely that this volatile loss took place through impact degassing and Porcelli et al. (2001) have proposed that this was the Moon-forming impact event, at ca. 30 Ma after t_0. Impacting on this scale would have led to extensive melting and volatile loss from all the outer

Earth reservoirs. If this event took place in the absence of a blanketing solar gas then there would have been substantial gas loss to space.

4. However, Genda and Abe (2005) and Kramers (2003) have shown that the Earth was likely to have had a very early liquid water ocean. Consequently, atmospheric loss probably took place in the presence of a liquid water ocean, leading to enhanced volatile loss.
5. Subsequent Earth history records that volatiles have been recycled from the Earth's surface back into the mantle through subduction and then returned again to the Earth's surface reservoirs (Honda et al., 2004).

5.3 The nature of the Archaean atmosphere

There is compelling evidence from both planetary science and the Earth sciences that our modern terrestrial atmosphere is very different in composition from the Earth's atmosphere in the Archaean. A useful starting point is to compare the composition of the Earth's present atmosphere with those of the neighboring terrestrial planets, Venus and Mars (Table 5.4). It is evident that the Earth is nitrogenous, oxygenic, hydrous and CO_2-poor and, in the language of James Lovelock, possesses a disequilibrium atmosphere. Why these differences exist between the terrestrial planets is much debated. One school of thought has argued that the initial atmospheric compositions of the three planetary atmospheres must have been identical. This means that the present differences reflect their subsequent planetary evolution, which must include differential volatile loss, chemical reactions between the atmosphere and planetary surface, and, on Earth, reflects the importance of biological process (Hunten, 1993). Thus, in this model Mars and Venus have lost their hydrous oceans.
The alternative view is that the different planetary atmospheres are original and are a function of their differing distances from the sun and contrasting accretion histories. On this basis Genda and Abe (2005) argued that Venus had never possessed a liquid water ocean.

However, our purpose here is to explore how the Earth's atmosphere has changed with time and examine the processes which have

TABLE 5.4 The compositions of the atmospheres of the terrestrial planets Venus, Earth, and Mars. Compositions expressed in parts per million by volume or volume percent. Data from compilation of Hunten (1993) and Lecuyer et al. (2000).

	Venus	Earth	Mars
Pressure of planetary atmosphere	**92.000 bars**	**1.013 bars +270 bars equiv H_2O in oceans +ca. 60 bars equiv CO_2 in crust**	**0.006 bars**
CO_2	96.5%	0.033%	95.3%
N_2	3.5%	78%	2.7%
H_2O	30–200 ppm	ca. 3%	ca. 100 ppm
He	12 ppm	5.2 ppm	100 ppm
Ne	7 ppm	18.2 ppm	2.5 ppm
Ar	70 ppm	9,340 ppm	16,000 ppm
Kr	0.05 ppm	1.14 ppm	0.03 ppm
Xe	0.04 ppm	0.087 ppm	0.08 ppm
D/H	2.5×10^{-2}	1.49×10^{-4}	9×10^{-4}

governed this change. Four main processes are thought to have been important and these can be summarized as a variety of different reactions between the atmosphere and the Earth system. They include:

- volcanic degassing – the influence of the mantle on the atmosphere;
- weathering – the influence of the crust (oceanic and continental) on the composition of the atmosphere;
- gas exchange – the influence of the oceans on the composition of the atmosphere;
- biogenic processes – the influence of life on the evolution of the atmosphere.

Each of these themes will keep appearing in the subsequent discussion.

5.3.1 The rise of atmospheric oxygen

The first systematic documentation of the changing levels of oxygen in the Earth's atmosphere was made by Cloud (1972) who showed that strongly oxidized iron-bearing sediments (red beds) only appeared in the geological record at about 2.0 Ga. Similarly, sediments containing abundant reduced iron – banded iron formations – ceased to be formed after about 1.8 Ga. These observations imply that there was a significant increase in the partial pressure of atmospheric oxygen during the early Proterozoic.

What is less certain is the proportion of oxygen in the atmosphere prior to this period, during the Archaean, although the evidence discussed below suggests that it was low. In this context it is helpful to think about oxygen sources and sinks. Free oxygen will only exist on Earth if production has first satisfied all possible oxygen sinks (Kump et al., 2001; Lasaga & Ohmoto, 2002). When seen in this way it is clear that there might be two different reasons for there being low oxygen levels in the Archaean. On the one hand oxygen production might have been low (e.g. Canfield, 2005). Alternatively, oxygen production might have been high but was buffered through reaction with surface rocks and reduced volcanic gases (e.g. Kump et al., 2001). The modern balance between oxygen sources and sinks was illustrated in the oxygen cycle diagram in Chapter 1 (Section 1.1.2.2).

First however, we must examine the evidence for ancient oxygen levels from paleo-oxygen barometers in the ancient sedimentary record.

5.3.1.1 Evidence from redox-sensitive detrital minerals in clastic sediments

An important source of information about the Earth's early atmosphere are the redox-sensitive minerals uraninite, pyrite, and siderite which are present in Archaean fluviodeltaic sediments but which are not found in similar modern sediments (Rasmussen & Buick, 1999; Frimmel, 2005). Notable occurrences are in 3.1–2.7 Ga clastic, fluviodeltaic sediments of the Witwatersrand basin in South Africa and in 3.25–2.75 Ga fluvial siliclastic sediments from the Pilbara Craton in western Australia. For example detrital uraninite (uranium oxide, UO_2) is found throughout the Witwatersrand basin succession and yet in the modern weathering environment uraninite is not stable and is removed as an oxyhydride. Similarly detrital pyrite is also present and thought to have been exposed for sufficient length of time to have equilibrated with the surface environment of its time. Pyrite oxidation takes place at even lower levels of oxygen activity than uraninite and so is hypersensitive to an oxygenic atmosphere. Calculations show that pyrite oxidation in Archaean seawater with a pH of 6.0 is very slow, even when the atmospheric fO_2 is 10^{-3}, and provide a further upper limit on atmospheric oxygen levels in the late Archaean (Frimmel, 2005). Detrital siderite is also unstable in an oxidizing environment, and yet is locally important in the Pilbara sediments where it appears to have been reworked through several episodes of erosion (Rasmussen & Buick, 1999). Taken together, the evidence from detrital redox-sensitive minerals points to low oxygen levels in the time interval 3.25–2.7 Ga.

5.3.1.2 Evidence from Fe-mobility in paleosols

Paleosols are fossil soils. They have formed by weathering in the subaerial environment and so have the potential to provide information about the composition of the atmosphere at the time of their formation. In a comprehensive critique of proposed paleosols in the geological literature Rye and Holland (1998) found that all paleosols formed prior to 2.44 Ga had experienced significant Fe-loss in the upper part of their weathering profile. However, this was not the case for younger paleosols formed after about 2.2 Ga, some of which have now been identified as former laterites (Gutzmer & Beukes, 1998; Beukes et al., 2002). This change in the mobility of Fe is attributed to the rise of atmospheric oxygen in the time interval 2.0–2.45.

The oldest recorded paleosols are developed on 2.8 Ga basalts at Mount Roe in western Australia. In these all the iron is stripped from the upper part of the weathering profile. Rye and Holland (1998) have used the compositions of these samples to calculate that the partial pressure of oxygen in the atmosphere at 2.8 Ga was very low and equal to or less than 3×10^{-4} atm. Support for this view comes from the identification of the mineral rhabdophane in 2.6–2.45 Ga paleosols from Pronto Mine Canada (Murakami et al., 2001). Rhabdophane is a REE-bearing phosphate mineral produced as a weathering product of apatite, which at Pronto Mine contains unoxidized Ce(III). The recognition of Ce(III) rather than the oxidized form Ce(IV) indicates that the weathering of the paleosol took place under oxygen deficient conditions.

Most workers agree that the paleosol evidence indicates that there was a change in the level of atmospheric oxygen between about 2.45 and 2.0 Ga, although Ohmoto (1996) dissents from this view. He believes that the loss of iron from paleosols was a late process and that initially ferric hydroxides were produced in the soil from an oxygenic atmosphere and subsequently removed in later reducing solutions.

5.3.1.3 Evidence from mass-independent sulfur fractionation

An exciting new development in the field of stable isotope geochemistry has been the recent recognition of the importance of mass-independent fractionation in the sulfur isotope system (see Text Box 5.1). This new field has important implications for assessing oxygen levels in the early atmosphere. Most stable isotope fractionations are controlled by the mass differences between isotopes. This is mass-dependent fractionation (MDF) and is expressed as a deviation from the appropriate standard as a "lower-case delta-value" in parts per thousand (e.g. $\delta^{33}S$ ‰). Mass-dependent fractionation follows specific rules based on the relative atomic mass differences between the fractionating isotopes (e.g. $\delta^{33}S \sim \frac{1}{2}\,\delta^{34}S$, and $\delta^{36}S \sim 2\,\delta^{34}S$). Fractionations of this type are governed by thermodynamic, kinetic, and diffusion processes and produce highly correlated relationships between $\delta^{33}S$ and $\delta^{34}S$ (Fig. 5.9a). MIF on the other hand, is where fractionations take place in a different manner from the simple proportionality predicted on the basis of atomic mass differences. Deviations from the MDF line are expressed using the convention "upper-case delta-value" in parts per thousand (e.g. $\Delta^{33}S$ ‰). One of the more commonly used diagrams to represent results of this kind is a plot of mass-dependent $\delta^{34}S$ on the *x*-axis versus mass-independent $\Delta^{33}S$ on the *y*-axis (Fig. 5.9b).

Laboratory studies have shown that the most likely explanation for mass-independent sulfur isotope fractionations within the Earth system is through reactions which take place within the gas phase and thereby provide an important geochemical fingerprint of atmospheric processes (Farquhar et al., 2000, 2002). A particularly important reaction is the photochemical oxidation of sulfur in the atmosphere. Today this reaction is prevented by the presence of ozone and oxygen in the atmosphere which shield the lower atmosphere from the ultraviolet radiation required for this reaction. Experimental studies show that only tiny amounts of atmospheric oxygen are needed to prevent the photochemical oxidation of sulfur, indicating that photochemical oxidation can only take place in an atmosphere with very low levels of oxygen.

What is important for studies of the early atmosphere is that there is a significant change

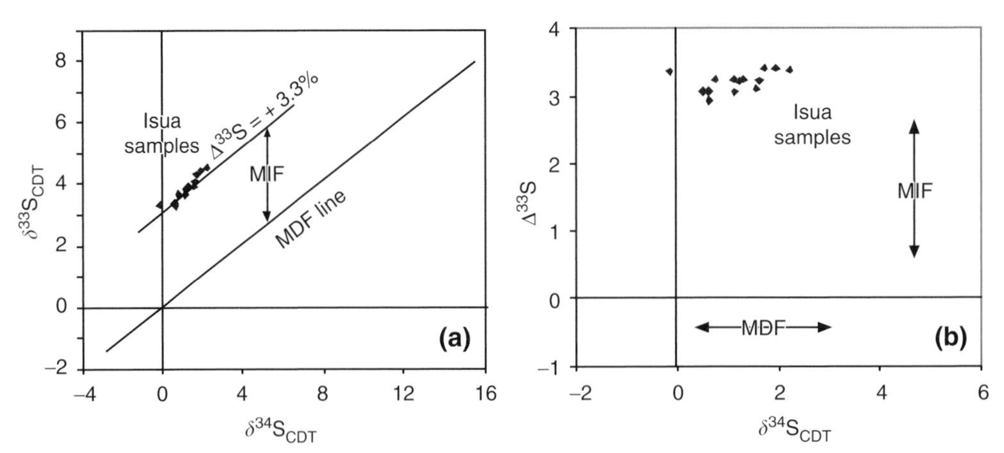

FIGURE 5.9 (a) A plot of $\delta^{34}S$ versus $\delta^{33}S$ showing the mass-dependent fractionation line (MDF) and samples from the >3.7 Ga Isua greenstone belt which plot above the MDF line indicating mass-independent fractionation (MIF). (b) A $\delta^{34}S$ versus $\Delta^{33}S$ plot for samples from Isua showing a small spread due to MDF and evidence of MIF (after Whitehouse et al., 2005). (See Text Box 5.1.)

in $\Delta^{33}S$ values in sulfide minerals before and after about 2.45 Ga (Fig. 5.10). This result first reported by Farquhar et al. (2000) has subsequently been confirmed by Mojzsis et al. (2003), Ono et al. (2003), and Whitehouse et al. (2005). Before 2.45 Ga there are measurable $\Delta^{33}S$ fractionations, whereas after this time they are not found in the geological record. This implies that before 2.45 Ga oxygen levels in the Earth's atmosphere were low and that there was insufficient oxygen and ozone in the upper atmosphere to block the ultraviolet radiation which drives the photochemical oxidation of volcanic sulfur gases in the atmosphere. In detail Farquhar and Wing (2003) have identified three stages of terrestrial $\Delta^{33}S$ fractionation (Fig. 5.10). Prior to ~2.45 Ga (Stage I) large positive and negative $\Delta^{33}S$ fractionations are recorded ($\Delta^{33}S$ > +3.0 to −2.0) requiring SO–SO_2 photolysis in the atmosphere. Pavlov and Kasting (2002) calculated that this reaction can only take place if oxygen levels were $<10^{-5}$ the level of the present atmosphere (PAL). The positive and negative

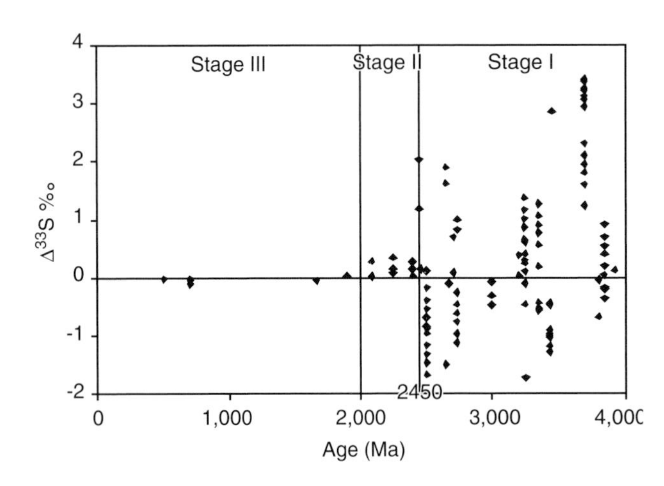

FIGURE 5.10 A plot of $\Delta^{33}S$ in sulfide and sulfate minerals versus time showing the three stages of sulfur isotope evolution. Stage I, between 3.8 and 2.45 Ga, is represented by extensive variability in $\Delta^{33}S$, stage II, between 2.45 and 2.0 Ga, shows subdued variability and stage III, after 2.0 Ga, is characterized by minimal variability with values of 0.0 ± 0.2‰ (after Farquhar & Wing, 2003 with additional data).

values of $\Delta^{33}S$ imply two sulfur reservoirs – a water soluble sulfate reservoir with negative values and a reduced insoluble reservoir with positive $\Delta^{33}S$ values. Between 2.45 and 2.0 Ga (Stage II) there is a small range of positive $\Delta^{33}S$ values. Farquhar and Wing's (2003) preferred interpretation is that this time interval records the onset of oxidative weathering. This requires a rise in oxygen levels from $<10^{-5}$ PAL before 2.45 Ga to $>10^{-2}$ PAL which would prevent the photochemical oxidation of sulfur in the atmosphere. After 2.0 Ga (Stage III) $\Delta^{33}S$ variations are close to zero indicating an absence of mass-independent sulfur isotope fractionation consistent with higher levels of oxygen in the atmosphere. These observations show that sulfur MIF also strongly supports a major change in the level of atmospheric oxygen between 2.45 and 2.0 Ga.

5.3.1.4 The oxidation state of the Earth's mantle

One of the major controls on oxygen levels in the modern atmosphere is the abundance of reducing volcanic gases, for these are an important oxygen sink. Lasaga and Ohmoto (2002) calculated that the modern flux of reduced carbon species (CO and CH_4) is $1{,}500 \times 10^{12}$ mol/Kyr, and the flux of other reduced gases (H_2, H_2S and SO_2) is $1{,}000 \times 10^{12}$ mol/Kyr. If this flux has changed with time, and in particular if it was more reducing in the past, as has been suggested by Kasting et al. (1993) and Kump et al. (2001), then this would exert an important influence on atmospheric oxygen levels.

Fundamental to this process is the oxidation state of the Earth's mantle, for equilibria between elemental carbon and C–O–H fluids control the proportions of oxidized and reducing gases degassed from the mantle. Particularly important here is knowing the extent to which the oxidation state of the mantle has changed with time. There have been a number of models which claim that the redox state of the mantle was different in the past due either to the separation of the Earth's core (Kuramoto & Matsui, 1996), the subduction of ferric iron (Lecuyer & Ricard, 1999), or intense mantle plume activity (Kump et al., 2001).

The principal tool for estimating the oxidation state of the mantle is the calculation of mantle oxygen fugacities. This is a measure of the chemical potential, that is, the reactivity of oxygen in the mantle system. A high mantle oxygen fugacity indicates a relatively oxidizing system. Mantle oxygen fugacities can be calculated from redox equilibria between mantle minerals. The results are normally compared to the quartz–fayalite–magnetite (QFM) reference curve, regarded as the "normal" oxidation state of the modern upper mantle and expressed in log units ± QFM. However, this enterprise is not simple, for in the modern mantle oxygen fugacity varies with tectonic setting (Parkinson & Arculus, 1999), and so the "normal" mantle can be hard to define. Furthermore, magmas can evolve with respect to their oxygen fugacity, as they rise to the surface and degas, placing additional limits on our detailed knowledge of their source (Mathez, 1984; Ballhaus & Frost, 1994).

A number of different studies have been carried out to determine the oxidation state of the Archaean mantle (Fig. 5.11). These indicate that Archaean komatiites have oxygen fugacities which approximate the QFM buffer curve (Nisbet et al., 1993; Canil, 1997, 1999; Rollinson, 1997). Similarly, Archaean basalts have oxygen fugacities close to that of modern MORB. Li and Lee (2004) used V/Sc ratios in Archaean basalts as a proxy for the oxygen fugacity of their mantle source. They show that 3.5 to 2.7 Ga basalts from a variety of Archaean greenstone belts have an almost identical V/Sc ratio to that of modern MORB. They calculate an oxygen fugacity of QFM -0.3 ± 0.5 log units for the modern MORB source and argue that the oxygen activity of the Archaean mantle is within 0.3 log units of the modern mantle. Support for this view comes from chromite compositions from a differentiated basaltic layered intrusion in the >3.8 Ga Itsaq gneisses of west Greenland which plot in the same field as chromites in modern MORB and OIB, leading Delano (2001) and Rollinson et al. (2002) to conclude that the redox state of the Earth's mantle has remained essentially constant for the past 3.8 Ga (Fig. 5.11).

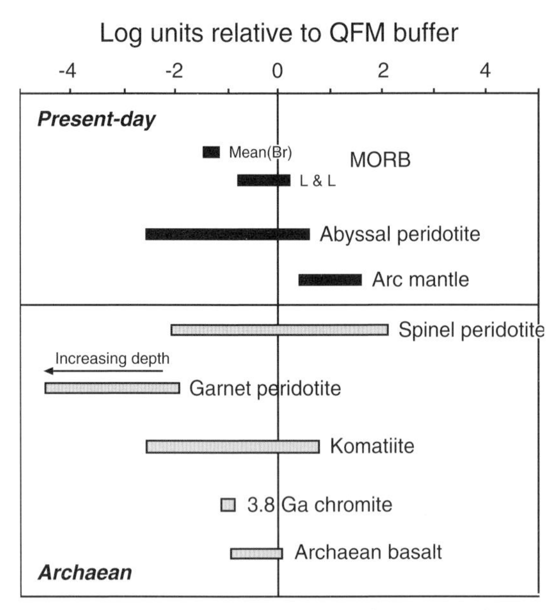

FIGURE 5.11 Calculated oxygen fugacity, relative to QFM for the present-day mantle and the Archaean mantle. The iron–wustite buffer, a likely lower limit for the Archaean mantle is located at QFM −3, for a pressure of 5 Ga and at 1,200°C. Data from Byndzia et al. (1989), Parkinson and Arculus (1999), Woodland and Koch (2003), Nisbet et al. (1993), Canil (1997, 1999), Rollinson (1997), Rollinson et al. (2002), Delano (2001), and Li and Lee (2004). The two sets of MORB data are a mean value from Byndzia et al. (1989) (mean (Br)) and the value from Li and Lee (2004) (L & L) for MORB source.

A number of studies have used xenoliths from Archaean subcontinental lithosphere to make inferences about the oxidation state of the early mantle. For example Woodland and Koch (2003) showed that the subcontinental lithosphere beneath the Kaapvaal Craton displays a systematic decrease in oxygen fugacity with depth (Fig. 5.11). However, it should be remembered that the highly depleted nature of the Archaean subcontinental lithosphere means that it is atypical of the mantle as a whole and may not therefore be useful as an indicator of mantle redox conditions (Chapter 3, Section 3.1.3.2).

In summary redox data for the Archaean mantle seem to show that the Earth's mantle is not substantially more oxidizing now than in the Archaean and that even in the early Archaean (3.8 Ga) the oxygen activity of the depleted mantle was broadly the same as it is today. Hence models for the evolution of the Earth's atmosphere, which require a changed proportion of oxidizing to reducing gases from volcanic sources over geological time, cannot be sustained.

5.3.1.5 Charting the rise of atmospheric oxygen over time

Currently the best fingerprint for the rise in atmospheric oxygen is provided by the mass-independent fractionation of sulfur isotopes. These studies show that the Earth's atmosphere experienced a significant oxidation event between 2.45 Ga and 2.0 Ga. This has been termed the Great Oxidation Event by Holland (2002). Pavlov and Kasting (2002) proposed that prior to 2.45 Ga oxygen levels were less than 10^{-5} the level of the present atmosphere (PAL), that is, <2 ppm, and after 2.0 Ga were greater than 10^{-2} PAL (>2,100 ppm). (Fig. 5.12) The argument for low oxygen levels prior to 2.45 is corroborated by the existence of the detrital minerals pyrite, uraninite and siderite in the 3.2–2.7 Ga Witwatersrand and Pilbara sediments, and from the composition of the 2.75 Ga Mount Roe paleosol. Evidence for high atmospheric oxygen levels at around 2.0 Ga comes from the presence of "red beds" in the stratigraphic record and the discovery of laterites, some of which are as old as 2.2 Ga (Beukes et al., 2002).

Recently Bekker et al. (2004) found syngenetic pyrite in the 2.32 Ga organic rich shales of the Rooihoogate formation and the conformably overlying Timeball Hill formation of the Transvaal Supergroup, South Africa, which shows no evidence of sulfur isotopes MIF. This result indicates that oxygen levels were in excess of $>10^{-5}$ PAL by 2.32 Ga, thus narrowing the interval of atmospheric oxygenation to between 2.47 Ga (Mojzsis et al., 2003) and 2.32 Ga. The rise of atmospheric oxygen may be further constrained by the observation that in the lowest part of the Rooihoogte–Timeball Hill sequence there are pebbles of sideritic facies iron formation, indicating a reducing environment, whereas in the upper part there are haematitic pisolites, indicating

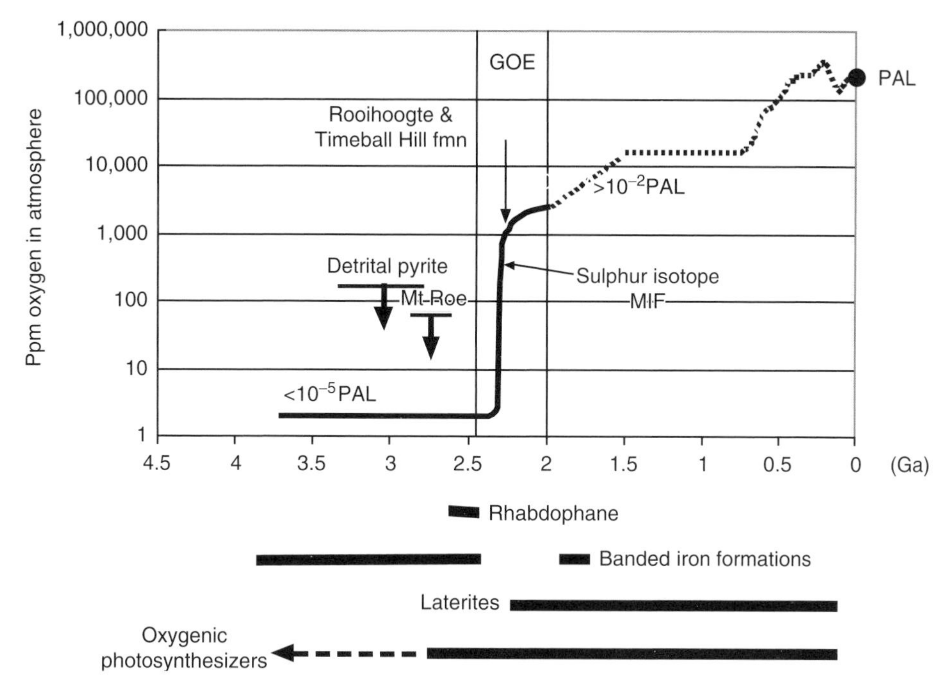

FIGURE 5.12 The probable evolution of atmospheric oxygen over time given the constraints discussed in the text. The solid curve is inferred from sulfur isotope MIF, the dotted curve is taken mostly from Canfield (2005). GOE is the great oxidation event of Holland (2002). Qv. Catling and Claire (2005).

an oxidizing environment (Bekker et al., 2004). It may be therefore that the first rise in atmospheric oxygen took place during the deposition of the Rooihoogte–Timeball Hill sequence at 2.32 Ga.

Collectively these data indicate that there was a change in the proportion of oxygen in the Earth's atmosphere over time. This change seems to have taken place at about 2.3 Ga. However, *why*, and precisely *how* the composition of the atmosphere changed at this time is still a matter of debate and will be discussed further below.

5.3.1.6 *Oxygen regulation over geological time*

In the modern Earth oxygen production through photosynthesis is balanced by the reduction of oxygen by organic carbon. Strictly therefore no oxygen should accumulate. However, over Earth history oxygen levels have been regulated by the compounds of carbon, sulfur and iron and the presence of free oxygen can be explained by the removal of a small amount of organic carbon and sulfur in the form of pyrite, during the burial of sediments.

The burial history of organic carbon over time can be deduced from the carbon isotope record (Fig. 5.13). Dissolved inorganic carbon is converted to organic carbon by autotrophic processes. In the course of these reactions carbon isotopes are fractionated such that ^{13}C is enriched in the inorganic reservoir and depleted in the organic carbon reservoir. Records of the composition of the organic reservoir over geological time can be obtained from the isotopic composition of buried organic carbon and of the inorganic carbon from the isotopic composition of carbonates precipitated from the oceans. Perhaps, rather surprisingly this isotopic record shows that the compositions of organic and inorganic carbon have been approximately constant since about 3.2–3.5 Ga (Schidlowski, 1988). This agrees with evidence from the fossil record which tells us that oxygen-producing cyanobacteria were present on Earth by at least 2.7 Ga and

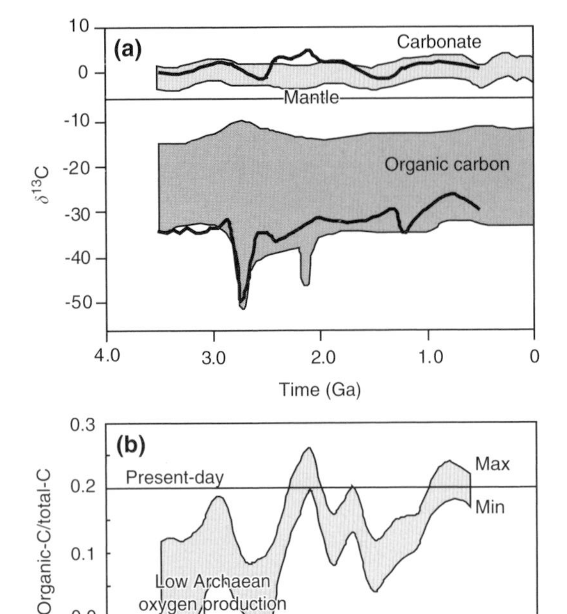

FIGURE 5.13 (a) The carbon isotope record for inorganic carbon and organic carbon from Schidlowski (1988) – shaded areas – and the average values from Canfield (2005) – continuous lines. (b) Calculated maximum and minimum values of the ratio of organic carbon to total carbon over geological time. The maximum curve is based upon the standard mass balance calculation, whereas the minimum curve assumes the additional carbon sink of carbonate in the ocean crust. The present-day value is 0.2 (after Canfield, 2005).

maybe as early as 3.5 Ga (Catling et al., 2001) (see Chapter 6, Section 6.5.1).

Bjerrum and Canfield (2004) have used the relationship in Fig. 5.13 to calculate the ratio of organic carbon burial to the total amount of carbon buried in the sedimentary record over time. They show (Fig. 5.13b) that organic carbon burial was less in the Archaean than at the present day. These authors also calculated the effects of a third carbon isotope reservoir – that of carbonate in the oceanic crust – and showed that in this case the degree of organic carbon burial was much less in the Archaean than at present. This result would imply that oxygen levels were low in the Archaean, not because of extensive consumption (large oxygen sinks) but because oxygen production was very low. However, some of the assumptions used in these calculations have been challenged, for example there is evidence that the average organic content of sediments has remained constant at 0.5 wt% over geological time (Catling & Claire, 2005) and the evidence for a depleted $\delta^{13}C$ reservoir of carbonates in the oceanic crust is not consistent with the carbon isotope measurements of Nakamura and Kato (2004) for 3.5 Ga carbonatized basalts.

A central question in this discussion is why atmospheric oxygen levels rose so sharply at about 2.3 Ga. A related question is why there was a delay of several hundred million years between the first oxygen production from cyanobacteria and the rise of atmospheric oxygen. A number of possible explanations have been offered. First, there may has been a dramatic change in the abundance of volcanic, reducing gases added to the Earth's atmosphere at 2.45 Ga, through the upwelling of an oxidized lower mantle plume (Kump et al., 2001; Barley et al., 2005). However, this model neither sits comfortably with the redox data for the Archaean mantle discussed earlier, nor with the reducing nature of the modern lower mantle (McCammon, 2005). Alternatively, oxygen sinks in the form of new continental and oceanic crust, formed in the late Archaean as the consequence of abundant plume magmatism, may have consumed oxygen, delaying its rise in the atmosphere until 2.4–2.3 Ga (Barley et al., 2005).

A second possibility, proposed by Catling et al. (2001) is that oxygen levels in the Earth's atmosphere result from a net loss of hydrogen from the Earth through the process of hydrogen escape. This model is based upon the assumption that a significant methane component had accumulated in the atmosphere in the late Archaean and that the hydrogen was produced through the oxidation of this methane. In this model, the rise in oxygen levels in the atmosphere marks the point at which crustal oxygen sinks become saturated, permitting the level of free atmospheric oxygen to rise.

A third model for the rise in atmospheric oxygen levels at ca. 2.45 Ga focuses on an increase in oxygen production, rather than, as in the previous models, a decrease in its consumption. Bjerrum and Canfield (2002, 2004) argue that oxygen production in the Archaean was limited by the supply of phosphorus as a nutrient, since during the Archaean most phosphorus was adsorbed onto Fe-oxides in BIFs (see Section 5.4.3.2). This changed between 2.4 and 2.5 Ga, with the increase in sulfate levels in the oceans leading to the sequestration of Fe as sulfide, hence freeing up more phosphorus. Thus after about 2.4 Ga the increased availability of phosphorus permitted an increased oxygenic photosynthesis, leading to higher levels of oxygen in the atmosphere.

Deciding between these competing options requires a quantitative knowledge of oxygen sources and sinks in the late Archaean/early Proterozoic (Catling & Claire, 2005). Initial attempts at such a quantification suggest that the redox states of the continental crust and/or the mantle may hold the key to this puzzle (Claire et al., 2005), although this more sophisticated approach is still in its infancy.

The implications of the rise in the level of atmospheric oxygen between 2.47 and 2.32 Ga are profound and affected more than just the composition of the Earth's atmosphere, oceans and surface. It was noted earlier that Th/U and Nb/U ratios in the Earth's mantle have changed with time (see Chapter 3, Section 3.2.3.3 and Chapter 4, Section 4.2.1.1). The increase in these ratios is thought to reflect the increasing mobility of U in an oxygenic surface environment and implies the return of oxidized materials from the Earth's surface to the mantle by subduction.

5.3.2 The reduction in atmospheric CO_2 levels over geological time

There are a number of reasons for believing that the Earth originally had a more CO_2-rich atmosphere than it does today. At a very basic level the simple comparison with its planetary neighbors Mars and Venus shows that they both now have atmospheres with more than 95% CO_2 (Table 5.4). Since it is likely that, in the absence of living organisms and an active tectonic cycle, these planets have retained much more of their original atmosphere than has the Earth, CO_2 was probably more abundant in the early Earth. A second reason for believing that the Earth originally had more CO_2 in its atmosphere is from estimates of the total amount of organic and carbonate carbon on Earth – thought to be at least 3.6×10^{21} mol (Fig. 5.14). (If this was all present as CO_2, although of course there is no certainty that it all was, then it would have made an atmosphere of about 60 bars (Table 5.4)).

The third reason for inferring a more CO_2-rich atmosphere in the past arises from what has become known as the "faint young sun paradox." This was first articulated by Sagan and Mullen (1972) who argued from solar models that in the early Archaean, the sun, as a relatively young star, would have had a 25–30% lower luminosity than it does now. This reduced luminosity, it is argued, would have led to much lower surface temperatures on Earth that at present, and a frozen Earth until about 2.0 Ga ago. The paradox arises because the Earth's oldest known sediments from the $>$ Ga Isua greenstone belt are clearly water-lain and show no evidence of a frozen Earth (Fedo et al., 2001).

One of the most widely accepted solutions to the faint young sun paradox is that the early Earth had a greenhouse atmosphere which compensated for the lower solar luminosity, and which kept surface temperatures above freezing throughout Earth history. The principal greenhouse gas, capable of producing a warm young Earth is CO_2, and it would have had to make a significant contribution to the composition of the atmosphere prior to about 2.0 Ga. In the sections which follow we discuss the geological evidence in support of the "CO_2-rich atmosphere" hypothesis, explore possible modifications to this hypothesis in the form of methane mixing, examine ways in which CO_2 levels have been reduced over time and seek to quantify CO_2 levels in the Earth's atmosphere over geological time.

5.3.2.1 Geological evidences for an early, CO_2-rich atmosphere

Mineral CO_2-barometers and other features of the Archaean sedimentary record confirm that there were higher levels of CO_2 in the Archaean atmosphere than today, and in some cases permit us to quantify them.

Nahcolite. The mineral nahcolite (imaginatively named after its formula $NaHCO_3$) is a primary evaporite mineral that has been found in 3.4 Ga sediments in the Barberton greenstone belt and in 3.2 Ga silicified sediments at Pilbara in western Australia (Lowe & Tice, 2004). Experimental studies show that, although nahcolite is not normally a stable phase in evaporites, it may become so if there is a high enough partial pressure of CO_2. On the basis of their observations Lowe and Tice (2004) calculated a minimum CO_2 level of between 1.4 and 10% in the atmosphere, for Archaean surface temperatures of between 55 and 85°C. It should be noted, however, that this value would be much lower if lower surface temperatures were used in the calculation.

Paleosols. Paleosols formed before 2.2 Ga tend to contain iron silicates, rather than siderite–iron carbonate. Rye et al. (1995) used carbonate-silicate mineral equilibria in the 2.75 Ga Mount Roe paleosol to estimate the partitioning of CO_2 between soil and air in the late Archaean. Their calculations suggest a maximum partial pressure of $10^{-1.4}$ (0.04) atmospheres CO_2 in the late Archaean atmosphere, significantly lower than the estimate of Lowe and Tice (2004) based on nahcolite. A lower limit for atmospheric CO_2 comes from the study of siderite–clay mineral equilibria for weathering rinds on clasts in river gravels from the 3.2 Ga Moodies Group of the Barberton greenstone belt, in south Africa (Hessler et al., 2004). In this study the minimum partial pressure for atmospheric CO_2 at 3.2 Ga was calculated to be 2.5×10^{-3} bars at 25°C.

Archaean weathering profiles. Further evidence for a CO_2-rich Archaean atmosphere comes from indications of an aggressive weathering regime in the late Archaean sedimentary record, driven by a high content of carbonic acid in the weathering environment. Two different arguments are used to support this claim. First, first-cycle sediments are much more refractory than their modern counterparts. An example of this is seen in the sediments of the 3.2 Ga Moodies Group in the Barberton greenstone belt, where Lowe and Tice (2004) show that only the most refractory components of the source terrain are present and all other mineral phases have been removed. Weathering rates, however, are also a function of temperature and high temperatures in the Archaean could have been caused by greenhouse gases other than CO_2. This means that arguments based upon weathering intensities cannot be used to argue for CO_2 as a unique weathering agent.

The second argument is based upon the chemical index of alteration (CIA) of Archaean sediments, which is another measure of the refractory nature of the sedimentary source. On the CIA scale fresh granitoids are 45–55 and Mesozoic to recent shales are 70–75. Archaean shales from the ca. 3.0 Ga Buhwa greenstone belt in southern Zimbabwe have very high CIA values of between 95 and 100, indicating that they were derived from a source region which had experienced the complete removal of Na, Ca and Sr and the conversion of virtually all plagioclase to aluminous clays (Fedo et al., 1996). Similarly weathering profiles in the 3.0 to 2.78 Ga Central Rand Group in the upper part of the Witwatersrand succession extend to several meters below erosional unconformities and have CIA values in the range 80–90 (Frimmel, 2005).

Interestingly, when this line of enquiry is extended back in time to the oldest known clastic sediments – the >3.7 Ga sediments of the Isua greenstone belt – there is no strong evidence for extreme source weathering. Bolhar et al. (2005a), found that most Isua samples have CIA values between 46 and 69, with only a few samples with higher values. At first glance this suggests that weathering in Isua times was not extreme, although it may be that weathering in an emergent oceanic setting (Isua) took a different form from that in a continental setting (Kaapvaal). Alternatively the CIA is not a good measure of atmospheric CO_2 levels. In this context the work of Holland (1984) is important. He examined the acid–base balance of Archaean rocks and calculated the amount of CO_2 needed per kilogram of igneous rock to obtain average shale. These calculations showed that there was no great difference between average Archaean shales and modern shales, suggesting that CO_2 levels in the Archaean atmosphere were not much greater in the Archaean than at the present day.

5.3.2.2 *The role of methane*

One of the important observations made by Rye et al. (1995) in their paleosol study of Archaean CO_2 levels is that the calculated levels are too low. The greenhouse properties of an atmosphere with 0.04 atmospheres of CO_2 are not enough at 2.75 Ga to compensate for the effects of the lower solar luminosity. This means that there has to have been an additional greenhouse gas present in the Archaean atmosphere. A similar line of reasoning has been followed by Zahnle and Sleep (2002) in their modeling of CO_2-drawdown (Section 5.3.2.3). They too find that there was not enough CO_2 present in the Archaean to warm the Earth sufficiently to maintain "modern" surface temperatures. The solution to the problem has to be that there was an additional greenhouse gas in the Archaean atmosphere. The preference is for methane, since it would have been stable in an anoxic atmosphere.

A likely source of this methane in the Archaean is methanogenic bacteria, probably one of Earth's earliest life-forms (Hayes, 1994; Kasting & Seifert, 2002). Today methane levels in the Earth's atmosphere are kept low (1.7 ppm) through reaction with atmospheric oxygen. In an anoxic Earth, however, methane levels could have risen to several orders of magnitude greater than today. Methanogens thrive at higher temperatures and, so the warming effect of higher methane levels could have resulted in increasing CH_4 production. There is some evidence that this may have happened during the late Archaean, for between 3.0 and 2.5 Ga there was a negative excursion in the ^{13}C isotope record (Hayes, 1994), which could be explained in terms of an overabundance of biogenic methane (Fig. 5.13).

A further consequence of a methane-rich early atmosphere is that it will photolyze in the upper atmosphere and undergo photochemical reactions to produce hydrogen (Catling et al., 2001). The net effect of this process is to liberate oxygen and allow it to accumulate in the Earth system.

5.3.2.3 *Controls on CO_2-drawdown*

In the modern Earth the principal agent of CO_2-drawdown is through the weathering of silicate rocks as part of what is known as the "Urey Cycle," after its identification by Urey (1952). The chemical reaction which summaries this process is

$$CO_2 + CaSiO_3 \rightarrow CaCO_3 + SiO_2$$

In the Urey cycle the CO_2 sequestered during silicate weathering eventually becomes immobilized as limestone. The process is then reversed during the burial and metamorphism of limestone, and CO_2 is released back into the atmosphere. The principal feedback in this cycle is through the strong temperature dependence of silicate weathering, so that in a warm Earth CO_2 is consumed at a greater rate than in a cool Earth. However, the more rapid uptake of CO_2 reduces the greenhouse effect, cooling the Earth so that CO_2 consumption is reduced. In this way the continental cycle of silicate weathering buffers atmospheric CO_2 levels and climate.

In a quasi-uniformitarian Urey cycle-based model Kramers (2002) sought to quantify the CO_2-drawdown capacity of the continental crust over time as a function of area of the silicate landmass. The change in area of the silicate landmass over time can be monitored using the crustal growth curve and erosion rates through the volume of crust-mantle recycling. This means that CO_2 is removed from the atmosphere proportionate to the volume of rock mass. Drawdown curves are shown for two different crustal drawdown capacities (Fig. 5.14), and the results compared to an estimate of the amount of CO_2 accumulated in the atmosphere as a result of mantle degassing over time. These calculations show that CO_2-drawdown exceeded CO_2 degassing at some time between 2.5 and 1.5 Ga, indicating that the Earth's CO_2 atmosphere was removed at some time between 2.5 and 1.5 Ga ago.

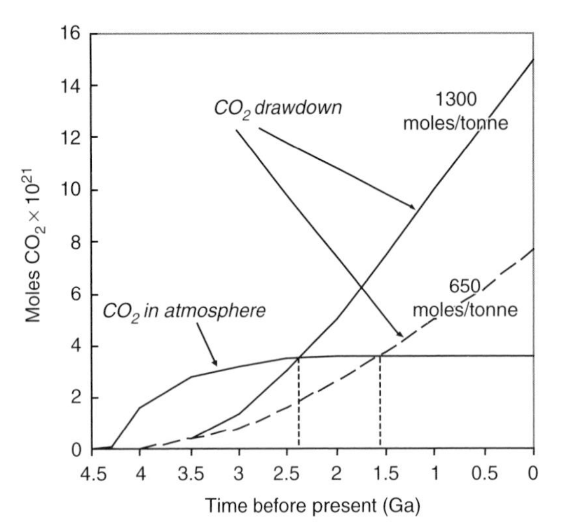

FIGURE 5.14 The mass of CO_2 in the Earth's atmosphere over time compared with the potential CO_2 drawdown capacity for two different cumulative CO_2 drawdown rates.

An alternative sink for CO_2 is the oceanic rather than the continental crust and a number of authors have argued that this was also an important process, perhaps the dominant process, in the Archaean. This approach to CO_2-drawdown was modeled by Zahnle and Sleep (2002). They show that CO_2 storage as carbonate in the ocean crust is controlled by the hydrothermal flux through the oceanic crust – the seafloor weathering flux. This in turn is governed by the exchange of CO_2 between the oceans and atmosphere, the supply of water to the ocean crust, and the availability of "reactable" cations in the oceanic crust. In a simple mass balance exercise Zahnle and Sleep (2002) demonstrate the importance of an oceanic crust reservoir by showing that present-day mantle outgassing of CO_2 at ocean ridges is the same as the mantle ingassing of carbonated oceanic crust during subduction. Their calculations are credible in as much as one of the outputs of their model is a steady state reservoir for carbon in the oceans and atmosphere, which looks very similar to that which exists today.

This model also works well for the Archaean, with its likely higher mantle heat flow and smaller continental mass and is consistent with calculations which show that there was a much higher flux of hydrothermal fluid into the early oceans compared to the present day (Section 5.4.2). In addition, it is consistent with the many observations of carbonate alteration in Archaean mafic and ultramafic rocks (Rose et al., 1996; Nakamura & Kato, 2004). Nakamura and Kato (2004) proposed a global carbon flux of 3.8×10^{13} mol/yr into the ocean floor at 3.46 Ga, an order of magnitude greater than the modern carbon flux of veined MORB (Table 5.3). If these Archaean fluxes are used in Zahnle and Sleep's (2002) ingassing-outgassing

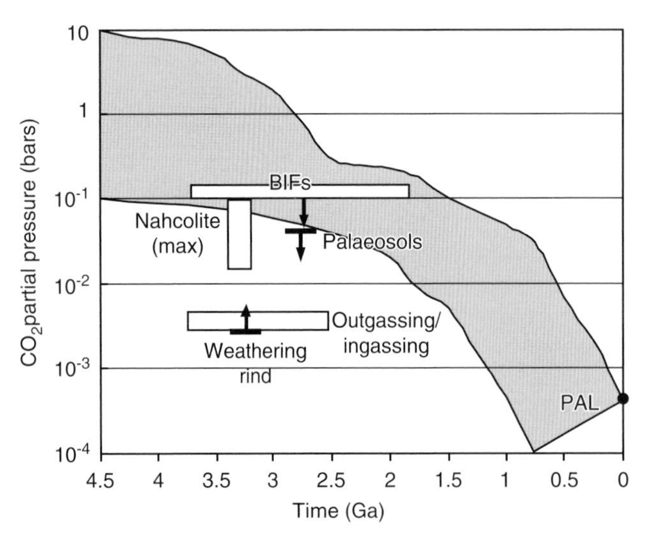

FIGURE 5.15 Changing CO_2 levels in the Earth's atmosphere over time expressed as partial pressure of CO_2. The grey shaded area is the range of concentrations required to compensate for the faint young sun (after Kasting, 1993). The other data points shown are discussed in the text.

mass balance model it predicts an Archaean atmosphere with a CO_2 level of between 8 and 14 PAL (0.002–0.004 bars).

If the Zahnle and Sleep (2002) model for CO_2-drawdown is correct, and CO_2 was principally stored in the oceanic crust during the Archaean, this does not necessarily negate the calculations of Kramers (2002), but it does shift the time of CO_2-drawdown back to perhaps the mid-Archaean. However, a problem with the Zahnle and Sleep (2002) model is that, unlike the Urey cycle, the Archaean oceanic weathering cycle has no inbuilt temperature feedback and needs a "thermostat" to maintain equable surface temperatures on the early Earth. This problem could be solved (just) if there were very high levels of CO_2 in the atmosphere, but the better solution is to include methane as a significant component of the Archaean atmosphere.

5.3.2.4 Quantitative estimates of CO_2 levels over time

The data presented above can be used to set some limits on CO_2 levels over geological time. The 2.75 Ga paleosol evidence sets an upper limit of 0.04 bars on CO_2 levels whereas the weathering rind study of Hassler et al. (2004) may set a lower limit of 0.0025 bars for 3.2 Ga. Calculations by Mel'nik (1982) show that CO_2 levels were not greater than 0.1–0.15 bars, for otherwise BIFs would be present as iron carbonates rather than oxides. The nahcolite study of Lowe and Tice (2004) lies within this range, although this result must be regarded as a maximum, as does the estimate of Zahnle and Sleep (2002) based purely on a mass balance between volcanic CO_2 and subducted, carbonated ocean crust (Fig. 5.15).

Figure 5.15 also shows that these data confirm what has already been discussed, that estimates of CO_2 levels in the Archaean are lower than those required to compensate for the faint young sun and an additional greenhouse gas is required. This additional gas was most probably methane.

5.3.3 How high were surface temperatures on the early Earth?

Ever since the faint young sun problem was identified by Sagan and Mullen (1972) there has been some uncertainty attached to the temperature of the Earth's surface during its earliest history. Initial calculations based upon a 25–30% lower solar luminosity indicated that the Earth should have been frozen until about 2.0 Ga resulting in global glaciation. This is no surprise for ices of various kinds are commonplace on other planets and their moons. However, this was not the case on Earth, as has been discussed above. Investigating this problem from the high-temperature end, Sleep et al. (2001) calculated that surface temperatures fell to below 100°C in as little as 2 Ma after the lunar impact magma ocean. Hence inferences of a hot early Earth – the Hell-like, "Hadean" period of Earth history – appear to be ill-founded, and it may be that Hell was quite cool, maybe even cold!

Geological evidences. The presence of liquid water on Earth over most of its history sets an approximate limit to Earth's surface temperature over time, indicating that, with the exception of a few ice ages, it has been maintained between greater than 0°C and less than 100°C. However, estimates of actual surface temperatures in the early Earth are few and are based upon oxygen isotope studies. Knauth and Lowe (1978, 2003) used the temperature dependence of oxygen isotope fractionation between chert and ocean water to show that cherts in the 3.2–3.5 Ga Barberton greenstone belt equilibrated under ocean temperatures of 70 ± 15°C (a hydrothermal ocean?). In a similar manner Veizer et al. (1989) found that mid- to late-Archaean limestones formed from ocean water about 30°C warmer than that at the present day. Evidence from the upper temperature limit of oxygen photosynthesis and the presence of photosynthesizing bacteria since about 2.7 Ga, limit surface temperatures to below 73°C (Lowe & Tice, 2004).

Early-Earth climate modeling. As already discussed (Section 5.3.2) the preferred solution to the faint young sun paradox is that the early Earth had a greenhouse atmosphere, which mitigated the effects of the lower solar luminosity. The preferred greenhouse warming gas is carbon dioxide, although there is good evidence that this must have been supplemented with methane (Zahnle & Sleep, 2002; Kasting, 2005).

A variety of climate models have been used to calculate the balance between the concentration of CO_2 in the atmosphere of the early Earth and surface temperatures. The model of Kasting (1993) illustrated in Fig. 5.15 explores the effect on temperature of between 0.1 and 10 bars CO_2 in the atmosphere at ca. 4.5 Ga, and assumes no albedo effect from cloud cover. The resulting temperatures were generally higher than the present-day mean of 15°C, apart from during glacial events as indicated by the inflexion at 2.5 Ga. More recently Zahnle and Sleep (2002) have shown that in climate models where CO_2 is the main greenhouse gas, temperatures may have been as low as 1°C, and in some special cases below freezing. However, in climate models which include methane, temperatures of about 15°C can be sustained.

The first glaciations. The earliest glacial events on Earth appear to be recorded in the diamictites of the Mozzan sediments of the 2.9 Ga Pongola supergroup in South Africa (Young et al., 1998) and in their possible correlatives in the West Rand Group of Witwatersrand succession (Frimmel, 2005). Slightly younger glacial sediments are reported from the early Proterozoic Huronian (Nesbitt & Young, 1982) and the 2.3 Ga sediments of Transvaal Supergroup, South Africa (Bekker et al., 2004). These latter diamictites are thought to have formed at low latitudes, for paleomagnetic estimates on contemporaneous lavas indicate a latitude of 11 ± 5° (Evans et al., 1997), prompting discussion of a Paleoproterozoic "snowball Earth."

These results suggest a largely ice-free world during the Archaean with excursions into freezing temperatures in the early Proterozoic and maybe at 2.9 Ga.

5.4 The nature of the Archaean oceans

When integrated over Earth history, probably the two most significant contributions of dissolved ions to the oceans have been from rivers – reflecting a continental contribution – and from ocean-ridge hydrothermal vents – representing a mantle contribution. The relative balance of these two has been the subject of much discussion. In addition to these contributions, the composition of the modern oceans is determined by a wide variety of other processes. These include outputs to sediment through scavenging and precipitation, equilibria of the surface layers with the atmosphere, and a range of biologically mediated processes.

Thus in the Archaean, when initially there was no life – or life of a different kind, when the atmosphere was more CO_2-rich and less oxygenic, and when landmasses were smaller, but weathering processes more aggressive, it is to be expected that sea water had a very different composition from the present day. In fact, in an early study Walker (1983) suggested that in the Archaean oceans had a lower pH and a lower carbonate and sulfate content, but higher concentrations of Ca, Fe^{2+}, Ba, Si, Na, Cl and bicarbonate.

In this section we first explore the likely timing of ocean formation and then investigate the nature of ocean water chemistry in the early Earth.

5.4.1 First indications of water – a cool early Earth?

Until a few years ago the best evidence for the antiquity of the Earth's oceans was from the > 3.7 Ga water-lain sediments of the Isua greenstone belt, west Greenland. Now we think that the oceans are probably much older than this, for there are several new, but indirect, lines of evidence which point to the existence of liquid water extremely early in Earth history. Perhaps the most striking of these is the observation that some carbonaceous chondrite meteorites display hydrothermal alteration. This means that there was water present on their parent asteroid and implies that liquid water was present even during the process of planetary accretion (Zolensky, 2005). This argument is consistent with the evidence from the distribution of volatile elements on the modern Earth (Section 5.2.2.3) which is most easily explained if there had been a liquid water ocean on Earth during impacting.

Further support for the notion of a very early water-ocean comes from the thermal modeling of Sleep et al. (2001) who showed that an early magma ocean, formed as a result of the Moon-forming event, would have cooled extremely rapidly, so that the Earth's surface, covered perhaps by only a thin crust, was at temperatures below 100°C at some time between 4.54 and 4.51 Ga. Liu (2004), in a study which supports this finding, calculated that under the high pressures of the primitive atmosphere – maybe 250–400 bars – a liquid ocean would begin to condense at temperatures even as high as 300–450°C, again supporting the view that a liquid water ocean condensed very rapidly after the formation of the Moon.

These calculations are also consistent with evidence from the oldest known Earth materials, the 4.4 Ga-old zircons from western Australia, which also indicate the presence of liquid water on the Earth at the time of their formation. Wilde et al. (2001) and Valley et al. (2002) found that these ancient zircons contain elevated oxygen isotope ratios, very similar to geologically more recent samples in which the oxygen isotopes indicate interaction with meteoric water. This oxygen isotope evidence provides a tantalizing glimpse at a possible hydrological cycle in ancient crust possibly as early as 4.325 Ga (Cavosie et al., 2005). Together these different lines of evidence provide strong evidence for cool surface temperatures on the young Earth, and support the idea of a liquid water ocean between 4.4. and 4.5 Ga.

Although the first water lain sediments have now been supplanted as the critical evidence for water on Earth, they are not without their own controversy. There has been much recent discussion of possible water lain sediments on Akilia Island in west Greenland older than 3.85 Ga (Nutman et al., 1997), although their sedimentary origin has been questioned by Fedo and Whitehouse (2002). There has also been some debate over the sedimentary status of the better known "sediments" in the Isua Greenstone Belt in west Greenland. Now the matter is settled for Fedo et al. (2001) have provided good evidence that water lain sediments and subaqueous pillow lavas formed at Isua before 3.71 Ga. However, what were once believed to be sedimentary carbonates are now thought to be metasomatized volcanic rocks (Rose et al., 1996). Even older sediments may be present in this area, for south of Isua there is a region of older crust (>3.8 Ga), from which BIF has been reported (see Nutman et al., 1997; Rollinson, 2002).

5.4.2 Proxies for the composition of ancient sea water

Chemical sediments such as evaporates, limestone, BIF and chert are rocks which have been precipitated directly from sea water and so are important as potential proxies for the composition of ancient sea water. Recent studies by Bolhar et al. (2004, 2005b) and Kamber and Webb (2001) have shown, from a comparison between ancient and modern carbonates, that the REEs and Y are very effective in fingerprinting the sea water signal in ancient chemical sediments (although see Johannesson et al., 2006). Modern ocean waters are characterized by positive La and Y anomalies, and a negative Ce anomaly and the REE and Y show a pattern of relative enrichment from light to heavy REE (Fig. 5.16). The Ce deficiency reflects the ease with which soluble Ce(III) is oxidized to the relatively insoluble Ce(IV) and subsequently scavenged from the water column (Bau, 1999). The presence of a negative Ce-anomaly therefore provides a useful monitor of oceanic redox conditions. An additional feature of REE + Y water chemistry is that it allows us to distinguish between marine and hydrothermal components in ocean water. Modern hydrothermal fluids have a very different composition from ocean waters, making them relatively easy to recognize. They have much higher abundances of the REE + Y, show relative depletion from light to heavy REE + Y, and have a prominent Eu anomaly (Fig. 5.16).

5.4.2.1 Archaean limestones

Archaean limestones make up a relatively minor part of the Archaean sedimentary record and are known from ca 3.5 Ga onwards. Nevertheless, they have the potential to record important information about the composition

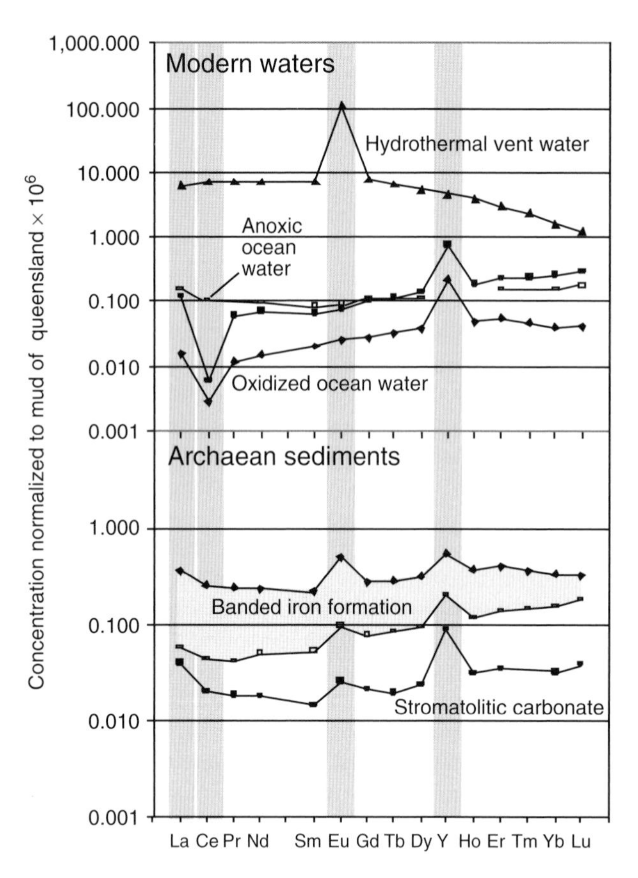

FIGURE 5.16 Rare Earth element + Y plots for modern waters and Archaean sediments, normalized to mud of Queensland (data taken from Bolhar et al., 2004, 2005b; normalization Kamber et al., 2005). The upper section of the diagram shows that hydrothermal vent fluids are characterized by a positive Eu anomaly, whereas modern ocean water is characterized by positive La and Y anomalies, a negative Ce anomaly in oxidized water and a progressive increase in heavy REE concentrations relative to light REE. The lower section of the diagram shows the composition of Archaean BIFs and an Archaean stromatolitic carbonate average, all with positive La, Eu and Y anomalies, but no negative Ce anomaly.

of Archaean sea water. This was recognized by Veizer et al. (1989) who showed that Archaean limestones have an order of magnitude more Fe and Mn than Phanerozoic limestones, and have mantle-like strontium and oxygen isotope ratios indicating a significant mantle (hydrothermal) flux into the Archaean oceans. In a more recent study, Kamber and Webb (2001) found that stromatolitic carbonates from the 2.5 Ga Campbellrand carbonate platform in South Africa have positive La and Y anomalies (Fig. 5.16) identical to those in modern ocean water. This confirms the view that ancient carbonates can serve as a proxy for the composition of ancient sea water, although the absence of a negative Ce anomaly indicates that, unlike the modern oceans, the water mass was not oxidizing.

5.4.2.2 BIFs

Recent precise measurements of REE and Y in 2.5 to 3.7 Ga-old BIFs show that they also contain a "sea water fingerprint." REE patterns in magnetite-, siderite- and jasper-rich bands display positive La and Y anomalies and show a progressive enrichment in REE from light to heavy (Bolhar et al., 2004), identical to the pattern found in modern sea water (Fig. 5.16). However, unlike modern sea water, there is no Ce anomaly, implying anoxic conditions, consistent with the abundance of Fe in solution. In addition, these lithologies have a positive Eu anomaly, indicating that they formed from ocean water which contained a significant hydrothermal component. A fuller discussion of the importance of BIFs for deciphering Archaean ocean chemistry is given later in Section 5.4.4.2.

5.4.3 Quantifying the hydrothermal input into the Archaean oceans

The two principal sources of dissolved ions in the modern oceans are the continents, via riverine input, and the mantle – the source of ocean-floor hydrothermal solutions. Given that the volume of the continents has increased over time, and that mantle heat flow has decreased, it is likely that the balance between continental and hydrothermal fluxes into the oceans was very different early in Earth history.

This problem has been investigated by a number of workers using stable and radiogenic isotopes who found that Archaean chemical sediments tend to have significant hydrothermal signature (e.g. Veizer et al., 1989). Kamber and Webb (2001) attempted to quantify the relative contributions of river water and the hydrothermal flux to the Archaean oceans using the REE, Nd- and Sr-isotope compositions of late Archaean stromatolites. They found that the Campbellrand microbialites possess a primitive Sr- and Nd-isotope signature and have a positive Eu-anomaly – both are features of a hydrothermal contribution to late Archaean ocean water chemistry. They show that isotopic variability within this sedimentary sequence implies mixing between mantle-derived fluids and a continental component, which in this case they show to be river water, rather than a clastic component. From this mixing relationship Kamber and Webb (2002) found that there was a 50-fold increase of the hydrothermal flux to river water flux ratio in the Archaean ocean relative to the present-day. This change is too great to be explained simply by waning mantle heat flow and must represent a lowered continental flux into the oceans during the late Archaean compared to the present-day. This is counter-intuitive given the more aggressive weathering expected in a late Archaean atmosphere, and so must imply that there was limited continental uplift during the late Archaean, and consequently much less erosion at this time.

5.4.4 The redox state of the Archaean oceans

It was shown earlier (Section 5.3.1.6) that the level of atmospheric oxygen has risen over time. Hence it is to be expected that the oceans too will show some evidence of progressive oxidation. Kasting et al. (1993) proposed a three-stage, three box model in which initially the atmosphere, surface ocean and deep ocean were all reducing. As the atmosphere was progressively oxidized, first the surface ocean and then the deep ocean became oxidized. This process can be monitored using sulfur and iron, the two great redox buffers of the marine environment.

5.4.4.1 Sulfide/sulfate equilibria

One of the major sources of sulfate in the modern oceans is through the oxidative weathering of sulfide in crustal rocks. This process returns the sulfur as soluble sulfate to the oceans. During the Archaean, when oxygen levels were thought to have been much lower, this process would have been suppressed, leading to lower sulfate levels in the Archaean oceans (Fig. 5.17). If true, this inference has a profound impact on the changing nature of oceanic redox processes over time, for today the microbial sulfate–sulfide redox reaction is the main

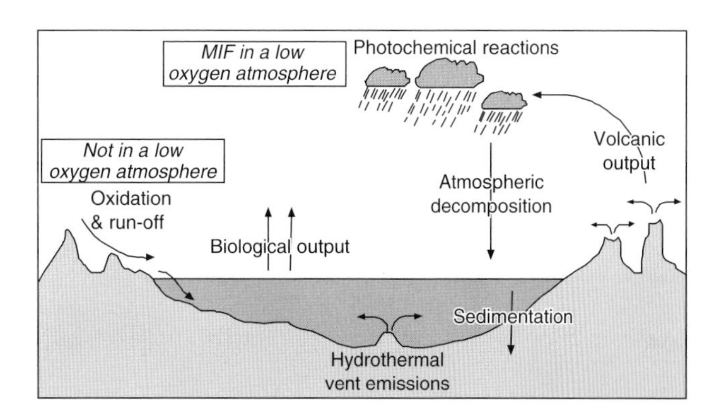

FIGURE 5.17 Simplified sulfur cycle showing the major differences between the modern and early Earth.

oceanic oxygen buffer. Thus there must have been a different oxygen buffer in the Archaean ocean.

Evidence to support the idea of low-sulfate levels in the Archaean oceans comes from the degree of mass-dependent sulfur isotope fractionation observed between sulfide and sulfate species in ancient sediments ($\delta^{34}S_{sulfate} - \delta^{34}S_{sulfide}$). This fractionation is driven by the microbial reduction of sulfate to sulfide, which in the modern produces a very large fractionation of between 45 to 70‰ $\delta^{34}S$. However, before 2.5–2.7 Ga the recorded fractionation is much smaller and is only about 10‰. A recent experimental study has made sense of these observations. Habicht et al. (2002) showed that, using bacterial cultures in the laboratory, the degree of fractionation is governed by the sulfate concentration of the aqueous environment, and that at low sulfate levels there is a low degree of isotopic fractionation. The critical value seems to be about 200 μmol.l^{-1} of sulfate, below which isotopic fractionation changes from modern to Archaean levels. Applying this result to the Archaean oceans indicates that before 2.5–2.7 Ga sulfate levels were about 200 μmol.l^{-1}, that is about 140 times less than at the present day (Table 5.5). Supporting evidence for a low-sulfate ocean in the Neoarchaean come from the presence of halite rather than gypsum in 2.58 Ga supratidal sediments (Eriksson et al., 2005).

An interesting consequence of these observations is that if sulfate reduction was not a common process in the oceans during the Archaean, then sulfate reducing bacteria were also not abundant. This means that organic matter must have been processed by some other metabolism. Habicht et al. (2002) have suggested that this was probably methanogenic bacteria, which would have given rise to

TABLE 5.5 Estimates of the pH, element ratios and composition of Archaean sea water compared to modern sea water. Modern sea water data from Holland (2003) and Bruland and Lohan (2003). Elemental concentrations that were higher in the Archaean oceans are indicated in italics.

	Present-day oceans (mmol.l^{-1})	Archaean oceans (mmol.l^{-1})	Reference
pH	7.5–8.5 (mean = 8.0)	≥6.5	
$CaCO_3$	Close to saturated	*Saturated/super saturated*	Walker (1983)
Mg^{2+}/Ca^{2+}	5.3	1 to 3	Late Archaean (Hardie, 2003)
Fe^{2+}/Ca^{2+}	ca. 5×10^{-8}	Max 4×10^{-3}	Walker (1983)
Na	469		
Mg	52.8	*55 to 95*	Late Archaean (Hardie, 2003)
K	10.2	*14 to 24*	Late Archaean (Hardie, 2003)
Ca	10.3	*28 to 50*	Late Archaean (Hardie, 2003)
Fe	5×10^{-7}	*0.1 to 1.0*	Ohmoto (2004) Mn
Sr	0.10		
Ba	1.1×10^{-4}	*Higher by factor of 10*	Walker (1983)
HCO_3	1.75		
CO_3	0.27		
SO_4	28.2	≤0.2	Habicht et al. (2002)
P	2.3×10^{-3}	1.5 to 6×10^{-4}	Bjerrum and Canfield (2002)
Cl	546		
Si	0.09	*Saturated (no biological removal)*	Walker (1983)
Mo	0.0001	lower	Siebert et al. (2005)
Re	4×10^{-8}	lower	Seibert et al. (2005)
U	1.4×10^{-5}	lower	Siebert et al. (2005)
Os	5×10^{-11}	lower	Siebert et al. (2005)
Ce	8.6×10^{-6}	*higher*	Bolhar et al. (2004)

significant quantities of atmospheric methane, as required by early atmospheric models (Section 5.3.2.2).

A puzzling exception to these inferences about low-sulfate levels in the Archaean oceans is the presence of bedded barite deposits formed during the time interval 3.2 to 3.5 Ga. The formation of evaporitic barite requires relatively high concentrations of sulfate, oxidizing conditions and of course elevated levels of Ba. These mid-Archaean deposits have a much lower $\delta^{34}S$ isotopic signature (ca. +5.0) than modern marine sulfates (ca. +20.0). Huston and Logan (2004) propose that mid-Archaean barite deposits are indicative of oxidized sulfur in the Archaean atmosphere, formed through the photolytic oxidation of volcanic gases (although this process can be anoxygenic and does not require free oxygen). In the absence of sulfate reducing bacteria, relatively high levels of sulfate accumulated in the shallow marine environment. Their disappearance from the geological record at ca. 3.2 Ga could indicate the rise of sulfate-reducing bacteria at this time. The alternative explanation for these anomalous sulfate occurrences is that they were evaporitic accumulations in areas of localized high oxygen productivity, and carry no information about the composition of the Archaean oceans.

5.4.4.2 Evidence from banded iron formations

Banded iron formations, as their name implies, are banded, iron-rich sedimentary rocks with >15 wt% Fe made up of alternating iron-rich and iron-poor (often silica-rich) layers a few millimeters thick. BIFs, as they are known, are found throughout the Archaean record and extend into the early Proterozoic. The oldest recognizable iron formations are from the Isua greenstone belt and have been dated at 3.71 Ga. BIFs were abundant in the late Archaean and persist until about 2.4 Ga. They then reappear at about 1.8 Ga before disappearing completely from the geological record. The largest BIF sequences formed at 2.5 Ga and are from the Hamersley Range in Australia and the Transvaal Supergroup in South Africa.

The very regular and continuous nature of the laminae in BIFs and the absence of a terrigenous component have led most authors to conclude that they formed as chemical precipitates in a quiet sedimentary environment, far off-shore. Most models for the formation of BIF are based upon a hydrothermal, oceanic source for the iron and a two-layer model of the ocean. The lower layer of the ocean is thought to have been anoxic, whereas the upper layer was oxidizing, although probably not as oxidizing as the modern oceans. In anoxic, sulfur-poor waters, iron exhaled from hydrothermal vents would remain in solution, and only when these waters rose above the chemocline into the oxidizing zone would iron precipitate as ferric oxyhydroxides, later to be converted to hematite (Fig. 5.18). Support for the hydrothermal origin of the iron comes from a variety of sources including sulfur isotope studies (see e.g. Bowins & Crocket, 1994) and the REE + Y data discussed in Section 5.4.2. It is likely that the silica-rich bands, which are interlayered with the iron-rich layers, were formed in quiescent periods between intervals of hydrothermal activity and reflect the silica saturation of sea water, in the absence of the biological utilization of silica.

Greater detail about these processes is beginning to emerge from the new field of iron-isotope geochemistry (Fig. 5.18). Iron isotopes are normally expressed as the $^{56}Fe/^{54}Fe$ isotope ratio ($\delta^{56}Fe$), and are remarkably unfractionated in most igneous rocks, but at low temperatures show measurable fractionations. What is important is that these fractionations are closely associated with redox changes, hence they provide powerful insights into biological and geochemical processes in the aqueous environment. Johnson et al. (2003) calculated the iron-isotopic compositions of the main iron-bearing minerals in the late Archaean iron formations of the Transvaal Supergroup from South Africa and used iron-isotope fractionation factors to interpret their mode of formation. Using the composition of modern hydrothermal fluids, with a slightly negative isotopic signature as the starting composition of the iron, they found that above the thermocline Fe(II) was oxidized by both inorganic processes and by photosynthetic bacteria, to produce ferric oxides with a heavy isotopic signature (more positive δ^{56}Fe). In this environment, residence times are low and iron isotopes have the capacity to record quite rapid changes in ocean water composition. In the deeper parts of the basin the iron-rich minerals are thought to have precipitated inorganically under relatively anoxic conditions. Here, with much longer residence times for iron, iron isotopes are much less responsive to changes in ocean chemistry.

A very particular finding of iron isotope studies comes from the recent discovery that there is a strong temporal control on the composition of iron in diagenetic pyrite. Rouxel et al. (2005) have found that δ^{56}Fe values for pyrites in black shales can be grouped into three categories which correspond very closely to the time intervals previously recognized for the oxygenation of the atmosphere (Fig. 5.10). They found that δ^{56}Fe in pyrites older than about 2.3 Ga show a wide spread of negative values, whereas between 2.3 and about 1.7 Ga the range is much reduced and more positive. In younger rocks the range is indistinguishable from that of igneous rocks and modern hydrothermal systems (Fig. 5.19). These rather striking changes in isotopic composition are thought to reflect the way in which, during the Archaean, the oxide minerals in BIFs were enriched in ^{56}Fe, so that the remaining iron in the ocean water became isotopically depleted, as recorded in diagenetic pyrites. At 2.3 Ga with the increase in sulfur supply to the oceans through the oxidation of crustal rocks, and the waning of the hydrothermal iron supply, the pyritic removal of iron from ocean water began to take over. At 1.8 Ga, a last "burst" of hydrothermal activity led to a short interval of BIFs, lowering the Fe-isotope ratio of the oceans (Kump, 2005).

The regular precipitation of iron oxides from the Archaean oceans is also thought to have had a strong control on phosphate levels. Phosphate is important because it is a major nutrient which controls oxygen productivity. Bjerrum and Canfield (2002) calculated the dissolved phosphate content of the Archaean

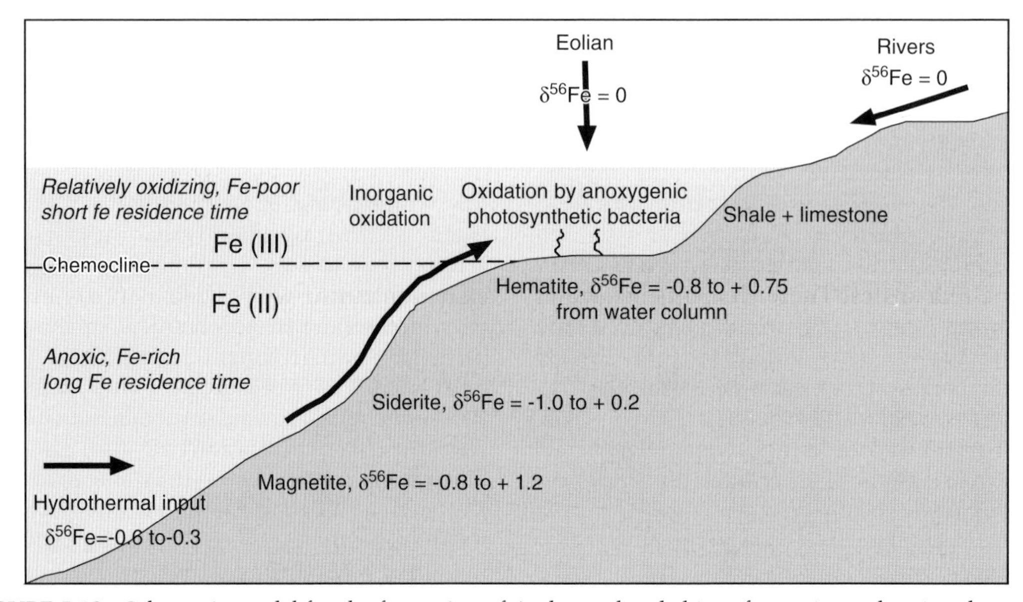

FIGURE 5.18 Schematic model for the formation of Archaean banded iron formations, showing the constraints imposed by iron isotope studies (modified from Johnson et al., 2003).

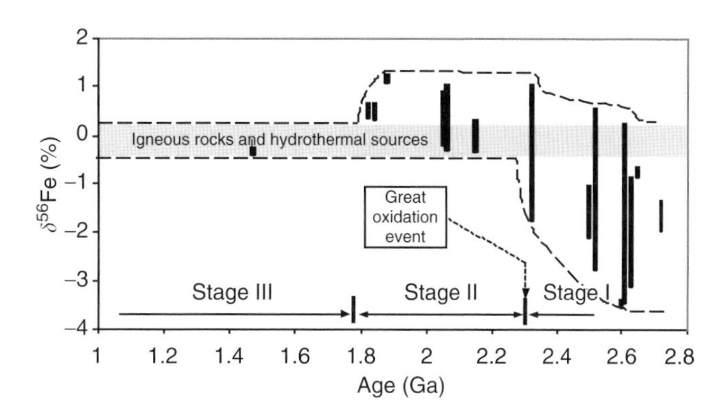

FIGURE 5.19 A plot of δ^{56}Fe over time for diagentic pyrite in black shales, showing three stages with different ranges of isotopic composition (after Rouxel et al., 2005). (See Text Box 5.1.)

oceans from the phosphorus content of BIFs, and found that it was 10–25% lower than in the present day. They suggest that dissolved orthophosphate was adsorbed onto precipitating iron oxide mineral surfaces and so removed from solution. They estimate that the low levels of phosphorus severely limited oxygen productivity and that, relative to the present-day, there was a 75–90% reduction in the rate of burial of organic carbon during the late Archaean (see Fig. 5.13).

5.4.4.3 Other redox-sensitive trace elements

In oxidizing waters the trace elements Mo, Re, Os, and U become soluble and the isotopes of Mo are fractionated, making them useful marine redox monitors (Holland, 2003; Siebert et al., 2005). This means that in oxidizing waters the elements Mo, Re, Os and U will be present in solution, and available to be scavenged during the sedimentation of organic matter, to be ultimately concentrated in organic-rich shales. These elements behave in the opposite way to Fe, which is soluble only in its reduced form, and also unlike Fe, they do not *control* the redox state of oceanic water, they simply respond to it.

On this basis Holland (2003) proposed that prior to the "Great Oxidation Event" at 2.3 Ga (see Section 5.3.1.6) the abundances of these elements should be low in organic shales, whereas after 2.3 Ga, when the oceans were more oxidizing their concentrations should be much higher. This was investigated by Siebert et al. (2005) in a study of black shales from sedimentary basins with ages between ca. 3.2 Ga and 2.2 Ga. They found evidence for slight Re and Mo mobilization at 3.15 Ga, more convincing evidence of oxidizing conditions at 2.7 Ga, but little evidence of oxidation in samples dated between 2.2 and 2.3 Ga. These results are not, at first sight, consistent with a Great Oxidation Event at 2.3 Ga and so may reflect the greater sensitivity of Re and Mo to low oxygen levels than is Fe, or they may be recording more localized redox phenomena.

5.4.4.4 Redox buffering of the Archaean oceans

Sulfur isotope fractionations show that the Archaean oceans had a much lower level of sulfate, probably more than a hundred times less, than the modern oceans. This observation is consistent with geological indications of low levels of atmospheric oxygen during the Archaean, resulting in low levels of sulfide oxidation during weathering and low levels of dissolved sulfate in Archaean rivers. An interesting extension of this observation is that it seems that ocean sulfate levels continued to remain low during the Proterozoic implying that atmospheric oxygen levels may not have risen substantially until the Neoproterozoic (Kah et al., 2004).

The bacterial reduction of oceanic sulfate to sulfide is the principal redox buffer of the modern oceans. In the Archaean, prior to the rise of atmospheric oxygen and when the oceans were low in sulfate, the redox state of the oceans was buffered by Fe, as is evidenced, at least in part, by the presence of BIFs. Hence the model proposed by Kasting et al. (1993) for the progressive oxidation of the atmosphere and oceans, in which there was oxidation first of the atmosphere, then of the shallow marine environment, and finally of the deep oceans, seems to be supported by geochemical evidence.

The oceanic record, as seen in the temporal distribution of BIFs and the Fe-isotope record of diagenetic pyrite, appears to reflect the progressive oxygenation of the atmosphere, with maybe a relatively late oxidation of the deep oceans (Canfield, 2005). What is even more important though, and suggested by the phosphorus study of Bjerrum and Canfield (2002), is that redox equilibria in the Archaean oceans were not simply responding to atmospheric conditions, but were *controlling* the rise of oxygen in the atmosphere.

5.4.5 The composition of Archaean sea water

Table 5.5 summarizes our current knowledge of the composition of Archaean sea water from a variety of sources. There are some important differences from modern seawater, as has already been discussed in the text. Where the concentrations of elements were higher in the Archaean these are italicised in the table. The estimates taken from Hardie (2003) are based upon the assumption that the oceans have maintained a constant volume since the Archaean and element concentrations are calculated from estimated ocean ridge/river water flux ratios.

These compositional differences reflect the different processes operating in the early Earth. In part they reflect the different redox conditions that existed but also indicate a different biology of the oceans and a different balance between hydrothermal and weathering inputs from that of the modern.

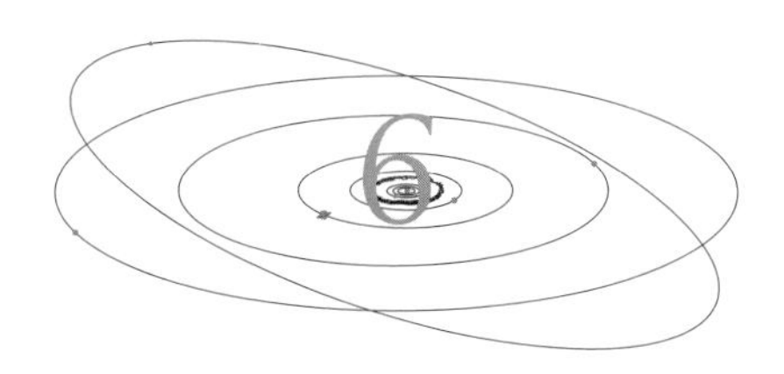

THE ORIGIN OF LIFE

The origin of life – the big picture

Life is improbable, but under the favorable physicochemical conditions of the early Earth it did begin. These conditions included an appropriate surface temperature, the presence of liquid water, the availability of appropriate chemical ingredients, the presence of a magnetic field, and an environment in which there were redox contrasts, all set in the context of a dynamic planet experiencing periodic resurfacing.

Life is also complex and the detailed chemistry of how the organic molecules which sustain and regulate life were formed is yet to be worked out. However, there was process in which simple molecules were synthesized to form complex polymers useful in the process of replication and the transmission of genetic information. Many believe that the nucleic acid RNA was a precursor to the modern DNA and that life first developed in an "RNA world." In addition the relevant energy sources were harnessed to facilitate life. Now these come mostly from sunlight, but in the early Earth they were probably chemical.

Recognizing life in the Archaean record depends mostly upon geochemical signals. The isotopes of essential elements in living organisms – carbon, sulfur, nitrogen, and iron – have all been useful in identifying biogenic signatures in ancient sediments. In addition the preservation of biomarker molecules in kerogen has allowed ancient living organisms to be identified with certainty. Stromatolites, laminated carbonate sediments precipitated in association with microbial activity, are also thought to provide evidence of life in the Archaean. These data show that in the late Archaean, by 2.7 Ga microbial life was diverse and abundant. However, tracing the record further back in time brings greater uncertainties, but there is good evidence of sulfate reducing bacteria as far back as 3.5 Ga and some evidence for photosynthesising cyanobacteria as old as 3.8 Ga in the Isua sediments.

In the past 30 years insights from molecular phylogenetics have allowed a new tree of life to be mapped. This comprises three domains – Bacteria, the Archaea (a new group), and Eucarya (the group to which all plants and animals belong). These three groups have their origin in a "last common ancestor," and there is much scientific interest in identifying the habitat and time of appearance of this precursor organism. Geological evidence from isotopic and biomarker evidence shows that the bacteria were in existence by 3.5 Ga and that the Eucarya had separated from the Archaea before 2.7 Ga. Photosynthesising cyanobacteria were in existence by or before 2.7 Ga, several hundred million years before the rise of atmospheric oxygen.

A favored habitat for the emergence of life on Earth is in an oceanic hydrothermal system. This is consistent with the thermophilic character of the most ancient members of the tree of life. Such systems were abundant in the Archaean and would have provided the energy, nutrients and catalysts needed for complex organic synthesis, and hence for life. They may even have provided a mineralogical mechanism for the formation of the first living cells.

Seeking to understand the origin of life is one of the greatest scientific questions we can ever pose, and for us as humans it is perhaps the most pertinent. For even a cursory examination of the complexities involved leads almost to incredulity. As Nisbet pointed out two decades ago in his book *The Young Earth*

> *Life is improbable, and it may be unique to this planet, but nevertheless it did begin and it is thus our task to discover how the miracle happened (Nisbet, 1987, p. 343).*

Rarely in science does the scientists' individual philosophy of life penetrate into their formal scientific reporting, but in this subject area such speculation is rife. For example, Richard Fortey writes in his *Life: An Unauthorized Biography*

> *It is not trivial to draw a parallel between the biography of life and an autobiography. From my early days, luck . . . has played a part in deciding which direction I have taken . . . . The greater story of life on our planet is partly a story of luck, [partly a story] of changes imposed upon the world by earthly and universal forces, and partly a story of genes, and finally a product of the changes life itself has wrought to modify the odds (Fortey, 1998).*

and Nobel prize winner Jacques Monod in his book *Chance and Necessity* writes

> *It is by chance that man has no importance in the Universe, that he counts for nothing, for his emergence is the result of chance; this the principal result of science is also the most unacceptable (Monod, 1972).*

Both authors, in their different ways, are reflecting on the enormity of the subject at hand.

Given such a huge scientific task, it is appropriate at the beginning of this discussion of the origin of life to ask how the Earth Sciences can illuminate this problem. Clearly much of the detailed genetics and molecular biology is outside its remit. Nevertheless, the early geological record contains important clues about the time at which life developed on Earth. There are chemical signals, preserved in ancient rocks which provide a record of life's developmental pathway. In addition, and this is where Earth scientists excel, it is possible to interpret the rock record in order to establish the former environmental setting of these early signals of life. Thus the habitat of the Earth's earliest life can also be established, providing a set of general constraints, which inform the more detailed enquiries of the molecular biologist and organic chemist.

6.1 Setting the scene for life

6.1.1 What makes a habitable planet?

As far as we know the Earth is the only planet inhabited by life. Why is this so? What are Earth's unique features that confer on it the status of a habitable planet? Here we identify six features which make Earth home to life.

- First, the Earth occupies the "habitable zone" of the solar system. This means that it lies at a distance from the sun which, over the past 4.5 Ga, as the sun has warmed, has consistently maintained a surface temperature conducive to life (Kasting et al., 1993; Kasting & Catling, 2003).
- The presence of liquid water on Earth is the second essential for life. Water is the dominant component of modern cells and a necessary medium for the majority of the relevant chemical reactions needed for the emergence of life. Life would have not evolved in the absence of water.
- Life also needs the "right chemistry," which in addition to the polymerized compounds of carbon, includes phosphorus, nitrogen, sulfur, and a variety of transition metals that are used in the redox chemistry of metabolism.
- The dynamic nature of planet Earth is important here for the periodic resurfacing of the planet means that fresh supplies of life-supporting elements are provided at the surface, preventing the premature exhaustion of the process of life. There is probably a close connection between the presence of water on the Earth and its dynamic nature, for planets that "dry out" cease to be active. This is perhaps the principal reason for the differences between Earth and its near neighbors, Venus and Mars. The presence of water and the dynamic nature

of the Earth, also indirectly influence the Earth's "self-regulatory" mechanisms, proposed by Lovelock (1979), and which have allowed the Earth to maintain a stable temperature over its life-span.

- A further factor which contributes to the habitable nature of planet Earth is that life on Earth today is protected by the van Allen belt, a product of the Earth's magnetic field. Hence, the dynamo effect of the Earth's core, may also be a significant factor in Earth sustaining life (Nisbet & Fowler, 2003).
- Finally, life needs a redox contrast in order to function. Prior to photosynthesis it is likely that organisms would have occupied habitats where naturally occurring redox contrasts were present. Such conditions are likely to have been available in the atmosphere, where there is and has been, a mixture of oxidized and reduced gases, in the oceans, and at atmosphere-rock and sea-water-rock interfaces

6.1.2 Steps toward synthesizing a cell

Life is complex, and a single cell contains complexities of the most incredible kind. If we knew how to assemble a cell, we would have solved the problem of the origin of life. For in comparison to a single cell, the complexities of the higher orders of life are trivial. Take for example, *Escherichia coli* a single celled organism a micron in size, which lives in the human gut. It swims, conjugates and replicates itself in 20 minutes. Its complexity is illustrated by its chromosomes which consists of a single double stranded chain of DNA with more than 4.6 million nucleotides (the individual building blocks of the DNA molecule and each a complex molecule), which together specify more than 4,000 different genes (Berg, 2000).

So what are the steps necessary to construct a single cell from nonliving starting materials? The first stage is the manufacture of the main "abiotic" molecules which are foundational to life. Abiotic molecules tend to be based upon a single carbon atom, they are monomers, and are the prelude to larger carbon-based molecules – polymers. Once polymers have formed the process of replication becomes important. Molecules capable of replication are thought to require at least 20–100 monomeric units. When the capacity for replication has been achieved these "loose" molecules must be kept together in order to capitalize on their chemical interactions. Hence the need for a membrane. When this is done we have created a primitive cell, and have taken a major step forward in solving the problem of the origin of life. Now we look at the processes in more detail.

6.1.2.1 Abiotic synthesis – making the molecules of life

The world before life – the prebiotic world – is a place where some very complex chemistry must have taken place. But how did it all start? There have been two fruitful lines of enquiry – one assuming that life started with only those molecules which were naturally occurring in the oceans and atmosphere. The other approach looks outside the Earth.

6.1.2.1.1 The Miller–Urey experiment

One of the most basic groups of molecules necessary for life are the amino acids. These are important because they are the building blocks of proteins, which themselves are essential to many of the functions of a cell. Over 50 years ago an attempt to synthesize amino acids from simpler molecules was carried out by Stanley Miller, working with Harold Urey at the University of Chicago. The "Miller–Urey" experiment as it has become known was based upon Urey's idea that the synthesis of organic molecules would have occurred in a reducing atmosphere and was driven by energy from lightning. Miller passed an electrical discharge, to simulate lightning, through a mixture of steam, hydrogen, ammonia, and methane, in a glass chamber in order to investigate which molecules might be generated by this process. The results were very positive and the experiment yielded a number of biologically significant molecules including amino acids, hydroxyacids, urea and an "oily material" (Miller, 1953; Miller & Urey, 1959). A similar set of experiments by Oro (1961) showed the importance of HCN, which when polymerized gives rise to the purines – adenine and guanine – essential components of both RNA and DNA. Whilst these experiments were able to make proteins, the much more difficult task of making nucleic acids, was not achieved.

Many workers have since replicated these results, but since the 1960s our understanding of the nature of the Archaean atmosphere has changed and it is now known to have been more CO_2-rich than assumed by Miller and Urey. Recently methane has been recognized as a likely component of the Archaean atmosphere, but still, this is only a mildly reducing component and insufficient to support Miller–Urey chemistry (Chapter 5, Section 5.3.2.2). Hence the highly reducing atmosphere assumed in these early experiments is now thought to be unlikely.

6.1.2.1.2 Meteoritic evidences

An alternative source of prebiotic molecules on the early Earth is from meteorites. Carbonaceous chondrite meteorites contain abundant carbon, present in a variety of molecular forms, including amino acids, polyols such as sugars and sugar-alcohols. These molecules are all important to life and form key components of nucleic acids and cell membranes (Cooper et al., 2001). Similar molecular evidence for amino acids is found in even more ancient fragments of the early Universe –IDPs. These molecules formed either in the solar nebula or in an interstellar environment (Flynn et al., 2003). Comets have also been considered a viable potential source of prebiotic molecules although currently the view is that most organic material was delivered by asteroids rather than by comets (Dauphas & Marty, 2002).

These lines of evidence demonstrate that the early Earth could have accumulated abiotic molecules in its oceans and smaller bodies of water from both extraterrestrial and terrestrial sources.

6.1.2.2 Prebiotic synthesis – proteins, nucleic acids, and replication

The next stage in complexity is the construction of polymers from abiotic monomeric molecules. This is the stage at which proteins and nucleic acids are formed. Proteins are long unbranched molecules, made up from amino acids and are the main structural and functional agents in a cell. In the game of molecular synthesis, proteins are extremely important because one group of proteins, known as enzymes, operate as biological catalysts. These have the function of delivering the right chemicals to the right place for organic synthesis.

However, amino acids are unlikely to form themselves into polymers without the help of some form of catalyst (Bada, 2004). Possible natural catalysts are mineral surfaces such as in the regular, repeating structure of clays although once bound to a clay the polymer has to be released. This is achieved in the laboratory with salt solutions and may, in nature, reflect an evaporative marine environment. An alternative venue is at hydrothermal vents where peptide bond formation is favored, and where catalysis may take place on sulfide mineral surfaces (Bada, 2004). Such a process has been described by Holm and Charlou (2001) who found linear saturated hydrocarbons with chain lengths of 16 to 29 carbon atoms in high-temperature hydrothermal fluids from a vent in the Mid-Atlantic Ridge.

If molecules are to be useful in the "life-business" they need to be able to copy themselves. Important here are the two nuclei acids RNA (ribonucleic acid) and DNA (deoxyribonucleic acid). DNA occurs in the nuclei of cells and is the principal component of chromosomes. It contains, in encrypted form, the instructions for the manufacture of proteins. Thus, encoded within the DNA of an organism is the order in which amino acids should be strung together to form all the necessary proteins. The clever feature of the DNA molecule is in its double helix structure, for it is this structure which allows the molecule to replicate accurately.

However, DNA needs RNA to catalyze protein manufacture and it is now thought that RNA preceded DNA in molecular evolution. This has led to the concept of the "RNA world" (see Section 6.4, and Joyce, 1989, 2002). The RNA molecule has the capacity to both store information and to catalyze reactions, but unlike DNA it does not have the capacity to synthesize proteins. Nevertheless, it does

TEXT BOX 6.1 Glossary of scientific terms used in this chapter

amino acid	An organic compound that is the building block of proteins. Amino acids may be joined together in any order by peptide bonds, to create a vast number of polypeptide chains with different amino acid sequences
assimilatory reduction	The reduction of an inorganic molecule to incorporate it into organic material. Energy is not made available during this process
autotrophy	The ability to synthesize organic compounds from inorganic compounds using chemical energy
biolipid	A reduced carbon compound, commonly synthesized by organisms. Cholesterol is an example
Calvin–Benson cycle	The synthesis of carbohydrate from CO_2, in particular during oxygenic photosynthesis
chloroplast	That part of plant cells and eukaryotic algae which conduct photosynthesis
dissimilatory reduction	The reduction of an inorganic molecule in association with the decomposition of an organic molecule. Energy is released
DNA	Deoxyribonucleic acid – the molecule that is the main repository of genetic information in living organisms. It occurs almost exclusively in the nuclei of eukaryotes
enzyme	A biological catalyst. The great majority of catalysts are proteins
eukaryote	Organisms whose cells have a nuclear membrane; includes all higher organisms such as plants and animals
extremophile	Microbes which exist in extreme environments on Earth – at great depth or at extremes of temperature
geolipid	The molecule produced from a biolipid after an organism has died. For example the biolipid cholesterol is converted at death to the geolipid cholestane
heterotrophy	The process whereby an organism uses existing preformed organic molecules for its energy needs
lithotrophy	The process whereby an organism uses inorganic substances as a source of energy
mesophile	Microbes which thrive in an aqueous environment at moderate temperatures – ca. 25–40°C.
methanogen	A cell which generates methane as a by-product of its metabolism
mitochondria	The part of a eukaryotic cell which is responsible for energy generation
monomer	A single-unit molecule from which a polymer can be constructed
nucleotide	The building block of a nucleic acid; the central structure of which is a 5- carbon sugar molecule (ribose)
peptide bond	The covalent bond that joins the carboxyl group of one amino acid to the amino group of another
photosynthesis	The reaction, driven by energy from sunlight, whereby CO_2 is converted into carbohydrate. This reaction produces oxygen. Anoxic photosynthesis uses light energy but does not produce oxygen
phototrophy	The fixation of CO_2 by photosynthesis, where the energy source is sunlight
polymer	A large molecule that is produced from smaller molecules (monomers) by connecting subunits to form a linear chain. Both DNA and RNA are linear polymers of nucleotides
ribosome	Part of a cell which is a complex of protein and RNA. It is the site of protein synthesis
RNA	Ribonucleic acid – a polymer that carries genetic information and can act as an enzyme
rRNA	Ribosomal RNA – the RNA that occurs in ribosomes
rubisco	Ribulose-1,5-bisphosphate carboxylase. An important enzyme used in the process of photosynthesis
saturated hydrocarbon	A hydrocarbon which has used all of its bonding electrons to make single bonds to other atoms
thermophile	Microbes which grow best, in an aqueous environment, at high temperatures >45–80°C.

have several features which make it attractive as a major stage in the development of life. It is a versatile molecule, readily soluble in water, simpler but more stable than DNA. But RNA is also a complex molecule and so is likely to have developed from a self replicating precursor. How exactly the RNA world developed is still uncertain. What is needed is a general mechanism of self-replication, in which the rate of polymer replication exceeds the rate of destruction, so allowing the polymer molecules to survive. Experimental studies suggest that low temperatures favor nucleic acid survival over higher temperatures and that salty solutions also favor their longevity. Possible pre-RNA candidates are the nucleic acid analogs peptide nucleic acid (PNA) and threose nucleic acid (TNA), or perhaps even a polymer which bears no relationship to a nucleic acid (Joyce, 2002).

Somehow, and at present, we know neither when nor how the RNA world was overtaken by DNA, a molecule which proved capable of carrying sufficient information to permit the synthesis of proteins. Initially RNA must have played the role of both replicator and catalyst, until its replacement by DNA saw it demoted to the role of catalyst.

6.1.2.3 Suitable energy sources – a variety of metabolisms

In addition to being able to replicate, life requires a source of energy. In the modern, the two processes – the genetic makeup and the metabolic processes in a cell – are inextricably linked. An organism cannot grow without a supply of energy and yet it cannot reproduce without the means to pass on essential information. Most Origin of Life theories emphasize one or other of the processes, either the replication process or the energy source as the essential component. It is very difficult to know which came first.

In the modern biosphere most energy comes from sunlight, but this skill had to be learned and it is likely that the first life obtained its energy from inorganic chemical reactions. There are a wide range of possible oxidation–reduction reactions which life utilizes to generate energy. These define a number of different metabolic pathways which can be classified as follows:

Autotrophic metabolisms are those which fix (or literally "eat") CO_2 by using inorganic compounds. There are two main classes. The first is *lithotrophy* (also known as chemoautotrophy, and chemolithoautotrophy) and is probably the most primitive. Here the energy source is chemical. Reduced inorganic compounds such as hydrogen, hydrogen sulfide, and ferrous iron are oxidized by oxygen, nitrate, sulfate, sulfur, or carbon dioxide. Likely metabolic processes involved hydrogen produced through water–rock reaction in a submarine environment (Nisbet & Fowler, 2003); thus in some Archaea (the methanogenic Archaea), CO_2 is reduced by hydrogen to form methane utilizing the reaction

$$CO_2(aq) + 4H_2 \rightarrow CH_4\,(aq) + 2H_2O$$

The second class of autotrophic metabolism is *phototrophy*, the process of photosynthesis. In this case the energy source is sunlight. There are two types of phototrophy – oxygenic and anoxic. *Oxygenic photosynthesis* creates oxygen and amongst the bacteria is performed by cyanobacteria. The basic reaction forms carbohydrate and oxygen and can be written

$$6CO_2 + 6H_2O \rightarrow C_6H_{12}O_6 + 6O_2$$

Anoxic photosynthesis utilizes light energy, but does not produce oxygen. Green and purple bacteria of the sulfur and nonsulfur kind are anoxic photosynthesizers. The sulfur bacteria oxidize sulfide to elemental sulfur and can even oxidize it further to sulfate.

Autotrophs are the primary producers, that is they are the organisms that make organic matter from an inorganic source. Also there are *heterotrophic metabolisms* (organotrophy) which take energy from existing organic compounds. These organisms depend therefore upon preexisting organic matter.

6.1.2.4 Holding it all together – the manufacture of membranes

A final step in the evolution of cellular life is the development of the cell membrane.

This then provides a stable environment for biochemical reactions to take place and protects and encloses the genome. In more advanced cellular development the cell nucleus is also contained within a membrane. Modern cell membranes are constructed from phospholipids, a group of organic molecules which have an affinity for water at one end but not at the other. These molecules are reinforced by other molecules, such as cholesterol, to give a rigid structure and thereby stiffen the membrane (Killops & Killops, 2005). Molecules of this type can be synthesized on a base of metal ions, and it has been suggested that the first membranes were synthesized on the surface of iron-sulfide bubbles in hydrothermal vents (Russell & Hall, 1997).

6.1.3 Possible habitats for the earliest life

Microbial life is now known to exist on Earth in all sorts of improbable settings. These are the extremophiles. Extremophiles are those microbes which exist in the hottest and coldest environments on Earth, and in places deeper than we imagined life could exist a decade or two ago. Thus the question of *where* life began is perhaps more open than it has been for a long time.

6.1.3.1 Primaeval soup

The notion of a primaeval (or prebiotic) soup implies that the early ocean was a "soup" of organic molecules. Within this melee chemical reactions took place, polymerized molecules formed and learned to self-replicate, finally to emerge as living cells. In reality, the ancient oceans probably had very active hydrothermal systems in which large volumes of water were circulated through the upper levels of the oceanic crust and so were unlikely to be an active meeting place for organic molecules (Nisbet & Fowler, 2003).

6.1.3.2 Hydrothermal vents

One place where life either began or at least spent its formative years is in deep-oceanic, hydrothermal vents. Hydrothermal vents are of a number of types, but the most spectacular are "black smokers," where mineral-laden, superheated water is discharged into cold ocean floor water creating a black, sulfide-rich, smoke-like plume. Both these ridge-flank, high-temperature vents, and more distal lower temperature vents are considered possible sites for early life.

There are a number of reasons why deep ocean hydrothermal vents might have been a suitable environment for the origin and nurturing of early life. These are briefly listed here. First there is evidence from the RNA "family tree of life" (Section 6.4.1) that the earliest life forms developed in a high-temperature environment. These microbes are known as thermophiles, first identified in hot springs and now known from the deep oceans. A second reason to favor a deep ocean site for the development of life is that it is protected from damaging solar radiation and from the volatilization of the surface of the oceans through asteroid impacting. Third, hydrothermal vents are rich in thermal energy, rich in nutrients in the form of metal-rich solutions and provide abundant mineral surfaces to catalyze chemical reactions. A more detailed discussion of this possible habitat is given in Section 6.5.2.

6.1.3.3 Clay minerals

One of the greatest difficulties in synthesizing complex organic molecules, such as RNA and DNA is the initiation of the self-replication process. An ingenious idea suggested by Cairns-Smith (1971) is that clay minerals play a part in the replication process. Clay minerals are not living but have a chemical structure which is characterized by endless repetition, and Cairns-Smith proposed that organic molecules might have "learned" the idea of self-replication from clays. Clays form in a variety of geological settings but interestingly are a common alteration product of ocean-floor basalt adjacent to hydrothermal vents. More recently, Ferris (2002) has suggested that the volcanic clay montmorillonite could have been the catalyst on which the chemical precursors to RNA were assembled to form the first short RNA polymers.

6.1.3.4 Extraterrestrial origins

It is well established that meteorites contain organic molecules, and it is possible that together with comets they delivered prebiotic molecules to the early Earth (Chyba, 1993; Pierazzo & Chyba, 1999). This "cosmic" origin for prebiotic carbon comes under the exciting new interdisciplinary field of astrobiology, popularized by NASA, and defined modestly as "the study of the origin, evolution, adaptation and distribution of past and present life in the Universe" (see Grady, 2003). Life's cosmic origins have previously been discussed under the grand title "panspermia" (Crick, 1981) and popularized by Hoyle and Wickramasinghe (1993). The possibility of life on the Martian meteorite ALH84001 announced in 1996 (McKay et al., 1996) has given new impetus to this theory, and although the claims for life within ALH84001 have not been proven, this research has given new focus to the present round of exploration of the Martian surface.

At the present time the balance of evidence is against a cometary origin for prebiotic carbon on Earth, for the same comets would also have delivered water to the Earth and yet the D/H ratio of the terrestrial oceans is different from that in comets. The more likely extraterrestrial input is from asteroids and meteorites, for there is evidence from both lunar and terrestrial samples that the late heavy bombardment event at 3.9 Ga (Section 6.4.1) contributed meteoritic material to the Earth at this time.

6.2 Geochemical signals of biological activity

Much of the evidence for the earliest life on Earth is based upon geochemical indicators of former living organisms. At best these indicators allow us to identify a specific organism or metabolic pathway. At worst, and this is particularly true for the early Archaean, geochemical signals are ambiguous, and biogenic and nonbiological signals easily confused.

There are two main types of geochemical proxy used to identify former life. These are biomarker molecules and isotopic fractionations. Biomarker molecules are specific molecules preserved in sediments as molecular fossils, which are unambiguous indicators of their biological precursors. Some of the most successful biomarkers are organic molecules called lipids. Biolipids are reduced carbon compounds, such as cholesterol, which are commonly synthesized by organisms. When a microorganism dies some of its molecular make-up may be converted through diagenetic processes into a related molecular form – a geolipid. So for example the "living" biolipid cholesterol ($C_{27}H_{46}O$) is converted to the "fossil" geolipid cholestane ($C_{27}H_{48}$). The geolipid bears the same carbon molecular structure, or "skeleton", as the biolipid. It is this skeletal form which can be recognized in the geological record and provides the diagnostic signal of former life (see Eglinton & Pancost, 2004).

The most frequently used isotopic proxies are those of the carbon and sulfur systems, although the study of nitrogen and iron isotopes is increasing (see Text Box 5.1). Isotopic biomarkers work in a different way from biomolecules. Take the case of carbon. During a chemical reaction associated with a particular metabolic pathway, the heavy and light isotopes of carbon can become fractionated relative to one another, principally as a result of the kinetic effects related to the reaction. Carbon-based molecules which result from this fractionation, and which are preserved in the geological record, will carry a memory of the isotopic fractionation. At its simplest, this isotopic memory can be related back to the original chemical reaction. In this way isotopic biomarkers are indicators of processes – particular chemical reactions, which are part of particular metabolic pathways, which in some instances may be specific to a particular group of organisms. Clearly the challenge is to identify those fractionations which are definitely biological in origin, and cannot be confused with other geological processes such as diagenesis, metamorphism, or fluid-rock reactions.

6.2.1 Carbon isotopes

Most biological reaction pathways discriminate against heavy carbon ^{13}C and become enriched in the lighter stable isotope ^{12}C. So for example in photosynthesis, ^{13}C is discriminated against during the fixation of CO_2 so that plants and autotrophic microbes tend to have low $\delta^{13}C$ values in the range −20 to −30. The net result of this fractionation is that there is a terrestrial reservoir of organic carbon which is depleted in ^{13}C ($\delta^{13}C$ = −20 to −30 ‰), relative to the mantle value $\delta^{13}C$ of −6‰, and a complementary enriched carbon reservoir in the form of limestones which has $\delta^{13}C$ values in the range −1 to +2‰ (Schidlowski, 1988). This fractionation of carbon isotopes is thought to be very ancient, dating from at least 3.5 Ga (see Fig 5.13).

Carbon isotope fingerprinting of biological processes is now a highly developed field within biogeochemistry and has a power of resolution far superior to that needed for the early Archaean organic sedimentary record (Hayes, 2001, 2004). Some of the relevant fractionations are identified in Table 6.1 below.

The isotopic fingerprinting of photosynthesis. Photosynthesis is currently an important means of fixing inorganic carbon and is used by plants, algae, bacteria, and Archaea. A particularly important enzyme (catalyst) in this process is Rubisco, the official shortened name for a molecule with a much more intricate systematic name. The reactions which take place in this process operate according to the Calvin–Benson Cycle and lead to isotopic fractionations of up to 30‰.

Methanogenesis. Methanogens produce much more extreme carbon isotope fractionations, although these can be very variable. A recent experimental study by Valentine et al. (2004) on moderately thermophilic methanogens showed that those which consume acetate show only a small carbon isotope fractionation

TABLE 6.1 Carbon isotopic fractionations in a variety of different biochemical pathways (data from Hayes, 2001; Orphan et al., 2001; Valentine et al., 2004; Horita, 2005).

Biogenic processes

Pathway	**Organism**	**Enzyme**	**Fractionation ‰**
Photosynthesis			
C3 pathway	Green plants and algae	Rubisco (form I)	−30
	Bacteria and cyanobacteria	Rubisco (form II)	−22
C4 pathway	Green plants and algae	Rubisco (form I)	−30
Acetyl-CoA		Carbon monoxide dehydrogenase	−52
Methanogenesis			
CO_2 reduction by H_2	*Methanothermobacter*	Rubisco (form III) in some methanogens	−22 to −58
Aceticlastic reaction	*Methsanosaeta thermophilia*		−7
CH_4 consumption	Methanotrophic Archaea		$\delta^{13}C$ values up to −96

Abiotic processes

Reactants	**Environment**	**Fractionation ‰**
CO, CO_2, CH_4	Atmospheric reactions	up to −25
CO–H_2 to CO_2, CH_4	Crust–mantle reactions	up to −64
CH_4 to C	Diamond reactions in deep mantle	up to −26
CO_2–H_2	Hydrothermal reactions in ocean crust	−35 to −60

of 7‰, whereas those which reduce carbon dioxide using hydrogen show a much stronger fractionation in the range of 22–58‰. At the extreme, methanotrophic archaea, organisms which consume methane, can be massively depleted in ^{13}C and may have values as low as −96 ‰ (Orphan et al., 2001).

The nonbiological fractionation of carbon isotopes. Horita (2005) has shown in a recent review that abiotic reactions can also produce carbon isotope fractionations and that these fractionations can be as large as those driven by biogenic processes. Particularly important is the production of methane (Sherwood Lollar et al., 2006). In the vicinity of ocean-floor hydrothermal vents, ultramafic rocks containing olivine may react with seawater, and dissolved bicarbonate can become converted to methane. In the presence of a Fe–Ni alloy catalyst carbon isotopes are fractionated as much as 50–60‰ at 200° C and 35 to 40‰ at 300° C (Horita & Berndt, 1999). This degree of fractionation is very similar to that produced by methanogenic bacteria.

There are therefore, problems in uniquely identifying biogenic signatures in the ancient sedimentary record, for there is an overlap in the degree of carbon isotope fractionation between biogenic and abiotic processes. Sometimes both biogenic and abiogenic carbon may be mixed together as recently suggested by Ueno et al. (2006) for methane in fluid inclusions. Characterizing biogenicity is currently the topic of much debate in the early Archaean sediments of west Greenland and Australia (see Section 6.3). One possible solution may be that, rather than simply relying on strongly negative absolute $\delta^{13}C$ values in the geological record, we must evaluate the extent of the fractionation by measuring the difference between the isotopic composition of the organic carbon and its source. Commonly this is inorganic carbon in the form of calcite (Sumner, 2001).

6.2.2 Sulfur isotopes

Sulfur is essential to modern life and is present in cells at about the 1% level (dry mass). However, most sulfur in the environment exists in its oxidized state, as sulfate, and must be reduced to H_2S to be useful in biosynthesis. The reduction of sulfate to sulfide is biologically mediated and the isotopic fractionations which take place during these reactions can be used as a biogenic fingerprint in the geological record. A summary of these processes is given below. For a fuller treatment the reader is referred to the reviews by Shen and Buick (2004) and Strauss (2003).

During microbial sulfate reduction microbes utilize ^{32}S over ^{34}S so that sulfides become depleted in ^{34}S relative to the original sulfate. This may take place via one of two metabolic pathways. During *assimilatory reduction*, sulfur is incorporated into the cell from the substrate, and there is a transfer of sulfate across the cell membrane. This is the means by which green plants, fungi, and most bacteria reduce sulfate to sulfide. During this process the overall isotopic fractionation is small, and is usually less than about 3‰. The second pathway is *dissimilatory sulfate reduction*. In this case sulfur is not incorporated into the cell, and sulfate reduction is closely linked to the oxidation of organic matter. Dissimilatory sulfate reduction is carried out by several different groups of microbes within the Archaea and the bacteria. In the marine environment, at normal marine sulfate levels, $\delta^{34}S$ fractionations of between 10–40‰ are recorded between sulfate and sulfide, although this decreases at lower sulfate concentrations. The most extreme fractionations have been recorded between H_2S and sulfate in pore waters, where Wortmann et al. (2001) found differences as great as 72‰.

The preservation of biogenic sulfur signatures in the geological record depends upon the mineralogical form in which the sulfur is preserved. Normally this is as iron pyrites – FeS_2, and the additional isotopic fractionation associated with this reaction is low, less than 1‰. This means that the bacterial fractionation of sulfur can be preserved relatively unmodified in the sedimentary record, in sedimentary pyrite. There are however a number of issues which can complicate the recognition of biogenic sulfur isotope signatures. First is the reoxidation of reduced sulfur species, which can occur in repeated cycles and give rise to significant $\delta^{34}S$ depletions. Complications of interpretation also arise from the isotopic fractionations that accompany the disproportionation of intermediate sulfites into sulfide–sulfate mixtures.

Inorganic sulfur isotope fractionation. The isotopic fractionation of sulfur during hydrothermal reduction of sea-water sulfate is governed by equilibrium fractionation processes. These fractionations vary as a function of temperature and give rise to isotopically heavy sulfates which are easy to distinguish from marine sulfate in the geological record. Isotopic depletions can also take place during the magmatic reduction of gaseous SO_2 to H_2S, giving rise to $\delta^{34}S$ depletions in the range 15–20‰. Abiotic sulfur isotope fractionations, therefore, tend to be the result of hydrothermal reactions, and so the cautious interpretation of sulfur isotope signatures should include an assessment of the likelihood of hydrothermal activity in the rocks under investigation.

6.2.3 Nitrogen isotopes

Nitrogen is central to the chemistry of life as a component of proteins and genetic material and in the modern world is abundant as gaseous nitrogen in the atmosphere (Chapter 5, Section 5.1.3). This is assimilated into the biological cycle through a variety of metabolic pathways ending up as a component of amino acids and nucleic acids. Hence both nitrogen and the presence of the ammonium ions have great potential as biomarkers (Boyd, 2001; Papineau et al., 2005).

Within the modern nitrogen cycle there is a variety of different metabolic processes, which give rise to a range of isotopic fractionations. These are summarized in Fig. 6.1 (after Papineau et al., 2005). Here we focus on two specific processes within the nitrogen cycle. First, is the production of ammonium ions, and second the process of denitrification.

Ammonium ions are produced by two different types of metabolic process. One is via the bacterial fixation of nitrogen. The other is

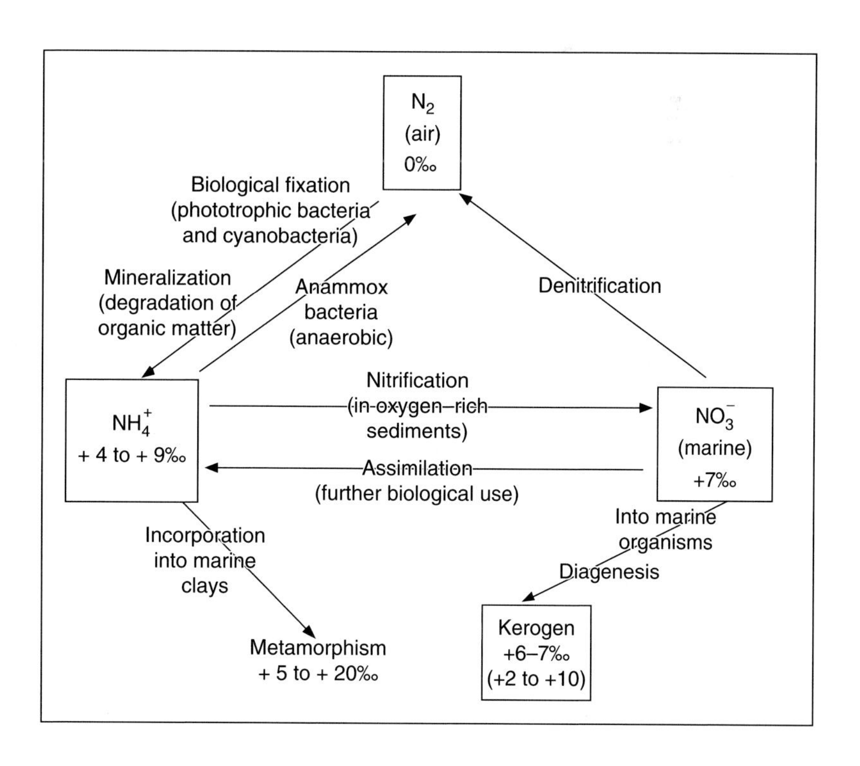

FIGURE 6.1 The modern nitrogen cycle showing the fractionations in $\delta^{15}N$ during the different bacterial metabolic pathways (data from Papineau et al., 2005 and Beaumont & Robert, 1999).

through the reduction of nitrate via bacterial nitrification. The net effect of these different processes is to produce ammonium ions enriched in heavy nitrogen, some of which is incorporated into the crystal structure of clays and buried as sediments. With increased depth of burial and metamorphism the NH_4^+ is structurally incorporated into biotite. At the higher grades of metamorphism, at amphibolite facies and above, biotites become further enriched in $\delta^{15}N$ (Boyd, 2001). Hence, although ammonium ions are common in the biotites of metasedimentary rocks, and these often have enriched $\delta^{15}N$ values of between 5 and 20‰, it is often difficult to uniquely define the metabolic origin of this signature. Even more problematical is the fact that there are also nonbiogenic processes which also fix nitrogen, the best known of these being lightning.

The process of denitrification, is the dominant metabolic process whereby marine nitrate is converted into atmospheric nitrogen, and is the principal control on the $\delta^{15}N$ composition of the oceans and modern sediment (Fig. 6.1). This leads to an enrichment of $\delta^{15}N$ in the oceans of +7 ‰. This enrichment is transmitted to marine organisms such as phytoplankton and ultimately to sedimentary kerogen, which is then preserved in the sedimentary record. There seems to be little further fractionation during the processes of diagenesis, for modern sedimentary organic matter has average $\delta^{15}N$ values between +6 and+7‰, close to that of ocean nitrate.

Nitrogen isotopes therefore provide a means of identifying denitrification in the geological record, through the +6 to +7‰, $\delta^{15}N$ enrichment of sedimentary kerogen (Fig. 6.1). Interestingly, Beaumont and Robert (1999) proposed that this enrichment is not found in early Archaean sediments, reflecting perhaps the absence of nitrates (and so denitrification) in the anoxygenic Earth, although this idea is not universally accepted (Chapter 5, Section 5.1.3.2). A further uncertainty is whether or not nitrogen isotope fractionations in ammonium, in metamorphosed Archaean sediments, reflect a true signature of biogenic nitrogen fixation.

6.2.4 Iron isotopes

Iron is thought to have been an important part of the biogeochemistry of the early Earth. However, fractionations within the iron-isotope system tend to be smaller than those observed for the lighter elements and are frequently only a few parts per mil. This means that it can be difficult to uniquely identify biogenic iron-isotope signatures in the geological record because abiotic fractionation is of comparable magnitude (Fig. 6.2).

Biogenic fractionations. There are two types of bacteria capable of fractionating iron isotopes and these represent two different types of Fe-metabolism. *Fe(III)-reducing* bacteria use Fe(III) as an electron acceptor during respiration. One genus, the mesophilic, dissimilatory *Shewanella alga* was found by Beard et al. (1999) to produce Fe(II) with $\delta^{56}Fe$ 1.3‰ *lighter* than the ferrihydrite substrate on which it was grown. In contrast, photosynthetic *Fe(II)-oxidizing* bacteria, operating under anaerobic conditions, and using $Fe(II)_{aq}$ as an electron donor, crystallize hydrous ferric oxide, *enriched* in $\delta^{56}Fe$ by about 1.5 ‰ (Croal et al., 2004).

Inorganic fractionations. Johnson et al. (2003) report inorganic $\delta^{56}Fe$ fractionations of up to 2.8‰ for Fe(III)–Fe(II) equilibria in dilute aqueous solutions and Anbar (2001) summarizes a range of inorganic iron-isotope fractionations, with a spread of up to 6‰ (Fig. 6.2).

It is possible that organisms used Fe(II) as an electron donor in anoxygenic photosynthesis very early in Earth history (Croal et al., 2004). Hence Johnson et al. (2003) interpreted the high $\delta^{56}Fe$ values in Archaean BIFs as a product of anoxygenic phototrophic Fe(II)-oxidizing organisms (Fig. 5.18), although Croal et al. (2004) point out that this can only be the case if there is independent evidence for an anoxygenic environment. Yamaguchi et al. (2005), on the other hand, argue that dissimilatory iron reduction also evolved very early and interpret negative $\delta^{56}Fe$ values in 2.9–2.6 Ga magnetite-bearing, organic-rich Archaean sediments (not BIFs) as a product of microbial dissimilatory Fe(III) reduction. This of course requires a source of oxidized iron, which is

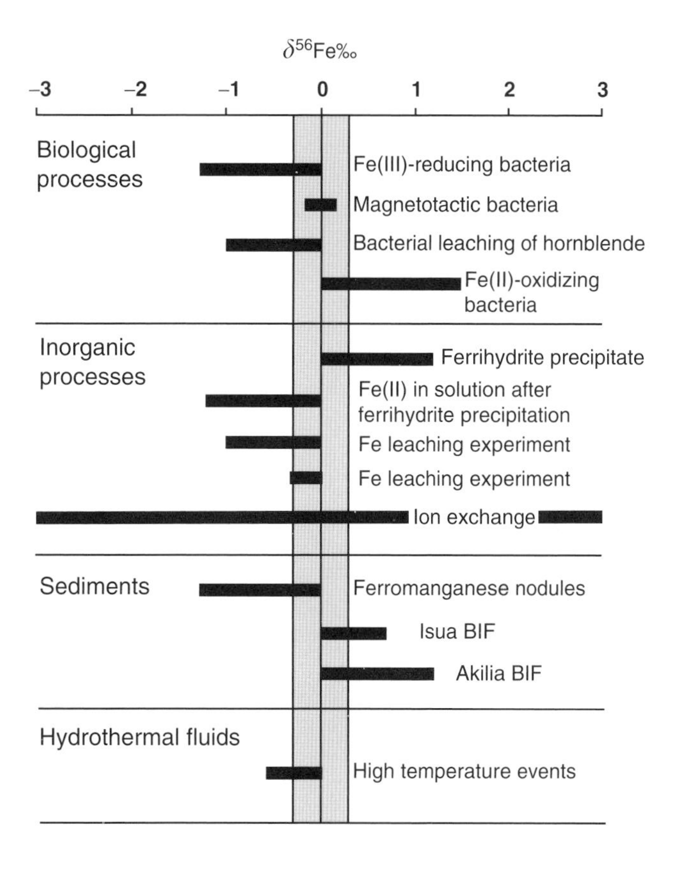

FIGURE 6.2 Iron-isotopic fractionations in biological and inorganic processes relative to the range of δ^{56}Fe found in igneous rocks (grey shaded area). Also shown are the range of Fe-isotopic fractionations recorded in modern ferromanganese nodules, early Archaean BIFs and modern high-temperature vents (after Anbar, 2001, with additional data from Croal et al., 2004; Johnson et al., 2003; and Dauphas et al., 2004).

a problem in the absence of an oxygenic atmosphere.

Given these conflicting interpretations of the iron-isotope record, coupled with organic and inorganic iron-isotope fractionations of similar magnitude, it is not possible at the present time to use iron isotopes to uniquely fingerprint early life on Earth.

6.2.5 Biomarker molecules

Most of the useful biomarker molecules are found in kerogen. Kerogen is the most abundant form of organic carbon in the Earth's crust and is made up of polymeric hydrocarbons, produced during the burial of organic matter in sedimentary rocks. It occurs as finely disseminated particles of organic matter and represents the selective preservation of biopolymers and the creation of new "geopolymer" molecules, through the process of diagenesis. Biomarker molecules can either be chemically bound into the structure of kerogen, or, in their unaltered form, may be enclosed within larger kerogen molecular structures.

The most stable biomarker molecules are those with a carbon base. Sulfate reducing microbes also produce biomarker molecules, but these do not survive the processes of sedimentary burial and diagenesis. Thus in ancient rocks the most likely biomarkers will be compounds of carbon. Currently, the oldest record of molecular hydrocarbon biomarkers comes from the 2.7 Ga organic-rich shales of the Fortescue Group in the Pilbara Block, western Australia (Brocks et al., 1999; Brocks & Summons, 2003). These rocks have experienced only low grade metamorphism and have been no hotter than 200–300°C since their formation. Contained within the kerogen of these sediments are the molecules belonging to 2α-methylhopanes and steranes with C_{26} to

C_{30} structures. The steranes are characteristic of eukaryotes, and the 2α-methylhopanes are geolipids, the degradation product of the biolipids 2-methyl bacteriohopanepolyols, synthesized only by cyanobacteria. Hence the oldest biomarker molecules indicate that the Eucarya and cyanobacteria were present at ca. 2.7 Ga.

6.3 The geological record of life's origins

Searching for life on the early Earth is a difficult enterprise, for most of the features which we use to identify life have their nonbiological equivalents. However, painstaking work over the past few decades has revealed that, certainly by the late Archaean, there was abundant life on early Earth, long before the appearance of the more conventional fossils.

6.3.1 Impacting – a difficult start

Life on Earth did not get off to an easy start. It is likely that during the final stages of accretion the Earth had already acquired a liquid water ocean (Chapter 5, Section 5.4.1), and it is possible that even at this stage life may have begun to develop in this early ocean. However, these early attempts at life would have been frustrated by impacting events and the volatilization of this early ocean.

The impacting history of the early Earth has long since been destroyed. However, we have a good idea of what it must have been like from the lunar record. In fact the Earth's stronger gravitational field and larger surface area mean that it would have been much more severely impacted than the Moon (Fig. 6.3), although the number of impactors would have declined massively in the first few hundred million years (Maher & Stefenson, 1988; Sleep et al., 1989).

The lunar record shows that within this period of impacting, there was a particularly intense event at about 3.9 Ga, lasting perhaps 100 Ma (Kring & Cohen, 2002). This event has become known as the "late heavy bombardment," and is thought to have been responsible for most of the impact craters on Mercury, the Moon, and the Martian highlands. Its effect on Earth would have been profound, although recognizing it in the terrestrial record is not straightforward. In fact the only terrestrial "memory" of the late heavy bombardment

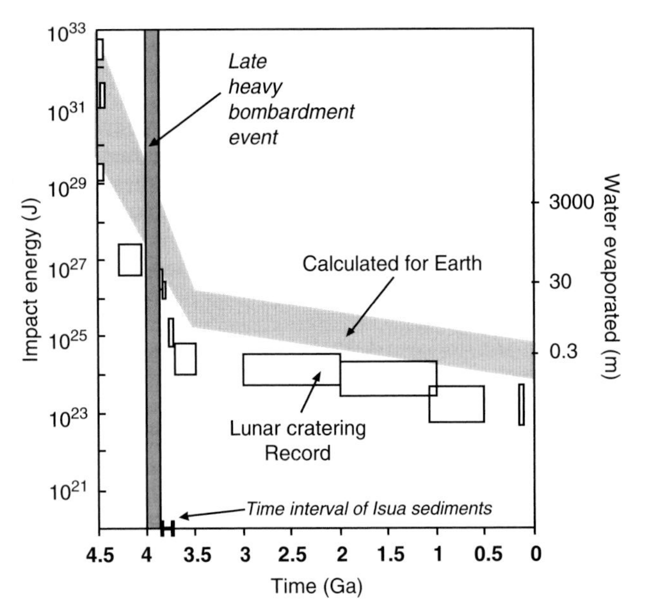

FIGURE 6.3 The impacting record for the Earth and Moon. The grey field for the Earth is calculated from the observed record on the Moon and expressed as impact energy (left) and depth of water evaporated (right). Superimposed upon the cratering record is the data from impact melts in lunar meteorites showing the time of the inner solar system late heavy bombardment (after Sleep et al. (1989) and Cohen et al. (2000)).

comes from tungsten isotope anomalies in the >3.7 Ga metasediments of the Isua greenstone belt (Schoenberg et al., 2002). It is possible that this late, intense impacting event explains why there are so few rocks on Earth older than 4.0 Ga.

A number of explanations have been offered for the origin of the impactors in the late heavy bombardment event. These can be categorized into either cometary or asteroid models. At the present time, isotopic and trace element data support an asteroidal rather than cometary origin (Kring & Cohen, 2002).

The implications of the late heavy bombardment event for the evolution of life on Earth are substantial. It has not gone unnoticed that there is a very short time interval between the end of the late heavy bombardment and the formation of sediments at Isua. Indeed the evidence for impacting at Isua recorded by Schoenberg et al. (2002) could signify that the two events overlapped. This means that either life did not start until after ca. 3.9 Ga, or if it had developed earlier, it had to survive a "high temperature stage" during which the oceans boiled.

6.3.2 The earliest record – W Greenland

In recent years the battle for the earliest signature of life on Earth has been fought out in west Greenland. One of the key localities is the Isua greenstone belt, arguably the oldest sequence of sedimentary rocks on Earth and an obvious place to search for the earliest record of life. The first indications of a biogenic signature in the Isua sediments came from studies by Schidlowski et al. (1979) and Schidlowski (1988) who reported light carbon isotope signatures in the metamorphosed sediments. They found $\delta^{13}C$ values as low as −25 ‰ in graphite in iron formation (values which are typical of graphitized kerogen) and $\delta^{13}C$ values in carbonates with a mean value of−2.5‰. On the basis of the difference in carbonate-organic carbon $\delta^{13}C$ values of greater than 20‰, Schidlowski (1988) proposed that carbon fixation was already taking place through photosynthesis at Isua times.

In 1996, Mojzsis et al. reported a new carbon isotope study of the Isua rocks. Instead of analyzing carbon from crushed samples of the whole rock, as had Schidlowski (1988), they used an ion microprobe to measure the carbon isotope ratios of individual graphite grains within the sediment. In this study they targeted graphitic inclusions located within the Ca-phosphate mineral apatite, an accessory phase in BIF, on the basis that phosphorus is likely to have been an essential component of ancient biological cycles. They found isotopically light carbon with a range of values ($\delta^{13}C$ = − 24 to −35‰, mean value − 30‰ ± 3), which they interpreted as biogenic. Further support for the presence of ancient life at Isua came from a study by Rosing (1999) who reported graphitic "globules" in clastic sediments with a $\delta^{13}C$ values as low as − 19‰. This result has now been extended to − 25.6‰ (Rosing & Frei, 2004), placing both studies within the range of the carbon isotope fractionations diagnostic of photosynthesis.

However, despite the apparently convincing biogenic signature, great care is needed in interpreting these carbon isotope results, for the rocks at Isua are deceptive. Although they show well-preserved primary structures in both sediments and igneous rocks (Appel et al., 1998), they have been extensively deformed and metamorphosed to amphibolite grade. In fact the samples examined by Mojzsis et al. (1996), and by Rosing (1999), have probably been metamorphosed twice (Rollinson, 2002, 2003). Unfortunately this is not the only difficulty with the Isua sediments, for the carbonate rocks at Isua, originally thought to be limestones are now known to be igneous rocks which have been metasomatically altered (Rose et al., 1996). This means that possible biogenic signatures need to be critically examined and metamorphic artifacts eliminated before biogenicity can be claimed (Fedo et al., 2006).

Considerable light has recently been shed on this dilemma by two very detailed petrographic studies of the graphitic rocks at Isua. Lepland et al. (2002) examined the mineral apatite, host to the graphite grains studied by Mojzsis et al. (1996). They found that there were two types of apatite in the Isua sediments, distinguished on the basis of their rare

Earth element patterns. There are primary and secondary apatites. The primary apatites are devoid of graphite, whereas the secondary apatites, which formed during the migration of later fluids, contain abundant graphitic carbon.

In a related study Van Zuilen et al. (2002, 2003) showed that there are also two types of graphites at Isua. There is secondary graphite which is the product of the metamorphic disproportionation of Fe(II)-carbonates, and characterized by whole rock $\delta^{13}C$ values of between − 10 and − 12‰, and primary graphite, not the product of metamorphic reactions, which has lighter $\delta^{13}C$ values, and which could be biogenic. Van Zuilen et al. (2002, 2003) concluded that the carbon globules in the clastic sediments studied by Rosing (1999) might be primary graphite and so biogenic in origin, whereas the sample examined by Mojzsis et al. (1996) contains secondary graphite and is part of a later, metamorphic siderite–graphite assemblage.

Given the uncertainty over the interpretation of the carbon isotope data from Isua, other isotopic methods have been used to assess the likelihood of life on Earth at Isua times. Even so, none of these is conclusive. Early studies of possible microfossils and biogenic amino acids have now been shown to be the product of contamination (Nagy et al., 1981; Appel et al., 2003). Sulfur isotope studies on sedimentary pyrite in iron formations (Strauss, 2003) and nitrogen isotope studies on nitrogen in graphitic carbon (Van Zuilen et al., 2005) have also proved inconclusive. Only the iron-isotope data of Dauphas et al. (2004) are within the range of iron oxidizing photoautotrophic bacteria, but this is also within the inorganic range and so, again, is inconclusive.

Isua therefore, illustrates how difficult the search for early life can be. Many Isua sediments contain carbon in the form of graphite, which, on the basis of its carbon isotope signature might have been kerogen. However, in most cases the graphite is a metamorphic artefact and not biogenic. The only study which has thus far withstood scrutiny is that of Rosing (1999), whose isotopically light graphitic globules might be biogenic in origin. However, even here Van Zuilen et al. (2005) urge caution, for "the degree of present graphite crystallinity, the amount of trapped argon, the low nitrogen content, and the $\delta^{15}N$ signature" render them similar to metasomatically derived graphite in other samples from the greenstone belt. Maybe this is not surprising given the metamorphic history of the Isua belt, for the graphite in the samples studied by Rosing (1999) has been multiply metamorphosed (Rollinson, 2003). For this reason, the high crystallinity need not be fatal for a biogenic origin.

6.3.3 The oldest microfossils

The confident recognition of Archaean microfossils is difficult. They tend to be miniscule, imperfectly preserved and have simple morphologies that can be mimicked by nonbiological structures (Schopf et al., 2002). Thus identifications of Archaean microfossils based upon morphology alone are open to serious criticism, and even when such identifications are supported by geochemical evidence, the interpretation can still be ambiguous. McLoughlin et al. (2005) recently reviewed the criteria for the biogenicity of Archaean microfossils. These include an established Archaean age for the samples, a biogenic morphology and geochemical evidence for biological processing which is syngenetic with the alleged "microfossils."

6.3.3.1 Fossil bacteria

Silicified spherules and "rod-shaped" bodies, a few microns in size, have been described from cherts in the 3.5 to 3.3 Ga Hooggenoeg and Kromberg Formations of the Onverwacht Group, the lowermost part of the Barberton Greenstone Belt, in South Africa (Westall et al., 2001). The shape, size, and colonial occurrence of these forms suggest that they are fossilized bacteria (Bacteria and Archaea). Spherules form rounded shapes about 1 μm in diameter and have the morphological characteristics of coccoid bacteria. They show some structural complexity reminiscent of cell division and occur in three-dimensional, cluster-like colonies. The "sausage-shaped" rods are between 2 and 4 μm in length and are thought to be of probable bacillar bacterial origin. Also present are smooth, ropy-textured, bedding plane films forming wispy laminae in the cherts. These are thought to represent biofilm – an extracellular excretion of micro- (and macro-) organisms, found on sediment surfaces, which is amenable to mineralization, and preservation in the sedimentary record. Within this "biofilm" are what appear to be gas escape structures, typical of bacterial colonies. The association of biofilm-like structures together with silicified spherules and "sausage-shapes" makes a strong case for these forms being fossil bacteria.

6.3.3.2 Microbial filaments

6.3.3.2.1 Organic filaments in the 3.5 Ga Apex chert, Pilbara

Kerogenous filaments in cherts from within the Apex basalt sequence in the Pilbara Craton, western Australia, became famous in the late 1980s as the world's oldest microfossils. They were described by Schopf and Packer (1987) and Schopf (1993) and have, until recently, been widely accepted as authentic, because of their supposed good preservation state. The kerogenous filaments are divided into cell-like compartments and have carbon isotope signatures as low as ($\delta^{13}C = -30‰$) – all the hallmarks of microfossils (Schopf, 1993; Schopf et al., 2002).

However, this claim has been contested by Brasier et al. (2002, 2005) who argue that these structures are not biological in origin. New geological mapping of the outcrop of the Apex cherts has shown that they are not deposited in a quiet marine environment as first supposed but instead formed as a hydrothermal vent breccia. Many of the supposed microfossils occur in fabrics now identified as late-stage fissure infillings. In addition, Brasier et al. (2002, 2005) showed that the filamentous forms described by Schopf (1999) are branched, a feature not otherwise seen until much later in the evolutionary record, indicating that they are secondary artefacts of crystal growth. The previously described cell-like structures are now thought to be the result of silica crystallization. The filaments are not hollow, as previously supposed but are composed of carbonaceous material wrapped around quartz crystals. The isotopically light carbon compounds, also previously regarded as a biogenic signature, are hypothesized to have been produced from volcanogenic carbon monoxide by a Fischer–Tropsch-type process in a hydrothermal vent environment (Brasier et al., 2002).

6.3.3.2.2 Organic filaments from the 3.42 Buck Reef Chert, South Africa

Similar filamentous structures to those found in the Apex chert have also been described from South Africa. In this case a biogenic origin seems likely. Tice and Lowe (2004) found carbonaceous filamentous structures up to 1.5 μm wide and 100 μm long forming black bands within the chert. These filaments have $\delta^{13}C$ values between − 20 and − 35‰ and are thought to be remnants of microbial mats. The particular unit examined, the Buck Reef Chert, extends for more than 50 km and can be traced across a range of depositional settings, from shelf to deep basin. Tice and Lowe (2004) argue that this would be unlikely if the chert was of hydrothermal origin and prefer an origin through precipitation from silica-saturated seawater. Controls from the likely environment of deposition and Fe(II) minerals in the sedimentary succession suggest that the microbes formed in shallow water, and so were probably photosynthesizers, but formed in an anoxygenic setting. Hence they were probably not cyanobacteria but anoxic photosynthesizers.

6.3.3.2.3 Pyritized filaments in 3.24 Ga cherts, Pilbara

Densely intertwined, pyritized threads, up to 2 μm wide and 300 μm long, are found in chert in the 3.24 Ga Sulfur Springs massive sulfide deposit, in the Pilbara Craton (Rasmussen, 2000). These filaments exhibit a similar morphology to microfossils described from younger rocks and for this reason are thought to be biogenic. Similar carbonaceous filaments in related rocks have isotopically light carbon signatures ($\delta^{13}C$ = − 27 to − 33‰), supporting a probable biogenic origin (Duck et al., 2004). The fact that the filaments described by Rasmussen (2000) are found within a massive sulfide deposit, suggests that they formed in a shallow, sea-floor hydrothermal vent system, and so were probably mesothermophilic, chemotrophic, prokaryotes.

6.3.3.3 *Stromatolites and microbialites*

Stromatolites are "laminated lithified sedimentary growth structures that form by accretion, through the addition of new laminae away from the point or surface of accretion." They normally are made up of carbonate minerals and thought to have formed in a shallow water environment. Modern freshwater stromatolites are made up of algal and cyanobacterial filaments. In the marine environment they form microbial mats and columns. These comparisons led many scientists to believe that stromatolites found in the ancient record are biogenic in origin and that they are in fact microbialites. That is, they are sedimentary structures in which minerals are precipitated as either a by-product of microbial metabolisms or as a by-product of microbial decay.

However, the biogenic view of stromatolites is not universally accepted. For example, Grotzinger and Knoll (1999) have pointed out that laminated growth structures can form from microbial, chemical, and physical processes and it is important to distinguish between these three. They suggest that in the early Earth stromatolites formed through in situ chemical precipitation, whereas in younger Proterozoic and more recent sediments they probably formed through microbial trapping and binding of sediment. However, it is increasingly being realized that carbonate precipitation processes which had previously been thought of as entirely inorganic are now known to be microbially controlled (Bosak & Newman, 2003).

6.3.3.3.1 *3.45 Ga Stromatolites from Pilbara*

The oldest known stromatolites are from 3.45 Ga-old rocks in the Pilbara Craton in western Australia. A recent study by Van Kranendonk et al. (2003) supports earlier claims of their biogenicity (Lowe, 1980; Walter et al., 1980). Van Kranendonk et al. (2003) present convincing geochemical evidence that the stromatolites of the Strelly Pool chert formed in an anoxic marine setting. Field observations of the stromatolitic structures indicate that a binding agent was required for the stromatolites to grow and this is presumed to have been biogenic. These observations suggest that the Strelly Pool Chert stromatolites were anoxic photosynthesizers occurring as a microbial mat.

6.3.3.3.2 *Late Archaean stromatolites*

Younger stromatolites have been described from 3.0 Ga rocks at Steep Rock in Canada (Wilks & Nisbet, 1985), from the 2.7 Ga Cheshire formation of the Belingwe Greenstone belt (Martin et al., 1980) and from 2.52 Ga shelf carbonate sediments in the Campbellrand subgroup, South Africa. In this latter locality there are also well-documented calcified microfossils of cyanobacteria (Kazmierczak & Altermann, 2002). It is also possible that at this locality the mineralization of the dead bacteria was the result of the action of heterotrophic bacteria.

Microbial mats have also been reported from 2.8 to 3.0 Ga clastic sediments in the Mozaan Group, South Africa (Noffke et al., 2003). These sediments formed in a high-energy environment and contain "wrinkly" microstructures which resemble filamentous bacteria. They are thought to have been benthic bacteria, analogous to those that stabilize sediment in modern, high-energy siliclastic environments. Carbonaceous residues from these rocks have a light carbon isotopic signature ($\delta^{13}C = -24‰$), consistent with a biogenic origin. Modern analogs indicate that these microbes could have been either cyanobacteria, or sulfur-oxidizing proteobacteria. It is possible that these sediments formed in the photic zone leading Noffke et al. (2003) to prefer photoautotrophic cyanobacteria, although the oxygen created by the cyanobacteria would also permit the coexistence of sulfur-oxidizing bacteria.

6.3.3.4 *Late Archaean microbial diversity*

By the time we reach the late Archaean there is a strong biogenic signal in the geological record. A particularly good example comes from the remarkably well-preserved sediments of the 2.7 Ga Belingwe Greenstone belt in Zimbabwe. Grassineau et al. (2001, 2002) have examined carbon and sulfur isotopes in cherts, shales, and stromatolites from the lowermost part of the Belingwe succession – the Manjiri Formation – and from the top of the succession– the Cheshire formation. Both of these units display a very wide range of $\delta^{13}C$ and $\delta^{34}S$ values, which Grassineau et al. (2001, 2002) elegantly interpret in terms of diverse metabolic processes.

Carbon in Belingwe carbonates has $\delta^{13}C$ values between +1.5 and −9.0‰. Organic carbon has values between −7 and −44‰ (Fig. 6.4b). Organic carbon–carbonate fractionations in the range −30 and −25‰ are thought to be the isotopic fingerprint of the enzyme rubisco and so indicate former photosynthesis and the presence of cyanobacteria. More extreme organic carbon-carbonate fractionations, with values of around −40‰, are thought to indicate the existence of methanogenic and methanotrophic bacteria.

The very wide range of sulfur isotope values found at Belingwe ($\delta^{34}S$ = +17 to −18) (Fig 6.4a) is unusual for Archaean sulfides, and, in part, is the result of the very careful, microscale sampling used in this study. Grassineau et al. (2001, 2002) argue that the wide range of sulfur isotope values implies that several different isotopic fractionation processes were taking place (Fig 6.4a). They proposed that the very light $\delta^{34}S$ values are best explained by a two-stage sulfate reduction process. In other words, seawater sulfate was reduced to sulfide by sulfate reducing bacteria, and then reoxidized by sulfide oxidizers, before being reduced again by sulfate reducers. If correct, this model requires that both sulfate reducing and sulfide oxidizing bacteria were present, and so the full sulfur oxidation–reduction cycle was in operation during the late Archaean.

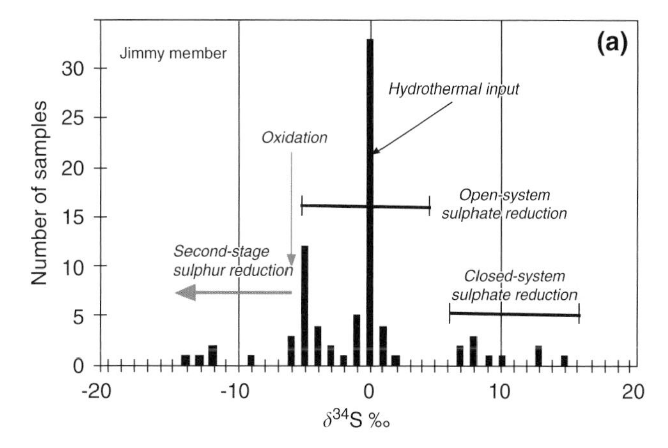

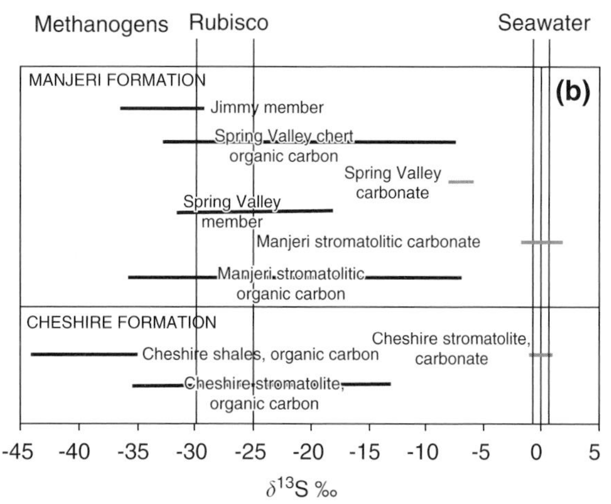

FIGURE 6.4 Carbon and sulfur isotope data for sediments from the 2.7 Ga Belingwe Greenstone belt, Zimbabwe. (a) Histogram of sulfur isotope data for the Jimmy member of the Manjiri Formation. The data are annotated showing the likely explanations for the sulfur isotope fractionations. The principal feature of these data is the hydrothermal source of the seawater sulfur. (b) Carbon isotope data for the Manjiri and Cheshire formations showing the range of isotopic signatures in the different units. Across the top of the diagram are the likely sources of carbon isotope fractionation. Data from Grassineau et al. (2001, 2002).

The carbon and sulfur isotope data provide evidence for a complex microbial community in Belingwe, at 2.7 Ga. There were photosynthesizing oxygenic bacteria, sulfate and sulfite reducers, sulfide oxidizers, methanogens, and methanotrophs. These may have formed an ecosystem in which cyanobacteria in shallow water reefs managed atmospheric CO_2 to provide oxygenated surface waters, whilst along strike there were microbial mats containing photosynthesizing green sulfur bacteria, together with sulfate and sulfur reducing bacteria. In the mud below these mats were methanogens overlain by methane oxidizing bacteria (Grassineau et al., 2002).

6.4 The microbial record of life's origins

A different insight into the history of life on Earth, and one which is completely independent of the fossil record, comes from an understanding of evolutionary relationships within the microbial world.

6.4.1 The RNA tree of life

Dramatic new insights into the evolution of life came through the application of molecular phylogenetics to the tree of life. The initial work was done in the 1970s by Carl Woese at the University of Illinois, who used ribosomal RNA (rRNA) gene sequences to construct a molecular-sequence-based phylogenetic tree. This was used to relate all organisms to each other, from which it was possible to construct a history of life (Woese et al., 1990). Woese's work challenged the contemporary view of five "kingdoms" of life – plants, animals, protists, fungi, and monera and instead proposed three "domains" of life. These were dominated by microbial life and were defined as the Eucarya (comprising eukaryotes – organisms whose cells have a nuclear membrane; this includes all higher organisms including plants and animals) and the Prokaryota. The Prokaryota are subdivided into the domains Bacteria and Archaea – single celled organisms which have no nuclear membrane. Bacteria are relatively well known as the cause of much human and animal disease, whereas Archaea were a new discovery (Fig. 6.5).

The methodology of molecular phylogeny involves the alignment of pairs of rRNA sequences from different organisms. The differences are counted and used as a measure of "evolutionary distance," although of course there is no absolute timescale embedded in this process. This approach allows phylogenetic trees to be identified and maps prepared to show the evolutionary paths which lead to modern rRNA sequences (Pace, 1997). From the entire tree, core lineages can be identified, and from these (the ultimate goal) a last common ancestor can be identified.

Pace (1997) identified two important outcomes of the RNA tree of life approach to phylogeny. First, it has become clear from the RNA sequencing that the major organelles of eukaryote cells – mitochondria and chloroplasts are derived from an original symbiotic relationship with bacteria. Mitochondria are representative of Proteobacteria (purple bacteria) and chloroplasts of cyanobacteria. A second outcome of this approach is that it shows the main links between the domains. For example, it has become clear that the Eucarya and Archaea have many properties in common, and appear to belong to a common lineage. The bacteria on the other hand are different. This means that some of the features of the Eucarya, such as the origin of the nuclear membrane, might be understood from examining their Archaeal roots (Pace, 1997).

More recent investigations have shown that the branching "tree of life" may look more like a braided river, for there seems to have been considerable merging and splitting of lines, including gene swapping (Doolittle, 1998, 1999; Nisbet & Sleep, 2001). This means that tracing phylogenetic relationships back through the tree of life may not be as easy as had previously been supposed, and it seems as though the molecular record is more noisy than had been initially expected (Nisbet & Fowler, 2003).

6.4.1.1 Metabolic themes

It is difficult to classify microbes on the basis of their metabolic properties, for similar metabolic pathways seem to have been used several times during the development of

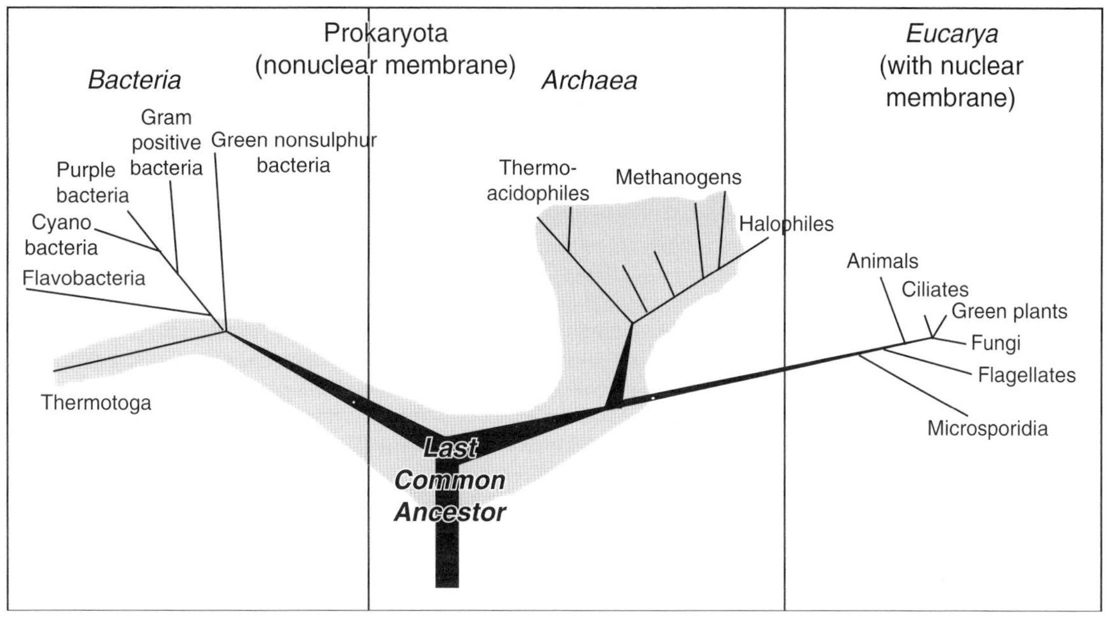

FIGURE 6.5 The relationships between the main groups of organisms based upon rRNA molecular phylogeny (after Knoll, 1999). This RNA tree of life shows the three main domains of life the Eucarya and the two domains of Prokaryota – Bacteria and Archaea. Thermophyllic organisms are outlined in grey.

microbial life. Nevertheless, it is possible to identify "domain-level tendencies" in the RNA tree of life (Pace, 1997). These can be summarized as follows:

Archaea are mostly lithotrophic and form two major groups, both of which rely on hydrogen:

- in the Euryarchaeota CO_2 is converted into methane;
- in the Crenarchaeota sulfur is converted into hydrogen sulfide.

The deepest members of the Bacteria are also lithotrophs and use hydrogen, together with sulfur compounds, or with low levels of oxygen. Chlorophyll-based photosynthesis was also developed by the bacteria although much later.

6.4.2 The last common ancestor of all life

The last common ancestor is the notional population from which all modern living cells are thought to be derived. It is probable that these organisms were akin to the bacteria, were DNA-based, lived in a hydrothermal setting (although Brochier & Philippe, 2002 have suggested an alternative), and metabolized hydrogen (Pace, 1997; Nisbet & Fowler, 2003). There is some debate about the precise temperature at which these organisms lived and some authors prefer a mesophile, rather than a hyperthermophile origin for the last common ancestor (Nisbet & Fowler, 2003).

The thermophilic character of the last common ancestor (Fig 6.5) has important implications, for it indicates that life, at some stage early in its development, passed through a high-temperature "bottle-neck." There are two possible scenarios for this thermophilic stage. First is the transient heating and volatilization of the oceans which took place during the late heavy bombardment at 3.9 Ga (see Section 6.3.1). In this case the heating of the oceans operated as a "filter," such that only thermophilic organisms survived these events. The second scenario is that life developed in the vicinity of sea-floor hydrothermal vent systems. This model is discussed in Section 6.5.2 and has much to commend it, not least because many of the elements fundamental to the construction of basic "life" molecules are readily available in this environment.

It has been suggested that nitrogen-based metabolisms must also be very ancient, as is the ability to incorporate ammonium into amino acids and nucleotide bases. In addition, the great variety of nitrogen-fixing organisms in both the Archaea and Bacteria imply that this capability too may be a feature of the last common ancestor (Papineau et al., 2005). Similarly, many hyperthermophilic organisms reduce elemental sulfur to H_2S using a metabolic pathway which is found in both the Bacterial and Archaeal lineages. The fact that this metabolic style is common to both lineages suggests either that it is also very ancient and inherited from the last common ancestor (Mojzsis et al., 2003) or that it is the result of a transfer.

6.4.3 A chronology for the tree of life

Although the rRNA tree of life is able to place organisms broadly into their correct relative positions, it does not have a reliable inbuilt clock (Graur & Martin, 2004), and so, until recently the timing of events on the tree of life has been poorly known. In a recent review Shen and Buick (2004) used fossil, rather than molecular evidence to place some initial time constraints on the evolutionary appearance of the Archaea, Bacteria and Eucarya (Fig. 6.6).

The principal biomarkers used to date the tree of life are those isotopic fractionations and relevant biomarker molecules that can be attributed to specific bacterial groups. For example, Brocks et al. (1999) used biomarker molecules to identify cyanobacteria in 2.6–2.7 Ga shales from the Pilbara Craton, western Australia. They also found steranes derived from eukaryotes in the same samples, implying that this domain is also at least 2.7 Ga old. An earlier study by Hayes (1994) had already established that methanogenic Archaea were present at 2.7 Ga.

Microbial sulfate reduction is known from 3.47 Ga barites, from North Pole, Australia (Shen & Buick, 2004). Petrological arguments constrain the temperature of sulfate reduction to below about 60°C and so limit the type of sulfate reducing bacteria to mesophiles. This argument is used to place sulfate reducing bacteria just above *Thermodesulfobacterium*.on the RNA tree (Shen & Buick, 2004). There are important implications in this finding, for sulfate reduction is a complex metabolic process. This means that even by 3.47 Ga microbes had developed many of the key cellular systems found in their modern relatives and implies that even at 3.47 Ga they had had a long history.

6.5 In the beginning

6.5.1 A brief history of life on Earth

The history of life in the Archaean can be divided into four stages – life before photosynthesis,

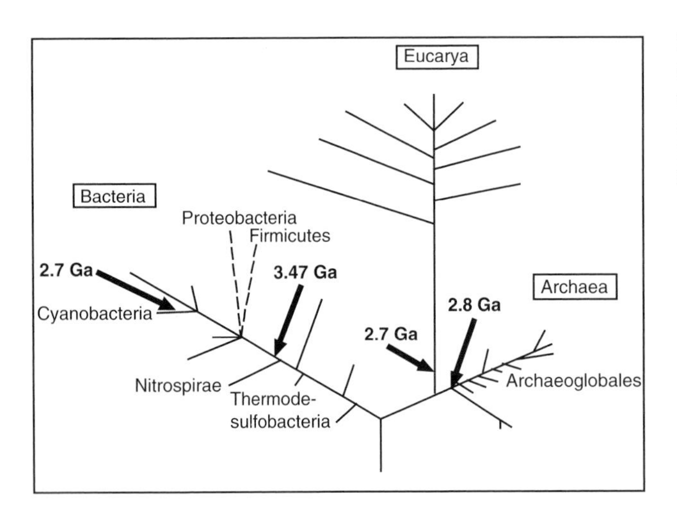

FIGURE 6.6 The RNA tree of life showing the three main domains and the members of the Bacteria and Archaea significant in dating the lower branches of the RNA tree (after Shen & Buick, 2004, with information from Brocks et al., 1999).

anoxygenic photosynthesis, oxygenic photosynthesis, and the evolution of the Eucarya (Nisbet, 2002). The first two stages correspond to the time before oxygen was produced on Earth, the third stage with the interval when oxygen was produced but it was not present in the atmosphere, and the final stage with the rise of atmospheric oxygen (Kerr, 2005). Here we attempt to set some time boundaries on these different stages of the evolution of life.

By the late Archaean the Earth was home to a wide range of microbial life. This is clear from the work of Grassineau et al. (2001, 2002) at Belingwe. There is good evidence for cyanobacteria and possible heterotrophs in the Cambellrand carbonate platform at 2.52 Ga and for cyanobacteria in the microbial mats of the Mozaan Group at 2.8 to 2.9 Ga. Other studies have identified cyanobacteria and biomarkers from Eucarya in 2.6–2.7 Ga shales from the Pilbara, and evidence of methanogenesis and methanogens at 2.7–2.8 Ga (Hayes, 1994; Brocks et al., 1999; Rye and Holland, 2000). Hence we can be reasonably certain that by the late Archaean oxygenic photosynthesis had been established and the Eucarya had evolved. Tracing life further back in time is more problematical.

6.5.1.1 *The rise of the Eucarya*

The accepted view of the Eucarya is that they derived their chloroplasts from cyanobacteria and their mitochondria from purple bacteria. Therefore they post-date these precursors. Biomarker molecules in 2.6–2.7 Ga Pilbara shales indicate that the Eucarya were in existence by this time, although Hartman and Fedorov (2002) have challenged the conventional view and believe that they could be much more ancient.

6.5.1.2 *Cyanobacteria and oxygenic photosynthesis*

There is good geological evidence from the Mozaan Group in South Africa for cyanobacteria at 2.8 to 2.9 Ga (Noffke et al., 2003). Older indications come from carbonaceous filaments in the 3.24 Ga Sulfur Springs cherts in Pilbara and in the 3.42 Ga Buck Reef Chert in South Africa, both of which have rubisco-like, light $\delta^{13}C$ isotopic signatures. There are also biogenic stromatolites in the 3.45 Ga Strelly Pool chert at Pilbara, although there is no claim that these were the product of cyanobacterial activity. Nevertheless, these data support the general argument of Schidlowski (1988) that the global difference between the $\delta^{13}C$ isotopic signatures of average carbonate carbon and kerogenous, organic carbon is constant and identical to the rubisco fractionation of about 30‰ since 3.5 Ga (Fig 6.7). Any earlier evidence for rubisco fractionation is tentative, but it is possible that the Isua kerogen globules, with a light $\delta^{13}C$ isotopic value of about −26‰ could originally have been photosynthesizing cyanobacteria (Rosing & Frei, 2004).

6.5.1.3 *Anoxygenic photosynthesis*

Photosynthesis probably evolved as organisms learned to use pigments to capture energy from light. Nisbet et al. (1995) suggested that early organisms associated with hydrothermal vent systems developed phototaxis using near-infrared radiation and that this 'skill' eventually gave rise to photosynthesis. In a study of the major groups of photosynthetic bacteria Xiong et al. (2000) found that chlorophyll-biosynthesis evolved from bacteriochlorophyll biosynthesis, indicating that the anoxygenic form of photosynthesis evolved before the oxygenic form. Dating the rise of anoxic photosynthesis, however, is not easy. Xiong's work indicates that six major bacterial lineages had developed before the cyanobacteria came into existence, at or before 2.8 Ga. On these grounds anoxic photosynthesis probably existed in the mid-Archaean and maybe earlier. Molecular phylogeny tells us that purple bacteria are probably the earliest photosynthetic lineage, although the geological record is silent on this subject.

Anoxygenic photosynthesis is found in purple and green sulfur and nonsulfur bacteria. In the sulfur bacteria there is photosynthetic oxidation of sulfide to elemental sulfur, although there is no recognizable isotopic fingerprint of this process. There may however be a record of anoxygenic photosynthesis from another source, for it is possible that some organisms used Fe(II) as an electron donor, in a very early form of anoxygenic photosynthesis. If this is the case then the high $\delta^{56}Fe$ values in Archaean BIFs could be evidence of anoxygenic phototrophic Fe(II)-oxidizing organisms (Johnson et al., 2003; Croal et al., 2004). Given the positive $\delta^{56}Fe$ value of +0.7 ‰ in the Isua iron formation (Dauphas et al., 2004) it is possible that anoxygenic photosynthesis can be traced back to this time.

6.5.1.4 Sulfate reducers

Shen and Buick (2004) demonstrated the great antiquity of sulfate reducing bacteria. The complexity of these organisms implies that they had a long evolutionary prehistory, prior to their first recorded appearance at 3.47 Ga. However, in preoxygenic times the early oceans were sulfate-poor, suggesting perhaps that these bacteria were localized to evaporitic environments.

6.5.1.5 Life before photosynthesis

The evolutionary lineages of the microbial world might be traceable to the oldest sediments, at ca. 3.8 Ga (Fig. 6.7). It is likely, however, that even before this time life existed on Earth and may have survived the late heavy bombardment. We know little about this earliest life, except that the last common ancestor was probably based upon a hydrogen-based metabolism. There are however a number of other possible metabolisms involving sulfur reduction, nitrogen reduction and carbon reduction which are all extremely ancient (Papineau et al., 2005).

6.5.2 The earliest habitat

The idea that modern hydrothermal systems represent "an ancestral niche" for life is a popular theme in contemporary writing on the origin of life. Here we investigate this idea more fully in order to explore the role played by hydrothermal systems in the early evolution of life.

Deep ocean hydrothermal vent systems are a relatively new geological discovery and over the past two to three decades have provided a wealth of information about the cooling of the oceanic crust, the formation of massive sulfide mineral deposits, and have revealed a whole new biota including extremophile microbes. We now know that oceanic hydrothermal vents

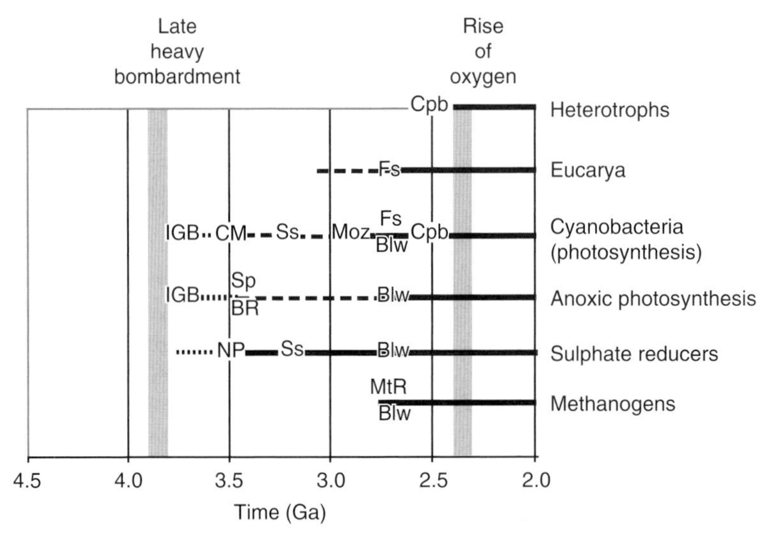

FIGURE 6.7 Time-lines for a range of different bacterial groups, relative to the late heavy bombardment and the rise of oxygen. The solid black lines indicate some certainty, the dashed lines indicate a high probability and the dotted lines the possibility of life at the time indicated. Data from Blw, Belingwe (Grassineau et al., 2001, 2002); Cpb, Campbellrand carbonate platform (Kazmierczak & Altermann, 2002); Moz, Mozaan group (Noffke et al., 2003); Ss, Sulfur Springs, Pilbara (Rasmussen, 2000); CM, general carbon management (Schidlowski, 1988); IGB, Isua greenstone belt (Rosing & Frei, 2004); Fs, Fortescue shales, Pilbara (Brocks et al., 1999); BR, Buck Reef Chert (Tice & Lowe, 2004); Sp, Strelly pool, Pilbara (Van Kranendonk et al., 2003); MtR, Mount Roe (Rye & Holland, 2000); NP, North Pole, Pilbara (Shen & Buick, 2004).

are of at least two types. There are the well-known high-temperature systems (>300°C), associated with ocean ridges, and the more recently discovered, cooler (40–75°C) off-axis systems (Kelley et al., 2001). What is unknown at present is whether it is the high- or low-temperature systems which have had the most influence on the development of early life.

The case that oceanic hydrothermal systems were particularly instrumental in the development of early life on Earth is strengthened when we consider how prevalent they were in the Archaean. Geochemical studies of Archaean oceanic chemical sediments show that they frequently have a strong geochemical signature of vent fluids, indicating that such fluids made a significant contribution (more than at present, see Kamber & Webb, 2002) to the composition of the Archaean oceans Section 5.4.3, this volume). Hydrothermal systems are also an important means of cooling the Earth. Given a hotter early Earth, in which the mantle potential temperature was perhaps 200°C greater than today (see Chapter 3, Section 3.2.2.3), the oceanic crust would have been thicker, and hydrothermal systems most likely more abundant.

Hydrothermal systems provide a suitable environment for the synthesis of complex organic molecules. For life to evolve, complex polymers have to be formed. One place where this happens at the present-day is in deep-sea hydrothermal vents. Holm and Charlou (2001) reported complex hydrocarbons formed by inorganic processes in high temperature fluids (364°C) from the Mid-Atlantic ridge. The high productivity of organic molecules in this particular setting is attributed to reactions between the hydrothermal fluids and ultramafic ocean-floor rocks. Interestingly, the reduced nature of oceanic hydrothermal fluids provides a microenvironment, albeit in an aquatic setting, not dissimilar to that investigated by Miller and Urey fifty years ago. Organic synthesis is aided by appropriate catalysts. In the case of deep oceanic vent systems both sulfide minerals and clays are present and capable of catalyzing organic reactions.

There has been some scepticism about the feasibility of high-temperature hydrocarbon synthesis, for it has been argued that fragile molecules such as amino acids are destroyed in high-temperature vent systems (Miller & Bada, 1988). However this concern appears to be unfounded for amino acids have now been reported in vents as hot as 308°C in the Izu-Bonin arc in the Pacific (Takano et al., 2004). In fact, when applied to DNA there is a positive side to this process, for the frequent damage to DNA in a high-temperature environment confers the ability of damage repair on the DNA of these organisms (Reysenbach & Shock, 2002).

A further aspect of molecular synthesis in a high-temperature environment is the recognition of "heat-shock proteins," proteins whose synthesis is promoted by heat stress. One particular group of heat-shock proteins is seen as particularly relevant to the development of photosynthesis. These are the chaperonins, which play an important role in binding and shaping the enzyme rubisco.

Hydrothermal systems are an important source of nutrients and energy. Hydrothermal systems are environments where there is a close interaction between geochemical and microbial processes. Their principal feature is that they represent surface fluids, drawn deep into the Earth, where they are reduced, heated, and laden with dissolved minerals and then forced back to the surface where they react with oxidizing fluids (Fig. 6.8). In the absence of sunlight in the deep ocean, the principal source of energy comes from redox reactions. In the absence of oxygen, as would be the case in the early Earth, chemolithoautotrophic primary production is driven by reactions which utilize H_2 to reduce CO_2, or sulfur. In addition to the chemical energy supplied in hydrothermal systems heat is also an important energy source. It has been proposed that a primitive heat-sensing capacity in early organisms, developed in a "hot" environment might be the precursor to the development of light sensitivity and photosynthesis (Nisbet & Fowler, 2003).

Living cells depend principally on a supply of carbon, oxygen, nitrogen, hydrogen, phosphorus, and sulfur. In addition their biochemistry

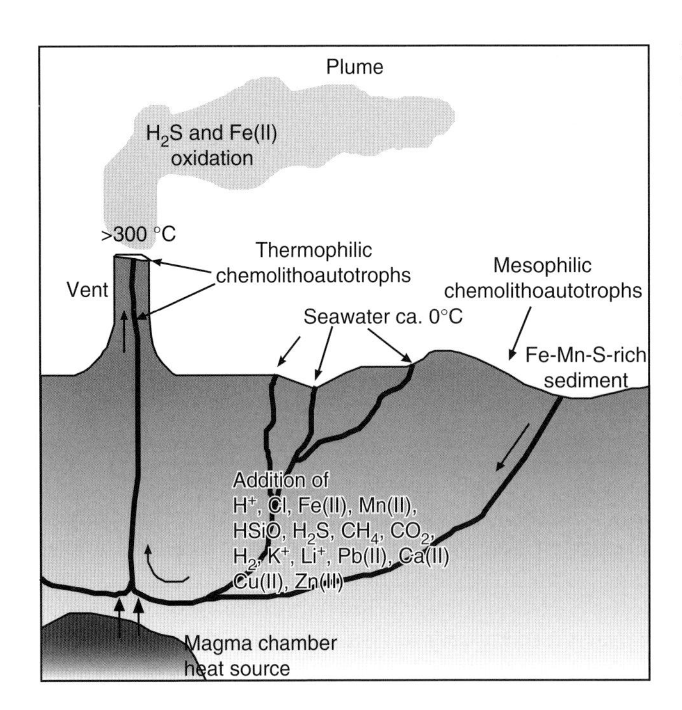

FIGURE 6.8 Schematic illustration of an ocean ridge hydrothermal system (after Reysenbach & Cady, 2001).

requires a wide range of trace metals. These either form an essential part of the molecule or act as a catalyst in molecular synthesis (e.g. Foustoukos & Seyfreid, 2004). For example Fe–S is important in the rubisco molecule and Mn in bacteriochlorophyll.

Particularly important here is the role of transition metal sulfides. In 1988 Wachtershauser proposed that pyrite, abundant in hydrothermal vent systems, provided an energy source for the first life. He suggested that pyrite provided the catalyst necessary to drive a number of essential chemical reactions which are important precursors to life. More recent studies have confirmed this view and have shown that the sulfides of Fe, Ni, Co, and Zn can play an important role in the fixation of carbon in a prebiotic world (Cody et al., 2004). Transition metal sulfides also play a role in more advance organic synthesis, and Huber and Wachtershauser (1998) showed how amino acids were converted into their peptides using a (NiFe)S catalyst.

Hydrothermal vents are home to thermophilic bacteria. One of the most important discoveries to arise out of the mapping of the universal phylogenetic tree (Woese et al., 1990) is that the most deep-rooted lineages are thermophilic (Fig. 6.5). Thermophiles are microbes which grow best between >45 and 80°C, and hyperthermophiles are those which comfortably live at temperatures above 80°C. Thermophilic organisms are found in both the Archaea and the Bacteria indicating that this property is extremely ancient and that the last common ancestor was thermophilic.

Thermophilic microbes have been known for some years from the hot springs of Yellowstone National Park, USA (Farmer, 2000) but now are also documented from deep ocean hydrothermal vents, the most extreme of which is an Fe(III)-reducing member of the Archaea that can grow at 121°C (Kashefi & Lovley, 2003). Ocean vent microbial communities are diverse and include Archaea and Bacteria (Hoek et al., 2003) which utilize hydrogen- and iron-oxidizing and sulfur-reducing metabolisms to produce microbial mats. They form rod-like, coccoid, and filamentous forms (Reysenbach and Cady, 2001). An equally diverse array of thermophilic microbes is also known from deep 65°C fluids, extracted from older, cooler ocean crust (Cowen et al., 2003).

The case for a mesophile, rather than a thermophilic or hyperthermophile origin for life has been argued by Forterre (1995). As discussed above, RNA is unstable at high temperatures and it unlikely than the RNA world began in a high-temperature vent environment. Mesophile organisms, on the other hand, would thrive at the warm (ca. 40°C) margins of a hydrothermal system, rather than in its high-temperature core. In this case it is possible that the hyperthermophiles are organisms which have strayed into and adapted to a higher temperature environment, from a cooler, less damaging environment (Forterre, 1995). In this model a mesophile organism would predate a thermophilic common ancestor.

It is possible that inorganic cell prototypes were created in hydrothermal systems. Iron sulfides are abundant in oceanic hydrothermal systems and are one of the main products of high-temperature vents. Russell and Hall (1997) and Martin and Russell (2003) proposed that iron monosulfide bubbles, created in an oceanic hydrothermal setting formed a template for the structure of the first cells. In this model the FeS membrane acted as a catalytic chamber within which organic synthesis could take place. As this process evolved, the sulfide membrane became coated with abiotic organic polymers which eventually took over from the sulfide and replaced their function.

As already mentioned, there is currently some debate about the type of hydrothermal system which was most likely to be responsible for the development of early life – terrestrial or oceanic, high-temperature hyperthermophilic or moderate-temperature mesophilic. On balance most arguments seem to support an oceanic setting, and the highly energetic, mineral laden system of deep ocean hydrothermal vents is the favored locality. In support of this hypothesis some of the sites of the most ancient recorded Archaean fossils are probably from this setting. Interestingly even at Isua the existence of methane- and saline-fluid inclusions indicate the operation of an oceanic high-temperature hydrothermal system (Appel et al., 2001). Nevertheless, one of the challenges of recognizing life from such systems is that they are capable of producing organic-looking signatures abiotically, and so confusing the search for ancient life. Recently a number of authors have drawn attention to the mid-Archaean carbonaceous cherts from the Pilbara Craton (Lindsay et al., 2006; van Kranendonk, 2006; see also section 3.3.2). These are most probably hydrothermal in origin and could represent a silica-rich slurry in which life first evolved. However, it is possible that in these rocks any early biosphere signal is masked by a predominant abiotic component.

If true, the hydrothermal vent hypothesis for the early development of life is a superior example of interactions within the Earth system. In this environment we have fluxes between sea-water and the ocean crust (and sometimes the mantle), and a range of biogenic processes taking place at the interface between the two.

6.5.3 Who is pulling the strings?

Lovelock (1979) argued that the modern atmosphere is a disequilibrium atmosphere, strongly modified by the presence of life on Earth. Extending this argument it is possible that over geological time there has been a long-term interaction between living organisms and the atmosphere so that the atmosphere is the product of, and is controlled by, "life." In turn the atmosphere has influenced the Earth's climate, maintaining equable temperatures for the last 4.4 Ga, and hence the presence of liquid water on Earth. Water is an important contributor to the dynamic nature of our planet, for without water the mantle would be more viscous and, in the absence of hydrous subduction, the continents unlikely (Rosing et al., 2006). So, it would appear that life has had a major influence not just on the Earth's atmosphere, as proposed by lovelock, but on the solid Earth as well.

What is less certain, but important to ponder (see e.g. Nisbet, 2002) is whether the Earth could have maintained its equable climate purely by inorganic means. What would an inorganic Earth look like? Nisbet (2002) has suggested a sluggish form of carbon management.

If CO_2-drawdown was in excess of volcanic CO_2 production then the Earth's temperature would fall, leading to a temporarily frozen Earth. At this point CO_2-drawdown would cease and volcanic outgassing would return the Earth to warmer temperatures. Alternatively, huge volcanic outpourings of CO_2 – for there is plenty of carbon in the mantle – might otherwise lead to a greenhouse Earth. In this case it is uncertain as to whether an inorganic carbon cycle would control surface temperatures, or whether the Earth would become a runaway Venusian greenhouse.

A more complex ocean–atmosphere cycle, discussed in Chapter 5, involves the elements phosphorus, sulfur, iron, and oxygen (Fig. 6.9). At low oxygen levels, typical of the Archaean, the lack of sulfide weathering results in a low-sulfate ocean, which permits high levels of dissolved Fe(II). The oxidation and precipitation of the iron would lead to low P-levels in the oceans, for the phosphorus is adsorbed onto the iron oxides. At about 2.3–2.4 Ga there were higher oxygen levels, which permitted higher concentrations of sulfate in the ocean, so that iron was removed from the ocean as a sulfide, releasing phosphorus to act as a nutrient. Increased phosphorus led to greater oxygen productivity and, through positive feedback led to an increase in oxygen levels.

What is not clear, however, is the precise cause of the rise in oxygen at between 2.3 and 2.4 Ga. It was not simply an increase in cyanobacterial oxygen production, because cyanobacteria had already been in existence for several hundred million years. Another, inorganic trigger is required – related perhaps to a change in the proportion of reducing volcanic gases produced, or to the way in which oxygen was utilized in oxdizing crustal rocks (Catling & Claire, 2005).

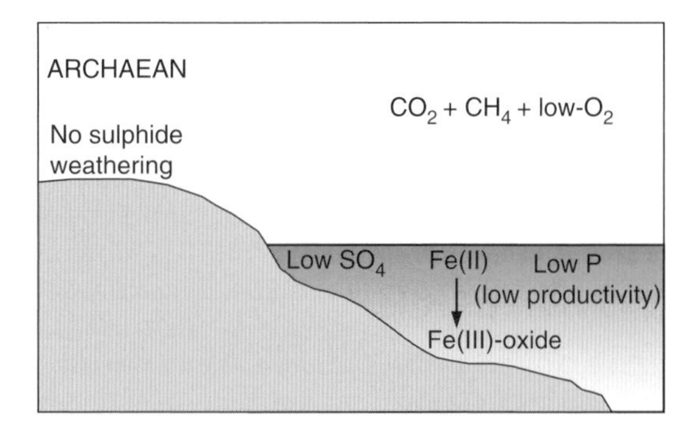

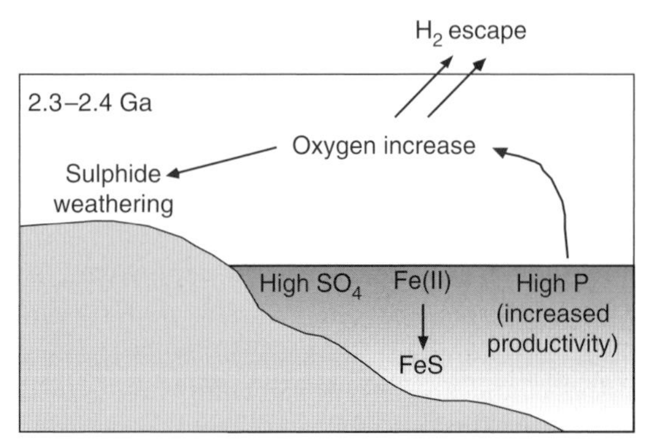

FIGURE 6.9 Atmosphere ocean interaction at the end of the Archaean as proposed by Catling et al. (2001) and Bjerrum and Canfield (2002, 2004).

So, what of Gaia? Our knowledge from the modern Earth carbon cycle suggests that inorganic processes could (probably) have maintained Earth's temperature within the bounds of 0–100°C. However, evidence from the rise of oxygen over time, displays a much more complex process, one which reflects the intricate interaction between organic and inorganic Earth processes.

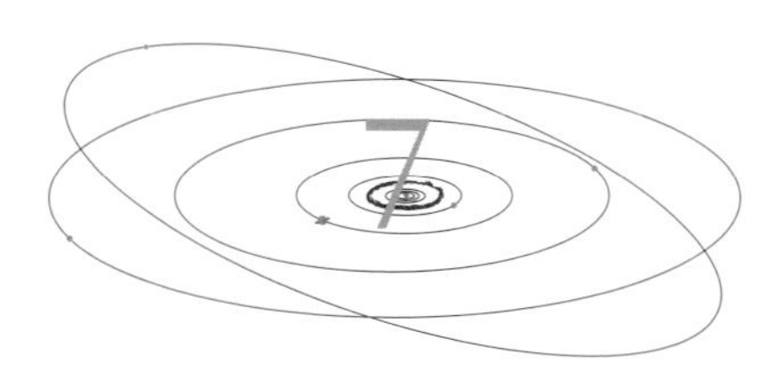

POST-SCRIPT

Much of the contemporary discussion on Earth systems is focused on the relatively recent geological past and on the Earth's surface environments. This has tended to keep the discussion of Earth systems within only a few sub-disciplines of the Earth Sciences and away from 'traditional hard-rock geology.' My goal, in this book, has been to show that if we consider the Earth on a long time scale, then Earth interactions involve all the major Earth reservoirs, including the solid Earth and the deep Earth. It has been my purpose here to draw attention to these interactions and to show that the solid Earth is an integral part of the Earth System which cannot be ignored, particularly in a discussion on the early Earth. Hence the focus of this book has been different from other Earth system science texts.

However, the reader will have detected that a detailed systems approach to the early Earth is not yet highly developed, especially when compared to the similar approach to the modern Earth. At best we are able to develop input–output models for the major Earth reservoirs in the Archaean. Only now are there attempts to begin to quantify these variables – see for example the model of Claire et al. (2005) for the rise of atmospheric oxygen.

In 2005, the journal *Science* celebrated its 125th anniversary by devoting an issue to the subject of "What don't we know?" Such is the thrust of modern science. However, it is important not to forget what we do know. So, it has also been my aim here to show that we actually know a great deal about the early Earth. Even though there are gaps in the geological record and ambiguities of interpretation in some difficult geological terrains, we certainly know enough now to map out a detailed sequence of Earth events since accretion and we broadly understand the major processes operating in the early Earth. In this context, a particularly important result is the growing recognition that many of the significant events in the history of our planet – the formation of the oceans, the core, the separation of the Moon, and the creation of the first basaltic crust – all took place within a few tens of millions of years of the formation of the solar system.

We also know that the Earth reservoirs have changed in composition over time. Such changes have been documented in this book. See for example – the isotopic evolution of the mantle (Chapter 3, Section 3.2.3), the secular evolution of the continental crust (Chapter 5, Section 5.3), the evolution of the composition of the atmosphere (Chapter 5, Section 5.3) and oceans (Table 5.5). Secular change in the biosphere, a process which we otherwise call evolution, is discussed in Chapter 6. Charting these changes and identifying the precise character of the systems of the early Earth is a task which is well underway.

Ahead lies the task of beginning to *quantify* interactions between the early Earth reservoirs.

REFERENCES

Abbott, A. and Mooney, W., 1995. The structural and geochemical evolution of the continental crust: support for the oceanic plateau model of continental growth. *Rev. Geophys., Suppl.*, **33**, 231–42.

Abbott, A., Sparks, D., Herzberg, C., Mooney, W., Nikishin, A., and Zhamg, Y.S., 2000. Quantifying Precambrian crustal extraction: the root is the answer. *Tectonophysics*, **322**, 163–90.

Abbott, D., Burgess, L., and Longhi, J., 1994. An empirical thermal history of the upper mantle. *J. Geophys. Res.*, **99**, B13835–50.

Abe, Y., 1997. Thermal and chemical evolution of the terrestrial magma ocean. *Phys. Earth Planet. Interior*, **100**, 27–39.

Abers, G.A., van Keken, P.E., Kneller, E.A., Ferris, A., and Stachnik, J.C., 2006. The thermal structure of subduction zones constrained by seismic imaging: implications for slab dehydration and wedge flow. *Earth Planet. Sci. Lett.*, **241**, 387–97.

Abouchami, W., Boher, M., Michard, A., and Albarede, F., 1990. A major 2.1 Ga event of mafic magmatism in west Africa: an early stage of crustal accretion. *J. Geophys. Res.*, **95**, 17605–29.

Agee, C.B., 1997. Melting temperatures of the Allende meteorite: implications for a Hadean magma ocean. *Phys. Earth Planet. Interior*, **100**, 41–7.

Agee, C.B. and Walker, D., 1988. Mass balance and phase density constraints on early differentiation of chondritic mantle. *Earth Planet. Sci. Lett.*, **90**, 144–56.

Albarede, F., 1998a. The growth of the continental crust. *Tectonophysics*, **296**, 1–14.

Albarede, F., 1998b. Time-dependent models of U–Th–He and K–Ar evolution and the layering of mantle convection. *Chem. Geol.*, **145**, 413–29.

Albarede, F., 2003. *Geochemistry: An Introduction*. Cambridge University Press, Cambridge, 248 pp.

Albarede, F. and van der Hilst, R.D., 1999. New mantle convecting model may reconcile conflicting evidence. *Eos*, **45**, 535–9.

Alexander, C.M.O'D., Boss, A.P., and Carlson, R.W., 2001. The early evolution of the inner solar system: a meteoritic perspective. *Science*, **293**, 64–8.

Alfe, D., Gillan, M.J., and Price, G.D., 2002. Composition and temperature of the earth's core constrained by combining ab initio calculations and seismic data. *Earth Planet. Sci. Lett.*, **195**, 91–8.

Alibert, C., Norman, M.D., and McCulloch, M.T., 1994. An ancient Sm–Nd age for a ferroan noritic anorthosite clast from lunar breccia 67016. *Geochim. Cosmochim. Acta*, **58**, 2921–6.

Allegre, C.J. and Turcott, D.L., 1986. Implications of a two component marble-cake mantle. *Nature*, **323**, 123–7.

Allegre, C.J., Dupre, B., and Lewin, E., 1986b. Thorium/uranium ratio of the Earth. *Chem. Geol.*, **56**, 219–27.

Allegre, C.J., Poirier, J.-P., Humler, E., and Hofmann, A.W., 1995. The chemical composition of the Earth. *Earth Planet. Sci. Lett.*, **134**, 515–26.

Allegre, C.J., Staudacher, T., and Sarda, P., 1986a. Rare gas systematics: formation of the atmosphere, evolution and structure of the Earth's mantle. *Earth Planet. Sci. Lett.*, **81**, 127–50.

Amelin, Y., 2005. Meteorite phosphates show constant ^{176}Lu decay rate since 4557 million years ago. *Science*, **310**, 839–41.

Amelin, Y., Krot, A.N., Hutcheon, I.D., and Ulyanov, A.A., 2002. Lead isotopic ages of chondrules and calcium–aluminium-rich inclusions. *Science*, **297**, 1678–83.

Amelin, Y., Krot, A., and Twelker, E., 2004. Pb isotopic age of the CB chondrite Gujiba and the duration of the chondrule formation interval (Abstract). *Geochim. Cosmochim. Acta*, **68(Suppl.)**, A759.

Amelin, Y., Lee, D.C., and Halliday, A.N., 2000. Early–middle Archaean crustal evolution deduced from Lu–Hf and U–Pb isotopic studies of single zircon grains. *Geochim. Cosmochim. Acta*, **64**, 4205–25.

Amelin, Y., Lee, D.C., Halliday, A.N., and Pidgeon, R.T., 1999. Nature of the Earth's earliest crust from Hafnium isotopes in single detrital zircons. *Nature*, **399**, 252–5.

Anbar, A.D., 2001. Iron isotope biosignatures: promise and progress. *Eos*, **82**, 173–9.

Anders, E. and Grevesse, 1989. Abundances of the elements: meteoritic and solar. *Geochim. Cosmochim. Acta*, **53**, 197–214.

Anderson, D.L., 2000. The thermal state of the upper mantle; no role for mantle plumes. *Geophys. Res. Lett.*, **27**, 3623–6.

Anderson, D.L., 2003. The plume hypothesis. *Geoscientist*, **13**, 16–17.

Anderson, O.L. and Isaak, D.G., 2002. Another look at the core density deficit of Earth's outer core. *Phys. Earth Planet. Interior*, **131**, 19–27.

Appel, P.W.U., Fedo, C.M., Moorbath, S., and Myers, J.S., 1998. Recogniseable primary volcanic and sedimentary features in a low strain domain of the highly deformed, oldest known (ca 3.7–3.8 Gyr) Greenstone belt, Isua, West Greenland. *Terra Nova*, **10**, 57–62.

Appel, P.W.U., Moorbath, S., and Myers, J.S., 2003. *Isuasphaera isua* (Pflug) revisited. *Precambrian Res.*, **126**, 309–12.

Appel, P., Rollinson, H.R., and Touret, J., 2001. 3.8 Ga hydrothermal fluids from fluid inclusions in a volcanic breccia from the Isua Greenstone belt, west Greenland. *Precambrian Res.*, **112**, 27–49.

Arculus, R.J., 1994. Aspects of magma genesis in arcs. *Lithos*, **33**, 189–208.

Armstrong, R.L., 1968. A model for the evolution of strontium and lead isotopes in a dynamic earth. *Rev. Geophys.*, **6**, 175–99.

Arndt, N.T., 1994. Archaean komatiites. In: Condie, K.C. (ed) *Archaean Crustal Evolution. Developments in Precambrian Geology 11*. Elsevier, Amsterdam, pp. 11–44.

Arndt, N.T., 2003. Komatiites, kimberlites, and boninites, *J. Geophys. Res.*, **108(B6)**, 2293, doi:10.1029/2002JB002157.

Arndt, N.T., Albarede, F., and Nisbet, E.G., 1997. Mafic and ultramafic magmatism. In: De Wit, M.J. and Ashwaal, L.D. (eds) *Greenstone Belts*. Oxford, Clarendon Press, pp. 233–54.

Arndt, N.T., Ginibre, C., Chauvel, C., Albarede, F., Cheadle, M., Herzberg, C., Jenner, G., and Lahaye, Y., 1998. Were komatiites wet? *Geology*, **26**, 739–42.

Artymowicz, P., 1997. Beta Pictoris: an early solar system. *Ann. Rev. Earth Planet. Sci.*, **25**, 175–219.

Asimow, P.D., and Langmuir, C.H., 2003. The importance of water to oceanic mantle melting regimes. *Nature*, **421**, 815–20.

Asphaug, E., Agnor, C.B., and Williams, Q., 2006. Hit-and-run planetary collisions. *Nature*, **439**, 155–60.

Atherton, M.P. and Petford, N., 1993. Generation of sodium-rich magmas from newly underplated basaltic crust. *Nature*, **362**, 144–6.

Ayers, J., 1998. Trace element modelling of aqueous fluid–peridotite interaction in the mantle wedge of subduction zones. *Contrib. Mineral. Petrol.*, **132**, 390–404.

Azbel, I.Ya., Tolsikhin, I.N., Kramers, J.D., Pechernikova, V., and Vityazev, A.V., 1993. Core growth and siderophile element depletion of the mantle during homogeneous accretion. *Geochim. Cosmochim. Acta*, **57**, 2889–98.

Bada, J.L., 2004. How life began on Earth: a status report. *Earth Planet. Sci. Lett.*, **226**, 1–15.

Bahcall, N., Ostriker, J.P., and Perlmutter, S., 1999. The cosmic triangle: revealing the state of the Universe. *Science*, **284**, 1481–8.

Baker, J. and Jensen, K.K., 2004. Coupled ^{186}Os–^{187}Os enrichments in the Earth's mantle – core–mantle interactions or recycling of ferromanganese crusts and nodules? *Earth Planet. Sci. Lett.*, **220**, 277–86.

Baker, J., Bizzarro, M., Wittig, N., Connelly, J., and Haack, H., 2005. Early planetesimal melting from an age of 4.5662 Gyr for differentiated meteorites. *Nature*, **436**, 1127–31.

Ballentine, C.J., Marty, B., Sherwood Lollar, B., and Cassidy, M., 2005. Neon isotopes constrain convection and volatile origin in the Earth's mantle. *Nature*, **433**, 33–8.

Ballhaus, C. and Frost, B.R., 1994. The generation of oxidized CO_2-bearing basaltic melts from reduced CH_4-bearing upper mantle sources. *Geochim. Cosmochim. Acta*, **58**, 4931–40.

Barley, M., Bekker, A., and Krapez, B., 2005. Late Archaean to early Proterozoic global tectonics, environmental change and the rise of atmospheric oxygen. *Earth Planet. Sci. Lett.*, **238**, 156–71.

Barth, M.G., McDonough, W.F., and Rudnick, R.L., 2000. Tracking the budget of Nb and Ta in the continental crust. *Chem. Geol.*, **165**, 197–213.

Bau, M., 1999. Scavenging of dissolved yttrium and rare earths by precipitating iron oxyhydroxide: experimental evidence for Ce oxidation, Y–Ho fractionation and lanthanide tetrad effect. *Geochim. Cosmochim. Acta*, **63**, 67–77.

Beard, B.L., Johnson, C.M., Cox, L., Sun, H., Nealson, K.H., and Aguilar, C., 1999. Iron isotope biosignatures. *Science*, **285**, 1889–92.

Beaumont, V. and Robert, F., 1999. Nitrogen isotope ratios of kerogens in Precambrian cherts: a record of the evolution of atmospheric chemistry? *Precambrian Res.*, **96**, 63–82.

Becker, H., Horan, M.F., Walker, R.J., Gao, S., Lorand, J.-P., and Rudnick, R.L., 2006. Highly siderophile element composition of the primitive upper mantle: Constraints from new data on peridotite massifs and xenoliths. *Geochim. Cosmochim. Acta*, **70**, 4528–50.

Becker, T.W., Kellogg, J.B., and O'Connell, R.J., 1999. Thermal constraints on the survival of primitive blobs in the lower mantle. *Earth Planet. Sci Lett.*, **171**, 351–65.

Bekker, A., Holland, H.D., Wang, P.-L., Rumble, D., Stein, H.J., Hannah, J.L., Coetzee, L.L., and Beukes, N.J., 2004. Dating the rise of atmospheric oxygen. *Nature*, **427**, 117–20.

Bell, K., Blenkinsop, J., Cole, T.J.S., and Menagh, D.P., 1982. Evidence from Sr isotopes for long-lived heterogeneities in the upper mantle. *Nature*, **298**, 251–3.

Bennett, C.L. and 20 others, 2003. First year Wilkinson Microwave Anisotropy Probe (WMAP) observations: preliminary maps and basic results. *Astrophys. J. Suppl.*, **148**, 1–27.

Bennett, V.C., Nutman, A.P., and Esat, T.M., 2002. Constraints on mantle evolution from $^{187}Os/^{188}Os$ isotopic compositions from Archaean ultramafic rocks from southern west Greenland (3.8 Ga) and western Australia (3.46 Ga). *Geochim. Cosmochim. Acta*, **66**, 2615–30.

Bennett, V.C., Nutman, A.P., and McCulloch, M.T., 1993. Nd isotopic evidence for transient, highly depleted mantle reservoirs in the early history of the earth. *Earth Planet. Sci. Lett.*, **119**, 299–317.

Benz, W. and Cameron, A.G.W., 1990. Terrestrial effects of the giant impact. In: Newsom, H.E. and Jones, J.H. (eds) *Origin of the Earth.* Oxford University Press, Oxford, pp. 61–7.

Bercovici, D. and Kurato, S.-I., 2003. Whole-mantle convection and the transition zone water filter. *Nature*, **425**, 39–44.

Berg, H.C., 2000. Motile behavior of bacteria. *Phys. Today*, **53**, 24–9.

Berger, M. and Rollinson, H.R., 1997. Isotopic and geochemical evidence for crust–mantle interaction during late Archaean crustal growth. *Geochim. Cosmochim. Acta*, **61**, 4809–29.

Beukes, N.J., Dorland, H., Gutzmer, J., Nedachi, M., and Ohmoto, H., 2002. Tropical laterites, life on land, and the history of atmospheric oxygen in the Palaeoproterozoic. *Geology*, **30**, 491–4.

Bickle, M.J., Nisbet, E.G., and Martin, A., 1994. Archaean greenstone belts are not oceanic crust. *J. Geol.*, **102**, 121–38.

Biellmann, C., Gillet, P., Guyot, F., Peyronneau, P., and Reynard, B., 1993. Experimental evidence for carbonate stability in the lower mantle. *Earth Planet. Sci. Lett.*, **118**, 31–41.

Birch, F., 1952. Elasticity and constitution of the earth's interior. *J. Geophys. Res.*, **57**, 227–86.

Bizzarro, M., Baker, J.A., and Haack, H., 2004. Mg isotope evidence for contemporaneous formation of chondrules and refractory inclusions. *Nature*, **431**, 275–8.

Bizzarro, M., Baker, J.A., Haack, H., Ulfbeck, D., and Rosing, M., 2003. Early history of earth's crust–mantle system inferred from hafnium isotopes in chondrites. *Nature*, **421**, 931–3.

Bizzarro, M., Simonetti, A., Stevenson, R.K., and David, J., 2002. Hf isotope evidence for a hidden mantle reservoir. *Geology*, **30**, 771–4.

Bjerrum, C.J. and Canfield, D.E., 2002. Ocean productivity before about 1.9 Gyr ago limited by phosphorus adsorption onto iron oxides. *Nature*, **417**, 159–62.

Bjerrum, C.J. and Canfield, D.E., 2004. New insights into the burial history of organic carbon on the early Earth. *Geochem. Geophys. Geosyst.*, **5**, doi 10:1029/2004GC000713.

Bjornerud, M.G. and Austrheim, H., 2004. Inhibited eclogite formation: the key to the rapid growth of strong and buoyant Archaean continental crust. *Geology*, **32**, 765–8.

Blake, T.S., Buick, R., Brown, S.J.A., and Barley, M.E., 2004. Geochronology of a late Archaean flood basalt province in the Pilbara Craton, Australia: constraints on basin evolution, volcanic and sedimentary accumulation, and continental drift rates. *Precambrian Res.*, **133**, 143–73.

Blenkinsop, T.G., Fedo, C.M., Bickle, M.J., Eriksson, K.A., Martin, A., Nisbet, E.G., and Wilson, J.F., 1993. An ensialic origin for the Ngezi group, Belingwe greenstone belt, Zimbabwe. *Geology*, **21**, 1135–8.

Blichert-Toft, J. and Albarede, F., 1997. The Lu–Hf isotope geochemistry of chondrites and the evolution of the mantle–crust system. *Earth Planet. Sci. Lett.*, **148**, 243–58.

Blichert-Toft, J. and Arndt, N.T., 1999. Hf isotope compositions of komatiites. *Earth Planet. Sci. Lett.*, **171**, 439–51.

Blichert-Toft, J. and Frei, R., 2001. Complex Sm–Nd and Lu–Hf isotope systematics in metamorphic garnets from Isua supracrustal belt, west Greenland. *Geochim. Cosmochim. Acta*, **65**, 3177–87.

Boher, M., Abouchami, W., Michard, A., Albarede, F., and Arndt, N., 1992. Crustal growth in west Africa at 2.1 Ga. *J. Geophys. Res.*, **97**, 345–69.

Bolfan-Casanova, N., Kepler, H., and Rubie, D.C., 2000. Water partitioning between nominally anhydrous minerals in the MgO–SiO_2–H_2O system up to 24 GPa: implications for the distribution of water in the Earth's mantle. *Earth Planet. Sci. Lett.*, **182**, 209–21.

Bolhar, R., Kamber, B.S., Moorbath, S., Fedo, C.M., and Whitehouse, M.J., 2004. Characterisation of early Archaean chemical sediments by trace element signatures. *Earth Planet. Sci. Lett.*, **222**, 43–60.

Bolhar, R., Kamber, B.S., Moorbath, S., Whitehouse, M.J., and Collerson, K.D., 2005a. Chemical characterization of earth's most ancient clastic sediments from the Isua greenstone belt, southern west Greenland. *Geochim. Cosmochim. Acta*, **69**, 1555–73.

Bolhar, R., van Kranendonk, M.J., and Kamber, B.S., 2005b. A trace element study of siderite–jasper banded iron formation in the 3.45 Warrawoona Group, Pilbara Craton – formation from hydrothermal fluids and shallow seawater. *Precambrian Res.*, **137**, 93–114.

Booker, J.R., Favetto, A., and Pomposiello, M.C., 2004. Low electrical resistivity associated with plunging of the Nazca flat slab beneath Argentina. *Nature*, **429**, 399–403.

Bosak, T. and Newman, D.K., 2003. Microbial nucleation of calcium carbonate in the Precambrian. *Geology*, **31**, 577–80.

Boss, A.P., 1998. Temperatures in protoplanetary disks. *Ann. Rev. Earth Planet. Sci.*, **26**, 53–80.

Boss, A.P., 2000. Massive planet formation through disk instability (extended abstract), In: Kessel-Deynet, O. and Burkert, A. (eds) *Ringberg Meeting: Star Formation 2000*, Max-Planck-Institut für Astronomie, Heidelberg, Germany.

Bowins, R.J. and Crocket, J.H., 1994. Sulphur and carbon isotopes in Archaean banded iron formations: implications for sulphur sources. *Chem. Geol.*, **111**, 307–23.

Bown, J.W. and White, R.S., 1994. Variation with spreading rate of oceanic crustal thickness and geochemistry. *Earth Planet. Sci. Lett.*, **121**, 435–49.

Bowring, S.A. and Housh, T.B., 1995. The earth's early evolution. *Science*, **269**, 1535–40.

Bowring, S.A. and Williams, I.S., 1999. Priscoan (4.00–4.03 Ga) orthogneisses from northwestern Canada. *Contrib. Mineral. Petrol.*, **134**, 3–16.

Bowring, S.A., Housh, T.B., and Isachsen, C.E., 1990. The Acasta Gneisses: remnant of Earth's early crust. In: Newsom, H.E. and Jones, J.H. (eds) *Origin of the Earth*, Oxford University Press, Oxford, pp. 319–43.

Bowring, S.A., Williams, I.S., and Compston, W. 1989. 3.96 Ga gneisses from the Slave Province, Northwest Territories, Canada. *Geology*, **17**, 971–5.

Boyd, F.R., 1973. A pyroxene geotherm. *Geochim. Cosmochim. Acta*, **37**, 2533–8.

Boyd, F.R., 1989. Compositional differences between oceanic and cratonic lithosphere. *Earth Planet. Sci. Lett.*, **96**, 15–26.

Boyd, F.R. and Nixon, P.H., 1975. Origins of the ultramafic nodules from some kimberlites of northern Lesotho and the Monastery Mine, South Africa. *Phys. Chem. Earth*, **9**, 431–54.

Boyd, O.S., Jones, C.H., and Sheehan, A.F., 2004. Foundering lithosphere imaged beneath the southern Sierra Nevada, California, USA. *Science*, **305**, 660–2.

Boyd, S.R., 2001. Ammonium as a biomarker in Precambrian metasediments. *Precambrian Res.*, **108**, 159–73.

Boyet, M. and Carlson, R.W., 2005. ^{142}Nd evidence for early (>4.53 Ga) global differentiation of the silicate Earth. *Science*, **309**, 576–81.

Boyet, M., Blichert-Toft, J., Rosing, M., Storey, M., Telouk, P., and Albarede, F., 2003. ^{142}Nd evidence for early earth differentiation. *Earth Planet. Sci. Lett.*, **214**, 427–42.

Brandon, A.D., Walker, R.J., Puchtel, I.S., Becker, H., Humayun, M., and Reveillon, S., 2003. ^{186}Os–^{187}Os systematics of Gorgona Island komatiites: implications for early growth of the inner core. *Earth Planet. Sci. Lett.*, **206**, 411–26.

Brasier, M.D., Green, O.R., Jephcoat, A.P., Kleppe, A.K., van Kranendonk, M.J., Lindsay, J.F., Steele, A., and Grassineau, N., 2002. Questioning the evidence for Earth's oldest fossils. *Nature*, **416**, 76–81.

Brasier, M.D., Green, O.R., Lindsay, J.F., McLoughlin, N., Steele, A., and Stoakes, C., 2005.

Critical testing of earth's oldest putative fossil assemblage from the 3.5 Ga Apex chert, Chinaman Creek, western Australia. *Precambrian Res.*, **140**, 55–102.

Braun, M.G. and Kelemen, P.B., 2002. Dunite distribution in the Oman ophiolite: implications for melt flux through porous dunite conduits. *Geochem. Geophys. Geosyst.*, **3**, 8603, doi:0.1029/2001GC000289.

Brochier, C. and Philippe, H., 2002. Phylogeny: an non-thermophilic ancestor for bacteria. *Nature*, **417**, 244.

Brocks, J.J. and Summons, R.E., 2003. Sedimentary hydrocarbons, biomarkers for early life. *Treatise Geochem.*, **8**, 63–115.

Brocks, J.J., Logan, G.A., Buick, R., and Summons, R.E., 1999. Archaean molecular fossils and the early rise of Eukaryotes. *Science*, **285**, 1033–6.

Brophy, J.G. and Marsh, B.D., 1986. On the origin of high-alumina arc basalt and the mechanics of melt extraction. *J. Petrol.*, **27**, 763–89.

Brown, G.C. and Mussett, A., 1993. *The Inaccessible Earth*, 2nd edition. Chapman and Hall, London, 276 pp.

Brown, L., Klein, J., Middleton, R., Sacks, I.S., and Tera, F., 1982. ^{10}Be in island-arc volcanoes and implications for subduction. *Nature*, **299**, 718–20.

Bruland, K.W. and Lohan, M.C., 2003. Controls of trace metals in seawater. *Treatise Geochem.* (Holland, H.D. and Turekian, K.K., eds), **6**, 23–47.

Buck, S.G. and Minter, W.E.L., 1985. Placer formation by fluvial degradation of an alluvial placer fan sequence: the Proterozoic Carbon leader placer, Witwatersrand Supergroup, South Africa. *J. Geol. Soc. Lond.*, **142**, 757–64.

Byndzia, L.T., Wood, B.J., and Dick, H.J.B., 1989. The oxidation state of the earth's sub-oceanic mantle from oxygen thermobarometry of abyssal spinel peridotites. *Nature*, **341**, 526–7.

Cairns-Smith, A.G., 1971. *The Life Puzzle: On Crystals and Organisms and On the Possibility of a Crystal as an Ancestor*. University of Toronto Press, Toronto, 165 pp.

Cameron, A.G.W., 1995. The first ten million years of the solar nebula. *Meteoritics*, **30**, 133–61.

Campbell, I.H. and Griffiths, R.W., 1993. The evolution of the mantle's chemical structure. *Lithos*, **30**, 389–99.

Campbell, I.H. and Taylor, S.R., 1983. No water, no granites – no oceans, no continents. *Geophys. Res. Lett.*, **10**, 1061–64.

Canfield, D.E., 2004. The evolution of the Earth surface sulfur reservoir. *Amer. J. Sci.*, **304**, 839–61.

Canfield, D.E., 2005. The early history of atmospheric oxygen: homage to Robert M. Garrels. *Ann. Rev. Earth Planet. Sci.*, **33**, 1–36.

Canil, D., 1997. Vanadium in peridotites, mantle redox and tectonic environments: Archaean to present. *Earth Planet. Sci. Lett.*, **195**, 75–90.

Canil, D., 1999. Vanadium partitioning between orthopyroxene, spinel and silicate melt and the redox states of mantle source regions for primary magmas. *Geochim. Cosmochim. Acta*, **63**, 557–72.

Canup, R.M., 2004. Simulations of a late lunar-forming impact. *Icarus*, **168**, 433–56.

Canup, R.M. and Asphaug, E., 2001. Origin of the Moon in a giant impact near the end of earth formation. *Nature*, **412**, 708–12.

Carlson, R.W., 1988. Mantle structure: layer cake or plum pudding? *Nature*, **334**, 380–1.

Carlson, R.W., 2005. Application of the Pt–Re–Os isotopic systems to mantle geochemistry and geochronology. *Lithos*, **82**, 249–72.

Caro, G., Bourdon, B., Birck, J.-L., and Moorbath, S., 2003. ^{146}Sm–^{142}Nd evidence from Isua metamorphosed sediments for the early differentiation of the earth's mantle. *Nature*, **423**, 428–32.

Caro, G., Bourdon, B., Birck, J.-L., and Moorbath, S., 2006. High precision ^{142}Nd/^{144}Nd measurements in terrestrial rocks: constraints on the early differentiation of the Earth's mantle. *Geochim. Cosmochim. Acta*, **70**, 164–91.

Cartigny, P. and Adler, M., 2003. A comment on "The nitrogen record of crust–mantle interaction and mantle convection from Archaean to present" by B. Marty and N. Dauphas [*Earth Planet. Sci. Lett.*, **206** (2003) 397–410]. *Earth Planet. Sci. Lett.*, **216**, 425–32.

Catling, D.C. and Claire, M.W., 2005. How Earth's atmosphere evolved to an oxic state: a status report. *Earth Planet. Sci. Lett.*, **237**, 1–20.

Catling, D.C., Zahnle, K.J., and McKay, C.P., 2001. Biogenic methane, hydrogen escape, and the irreversible oxidation of early Earth. *Science*, **293**, 839–43.

Catuneanu, O., 2001. Flexural partitioning of the late Archaean Witwatersrand foreland system, South Africa. *J. Sediment. Geol.*, **141**, 95–112.

Cavosie, A.J., Valley, J.W., Wilde, S.A., and EIMF, 2005. Magmatic δ^{18}O in 4400–3900 Ma detrital zircons: a record of the alteration and recycling of crust in the early Archaean. *Earth Planet. Sci. Lett.*, **235**, 663–81.

Cayrel, R., Hill, V., Beers, T.C., Barbuy, B., Spite, M., Spite, F., Plez, B., Andersen, J., Bonifacio, P., Francois, P., Molaro, P., Nordstrom, B., and

Primas, F., 2001. Measurement of stellar age from uranium decay. *Nature*, **409**, 691–2.
Chabot, N.L. and Agee, C.B., 2003. Core formation in the Earth and Moon: new experimental constraints from V, Cr and Mn. *Geochim. Cosmochim. Acta*, **67**, 2077–91.
Chabot, N.L., Draper, D.S., and Agee, C.B., 2005. Conditions of core formation in the Earth: constraints from nickel and cobalt partitioning. *Geochim. Cosmochim. Acta*, **69**, 2141–51.
Christensen, N.I. and Mooney, W.D., 1995. Seismic velocity structure and composition of the continental crust: a global view. *J. Geophys. Res.*, **100**, 9761–88.
Christensen, P.R. and 11 others, 2005. Evidence for magmatic evolution and diversity on Mars from infrared observations. *Nature*, **436**, 504–9.
Christensen, U.R., 1985. Thermal evolution models for the Earth. *J. Geophys. Res.*, **90**, 2995–3008.
Chyba, C.F., 1987. The cometary contribution to the oceans of the primitive Earth. *Nature*, **330**, 632–5.
Chyba, C.F., 1990. Impact delivery and erosion of planetary oceans in the early inner solar system. *Nature*, **343**, 129–33.
Chyba, C.F., 1993. The violent environment of the origin of life: progress and uncertainties. *Geochim. Cosmochim. Acta*, **57**, 3351–8.
Claire, M.W., Catling, D.C., and Zahnle, K.J., 2005. Modeling the rise of oxygen. *GSA Speciality Meetings Abstr. Programs*, **1**, 40.
Class, C. and Goldstein, S.L., 2005. Evolution of helium isotopes in the Earth's mantle. *Nature*, **436**, 1107–12.
Clayton, R.N. and Mayeda, T.K., 1999. Oxygen isotope studies of carbonaceous chondrite. *Geochim. Cosmochim. Acta*, **63**, 2089–104.
Clayton, R.N., Onuma, N., and Mayeda, T.K., 1976. A classification of meteorites based on oxygen isotopes. *Earth Planet. Sci. Lett.*, **30**, 10–18.
Cloud, P., 1972. A working model of the primitive earth. *Amer. J. Sci.*, **272**, 537–48.
Cody, G.D., Boctor, N.Z., Brandes, J.A., Filley, T.R., Hazen, R.M., and Yoder, H.S., 2004. Assaying the catalytic potential of transition metal sulphides for abiotic carbon fixation. *Geochim. Cosmochim. Acta*, **68**, 2185–96.
Cohen, B.A., Hewins, R.H., and Alexander, C.M.O'D., 2004. The formation of chondrules by open system melting of nebular condensates. *Geochim. Cosmochim. Acta*, **68**, 1661–75.
Cohen, B.A., Swindle, T.D., and Kring, D.A., 2000. Support for the lunar cataclysm hypothesis from Lunar meteorite impact melt ages. *Science*, **290**, 1754–6.
Collerson, K.D. and Kamber, B., 1999. Evolution of the continents and the atmosphere inferred from Th–U–Nb systematics of the depleted mantle. *Science*, **283**, 1519–22.
Collerson, K.D., Campbell, L.M., Weaver, B.L., and Palacz, Z.A., 1991. Evidence for extreme mantle fractionation in early Archaean ultramafic rocks from northern Labrador. *Nature*, **349**, 209–14.
Collerson, K.D., Hapugoda, S., Kamber, B.S., and Williams, Q., 2000. Rocks from the mantle transition zone: majorite-bearing xenoliths from Malaita, southwest Pacific. *Science*, **288**, 1215–23.
Compston, W. and Pidgeon, R.T., 1986. Jack Hills, evidence of more very old detrital zircons in western Australia. *Nature*, **321**, 766–9.
Compston, W., Williams, I.S., and Clement, S.W.J., 1982. U–Pb ages with single zircon crystals using a sensitive high mass-resolution ion microprobe. *30th Am. Soc. Mass Spectrom. Conf.*, Honolulu (*Am. Soc. Mass Spectrom.*), pp. 593–5.
Condie, K.C., 1997. *Plate Tectonics and Crustal Evolution*, 4th edition. Butterworth-Heinemann, London, 282 pp.
Condie, K.C., 1998. Episodic continental growth and supercontinents: a mantle avalanche connection. *Earth Planet. Sci. Lett.*, **163**, 97–108.
Condie, K.C., 2000. Episodic continental growth models: afterthoughts and extensions. *Tectonophysics*, **322**, 153–62.
Condie, K.C., 2005. High field strength element ratios in Archaean basalts: a window into evolving sources of mantle plumes. *Lithos*, **79**, 491–504.
Coogan, L.A., Banks, G.J., Gillis, K.M., MacLeod, C.J., and Pearce, J.A., 2003. Hidden melting signatures recorded in the Troodos ophiolite plutonic suite: evidence for widespread generation of depleted melts and intra-crustal aggregation. *Contrib. Mineral. Petrol.*, **144**, 484–505.
Cooper, G., Kimmich, N., Belisle, W., Sarinana, Brabham, K., and Garrel, L., 2001. Carbonaceous meteorites as a source of sugar-related organic compounds for the early earth. *Nature*, **414**, 879–83.
Corfu, F., 1987. Inverse age stratification in the Archaean crust of the Superior Province: evidence for infra- and subcrustal accretion from high resolution U–Pb zircon and monazite ages. *Precambrian Res.*, **36**, 259–75.
Corgne, A., Liebske, C., Wood, B.J., Rubie, D.C., and Frost, D.J., 2005. Silicate perovskite-melt partitioning of trace elements and geochemical signature of a deep perovskite reservoir. *Geochim. Cosmochim. Acta*, **69**, 485–96.

Courtillot, V., Davaille, A., Besse, J., and Stock, J., 2003. Three distinct types of hotspots in the Earth's mantle. *Earth Planet. Sci. Lett.*, **205**, 295–308.

Coward, M.P., Spencer, R.M., and Spencer, C.E., 1995. Development of the Witwatersrand basin, South Africa. In: Coward, M.P. and Reis, A.C. (eds) *Early Precambrian Processes. Spl. Publ. Geol. Soc. Lond.*, **95**, 243–69.

Cowen, J.P., Giovannoni, S.J., Kenig, F., Johnson, H.P., Butterfield, D., Rappe, M.S., Hutnak, M., and Lam, P., 2003. Fluids from aging ocean crust that support microbial life. *Science*, **299**, 120–3.

Crawford, A.J., Falloon, T.J., and Green, D.H., 1989. Classification, petrogenesis and tectonic setting of boninites. In: Crawford, A.J. (ed), *Boninites and Related Rocks*. Unwin Hyman, London, pp. 1–49.

Crick, F., 1981. *Life Itself: Its Origin and Nature*. Simon and Schuster, New York, 192 pp.

Crisp, J.A., 1984. Rates of magma emplacement and volcanic output. *J. Volcan. Geotherm. Res.*, **20**, 177–211.

Croal, L.R., Johnson, C.M., Beard, B.L., and Newman, D.K., 2004. Iron isotope fractionation by Fe(II)-oxidising photoautotrophic bacteria. *Geochim. Cosmochim. Acta*, **68**, 1227–42.

Dasgupta, R., Hirschmann, M.M., and Withers, A.C., 2004. Deep global cycling of carbon constrained by the solidus of anhydrous carbonated eclogite under upper mantle conditions. *Earth Planet. Sci. Lett.*, **227**, 73–85.

Dauphas, N. and Marty, B., 2002. Inference on the nature and mass of the Earth's late veneer from noble gases and metals. *J. Geophys. Res.*, **107**, doi: 10.1029/2001JE001617.

Dauphas, N., van Zuilen, M., Wadhwa, M., Davies, A.M., Marty, B., and Janney, P., 2004. Clues from Fe isotope variations on the origin of early Archaean BIFs from Greenland. *Science*, **306**, 2077–80.

Davidson, J.P., 1983. Lesser Antilles isotopic evidence of the role of subducted sediment in island arc magma genesis. *Nature*, **306**, 253–6.

Davies, G.F., 1993. Cooling of the core and mantle by plume and plate flows. *Geophys. J. Int.*, **115**, 132–46.

Davies, G.F., 1998. Topography: a robust constraint on mantle fluxes. *Chem. Geol.*, **145**, 479–89.

Davies, G.F., 1999. Geophysically constrained mantle mass flows and the ^{40}Ar budget: a degassed lower mantle? *Earth Planet. Sci. Lett.*, **166**, 149–62.

Davies, G.F., 2006. Gravitational depletion of the early Earth's upper mantle and the viability of early plate tectonics. *Earth Planet. Sci. Lett.*, **243**, 376–82.

De Bernadis, P. and 35 others, 2000. A flat universe from high resolution maps of the cosmic microwave background radiation. *Nature*, **404**, 955–9.

Deines, P., 2002. The carbon isotope geochemistry of mantle xenoliths. *Earth Sci. Rev.*, **58**, 247–78.

Delano, J.W., 2001. Redox history of the Earth's interior since ca 3900 Ma: implications for prebiotic molecules. *Origins Life Evol. Bios.*, **31**, 311–41.

De Ronde, C.E.J. and De Wit, M.J., 1994. Tectonic history of the Barberton greenstone belt, South Africa: 490 million years of Archaean crustal evolution. *Tectonics*, **13**, 983–1005.

Desrochers, J.-P., Hubert, C., Ludden, J.N., and Pilote, P., 1993. Accretion of Archaean oceanic plateau fragments in the Abitibi greenstone belt, Canada. *Geology*, **21**, 451–4.

De Wit, M.J. and Ashwal, L.D., 1995. Greenstone Belts: what are they? *S. Afr. J. Geol.*, **98**, 505–20.

De Wit, M.J. and Ashwal, L.D., 1997. *Greenstone Belts*. Oxford University Press, Oxford, 809 pp.

De Wit, M.J., Hart, R.A., and Hart, R.J., 1987. The Jamestown ophiolite complex, Barberton Mountain belt: a section through 3.5 Ga oceanic crust. *J. Afr. Earth Sci.*, **6**, 681–730.

Dirks, P.H.G.M., Jelsma, H.A., and Hofmann, A., 2002. Thrust-related accretion of an Archaean greenstone belt in the Midlands of Zimbabwe. *J. Struct. Geol.*, **24**, 1707–27.

Doolittle, W.F., 1998. A paradigm gets shifty. *Nature*, **392**, 15–16.

Doolittle, W.F., 1999. Phylogenetic classification and the universal tree. *Science*, **284**, 2124–8.

Drake, M.J. and Righter, K., 2002. Determining the composition of the Earth. *Nature*, **416**, 39–44.

Drummond, M.S. and Defant, M.J., 1990. A model for trondhjemite–tonalite–dacite genesis and crustal growth via slab melting: Archaean to modern comparisons. *J. Geophys. Res.*, **95**, 21503–21.

Ducea, M. and Saleeby, J., 1998. A case for delamination of the deep batholithic crust beneath the Sierra Nevada, California. *Int. Geol. Rev.*, **40**, 78–93.

Duck, L.J., Glikson, M., Golding, S.D., and Webb, R.E., 2004. Characteristics of organic matter in sediments from 3.24 Sulphur Springs, VHMS deposit, western Australia. *Geochim. Cosmochim. Acta*, **68(Suppl. 1)**, Goldschmidt Abstracts, A798.

Durrheim, R.J. and Mooney, W.D., 1991. Archaean and Proterozoic crustal evolution: evidence from crustal seismology. *Geology*, **19**, 606–9.

Durrheim, R.J. and Mooney, W.D., 1994. Evolution of the Precambrian lithosphere: seismological and geochemical constraints. *J. Geophys. Res.*, **99**, 15359–74.

Dziewonski, A.M. and Anderson, D.L., 1981. Preliminary reference Earth model. *Phys. Earth Planet. Interior*, **25**, 297–356.

Eglinton, G. and Pancost, R., 2004. Immortal molecules. *Geoscientist, Geol. Soc. Lond.*, **14**, 4–16.

Ellam, R.M. and Hawkesworth, C.J., 1988a. Elemental and isotopic variations in subduction related basalts: evidence for a three component model. *Contrib. Mineral. Petrol.*, **98**, 72–80.

Ellam, R.M. and Hawkesworth, C.J., 1988b. Is average continental crust generated at subduction zones? *Geology*, **16**, 314–17.

Elliott, T., Plank, T., Zindler, A., White, W., and Bourdon, B., 1997. Element transport from slab to volcanic front at the Mariana arc. *J. Geophys. Res.*, **B102**, 14991–5019.

Elliott, T., Zindler, A., and Bourdon, B., 1999. Exploring the kappa conundrum: the role of recycling in the lead isotope evolution of the mantle. *Earth Planet. Sci. Lett.*, **169**, 129–45.

Elthon, D., 1986. Comments on "Composition and depth of origin of primary mid-ocean ridge basalts," by D.C. Presnall, and J.D. Hoover. *Contrib. Mineral. Petrol.*, **94**, 253–6.

England, P. and Houseman, G., 1989. Extension during continental convergence with applications to the Tibetan Plateau. *J. Geophys. Res.*, **94**, 17561–579.

Eriksson, K.A., Krapez, B., and Fralick, P.W., 1994. Sedimentology of Archaean greenstone belts: signatures of tectonic evolution. *Earth Sci. Rev.*, **37**, 1–88.

Eriksson, K.A., Simpson, E.L., Master, S., and Henry, G., 2005. Neoarchaean (c. 2.58 Ga) halite clasts: implications for palaeoceanic chemistry. *J. Geol. Soc. Lond.*, **162**, 789–99.

Eriksson, P.G., Martins-Neto, M.A., Nelson, D.R., Aspler, L.B., Chiarenzelli, J.R., Catenuneanu, O., Sarker, S., Altermann, W., and Rautenbach, C.J. de W., 2001. An introduction to Precambrian basins: their characteristics and genesis. *Sediment. Geol.*, **141**, 1–35.

Ernst, R.E., Buchan, K.L., and Campbell, I.H., 2005. Frontiers in large igneous province research. *Lithos*, **79**, 271–97.

Evans, D.A., Beukes, N.J., and Kirschvink, J.L., 1997. Low-latitude glaciation in the palaeoproterozoic earth. *Nature*, **386**, 262–6.

Fagan, T.J., Krot, A.N., Keil, K., and Yurimoto, H., 2004. Oxygen isotopic evolution of amoeboid olivine aggregates in the reduced CV3 chondrites Efremovka, Vigarano and Leoville. *Geochim. Cosmochim. Acta*, **68**, 2591–611.

Farmer, J.D., 2000. Hydrothermal systems: doorways to early biosphere evolution. *GSA Today*, July, 1–9.

Farquhar, J. and Wing, B.A., 2003. Multiple sulfur isotopes and the evolution of the atmosphere. *Earth Planet. Sci. Lett.*, **213**, 1–13.

Farquhar, J., Bao, H., and Thiemens, M., 2000. Atmospheric influence of earth's earliest sulfur cycle. *Science*, **289**, 756–8.

Farquhar, J., Wing, B.A., McKeegan, K.D., Harris, J.W., Cartigny, P., and Thiemens, M., 2002. Mass-independent sulfur of inclusions in diamond and sulfur recycling on early earth. *Science*, **298**, 2369–72.

Fedo, C.M. and Whitehouse, M.J., 2002. Metasomatic origin of Quartz-pyroxene rock, Akilia, Greenland, and implications for Earth's earliest life. *Science*, **296**, 1448–52.

Fedo, C.M., Eriksson, K.A., and Krogstad, E., 1996. Geochemistry of shales from the Archaean (~3.0 Ga) Buhwa greenstone belt, Zimbabwe: implications for provenance and source-area weathering. *Geochim. Cosmochim. Acta*, **60**, 1751–63.

Fedo, C.M., Myers, J.S., and Appel, P.W.U., 2001. Depositional setting and palaeogeographic implications of earth's oldest supracrustal rocks, the >3.7 Ga Isua greenstone belt, West Greenland. *Sediment. Geol.*, **141**, 61–77.

Fedo, C.M., Whitehouse, M.J., and Kamber, B.S., 2006. Geological constraints on detecting the earliest life on earth: a perspective from the early Archaean (>3.7 Ga) of southwest Greenland. *Phil. Trans. Roy. Soc., B*, **361**, 851–67.

Ferris, J.P., 2002. Montmorillonite catalysis of 30–50 mer oligonucleotides: laboratory demonstration of potential steps in the origin of the RNA world. *Origins Life Evol. Bios.*, **32**, 311–32.

Fischer, T.P., Hilton, D.R., Zimmer, M.M., Shaw, A.M., Sharp, Z.D., amd Walker, J.A., 2002. Subduction and recycling of nitrogen along the central American margin. *Science*, **297**, 1154–7.

Flynn, G.J., Keller, L.P., Feser, M., Wirick, S., and Jacobsen, C., 2003. The origin of organic matter in the solar system: evidence from the interplanetary dust particles. *Geochim. Cosmochim. Acta*, **67**, 4791–806.

Foley, S., Tiepolo, M., and Vannucci, R., 2002. Growth of early continental crust controlled by melting of amphibolite in subduction zones. *Nature*, **417**, 837–40.

Forterre, P., 1995. Thermoreduction, a hypothesis for the origin of prokaryotes. *C. R. Acad. Sci. Paris; Sci de la vie*, **318**, 415–22.

Fortey, R., 1998. *Life: A Natural History of the First Four Billion Years of Life on Earth*. Alfred A Knopf, New York, 368 pp.

Foster, J.G., Lambert, D.D., Frick, L.R., and Maas, R., 1996. Re–Os isotopic evidence for genesis of Archaean nickel ores from uncontaminated komatiites. *Nature*, **382**, 703–6.

Foulger, G.R. and Natland, J.H., 2003. Is "hotspot" volcanism a consequence of plate tectonics? *Science*, **300**, 921–2.

Foulger, G.R., Natland, J.H., Presnall, D.C., and Anderson, D.L. (eds), 2005. *Plates, Plumes and Paradigms. Geol. Soc. Amer. Special Paper* **388**, 861 pp.

Foustoukos, D.I. and Seyfreid, W.E., 2004. Hydrocarbons in hydrothermal vent fluids: the role of chromium-bearing catalysts. *Science*, **302**, 1002–5.

Franck, S. and Bounamam, C., 2001. Global water cycle and Earth's thermal evolution. *J. Geodynam.*, **32**, 231–46.

Frei, R. and Jensen, B.T., 2003. Re–Os, Sm–Nd isotope and REE systematics on ultramafic rocks and pillow basalts from the Earth's oldest oceanic crustal fragments (Isua supracrustal belt and Ujaragssuit nunat area, W. Greenland). *Chem. Geol.*, **196**, 163–91.

Frei, R. and Rosing, M.T., 2001. The least radiogenic terrestrial leads: implications for the early Archaean crustal evolution and hydrothermal-metasomatic processes in the Isua supracrustal belt (west Greenland). *Chem. Geol.*, **181**, 47–66.

Frei, R. and Rosing, M.T., 2005. Search for traces of the late heavy bombardment on Earth – results from high precision chromium isotopes. *Earth Planet. Sci. Lett.*, **236**, 28–40.

Frei, R., Polat, A., and Meibom, A., 2004. The Hadean upper mantle conundrum: evidence for source depletion and enrichment from Sm–Nd, Re–Os and Pb-isotopic compositions in 3.71 Ga boninite-like metabasalts from the Isua Supracrustal Belt, Greenland. *Geochim. Cosmochim. Acta*, **68**, 1645–60.

Frey, F.A. and Prinz, M., 1978, Ultramafic inclusions from San Carlos, Arizona: petrological and geochemical data being on their petrogenesis. *Earth Planet. Sci. Lett.*, **38**, 129–76.

Friend, C.R.L. and Nutman, A.P., 2005. New pieces to the Archaean terrane jigsaw puzzle in the Nuk region, southern west Greenland: steps in transforming a simple insight into a complex regional tectonothermal model. *J. Geol. Soc. Lond.*, **162**, 147–62.

Friend, C.R.L., Bennett, V.C., and Nutmann, A.P., 2002. Abyssal peridotites >3,800 Ma from southern west Greenland: field relationships, petrography, geochronology, whole-rock and mineral chemistry of dunite and harzburgite inclusions in the Itsaq Gneiss Complex. *Contrib. Mineral. Petrol.*, **143**, 71–92.

Friend, C.R.L., Nutman, A.P., and McGregor, V.R., 1988. Late Archaean terrane accretion in the Godthab region, southern west Greenland. *Nature*, **335**, 535–8.

Frimmel, H.E., 1997. Detrital origin of hydrothermal Witwatersrand gold – a review. *Terra Nova*, **9**, 192–7.

Frimmel, H.E., 2005. Archaean atmospheric evolution: evidence from the Witwatersrand gold fields, South Africa. *Earth Sci. Rev.*, **70**, 1–46.

Fripp, R.E.P. and Jones, M.G., 1997. Sheeted intrusions and peridotite-gabbro assemblages in the Yilgarn Crato, western Australia: elements of Archaean ophiolites. In: De Wit, M.J. and Ashwaal, L.D. (eds), *Greenstone Belts*. Oxford University Press, Oxford, pp. 422–37.

Froude, D.O., Ireland, T.R., Kinny, P.D., Williams, I.S., Compston, W., Williams, I.R., and Myers, J.S., 1983. Ion microprobe identification of 4,100–4,200 Myr-old terrestrial zircons. *Nature*, **304**, 616–18.

Fukao, Y., Widiyantoro, S., and Obayashi, M., 2001. Stagnant slabs in the upper and lower mantle transition zone. *Rev. Geophys.*, **39**, 291–323.

Galer, S.J.G. and Goldstein, S.L., 1988. Early mantle depletion and thermal consequences. *Chem. Geol.*, **70**, 143.

Galer, S.J.G. and Goldstein, S.L., 1991. Early mantle differentiation and thermal consequences. *Geochim. Cosmochim. Acta*, **55**, 227–39.

Galer, S.J.G. and O'Nions, R.K., 1985. Residence time for thorium, uranium and lead in the mantle with implications for mantle convection. *Nature*, **316**, 778–82.

Gallet, S., Jahn, B.-M., van Vliet-lanoe, B., Dia, A., and Rossello, E., 1998. Loes geochemistry and its implications for particle origin and composition of the upper continental crust. *Earth Planet. Sci. Lett.*, **156**, 157–72.

Gangopadhyay, A. and Walker, R.J., 2003. Ro–Os systematics of the ca. 2.7-Ga komatiites from Alexo, Ontario, Canada. *Chem. Geol.*, **196**, 147–62.

Gao, S., Rudnick, R.L., Yuan, H.-L., Liu, X.-M., Liu, Y.-S., Xu, W.-L., Ling, W.-L., Ayers, J., Wang, X.-C., and Wang, Q.-H., 2004. Recycling lower continental crust in the North China Craton. *Nature*, **432**, 892–7.

Gao, S., Zhang, B.-R., Jin, Z.-M., Kern, H., Luo, T.-C., and Zhao, Z.-D., 1998. How mafic is the lower continental crust? *Earth Planet. Sci. Lett.*, **161**, 101–17.

Genda, H. and Abe, Y., 2005. Enhanced atmospheric loss on protoplanets at the giant impact phase in the presence of oceans. *Nature*, **433**, 842–4.

Gessmann, C.K. and Wood, B.J., 2002. Potassium in the Earth's core? *Earth Planet. Sci. Lett.*, **200**, 63–78.

Gessmann, C.K., Wood, B.J., Rubie, D.C., and Kilburn, M.R., 2001. Solubility of silicon in liquid metal at high pressure: implications for the composition of the Earth's core. *Earth Planet. Sci. Lett.*, **184**, 367–76.

Gold, D.J.C. and von Veh, M.W., 1995. Tectonic evolution of the late Archaean Pongola-Mozaan basin, South Africa. *J. Afr. Earth Sci.*, **21**, 200–12.

Goldstein, S.L., 1988. Decoupled evolution of Nd and Sr isotopes in the continental crust and the mantle. *Nature*, **336**, 733–8.

Gradstein, F.M. and too many others to count, 2005. *A Geologic Time Scale 2004*. Cambridge University Press, Cambridge, 589 pp.

Grady, M.M., 2000. *Catalogue of Meteorites*, 5th edition. Cambridge University Press, Cambridge, 696 pp.

Grady, M., 2003. Astrobiology: the search for extraterrestrial life beyond the Earth. *Geol. Today*, **19**, 99–102.

Graham, D.W., 2005. Neon illuminates the mantle. *Nature*, **433**, 25–6.

Grassineau, N.V., Nisbet, E.G., Bickel, M.J., Fowler, C.M.R., Lowry, D., Mattey, D.P., Abell, P., and Martin, A., 2001. Antiquity of the biological sulphur cycle: evidence from sulphur and carbon isotopes in 2700 million-year-old rocks of the Belingwe Belt, Zimbabwe. *Proc. Roy. Soc. Lond., B*, **268**, 113–19.

Grassineau, N.V., Nisbet, E.G., Fowler, C.M.R., Bickel, M.J., Lowry, D., Chapman, H.J., Mattey, D.P., Abell, P., Yong, J., and Martin, A., 2002. Stable isotopes in the Archaean Belingwe belt, Zimbabwe: evidence for a diverse microbial mat ecology. *Special Publ. Geol. Soc. Lond.* (Fowler, C.M.R. Ebinger, C.J., and Hawkesworth, C.J., eds), **199**, 309–28.

Graur, D. and Martin, W., 2004. Reading the entrails of chickens: molecular timescales of evolution and the illusion of precision. *Trends Genetics*, **20**, 80–6.

Gray, G.J., Lawrence, S.R., Kenyon, K., and Cornford, C., 1998. Nature and origin of "carbon" in the Archaean Witwatersrand Basin, South Africa. *J. Geol. Soc. Lond.*, **155**, 39–59.

Green, M.G., Sylvester, P.J., and Buick, R., 2000. Growth and recycling of early Archaean continental crust: geochemical evidence from Coonterunah and Warrawoona Groups, Pilbara craton, Australia. *Tectonophysics*, **322**, 69–88.

Griffin, W.L., O'Reilly, S.Y., Abe, N., Aulbach, S., Davies, R.M., Pearson, N.J., Doyle, B.J., and Kivi, K., 2003. The origin and evolution of Archaean lithospheric mantle. *Precambrian Res.*, **127**, 19–41.

Grossman, L., 1972. Condensation in the primitive solar nebula. *Geochim. Cosmochim. Acta*, **36**, 597–619.

Grotzinger, J.P. and Knoll, A.H., 1999. Stromatolites in Precambrian carbonates: evolutionary mileposts or environmental dipsticks? *Ann. Rev. Earth Planet. Sci.*, **27**, 313–58.

Grove, T.L. and Parman, S.W., 2004. Thermal evolution of the earth as recorded by komatiites. *Earth Planet. Sci. Lett.*, **219**, 173–87.

Groves, D.I., Korkiaaakoski, E.A., McNaughton, N.J., Lesher, C.M., and Cowden, A., 1986. Thermal erosion by komatiites at Kambalda, western Australia and the genesis of nickel ores. *Nature*, **319**, 136–9.

Guan, Y., McKeegan, K.D., and MacPherson, G.J., 2000. Oxygen isotopes in calcium–aluminium inclusions from enstatite chondrites: new evidence for a single CAI source in the solar nebula. *Earth Planet. Sci. Lett.*, **181**, 271–7.

Gust, D.A. and Perfit, M.R., 1987. Phase relations of a high-Mg basalt from the Aleutian Island Arc: implications for primary arc basalts and high-Al basalts. *Contrib. Mineral. Petrol.*, **97**, 7–18.

Gutzmer, J. and Beukes, N.J., 1998. Earliest laterites and possible evidence for terrestrial vegetation in the early Proterozoic. *Geology*, **26**, 263–6.

Haack, H., Bizzarro, M., Lundgaard, K.L., and Baker, J.A., 2004. Timescale of chondrule formation, accretion and differentiation – the tale of Al-26 (Abstract). *Geochim. Cosmochim. Acta*, **68 (Suppl.)**, A762.

Habicht, K.S., Gade, M., Thamdrup, B., Berg, P., and Canfield, D.E., 2002. Calibration of sulphate levels in the Archaean ocean. *Science*, **298**, 2372–4.

Halla, J., 2005. Late Archaean high-Mg granitoids (sanukitoids) in the southern Karelian domain, eastern Finland: Pb and Nd isotopic constraints on crust–mantle interactions. *Lithos*, **79**, 161–78.

Halliday, A.N., 2000. Terrestrial accretion rates and the origin of the moon. *Earth Planet. Sci. Lett.*, **176**, 17–30.

Halliday, A.N., 2004. Mixing, volatile loss and compositional change during impact-driven accretion of the Earth. *Nature*, **427**, 505–9.

Halliday, A.N. and Porcelli, D., 2001. In search of lost planets – the palaeocosmochemistry of the inner solar system. *Earth Planet. Sci. Lett.*, **192**, 545–59.

Hama, J. and Suito, K., 2001. Thermoelastic models of minerals and the composition of the earth's lower mantle. *Phys. Earth Planet. Interior*, **125**, 147–66.

Hammouda, T., 2003. High pressure melting of carbonated eclogite and experimental constraints on carbon recycling and storage in the mantle. *Earth Planet. Sci. Lett.*, **214**, 357–68.

Hardie, L.A., 2003. Secular variations in Precambrian seawater chemistry and the timing of Precambrian aragonite seas and calcite seas. *Geology*, **31**, 785–8.

Harper, C.L. and Jacobsen, S.B., 1992. Evidence from coupled ^{147}Sm–^{143}Nd and ^{146}Sm–^{142}Nd systematics for very early (4.5 Gyr) differentiation of the Earth's mantle. *Nature*, **360**, 728–32.

Harper, C.L. and Jacobsen, S.B., 1996. Evidence for ^{182}Hf in the early solar system and constraints on the timescale of terrestrial accretion and core formation. *Geochim. Cosmochim. Acta*, **60**, 1131–53.

Harper, G.D., 1985. Dismembered Archaean ophiolite, Wind River Mountains, Wyoming, USA. *Ofioliti*, **10**, 297–306.

Harrison, T.M., Blichert-Toft, J., Muller, W., Albarede, F., Holdern, P., and Mojzsis, S.J., 2005. Heterogeneous Hadean hafnium: evidence of continental crust at 4.4 to 4.5 Ga. *Science*, **310**, 1947–50.

Hart, S.R., 1988. Heterogeneous mantle domains: signatures, genesis and mixing chronologies, *Earth Planet. Sci. Lett.*, **90**, 273–96.

Hart, S.R. and Zindler, A., 1986. In search of a bulk earth composition. *Chem. Geol.*, **57**, 247–67.

Harte, B., 1983. Mantle peridotites and processes – the kimberlite sample. In: Hawkesworth, C.J. and Norry, M.J. (eds) *Continental Basalts and Mantle Xenoliths*. Shiva, Nantwich, pp. 46–91.

Harte, B. and Harris, J.W., 1994. Lower mantle associations preserved in diamonds. *Mineral. Mag.*, **58A**, 384–5.

Hartman, H. and Fedorov, A., 2002. The origin of the eukaryotic cell: a genomic investigation. *Proc. Natl. Acad. Sci. USA*, **98**, 14, 778–83.

Hartmann, W.K., Malin, M., McEwen, A., Carr, M., Soderblom, L., Thomas, P., Danielson, E., James, P., and Veverka, J., 1999. Evidence for recent volcanism on Mars from crater counts. *Nature*, **397**, 586–9.

Hauri, E., 2002. SIMS analysis of volatiles in silicate glasses, 2: isotopes and abundances in Hawaiian melt inclusions. *Chem. Geol.*, **183**, 115–41.

Hawkesworth, C.J. and Kemp, A., 2006. Using hafnium and oxygen isotopes to unravel the record of crustal evolution. *Chem. Geol.*, **226**, 144–62.

Hayes, J.M., 1994. Global methanotrophy at the Archaean–Proterozoic transition. In: Bengtson, S. (ed) *Early Life on Earth*. Columbia University Press, New York, pp. 220–36.

Hayes, J.M., 2001. Fractionation of the isotopes of carbon and hydrogen in biosynthetic processes. In: Valley, J.W. and Cole, D.R. (eds) *Stable Isotope Geochemistry, Reviews in Mineralogy and Geochemistry*. Mineralogical Society of America, Chantilly, VA, Vol. 43, pp. 225–78.

Hayes, J.M., 2004. Isotopic order, biogeochemical processes and earth history. *Geochim. Cosmochim. Acta*, **68**, 1691–700.

Hayman, P.C., Kopylova, M.G., and Kaminsky, F.V., 2005. Lower mantle diamonds from Rio Soriso (Juina area, Mato Grosso, Brazil). *Contrib. Mineral. Petrol.*, **149**, 430–45.

Helffrich, G. and Kaneshima, S., 2004. Seismological constraints on core composition from Fe–O–S liquid immiscibility. *Science*, **306**, 2239–42.

Helffrich, G.R. and Wood, B.J., 2001. The earth's mantle. *Nature*, **412**, 501–7.

Helmstedt, H., Padgham, W.A., and Brophy, J.A., 1986. Multiple dykes in the Lower Kam Group, Yellowknife greenstone belt: evidence for Archaean sea-floor spreading. *Geology*, **14**, 562–6.

Henry, P., Stevenson, R.K., and Gariepy, C., 1998, Late Archaean mantle composition and crustal growth in the western Superior Province of Canada: neodymium and lead isotopic evidence from the Wawa, Quetico and Wabigood subprovinces. *Geochim. Cosmochim. Acta*, **62**, 143–57.

Herzberg, C., 1992. Depth and degree of melting of komatiites. *J. Geophys. Res.*, **97**, B4521–40.

Hessler, A.M., Lowe, D.R., Jones, R.L., and Bird, D.K., 2004. A lower limit for atmospheric carbon dioxide levels 3.2 billion years ago. *Nature*, **428**, 736–8.

Hirth, G. and Kohlstedt, D., 1996. Water in the oceanic upper mantle: implications for rheology, melt extraction and the evolution of the lithosphere. *Earth Planet. Sci. Lett.*, **144**, 93–108.

Hochstaedter, A.G., Kepezhinskas, P., and Defant, M., 1996. Insights into the volcanic arc mantle wedge from magnesian lavas from the Kamchatka arc. *J. Geophys. Res.*, **B101**, 697–712.

Hoek, J., Banta, A., Hubler, F., and Reysenbach, A.-L., 2003. Microbial diversity of a sulphide spire located in the Edmond deep-sea hydrothermal vent field on the Central Indian Ridge. *Geobiology*, **1**, 119–27.

Hofmann, A., 1998. Chemical differentiation of the Earth: the relationship between mantle, continental crust and oceanic crust. *Earth Planet. Sci. Lett.*, **90**, 297–314.

Hofmann, A., 2003. Just add water. *Nature*, **425**, 24–5.

Hofmann, A. and White, W.M., 1982. Mantle plumes from ancient oceanic crust. *Earth Planet. Sci. Lett.*, **57**, 421–36.

Hofmann, A.W., 2001. Lead isotopes and the age of the Earth – a geochemical accident. In: Lewis, C.L.E. and Knell, S.J. (eds) *The Age of the Earth from 4004 BC to AD 2002*. Geological Society of London Publishing House, Bath, *Spl. Publ. Geol. Soc. Lond.*, **190**, 223–36.

Hofmann, A.W., Jochum, K.P., Seufert, M., and White, W.M., 1986. Nb and Pb in oceanic basalts: new constraints on mantle evolution. *Earth Planet. Sci. Lett.*, **79**, 33–45.

Huang, X., Xu, Y., and Karato, S.-I., 2005. Water content in the transition zone from electrical conductivity of wadsleyite and ringwoodite. *Nature*, **434**, 746–9.

Holland, H.D., 1984. *The Chemical Evolution of the Atmosphere and Oceans*. Princeton University Press, Princeton.

Holland, H.D., 2002. Volcanic gases, black smokers and the Great Oxidation Event. *Geochim. Cosmochim. Acta*, **66**, 3811–26.

Holland, H.D., 2003. The geological history of seawater. In: Holland, H.D. and Turekian, K.K. (eds) *Treatise on Geochemistry*, Vol. 6, pp. 583–625.

Holloway, J.R., 1998. Graphite-melt equilibria during mantle melting: constraints on CO_2 in MORB magmas and the carbon content of the mantle. *Chem. Geol.*, **147**, 89–97.

Holm, N.G. and Charlou, J.L., 2001. Initial indications of abiotic formation of hydrocarbons in the rainbow ultramafic hydrothermal system, mid-atlantic ridge. *Earth Planet. Sci. Lett.*, **191**, 1–8.

Honda, M., Phillips, D., Harris, J.W., and Yatsevich, I., 2004. Unusual noble gas compositions in polycrystalline diamonds: preliminary results for the Jwaneng kimberlite, Botswana. *Chem. Geol.*, **203**, 347–58.

Horita, J., 2005. Some perspectives on isotope biosignatures for early life. *Chem. Geol.*, **218**, 171–86.

Horita, J. and Berndt, M.E., 1999. Abiogenic methane formation and isotopic fractionation under hydrothermal conditions. *Science*, **285**, 1055–57.

Hoyle, F. and Wickramasinghe, C., 1993. *Our Place in the Cosmos: The Unfinished Revolution*. Dent, London, 190 pp.

Huang, X., Xu, Y., and Karato, S.-I., 2005. Water content in the transition zone from electrical conductivity of wadsleyite and ringwoodite. *Nature*, **434**, 746–9.

Hubble, E.A., 1929. A relation between distance and radial velocity among extra-galaxtic nebulae. *Proc. Natl. Acad. Sci. USA*, **15**, 168–73.

Huber, C. and Wachtershauser, G., 1998. Peptides by activation of amino acids with CO on (Ni,Fe)S surfaces: implications for the origin of life. *Science*, **281**, 670–2.

Hunten, D.M., 1993. Atmospheric evolution of the terrestrial planets. *Science*, **259**, 915–20.

Huppert, H.E., Sparks, S.J., Turner, J.S., and Arndt, N.T., 1984. Emplacement and cooling of komatiite lavas. *Nature*, **309**, 19–22.

Hurley, P.M. and Rand, J.R., 1969. Pre-drift continental nucleii. *Science*, **164**, 1229–42.

Huston, D. and Logan, G.A., 2004. Barite, BIFs and bugs: evidence for the evolution of the Earth's early hydrosphere. *Earth Planet. Sci. Lett.*, **220**, 41–55.

Hyndman, R.D. and Peacock, S.M., 2003. Serpentinization of the forearc mantle. *Earth Planet. Sci. Lett.*, **212**, 417–32.

Iizuka, T., Horie, K., Komiya, T., Maruyama, S., Hirata, T., Hidaka, H., and Windley, B., 2006. 4.2 Ga zircon xenocryst in an Acasta gneiss from northwestern Canada: evidence for early continental crust. *Geology*, **34**, 245–8.

Inoue, T., Rapp, R.P., Zhang, J., Gasparik, T., Wedner, D.J., and Irifune, T., 2000. Garnet fractionation in a hydrous magma ocean and the origin of Al-depleted komatiites: melting experiments of hydrous pyrolite with REEs at high pressure. *Earth Planet. Sci. Lett.*, **177**, 81–7.

Ito, E. and Takahashi, E., 1987. Melting of peridotite at uppermost lower-mantle conditions. *Nature*, **328**, 514–16.

Jackson, M.D., Gallagher, K., Petford, N., and Cheadle, M.J., 2005. Towards a coupled physical and chemical model for tonalite–trondhjemite–granodiorite magma formation. *Lithos*, **79**, 43–60.

Jackson, M.J. and Pollack, H.N., 1984. On the sensitivity of parameterized convection to the rate of decay of internal heat sources. *J. Geophys. Res.*, **89**, 10103–8.

Jacob, D.E., Bizimis, M., and Salters, V.J.M., 2005. Lu–Hf and geochemical systematics of recycled ancient oceanic crust: evidence from Roberts Victor eclogites. *Contrib. Mineral. Petrol.*, **148**, 707–20.

Jacobsen, S.B., 2005. The Hf–W isotopic system and the origin of the Earth and Moon. *Ann. Rev. Earth Planet. Sci.*, **33**, 531–70.

Jagoutz, E., Palme, H., Baddenhausen, H., Blum, K., Cendales, M., Dreibus, G., Spettel, B., Lorenz, V., and Wanke, H., 1979. The abundances of major, minor and trace elements in the Earth's mantle as derived from primitive ultramafic nodules. *Proc. 10th Lunar Planet. Sci. Conf.*, **10**, 2031–50.

Jahn, B.-M., 2004. Generation of juvenile crust in central Asia. *32nd IGC Florence, Abstr.*, 1344.

James, D.E. and Fouch, M.J., 2002. Formation and evolution of Archaean cratons: insights from southern Africa. In: Fowler, C.M.R., Ebinger, C.J., and Hawkesworth, C.J. (eds) *The Early Earth: Physical, Chemical and Biological Development*. Geological Society of London Publishing House, Bath, *Spl. Publ. Geol. Soc. Lond.* **199**, 1–26.

Jaques, A.L. and Green, D.H., 1980. Anhydrous melting of peridotite at 0–15 kb pressure and the genesis of tholeiitic basalts. *Contrib. Mineral. Petrol.*, **73**, 287–310.

Jarrard, R.D., 2003. Subduction fluxes of water, carbon dioxide, chlorine and potassium. *Geochem. Geophys. Geosyst.*, **4**, 8905, doi:10.1029/2002GC000392.

Jaupart, C. and Mareschal, J.C., 1999. The thermal structure and thickness of continental roots. *Lithos*, **48**, 93–114.

Javoy, M. and Pineau, 1991. The volatiles record of a "popping" rock from the mid-Atlantic ridge at 14°N: chemical and isotopic composition of gas trapped in the vesicules. *Earth Planet. Sci. Lett.*, **107**, 598–611.

Johannesson, K.H., Hawkins, D.L. Jr., and Cortes, A., 2006. Do Archaean chemical sediments record ancient seawater rare earth element patterns? *Geochim. Cosmochim. Acta*, **70**, 871–90.

Johnson, C.M., Beard, B.L., Beukes, N.J., Klein, C., and O'Leary, J.M., 2003. Ancient geochemical cycling in the earth as inferred from Fe isotope studies of banded iron formations from the Kaapvaal craton. *Contrib. Mineral. Petrol.*, **144**, 523–47.

Jones, J.H. and Hood, L.L., 1990. Does the Moon have the same chemical composition as the earth's upper mantle. In: Newsom, H.E. and Jones, J.H. (eds) *Origin of the Earth*. Oxford University Press, Oxford, pp. 85–98.

Jones, R.H., 1996. FeO-rich porphyritic pyroxene chondrules in unequilibrated ordinary chondrites. *Geochim. Cosmochim. Acta*, **60**, 3115–38.

Jordan, T.H., 1988. Structure and formation of the continental tectosphere. *J. Petrol.* (Special Lithosphere Issue), 11–37.

Joyce, G.F., 1989. RNA evolution and the origins of life. *Nature*, **338**, 217–24.

Joyce, G.F., 2002. The antiquity of RNA-based evolution. *Nature*, **418**, 214–21.

Jull, M. and Kelemen, P.B., 2001. On the conditions for lower crustal convective instability. *J. Geophys. Res.*, **B106**, 6423–46.

Kah, L.C., Lyons, T.W., and Frank, T.D., 2004. Low marine sulphate and protracted oxygenation of the Proterozoic biosphere. *Nature*, **432**, 834–8.

Kalas, P. and Jewitt, P., 1995. Assymetries in the Beta Pictoris dust disk. *Astron. J.*, **110**, 794–804.

Kallemeyn, G.W. and Wasson, J.T., 1986. Compositions of enstatite (EH3, EH4,5 and EL6) chondrites: implications regarding their formation. *Geochim. Cosmochim. Acta*, **50**, 2153–64.

Kamber, B.S. and Collerson, K.D., 1999. Origin of ocean island basalts: a new model based upon lead and helium isotope systematics. *J. Geophys. Res.*, **104**, 25479–92.

Kamber, B.S. and Collerson, K.D., 2000. Role of "hidden" deeply subducted slabs in mantle depletion. *Chem. Geol.*, **166**, 241–54.

Kamber, B.S. and Webb, G.E., 2001. The geochemistry of late Archaean microbial carbonate: implications for ocean chemistry and continental erosion history. *Geochim. Cosmochim. Acta*, **65**, 2509–25.

Kamber, B.S., Collerson, K.D., Moorbath, S., and Whitehouse, M.J., 2003. Inheritance of early Archaean Pb-isotope variability from long-lived Hadean protocrust. *Contrib. Mineral. Petrol.*, **145**, 25–46.

Kamber, B.S., Ewart, A., Collerson, K.D., Bruce, M.C., and McDonald, G.D., 2002. Fluid-mobile trace element constraints on the role of slab melting and implications for Archaean crustal growth models. *Contrib. Mineral. Petrol.*, **144**, 38–56.

Kamber, B.S., Greig, A., and Collerson, K.D., 2005a. A new estimate for the composition of weathered, young, upper continental crust from alluvial sediments, Queensland, Australia. *Geochim. Cosmochim. Acta*, **69**, 1041–58.

Kamber, B.S., Whitehouse, M.J., Bolhar, R., and Moorbath, S., 2005b. Volcanic resurfacing and the early terrestrial crust: zircon U–Pb and REE constraints from the Isua greenstone belt, southern west Greenland. *Earth Planet. Sci. Lett.*, **240**, 276–90.

Kashefi, K. and Lovley, D.R., 2003. Extending the upper temperature of life. *Science*, **301**, 934.

Kasting, J.F., 1993. Earth's early atmosphere. *Science*, **259**, 920–6.

Kasting, J.F., 2005. Methane and climate during the Precambrian era. *Precambrian Res.*, **137**, 119–29.

Kasting, J.F. and Catling, D., 2003. Evolution of a habitable planet. *Ann. Rev. Astron. Astrophys.*, **41**, 429–63.

Kasting, J.F. and Seifert, J.L., 2002. Life and the evolution of Earth's atmosphere. *Science*, **296**, 1066–8.

Kasting, J.F., Eggler, D.H., and Raeburn, S.P., 1993. Mantle redox evolution and the oxidation state of the Archaean atmosphere. *J. Geol.*, **101**, 245–57.

Kasting, J.F., Whitemire, D.P., and Reynolds, R.T., 1993. Habitable zones around main sequence stars. *Icarus*, **101**, 108–28.

Kawamoto, T., Hervig, R.L., and Holloway, J.R., 1996. Experimental evidence for a hydrous transition zone in the early Earth's mantle. *Earth Planet. Sci. Lett.*, **142**, 587–92.

Kay, R.W., 1978. Aleutian magnesian andesites: melts from subducted Pacific Ocean crust. *J. Volcanol. Geotherm. Res.*, **4**, 117–32.

Kay, R.W. and Kay, S.M., 1993. Delamination and delamination magmatism. *Tectonophysics*, **219**, 177–89.

Kazmierczak, J. and Altermann, W., 2002. Neoarchaean biomineralisation by benthic cyanobacteria. *Science*, **298**, 2351.

Kelemen, P.B., 1995. The genesis of high mg# andesites and the continental crust. *Contrib. Mineral. Petrol.*, **120**, 1–19.

Kelemen, P.B., Braun, M., and Hirth, G., 2000. Spatial distribution of melt conduits in the mantle beneath oceanic spreading ridges: observations from the Ingalls and Oman ophiolites. *Geochem. Geophys. Geosyst.*, **1**, Paper 1999GC000012.

Kelemen, P.B., Yogodzinski, G.M., and Scholl, D.W., 2003. Along-strike variation in the aleutian island arc: genesis of high-Mg# andesite and implications for continental crust. In: Eiler, J. (ed) *Inside the Subduction Factory. Geophysical Monograph*, **138**, *American Geophysical Union*, Washington DC, 223–76.

Keller, L.P., Messenger, S., Flynn, G.J., Clemett, S., Wirick, S., and Jacobsen, C., 2004. The nature of molecular cloud material in interplanetary dust. *Geochim. Cosmochim. Acta*, **68**, 2577–89.

Kellogg, L.H., Hager, B.H., and van der Hilst, R.D., 1999. Compositional stratification in the deep mantle. *Science*, **283**, 1881–4.

Kelley, D.S., Karson, J.A., Blackman, D.K., Fruh-Green, G.L., Butterfield, D.A., Lilley, M.D., Olson, E.J., Schrenk, M.D., Roe, K.K., Lebon, G.T., Rivizzigno, P., and the AT3–60 shipboard party, 2001. An off-axis hydrothermal vent field near the mid-Atlantic Ridge at 30 °N. *Nature*, **412**, 145–9.

Kemp, A.I.S. and Hawkesworth, C.J., 2003. Granitic perspectives on the generation and secular evolution of the continental crust. In: Holland, H.D. and Turekian, K.K. (eds) *Treatise on Geochemistry*, Elsevier, Amsterdam, Vol. 3.11, pp. 349–410.

Kemp, A.I.S., Hawkesworth, C.J., Paterson, B.A., and Kinny, P.D., 2006. Episodic growth of the Gondwana supercontinent from hafnium and oxygen isotopes in zircon. *Nature*, **439**, 580–3.

Keppler, H., Weidenbeck, M., and Shcheka, S.S., 2003. Carbon solubility in olivine and the mode of carbon storage in the Earth's mantle. *Nature*, **424**, 414–16.

Kerr, R.A., 2005. The story of O_2. *Science*, **308**, 1730–2.

Kerrich, R., Wyman, D.A., Fan, J., and Bleeker, W., 1998. Boninite series: low Ti-tholeiite associations from the 2.7 Ga Abitibi Greenstone Belt. *Earth Planet. Sci. Lett.*, **164**, 303–16.

Killops, S. and Killops, V., 2005. *Introduction to Organic Geochemistry*, 2nd edition. Blackwell Publishing, Oxford, 393 pp.

Klein, E. and Langmuir, C.H., 1987. Global correlations of ocean ridge basalt chemistry with axial depth and crustal thickness. *J. Geophys. Res.*, **92**, 8089–115.

Kleine, T., Mezger, K., Palme, H., and Munker, C., 2004. The W isotope evolution of the bulk silicate Earth: constraints on the timing and mechanisms of core formation and accretion. *Earth Planet. Sci. Lett.*, **228**, 109–23.

Kleine, T., Munker, C., Mezger, K., and Palme, H., 2005a. Rapid accretion and early core-formation on asteroids and the terrestrial planets from Hf–W chronometry. *Nature*, **418**, 952–5.

Kleine, T., Palme, H., Mezger, K., and Halliday, A.N., 2005b. Hf–W chronometry of lunar metals and the age and early differentiation of the moon. *Science*, **310**, 1671–4.

Knauth, L.P. and Lowe, D.R., 1978. Oxygen isotope geochemistry of cherts from the Onverwacht group (3.4 billion years), Transvaal, South Africa, with implications for secular variations in the isotopic composition of cherts. *Earth Planet. Sci. Lett.*, **41**, 209–22.

Knauth, L.P. and Lowe, D.R., 2003. High Archaean climatic temperatures inferred from oxygen isotope geochemistry of cherts in the 3.5 Ga Swaziland Supergroup, South Africa. *Bull. Geol. Soc. Amer.*, **115**, 566–80.

Knoll, A.H., 1999. A new molecular window on early life. *Science*, **285**, 1025–6.

Kortenkamp, S.J., Wetherill, G.W., and Inaba, S., 2001. Runaway growth of planetary embryos facilitated by massive bodies in a protoplanetary disk. *Science*, **293**, 1127–9.

Kosigo, T., Hirschmann, M.M., and Reiners, P.W., 2004. Length scales of mantle heterogeneities and their relationship to ocean island basalt geochemistry. *Geochim. Cosmochim. Acta*, **68**, 345–60.

Kosticin, N. and Krapez, B., 2004. Relationship between detrital zircon age-spectra and the tectonic evolution of the late Archaean Witwatersrand Basin, South Africa. *Precambrian Res.*, **129**, 141–68.

Kramers, J.D., 1987. Link between Archaean continent formation and anomalous sub-continental mantle. *Nature*, **325**, 47–50.

Kramers, J.D., 1988. An open-system fractional crystallisation model for very early continental crust formation. *Precambrian Res.*, **38**, 281–95.

Kramers, J.D., 1998. Reconciling siderophile element data in the earth and Moon, W isotopes and the upper lunar age limit in a simple model of homogeneous accretion. *Chem. Geol.*, **145**, 461–78.

Kramers, J.D., 2002, Global modeling of continent formation and destruction through geological time and implications for CO_2 drawdown in the Archaean Eon. In: Fowler, C.M.R., Ebinger, C.J., and Hawkesworth, C.J. (eds) *The Early Earth: Physical, Chemical and Biological Development*. Geological Society of London Publishing House, Bath, *Geol. Soc. Lond. Spl. Publ.*, **199**, 231–57.

Kramers, J.D., 2003. Volatile element abundance patterns and an early liquid water ocean on Earth. *Precambrian Res.*, **126**, 379–94.

Kramers, J.D. and Tolstikhin, I.N., 1997. Two terrestrial lead isotope paradoxes, forward transport modeling, core formation and the history of the continental crust. *Chem. Geol.*, **139**, 75–110.

Kreissig, K. and Elliott, T., 2005. Ca isotope fingerprints of early crust–mantle evolution. *Geochim. Cosmochim. Acta*, **69**, 165–76.

Kring, D.A., 1991. High temperature rims around chondrules in primitive chondrites: evidence for fluctuating conditions in the solar nebula. *Earth Planet. Sci. Lett.*, **105**, 65–80.

Kring, D.A. and Cohen, B.A., 2002. Cataclysmic bombardment throughout the inner solar system 3.9–4.0 Ga. *J. Geophys. Res.*, **107**, doi 10.1029/2001/JE001529.

Krot, A.N., Yuriomoto, H., Hutcheon, I.D., and MacPherson, G.J., 2005. Chronology of the early solar system from chondrule-bearing calcium–aluminium-rich inclusions. *Nature*, **434**, 998–1001.

Kump, L.R., 2005. Ironing out biosphere oxidation. *Science*, **307**, 1058–9.

Kump, L.R., Kasting, J.F., and Barley, M., 2001. Rise of atmospheric oxygen and the "upside-down" Archaean mantle. *Geochem. Geophys. Geosyst.*, **2**, Paper number 2000GC000114.

Kump, L.R., Kasting, J.F., and Crane, R.G., 1999. *The Earth System*. Prentice Hall, London, 351 pp.

Kuramoto, K. and Matsui, T., 1996. Partitioning of H and C between the mantle and core during the core formation in the Earth: its implications for the atmosphere evolution and the redox state of early mantle. *J. Geophys. Res.*, **101**, 14909–32.

Kusky, T.M. (ed), 2004. *Precambrian Ophiolites and Related Rocks*. Elsevier Science, Amsterdam, 739 pp.

Kusky, T.M. and Kidd, W.S.F., 1992. Remnants of an Archaean Oceanic plateau, Belingwe greenstone belt, Zimbabwe. *Geology*, **20**, 43–6.

Kusky, T.M., Li, J.-H., and Tucker, R.D., 2001. The Archaean Dongwanzi ophiolite complex, north China Craton: 2.505 billion-year-old oceanic crust and mantle. *Science*, **292**, 1142–5.

Labrosse, S., 2002. Hot spots, mantle plumes and core heat loss. *Earth Planet. Sci. Lett.*, **199**, 147–56.

Larimer, J.W., 1988. The cosmochemical classification of the elements. In: Kerridge, J.F. and Matthews, M.S. (eds) *Meteorites and the Early Solar System*, University of Arizona Press, Tuscon, AZ, pp. 375–89.

Lasaga, A.C. and Berner, R.A., 1998. Fundamental aspects of quantitative models for geochemical cycles. *Chem. Geol.*, **145**, 161–75.

Lasaga, A.C. and Ohmoto, H., 2002. The oxygen cycle: dynamics and stability. *Geochim. Cosmochim. Acta*, **66**, 361–81.

Lawton, J., 2001. Earth system science. *Science*, **292**, 1965.

Leat, P.T., Riley, T.R., Wareham, C.D., Millar, I.L., Kelley, S.P., and Storey, B.C., 2002. Tectonic setting of primitive magmas in volcanic arcs: an example from the Antarctic peninsula. *J. Geol. Soc. Lond.*, **159**, 31–44.

Lecluse, C., Robert, F., Kaiser, R.-I., Roessler, K., Pillinger, C.T., and Javoy, M., 1998. Carbon isotope ratios of irradiated methane ices: implications for cometary $^{13}C/^{12}C$ ratio. *Astron. Astrophys.*, **330**, 1175–9.

Lecuyer, C. and Ricard, Y., 1999. Long-term fluxes and budget of ferric iron: implication for the

redox states of the Earth's mantle and atmosphere. *Earth Planet. Sci. Lett.*, **165**, 197–211.
Lecuyer, C., Gillet, P., and Robert, F., 1998. The hydrogen isotope composition of seawater and the global water cycle. *Chem. Geol.*, **145**, 249–61.
Lecuyer, C., Simon, L., and Guyot, F., 2000. Comparison of carbon, nitrogen and water budgets on Venus and the Earth. *Earth Planet. Sci. Lett.*, **181**, 33–40.
Lepland, A., Arrhenius, G., and Cornell, D., 2002. Apatite in early Archaean Isua supracrustal rocks, southern west Greenland: its origin, association with graphite and potential as a biomarker. *Precambrian Res.*, **118**, 221–41.
Lewis, J.S., 2004. *Physics and Chemistry of the Solar System*, 2nd edition. Elsevier, Oxford, 655 pp.
Li, B. and Zhang, J., 2005. Pressure and temperature dependence of elastic wave velocity of $MgSiO_3$ perovskite and the composition of the lower mantle. *Phys. Earth Planet. Interior*, **151**, 143–54.
Li, J. and Agee, C.B., 2001. The effect of pressure, temperature, oxygen fugacity and composition on partitioning of nickel and cobalt between liquid Fe–Ni–S alloy and liquid silicate: implications for the Earth's core formation. *Geochim. Cosmochim. Acta*, **65**, 1821–32.
Li, J., Fei, Y., Mao, H.K., Hirose, K., and Shieh, S.R., 2001. Sulfur in the Earth's inner core. *Earth Planet. Sci. Lett.*, **193**, 509–14.
Li, Z.-X.A. and Lee, C.-T.A., 2004. The constancy of upper mantle fO_2 through time inferred from V/Sc ratios in basalts. *Earth Planet. Sci. Lett.*, **228**, 483–93.
Lindsay, J.F., Brasier, M.D., McLoughlin, N., Green, O.R., Fogel, Steele, A., and Mertzmann, S.A., 2006. The problem of deep carbon – an Archaean paradox. *Precambrian Res.*, **143**, 1–22.
Lineweaver, C.H., 1999. A younger age for the Universe. *Science*, **284**, 1503–7.
Lissauer, J.J., 1993. Planet formation. *Ann. Rev. Astron. Astrophys.*, **31**, 129–74.
Lissauer, J.J., 2002. Extrasolar planets. *Nature*, **419**, 355–8.
Liu, L.-G., 2004. The inception of the oceans and CO_2-atmosphere in the early history of the Earth. *Earth Planet. Sci. Lett.*, **227**, 179–84.
Longair, M.S., 1998. *Galaxy Formation*. Springer-Verlag, New York, 536 pp.
Lovelock, J.E., 1979. *Gaia. A New Look at Life on Earth*. Oxford University Press, Oxford, 157 pp.
Lovelock, J.E., 1988. *The Ages of Gaia: A Biography of Our Living Earth*. Oxford University Press, Oxford, 305 pp.
Lovelock, J.E., 1991. *Gaia: The Practical Science of Planetary Medicine*. Oxford University Press, Oxford, 192 pp.
Lowe, D.R., 1980. Stromatolites 3,400 Myr old from the Archaean of western Australia. *Nature*, **284**, 441–3.
Lowe, D.R. and Tice, M.M., 2004. Geologic evidence for Archaean atmospheric and climatic evolution: fluctuating levels of CO_2, CH_4 and O_2 with an overriding tectonic control. *Geology*, **32**, 493–6.
Ludden, J. and Hubert, C., 1986. Geologic evolution of the late Archaean Abitibi greenstone belt of Canada. *Geology*, **14**, 707–11.
Maas, R., Kinny, P.D., Williams, I.S., Froude, D.O., and Compston, W., 1992. The Earth's oldest known crust: a geochronological and geochemical study of 3,900–4,200 Ma old detrital zircons from Mt Narryer and Jack Hill, western Australia. *Geochim. Cosmochim. Acta*, **56**, 1281–300.
Macdonald, R., Hawkesworth, C.J., and Heath, E., 2000. The Lesser Antilles volcanic chain: a study in arc magmatism. *Earth Sci. Rev.*, **49**, 1–76.
MacKenzie, K., 2001. Evidence for a plate tectonics debate. *Earth Sci. Rev.*, **55**, 235–336.
Maher, K.A. and Stevenson, D.J., 1988. Impact frustration of the origin of life. *Nature*, **331**, 612–14.
Malavergne, V. and 8 others, 2004. Si in the core? New high-pressure and high temperature experimental data. *Geochim. Cosmochim. Acta*, **68**, 4201–11.
Martin, A., Nisbet, E.G., and Bickle, M.J., 1980. Archaean stromatolites of the belingwe greenstone belt, Zimbabwe (Rhodesia). *Precambrian Res.*, **13**, 337–62.
Martin, H., 1986. Effects of steeper Archaean geothermal gradient on geochemistry of subduction-zone magmas. *Geology*, **14**, 753–6.
Martin, H., 1994. The Archaean grey gneisses and the genesis of the continental crust. In: Condie, K.C. (ed) *Archaean Crustal Evolution, Developments in Precambrian Geology*. Elsevier, Amsterdam, vol. 11, pp. 205–59.
Martin, H., 1999. Adakitic magmas: modern analogues of Archaean granitoids. *Lithos*, **46**, 411–29.
Martin, H. and Moyen, J.-F., 2002. Secular changes in tonalite–trondhjemite–granodiorite composition as markers of the progressive cooling earth. *Geology*, **30**, 319–22.
Martin, H., Smithies, R.H., Rapp, R., Moyen, J.-F., and Champion, D., 2005. An overview of adakite, tonalite–trondhjemite–granodiorite (TTG) and sanukitoid relationships and some implications for crustal evolution. *Lithos*, **79**, 1–24.

Martin, W. and Russell, M.J., 2003. On the origin of cells: a hypothesis for the evolutionary transitions from abiotic geochemistry to chemoautotrophic prokaryotes and from prokaryotes to nucleated cells. *Phil. Trans. R. Soc. Lond.*, **385B**, 27–85.

Marty, B. and Dauphas, N., 2003. The nitrogen record of crust–mantle interaction and mantle convection from Archaean to Present. *Earth Planet. Sci. Lett.*, **206**, 397–410.

Marty, B. and Tolstikhin, I.N., 1998. CO_2 fluxes from mid-ocean ridges, arcs and plumes. *Chem. Geol.*, **145**, 233–48.

Maruoka, T., Kurat, G., Dobosi, G., and Koeberl, C., 2004. Isotopic composition of carbon in diamonds of diamondites: record of mass fractionation in the upper mantle. *Geochim. Cosmochim. Acta*, **68**, 1635–44.

Mather, T.A., Pyle, D.M., and Allen, A.G., 2004. Volcanic source for fixed nitrogen in the early Earth's atmosphere. *Geology*, **32**, 905–8.

Mathez, E.A., 1984. Influence of degassing on oxidation states of basaltic magmas. *Nature*, **310**, 371–5.

McCammon, C., 2005. The paradox of mantle redox. *Science*, **308**, 807–8.

McCammon, C.A., Stachel, T., and Harris, J.W., 2004. Iron oxidation state in lower mantle mineral assemblages II. Inclusions in diamonds from Kankan, Guinea. *Earth Planet. Sci. Lett.*, **222**, 423–34.

McCulloch, M.T., 1993. The role of subducting slabs in an evolving Earth. *Earth Planet. Sci. Lett.*, **115**, 89–100.

McCulloch, M.T. and Bennett, V.C., 1993. Evolution of the early Earth: constraints from ^{143}Nd–^{142}Nd isotopic systematics. *Lithos*, **30**, 237–55.

McCulloch, M. and Bennett, V.C., 1994. Progressive growth of the Earth's continental crust and depleted mantle: geochemical constraints. *Geochim. Cosmochim. Acta*, **58**, 4717–38.

McDonough, W.F., 1991. Partial melting of subducted oceanic crust and isolation of its residual eclogitic lithology. *Phil. Trans. R. Soc. Lond. A*, **335**, 407–18.

McDonough, W.F. and Sun, S.-S., 1995. The composition of the Earth. *Chem. Geol.*, **120**, 223–53.

McDonough, W.F., Sun, S.-S., Ringwood, A.E., Jagoutz, E., and Hofmann, A.W., 1992. Potassium, rubidium and cesium in the Earth and Moon and the evolution of the mantle of the Earth. *Geochim. Cosmochim. Acta*, **56**, 1001–12.

McKay, D.S., Gibson, E.K. Jr., Thomas-Keprta, K.L., Vali, H., Romanek, C.S., Clemett, S.J., Chillier, X.D.F., Maechling, C.R., and Zare, R.N., 1996. Search for past life on Mars: possible relic biogenic activity in Martian meteorite ALH84001. *Science*, **273**, 103–24.

McKenzie, D.P., 1989. Some remarks on the removal of small melt fractions in the mantle. *Earth Planet. Sci. Lett.*, **95**, 53–72.

McKenzie, D. and Bickle, M.J., 1988. The volume and composition of melt generated by extension of the lithosphere. *J. Petrol.*, **29**, 625–79.

McLennan, S.M., Taylor, S.R., and Hemming, S.R., 2006. Composition, differentiation and evolution of continental crust: constraints from sedimentary rocks and heat flow. In: Brown, M. and Rushmer, T. (eds) *Evolution and Differentiation of the Continental Crust*. Cambridge University Press, Cambridge, pp. 92–134.

McLoughlin, N., Brasier, M.D., Wacey, D., Perry, R.S., Green, O.S., Kilburn, M., and Grovenor, C., 2005. Establishing biogenicity criteria for Archaean endoliths. *GSA Speciality Meetings Abstracts Programs*, **1**, 28.

Meibom, A. and Anderson, D.L., 2003. The statistical upper mantle assemblage. *Earth Planet. Sci. Lett.*, **217**, 123–39.

Meibom, A., Anderson, D.L., Sleep, N.H., Frei, R., Chamberlain, C.P., Hren, M.T., and Wooden, J.L., 2003. Are high ^{3}He/^{4}He ratios in oceanic basalts indicators of deep-mantle plume components. *Earth Planet. Sci. Lett.*, **208**, 197–204.

Meibom, A., Sleep, N.H., Chamberlain, C.P., Coleman, R.G., Frei, R., Hren, M.T., and Wooden, J.L., 2002. Re–Os isotopic evidence for long-lived heterogeneity and equilibration processes in the Earth's upper mantle. *Nature*, **419**, 705–8.

Meisel, T., Walker, R.J., Irving, A.J., and Lorand, J.-P., 2001. Osmium isotopic compositions of mantle xenoliths: a global perspective. *Geochim. Cosmochim. Acta*, **65**, 1311–23.

Mel'nik, Y.P., 1982. *Banded Iron Formations: Physicochemical Conditions of Formation*. Elsevier, Amsterdam, 310 pp.

Melosh, J., 2001. A new model Moon. *Nature*, **412**, 694–5.

Menzies, M., 1983. Mantle ultramafic xenoliths in alkaline magmas: evidence for mantle heterogeneity modified by magmatic activity. In: Hawkesworth, C.J. and Norry, M.J. (eds) *Continental Basalts and Mantle Xenoliths*. Shiva, Nantwich, pp. 92–110.

Menzies, M.A., Klemperer, S.L., Ebinger, C.J., and Baker, J., 2002. Characteristics of volcanic rifted margins. In: Menzies, M.A., Klemperer, S.L., Ebinger, C.J., and Baker, J. (eds) *Volcanic Rifted Margins*. Geological Society of America, Boulder, CO, *Geol. Soc. Amer. Spl. Paper*, **362**, 1–14.

Messenger, S., Keller, L.P., Stadermann, F.J., Walker, R.M., and Zinner, E., 2003. Samples of stars beyond the solar system: silicate grains in interplanetary dust. *Science*, **300**, 105–8.

Midgley, M., 2001. *Gaia: The Next Big Idea*. Demos, London, 52 pp.

Miller, S.L., 1953. Production of amino acids under possible primitive Earth conditions. *Science*, **117**, 528.

Miller, S.L. and Bada, J.L., 1988. Submarine hot springs and the origin of life. *Nature*, **334**, 609–11.

Miller, S.L. and Urey, H.C., 1959. Organic compound synthesis on the primitive Earth. *Science*, **130**, 245–51.

Minter, W.E.L., 1999. Irrefutable detrital origin of Witwatersrand gold and evidence of eolian signatures. *Eco. Geol.*, **94**, 665–70.

Miyazaki, A., Hiyagon, H., Sugiura, N., Hirose, K., and Takahashi, E., 2004. Solubilities of nitrogen and noble gases in silicate melts under various oxygen fugacities: implications for the origin and degassing history of nitrogen and noble gases in the Earth. *Geochim. Cosmochim. Acta*, **68**, 387–401.

Mojzsis, S.J., Arrhenius, G., McKeegan, K.D., Harrison, T.M., Nutman, A.P., and Friend, C.R.L., 1996. Evidence for life on earth before 3,800 million years ago. *Nature*, **384**, 55–9.

Mojzsis, S.J., Coath, C.D., Greenwood, J.P., McKeegan, K.D., and Harrison, T.M., 2003. Mass-independent isotope effects in Archaean (2.5 to 3.8 Ga) sedimentary sulfides determined by ion microprobe anlysis. *Geochim. Cosmochim. Acta*, **67**, 1635–58.

Mojzsis, S.J., Harrison, M.T., and Pidgeon, R.T., 2001. Oxygen isotope evidence from very ancient zircons for liquid water at the Earth's surface 4,300 Myr ago. *Nature*, **409**, 178–81.

Monod, J., 1972. *Chance and Necessity: An Essay on the Natural Philosophy of Modern Biology*. Collins, London, pp. 187.

Montagner, J.P. and Anderson, D.L., 1989. Constrained reference mantle model. *Phys. Earth Planet. Interior*, **58**, 205–27.

Montelli, R., Nolet, G., Dahlen, F.A., Masters, G., Engdahl, R., and Hung, S.-H., 2004. Finite frequency tomography reveals a variety of plumes in the mantle. *Science*, **303**, 338–43.

Moorbath, S., O'Nions, R.K., Pankhurst, R.J., Gale, N.H., and McGreggor, V.R., 1972. Further rubidium–strontium age determinations on the very early Precambrian rocks of the Godthaab district, west Greenland. *Nat. Phys. Sci.*, **240**, 78–82.

Moorbath, S., Whitehouse, M.J., and Kamber, B.S., 1997. Extreme Nd-isotope heterogeneity in the early Archaean – fact or fiction? Case histories from northern Canada and west Greenland. *Chem. Geol.*, **135**, 213–31.

Moore, R.O. and Gurney, J.J., 1985. Pyroxene solid solution in garnets included in diamonds. *Nature*, **318**, 553–5.

Morbidelli, A., Chambers, J., Lunine, J.I., Petit, J.M., Robert, F., Valsecchi, G.B., and Cyr, K.E., 2000. Source regions and timescales for the delivery of water to Earth. *Meteoritics Planet. Sci.*, **35**, 1309–20.

Morgan, P., 1985. Crustal radiogenic heat production and the selective survival of ancient continental crust. *Proc. 15th Lunar Planet. Sci. Conf., J. Geophys. Res.*, **90(Suppl.)**, C561–70.

Morgan, Z. and Liang, Y., 2003. An experimental and numerical study of the kinetics of harzburgite reactive dissolution with applications to dunite dyke formation. *Earth Planet. Sci. Lett.*, **214**, 59–74.

Moyen, J.-F. and Stevens, G., 2005. Experimental constraints on TTG petrogenesis: implications for Archaean geodynamics. In: Benn, K., Mareschal, J.-C., and Condie, K. (eds) *Archaean Geodynamics and Environments*, AGU Monograph **164**, Washington, Chapter 10, pp. 149–75.

Murakami, M., Hirose, K., Yurimoto, H., Nakashima, S., and Takafuji, N., 2002. Water in the Earth's lower mantle. *Science*, **295**, 1885–7.

Murakami, T., Utsunomiya, S., Imazu, Y., and Prasad, N., 2001. Direct evidence of late Archaean to early Proterozoic anoxic atmosphere from a product of 2.5 Ga old weathering. *Earth Planet. Sci. Lett.*, **184**, 523–8.

Murphy, D.T., Kamber, B.S., and Collerson, K.D., 2002. A refined solution to the first terrestrial Pb-paradox. *J. Petrol.*, **44**, 39–53.

Myers, J.S., 2001. Protoliths of the 3.8–3.7 Ga Isua greenstone belt, West Greenland. *Precambrian Res.*, **105**, 129–41.

Nagler, T.F. and Kramers, J.D., 1998. Nd isotopic evolution of the upper mantle during the Precambrian: models, data and the uncertainty of both. *Precambrian Res.*, **91**, 233–52.

Nagler, T.F., Kramers, J.D., Kamber, B.S., Frei, R., and Prendergast, M.D.A., 1997. Growth of sub-continental lithospheric mantle beneath Zimbabwe started at or before 3.8 Ga: Re–Os study on chromites. *Geology*, **25**, 983–6.

Nagy, B., Engel, M.H., Zumberge, J.E., Ogino, H., and Chang, S.Y., 1981. Amino acids and hydrocarbons

3,800-Myr old in the Isua rocks, southwestern Greenland. *Nature*, **289**, 53–6.

Nakamura, K. and Kato, Y., 2004. Carbonatisation of oceanic crust by the seafloor hydrothermal activity and its significance as a CO_2 sink in the early Earth. *Geochim. Cosmochim. Acta*, **68**, 4595–618.

Nelson, D.R., Robinson, B.W., and Myers, J.S., 2000. Complex histories extending for >4.0Ga deciphered from xenocrystic zircon microstructures. *Earth Planet. Sci. Lett.*, **181**, 89–102.

Nesbitt, H.W. and Young, G.M., 1982. Early Proterozoic climates and plate motions inferred from major element chemistry of lutites. *Nature*, **299**, 715–17.

Newsom, H.E. and Jones, J.H., 1990. *Origin of the Earth*. Oxford University Press, Oxford, 378 pp.

Ni, S. and Helmberger, D.V., 2003. Further constraints on the African superplume structure. *Phys. Earth Planet. Interior*, **140**, 243–51.

Nicolas, A., 1989. *Structures of Ophiolites and Dynamics of Ocean Lithosphere*. Kluwer Academic, Dordrecht, 367 pp.

Nicolas, A., Boudier, F., Ildefonse, B., and Ball, E., 2000. Accretion of the Oman and United Arab Emirates ophiolite – discussion of new structural map. *Marine Geophys. Res.*, **21**, 147–79.

Nisbet, E.G., 1987. *The Young Earth: An Introduction to Archaean Geology*. Allen and Unwin, London, 402 pp.

Nisbet, E.G., 2002. The influence of life on the face of the Earth: garnets and moving continents. *Special Publ. Geol. Soc. Lond.* (Fowler, C.M.R., Ebinger, C.J., and Hawkesworth, C.J., eds), **199**, 275–307.

Nisbet, E.G. and Fowler, C.M.R., 2003. The early history of life. *Treatise Geochem.* (Holland, H.D. and Turekian, K.K., eds), **8.01**, 1–39.

Nisbet, E.G. and Sleep, N.H., 2001. The habitat and nature of early life. *Nature*, **409**, 1083–91.

Nisbet, E.G. and Walker, D., 1982. Komatiites and the structure of the Archaean mantle. *Earth Planet. Sci. Lett.*, **60**, 105–13.

Nisbet, E.G. and 10 others, 1987. Uniquely fresh komatiites from the Belingwe greenstone belt, Zimbabwe. *Geology*, **15**, 1147–50.

Nisbet, E.G., Cann, J.R., and van Dover, C.L., 1995. Origins of photosynthesis. *Nature*, **373**, 479–80.

Nisbet, E.G., Cheadle, M.J., Arndt, N.T., and Bickle, M.J., 1993. Constraining the potential temperature of the Archaean mantle: a review of the evidence from komatiites. *Lithos*, **30**, 291–307.

Nittler, L.R., 2003. Presolar stardust in meteorites: recent advances and scientific frontiers. *Earth Planet. Sci. Lett.*, **209**, 259–73.

Niu, F. and James, D.E., 2002. Fine structure of the lowermost crust beneath the Kaapvaal craton and its implications for crustal formation and evolution. *Earth Planet. Sci. Lett.*, **200**, 121–30.

Niu, Y. and O'Hara, M.J., 2003. Origin of ocean island basalts: a new perspective from petrology, geochemistry and mineral physics considerations. *J. Geophys. Res. B*, **108**. doi: 10.1029/2002/JB002048.

Noffke, N., Hazen, R., and Nhleko, N., 2003. Earth's earliest microbial mats in a siliclastic marine environment (2.9 Ga Mozzan Group, South Africa). *Geology*, **31**, 673–6.

Nutman, A.P., Allart, J.P., Bridgwater, D., Dimroth, E., and Rosing, M., 1984. Stratigraphic and geochemical evidence for the depositional environment of the early Archaean Isua supracrustal belt, southern west Greenland. *Precambrian Res.*, **25**, 365–96.

Nutman, A.P., McGregor, V.R., Friend, C.R.L., Bennett, V.C., and Kinny, P.D., 1996. The Itsaq Gneiss complex of southern west Greenland: the world's most extensive record of early crustal evolution (3,900–3,600 Ma). *Precambrian Res.*, **78**, 1–39.

Nutman, A.P., Mojzsis, S.J., and Friend, C.R.L., 1997. Recognition of >3.85 Ga old water-lain sediments in west Greenland and their significance for the early Archaean Earth. *Geochim. Cosmochim. Acta*, **67**, 2093–108.

Nyblade, A.A., 1999. Heat flow and the structure of Precambrian lithosphere. *Lithos*, **48**, 81–91.

Nyblade, A.A. and Pollack, H.N., 1993. A global analysis of heat flow from Precambrian terrains: implications for the thermal structure of Archaean and Proterozoic lithosphere. *J. Geophys. Res.*, **98**, 12207–18.

Oganov, A.R. and Ono, S., 2004. Theoretical and experimental evidence for a post-perovskite phase of $MgSiO_3$ in Earth's D″ layer. *Nature*, **430**, 445–8.

O'Hara, M.J., 1968. Are ocean-floor basalts primary magmas? *Nature*, **220**, 683–6.

Ohmoto, H., 1996. Evidence in pre-2.2 palaeosols for the early evolution of atmospheric oxygen and terrestrial biota. *Geology*, **24**, 1135–8.

Ohmoto, H., 2004. The Archaean atmosphere, hydrosphere and biosphere. In: Eriksson, P.G., Altermann, W., Nelson, D.R., and Cataneanu, O. (eds) *The Precambrian Earth: Tempos and Events*. Elsevier, Amsterdam, pp. 361–88.

Ohtani, E., 2005. Water in the mantle. *Elements*, **1**, 25–30.

Ohtani, E., Kato, T., and Sawamoto, H., 1986. Melting of a model chondritic mantle to 20 GPa. *Nature*, **322**, 352–3.

Okamoto, Y.K. and 9 others, 2004. An early extrasolar planetary system revealed by planetesimal belts in β Pictoris. *Nature*, **431**, 660–3.

O'Neill, H.St.C., Dingwall, D.B., Borisov, A., Spettel, and Palme, H., 1995. Experimental petrochemistry of some highly siderophile elements at high temperatures and some implications for core formation and the mantle's earliest history. *Chem. Geol.*, **120**, 255–73.

O'Nions, R.K. and McKenzie, D.P., 1988. Melting and continent generation. *Earth Planet. Sci. Lett.*, **90**, 449–56.

Ono, S., Eigenbrode, J.L., Pavlov, A.A., Kharecha, P., Rumble, D., Kasting, J.F., and Freeman, K.H., 2003. New insights into Archaean sulfur cycle from mass-independent sulfur records from the Hammersley basin, Australia. *Earth Planet. Sci. Lett.*, **213**, 15–30.

Oro, J., 1961. Mechanism of synthesis of adenine from hydrogen cyanide under possible primitive Earth conditions. *Nature*, **191**, 1193–4.

Orphan, V.J., House, C.H., Hinrichs, K.-U., McKeegan, K.D., and DeLong, E.F., 2001. Methane-consuming archaea revealed by directly coupled isotopic and phylogenetic analysis. *Science*, **293**, 484–7.

Ozima, M. and Igarashi, G., 2000. The primordial noble gases in the Earth: a key constraint on Earth evolution models. *Earth Planet. Sci. Lett.*, **176**, 219–32.

Pace, N.R., 1997. A molecular view of microbial diversity and the biosphere. *Science*, **276**, 734–40.

Palme, H. and O'Neill, H.St.C., 2003. Cosmochemical estimates of mantle composition. *Treatise Geochem.*, **2.02**, 1–38.

Panning, M. and Romanowicz, B., 2004. Inferences on flow at the base of the Earth's mantle based on seismic anisotropy. *Science*, **303**, 351–3.

Papineau, D., Mojzsis, S.J., Karhu, J.A., and Marty, B., 2005. Nitrogen isotopic composition of ammoniated phyllosilicates: case studies from Precambrian metamorphosed sedimentary rocks. *Chem. Geol.*, **216**, 37–58.

Parkinson, I.J. and Arculus, R.J., 1999. The redox state of subduction zones: insights from arc-peridotites. *Chem. Geol.*, **160**, 409–23.

Parman, S.W., Dann, J.C., Grove, T.L., and de Wit, M.J., 1997. Emplacement conditions of komatiite magmas from the 3.49 Ga komati formation, Barberton Greenstone Belt, South Africa. *Earth Planet. Sci. Lett.*, **150**, 303–23.

Parman, S.W., Grove. T.L., and Dann, J.C., 2001. The production of Barberton komatiites in an Archaean subduction zone. *Geophys. Res. Lett.*, **28**, 2513–16.

Parman, S.W., Grove, T.L., Dann, J.C., and deWit, M.J., 2004. A subduction origin for komatiites and cratonic lithospheric mantle. *S. Afr. J. Geol.*, **107**, 107–18.

Parman, S.W., Schimizu, N., Grove. T.L., and Dann, J.C., 2003. Constraints on the pre-metamorphic element composition of Barberton komatiites from ion probe analyses of preserved clinopyroxenes. *Contrib. Mineral. Petrol.*, **144**, 383–96.

Patchett, P.J. and Arndt, N.T., 1986. Nd isotopes and tectonics of 1.9–1.7 Ga crustal genesis. *Earth Planet. Sci. Lett.*, **78**, 329–38.

Patchett, P.J. and Samson, S.D., 2003. Ages and growth of the continental crust from radiogenic isotopes. In: Holland, H.D. and Turekian, K.K. (eds) *Treatise on Geochemistry*. Elsevier, Amsterdam, vol. 3.10, pp. 1–28.

Patchett, P.J., Vervoort, J.D., Soderlund, U., and Salters, V.J.M., 2004. Lu–Hf and Sm–Nd isotopic systematics in chondrites and their constraints on the Lu–Hf properties of the Earth. *Earth Planet. Sci. Lett.*, **222**, 29–41.

Pavlov, A.A. and Kasting, J.F., 2002. Mass-independent fractionation of sulphur isotopes in Archaean sediments: strong evidence for an anoxic Archaean atmosphere. *Astrobiology*, **2**, 27–41.

Peacock, S.M., 1990. Fluid processes in subduction zones. *Science*, **248**, 329–37.

Peacock, S.M., Rushmer, T., and Thompson, A.B., 1994. Partial melting of subducting oceanic crust. *Earth Planet. Sci. Lett.*, **121**, 227–44.

Pearce, J.A. and Parkinson, I.J., 1993. Trace-element models for mantle melting: application to volcanic arc petrogenesis. *Geol. Soc. Lond. Spl. Publ.*, **76**, 373–403.

Pearson, D.G., 1999. The age of continental roots. *Lithos*, **48**, 171–94.

Pearson, N.J., Alard, O., Griffin, W.L., Jackson, S.E., and O'Reilly, S.Y., 2002. In situ measurement of Re–Os isotopes in mantle sulfides by laser ablation multicollector-inductively coupled plasma mass spectrometry: analytical methods and preliminary results. *Geochim. Cosmochim. Acta*, **66**, 1037–50.

Peck, W.H., Valley, J.W., Wilde, S.A., and Graham, C., 2001. Oxygen isotope ratios and rare earth elements in 3.3 to 4.4 Ga zircons: ion microprobe evidence for high $\delta^{18}O$ continental crust and oceans in the early Earth. *Geochim. Cosmochim. Acta*, **65**, 4215–29.

Penzias, A.A. and Wilson, R.W., 1965. A measurement of excess antenna temperature at 4080 Mc/s. *Astrophys. J.*, **142**, 419–21.

Pepin, R.O., 2000. On the isotopic composition of primordial xenon in terrestrial planet atmospheres. *Space Sci. Rev.*, **92**, 371–95.

Percival, J.A. and West, G.F., 1994. The Kapuskasing uplift: a geological and geophysical synthesis. *Can. J. Earth Sci.*, **31**, 1256–86.

Pichavant, M., Mysen, B.O., and Macdonald, R., 2002. Source and H_2O content of high-MgO magmas in island arc settings: an experimental study of a primitive calc-alkaline basalt from St Vincent, Lesser Antilles arc. *Geochim. Cosmochim. Acta*, **66**, 2193–209.

Pierazzo, E. and Chyba, C.F., 1999. Amino acid survival in large cometary impacts. *Meteoritics Planet. Sci.*, **34**, 909–18.

Pinti, D.L. and Hashizume, K., 2001. ^{15}N-depleted nitrogen in early Archaean kergens: clues on ancient marine chemosynthetic-based ecosystems? A comment to Beaumont, V., Robert, F., 1999. *Precambrian Res.*, **96**, 62–82; **105**, 85–8.

Plank, T., 2005. Constraints from Th/La on sediment recycling at subduction zones and the evolution of the continents. *J. Petrol.*, **46**, 921–44.

Plank, T. and Langmuir, C.H., 1988. An evaluation of the global variations in the major element chemistry of arc basalts. *Earth Planet. Sci. Lett.*, **90**, 349–70.

Plank, T. and Langmuir, C.H., 1992. Effects of melting regimes on the composition of ocean crust. *J. Geophys. Res.*, **97**, 19749–70.

Plank, T. and Langmuir, C.H., 1998. The chemical composition of subducting sediment and its consequences for the crust and mantle. *Chem. Geol.*, **145**, 325–94.

Polat, A., Hofmann, A.W., and Rosing, M.T., 2002. Boninite-like volcanic rocks in the 3.7–3.8 Ga Isua greenstone belt, west Greenland: geochemical processes for intra-oceanic subduction zone processes in the early earth. *Chem. Geol.*, **184**, 231–54.

Poli, S. and Schmidt, M.W., 1995. H_2O transport and release in subduction zones: experimental constraints on basaltic melts and andesitic systems. *J. Geophys. Res.*, **B100**, 22299–314.

Pollack, H.N., 1997. Thermal characteristics of the Archaean. In: De Wit, M.J. and Ashwaal, L.D. (eds) *Greenstone Belts*. Oxford University Press, Oxford, pp. 223–32.

Porcelli, D., Woolum, D., and Cassen, P., 2001. Deep Earth rare gases: initial inventories, capture from the solar nebula, and losses during moon formation. *Earth Planet. Sci. Lett.*, **193**, 237–51.

Poupinet, G., Arndt, N., and Vacher, P., 2003. Seismic tomography beneath stable tectonic regions and the origin and composition of the continental lithosphere. *Earth Planet. Sci. Lett.*, **212**, 89–101.

Presnall, D.C. and Hoover, J.D., 1984. Composition and depth of origin of primary mid-ocean ridge basalts. *Contrib. Mineral. Petrol.*, **87**, 170–8.

Presnall, D.C. and Hoover, J.D., 1986. Composition and depth of origin of primary mid-ocean ridge basalts – reply to D. Elthon. *Contrib. Mineral. Petrol.*, **94**, 257–61.

Puchtel, I.S., Hofmann, A.W., Jochum, K.P., Mezger, K., Shchipansky, A.A., and Samsonov, A.V., 1997. The Kostomuksha greenstone belt, NW Baltic Shield: remnant of a late Archaean oceanic plateau? *Terra Nova*, **9**, 87–90.

Rama Murthy, V. and Karato, S., 1997. Core formation and the chemical equilibrium in the Earth – II. Chemical consequences for the mantle and core. *Phys. Earth Planet. Interior*, **100**, 81–96.

Rapp, R.P., 1997. Heterogeneous source regions for Archaean granitoids: experimental and geochemical evidence. In: De Wit, M. and Ashwal, L.D. (eds) *Greenstone Belts*. Oxford University Press, Oxford, pp. 256–66.

Rapp, R.P., Shimizu, N., and Norman, M.C., 2003. Growth of early continental crust by partial melting of eclogite. *Nature*, **425**, 605–9.

Rapp, R.P., Watson, E.B., and Miller, C.F., 1991. Partial melting of amphibolite/eclogite and the origin of Archaean trondhjemites and tonalites. *Precambrian Res.*, **51**, 1–25.

Rasmussen, B., 2000. Filamentous microfossils in a 3,235-million-year-old volcanogenic massive sulphide deposit. *Nature*, **405**, 676–9.

Rasmussen, B. and Buick, R., 1999. Redox state of the Archaean atmosphere: evidence from detrital heavy minerals in ca. 3,250–2,750 Ma sandstones from the Pilbara Craton. *Geology*, **27**, 115–18.

Rees, M., 2000. *Just Six Numbers: The Deep Forces That Shape the Universe*. Basic Books, New York, 173 pp.

Regenauer-Lieb, K. and Kohl, T., 2003. Water solubility and diffusivity in olivine: its role in planetary tectonics. *Mineral. Mag.*, **67**, 697–715.

Reisberg, L. and Lorand, J.-P., 1995. Longevity of sub-continental mantle lithosphere from osmium isotope systematics in orogenic peridotite massifs. *Nature*, **376**, 159–62.

Reiss, A. and 18 others, 2001. The farthest known supernova: support for an accelerating Universe and a glimpse of the epoch of deceleration. *Astrophys. J.*, **560**, 49–71.

Reymer, A. and Schubert, G., 1984. Phanerozoic additions to the continental crust and crustal growth. *Tectonics*, **3**, 63–77.

Reymer, A. and Schubert, G., 1986. Rapid growth of some major segments of continental crust. *Geology*, **14**, 299–302.

Reysenbach, A.-L. and Cady, S.L., 2001. Microbiology of ancient and modern hydrothermal systems. *Trends Microbiol.*, **9**, 79–85.

Reysenbach, A.-L. and Shock, E., 2002. Merging genomes with geochemistry in hydrothermal ecosystems. *Science*, **296**, 1077–82.

Richardson, S.H., Erlank, A.J., and Hart, S.R., 1985. Kimberlite-borne garnet peridotite xenoliths from old enriched subcontinental lithosphere. *Earth Planet. Sci. Lett.*, **75**, 116–28.

Richardson, S.H., Gurney, J.J., Erlank, A.J., and Harris, J.W., 1984. Origin of diamonds in old enriched mantle. *Nature*, **310**, 198–202.

Richardson, S.H., Shirey, S.B., Harris, J.W., and Carlson, R.W., 2001. Archaean subduction recorded by Re–Os isotopes in eclogitic sulfide inclusions in Kimberly diamonds. *Earth Planet. Sci. Lett.*, **191**, 257–66.

Richter, F.M., 1988. A major change in the thermal state of the earth at the Archaean–Proterozoic boundary: consequences for the nature and preservation of continental lithosphere. *J. Petrol.* (Special Lithosphere Issue), 39–52.

Righter, K., Drake, M.J., and Yaxley, G., 1997. Prediction of siderophile element metal-silicate partition coefficients to 20 GPa and 2,800°C: the effects of pressure, temperature, oxygen fugacity and silicate and metallic melt compositions. *Phys. Earth Planet. Interior*, **100**, 115–34.

Ringwood, A.E., 1962. A model for the upper mantle. *J. Geophys. Res.*, **67**, 857–67.

Ringwood, A.E., 1966. The chemical composition and origin of the Earth. In: Hurley, P.M. (ed) *Advances in Earth Sciences*. MIT Press, Cambridge, MA, pp. 287–356.

Ringwood, A.E., 1991. Phase transitions and their bearing on the constitution and dynamics of the mantle. *Geochim. Cosmochim. Acta*, **55**, 2083–110.

Rino, S., Komiya, T., Windley, B.F., Katayama, I., Motoki, A., and Hirata, T., 2004. Major episodic increases of continental crustal growth determined from zircon ages of river sands: implications for mantle overturns in the early Precambrian. *Phys. Earth Planet. Interior*, **146**, 369–94.

Robb, L.J. and Meyer, F.M., 1995. The Witwatersrand Basin, South Africa: geological framework and mineralisation processes. *Ore Geology Rev.*, **10**, 67–94.

Rocholl, A. and Jochum, K.P., 1993. Th, U and other trace elements in carbonaceous chondrites: implications for the terrestrial and solar system Th/U ratios. *Earth Planet. Sci. Lett.*, **117**, 265–78.

Rollinson, H.R., 1993. *Using Geochemical Data: Evaluation, Presentation, Interpretation*. Longman, Harlow, 352 pp.

Rollinson, H.R., 1996. Tonalite–trondhjemite–granodiorite magmatism and the genesis of the Lewisian crust during the Archaean. In: Brewer, T.S. (ed) *Precambrian Crustal Evolution in the North Atlantic Region*. Geological Society of London Publishing House, Bath, *Geol. Soc. Spl. Publ.*, **112**, 25–42.

Rollinson, H.R., 1997a. The Archaean, komatiite-related Inyala chromitite, southern Zimbabwe. *Eco. Geol.*, **92**, 98–107.

Rollinson, H.R., 1997b. Eclogite xenoliths in West African kimberlites are residues from Archaean granitoids. *Nature*, **389**, 173–6.

Rollinson, H.R., 1999. Petrology and geochemistry of metamorphosed komatiites and basalts from the Sula Mountains greenstone belt, Sierra Leone. *Contrib. Mineral. Petrol.*, **134**, 86–101.

Rollinson, H.R., 2002. The metamorphic history of the Isua Greenstone belt, west Greenland. In: Fowler, C.M.R., Ebinger, C.J., and Hawkesworth, C.J. (eds) *The Early Earth: Physical, Chemical and Biological Development*. Geological Society of London Publishing House, Bath, *Special Publ. Geol. Soc. Lond.*, **199**, 329–50.

Rollinson, H.R., 2003. Metamorphic history suggested by garnet-growth chronologies in the Isua Greenstone Belt, west Greenland. *Precambrian Res.*, **126**, 181–96.

Rollinson, H.R., 2006. Crustal generation in the Archaean. In: Brown, M. and Rushmer, T. (eds) *Evolution and Differentiation of the Continental Crust*. Cambridge University Press, Cambridge, pp. 173–230.

Rollinson, H.R. and Blenkinsop, T.G., 1995. The magmatic, metamorphic and tectonic evolution of the northern marginal zone of the Limpopo belt in Zimbabwe. *J. Geol. Soc. Lond.*, **151**, 65–75.

Rollinson, H.R. and Fowler, M.B., 1987. The magmatic evolution of the Scourian Complex at Gruinard Bay. In: Park, R.G. and Tarney, J. (eds) *The Evolution of the Lewisian and Comparable Precambrian High-Grade Terrains*. Geological Society of London Publishing House, Bath, *Geol. Soc. Spl. Publ.*, **27**, 57–71.

Rollinson, H.R. and Tarney, J., 2005. Adakites – the key to understanding LILE depletion in granulites. *Lithos*, **79**, 61–81.

Rollinson, H.R., Appel, P.W.U., and Frei, R., 2002. A metamorphosed, early Archaean chromitite from west Greenland: implications for the genesis of Archaean anorthositic chromitites. *J. Petrol.*, **43**, 2143–70.

Rose, N.M., Rosing, M.T., and Bridgwater, D., 1996. The origin of metacarbonate rocks in the Archaean Isua supracrustal belt, West Greenland. *Am. J. Sci.*, **296**, 1004–44.

Rosing, M.T., 1999. ^{13}C-depleted carbon microparticles in >3,700-Ma sea-floor sedimentary rocks from west Greenland. *Science*, **283**, 674–6.

Rosing, M.T. and Frei, R., 2004. U-rich Archaean sea-floor sediments from Greenland – indications of >3,700 Ma oxygenic photosynthesis. *Earth Planet. Sci. Lett.*, **217**, 237–44.

Rosing, M.T., Bird, D.K., Sleep, N.H., Glassley, W., and Albarede, F., 2006. The rise of continents – an essay on the geologic consequences of photosynthesis. *Palaeogeography, Palaeoclimatol., Palaeoecol.*, **232**, 99–113.

Rouxel, O.J., Bekker, A., and Edwards, K.J., 2005. Iron isotope constraints on the Archaean and Palaeoproterozoic oceanic redox state. *Science*, **307**, 1088–91.

Rubie, D.C., Gessmann, C.K., and Frost, D.J., 2004. Partitioning of oxygen during core formation on the earth and Mars. *Nature*, **429**, 58–61.

Rubie, D.C., Melosh, H.J., Reid, J.E., Liebske, C., and Righter, K., 2003. Mechanisms of metal-silicate equilibration in the terrestrial magma ocean. *Earth Planet. Sci. Lett.*, **205**, 239–55.

Rudnick, R.L., 1995. Making continental crust. *Nature*, **378**, 571–8.

Rudnick, R.L. and Fountain, D.M., 1995. Nature and composition of the continental crust: a lower crustal perspective. *Rev. Geophys.*, **33**, 267–309.

Rudnick, R.L. and Gao, S., 2003. Composition of the continental crust. In: Holland, H.D. and Turekian, K.K. (eds) *Treatise on Geochemistry*. Elsevier, Oxford, Vol. 3.01, pp. 1–63.

Rudnick, R.L. and Taylor, S.R., 1987. The composition and petrogenesis of the lower crust: a xenolith study. *J. Geophys. Res.*, **92**, 13981–4005.

Rudnick, R.L., Barth, M., Horn, I., and McDonough, W., 2000. Rutile-bearing refractory eclogites: the missing link between continents and depleted mantle. *Science*, **287**, 278–81.

Rudnick, R.L., McDonough, W., and O'Connell, R.J., 1998. Thermal structure, thickness and composition of continental lithosphere. *Chem. Geol.*, **145**, 395–411.

Russell, M.J. and Hall, A.J., 1997. The emergence of life from iron monosulphide bubbles at a submarine hydrothermal redox and pH front. *J. Geol. Soc. Lond.*, **154**, 377–402.

Russell, S.A., Lay, T., and Garnero, E.J., 1998. Seismic evidence for small-scale dynamics in the lowermost mantle at the root of the Hawaiian hotspot. *Nature*, **396**, 255–8.

Rye, R. and Holland, H.D., 1998. Palaeosols and the evolution of atmospheric oxygen: a critical review. *Amer. J. Sci.*, **298**, 621–72.

Rye, R. and Holland, H.D., 2000. Life associated with a 2.76 Ga ephemeral pond? Evidence from Mount Roe #2 palaeosol. *Geology*, **28**, 483–486.

Rye, R., Kuo, P.H., and Holland, H.D., 1995. Atmospheric carbon dioxide concentrations before 2.2 billion years ago. *Nature*, **378**, 603–5.

Saal, A.E., Hauri, E.H., Langmuir, C.H., and Perfit, M.R., 2002. Vapour undersaturation in primitive mid-ocean-ridge basalt and the volatile content of Earth's upper mantle. *Nature*, **419**, 451–5.

Sagan, C. and Mullen, G., 1972. Earth and Mars: evolution of atmospheres and surface temperatures. *Science*, **177**, 52–6.

Salters, V.J.M. and White, W.M., 1998. Hf isotope constraints on mantle evolution. *Chem. Geol.*, **145**, 447–60.

Sasaki, S., 1990. The primary solar-type atmosphere surrounding the accreting earth: H_2O-induced high surface temperature. In: Newsom, H.E. and Jones, J.H. (eds) *Origin of the Earth*. Oxford University Press, Oxford, pp. 195–209.

Saunders, A.D., 2003. Mantle plumes: an alternative to the alternative. *Geoscientist*, **13**, 20–2.

Saunders, A.D., Tarney, J., Kerr, A.C., and Kent, R.W., 1996. The formation and fate of large oceanic igneous provinces. *Lithos*, **37**, 81–95.

Schersten, A., Elliott, T., Hawkesworth, C., and Norman, M., 2004. Tungsten isotope evidence that mantle plumes contain no contribution from the Earth's core. *Nature*, **427**, 234–7.

Schidlowski, M., 1988. A 3,800-million-year isotopic record of life from the carbon in sedimentary rocks. *Nature*, **333**, 313–18.

Schidlowski, M., Appel, P.W.U., Eichmann, R., and Junge, C.E., 1979. Carbon isotope geochemistry of the 3.7×10^9-yr-old Isua sediments, west Greenland: implicatons for the Archaean carbon and oxygen cycles. *Geochim. Cosmochim. Acta*, **43**, 189–99.

Schmidberger, S.S. and Francis, D., 1999. Nature of the mantle roots beneath the North American Craton: mantle xenolith evidence from Somerset Island kimberlites. *Lithos*, **48**, 195–216.

Schmidt, M.W. and Poli, S., 1998. Experimentally based water budgets for dehydrating slabs and consequences for arc magma generation. *Earth Planet. Sci. Lett.*, **163**, 361–79.

Schoenberg, R., Kamber, B.S., Collerson, K.D., and Eugster, O., 2002. New W-isotope evidence for rapid terrestrial accretion and very early core formation. *Geochim. Cosmochim. Acta*, **66**, 3151–60.

Schopf, J.W., 1993. Microfossils of the early Archaean Apex Chert: new evidence of the antiquity of life. *Science*, **260**, 640–6.

Schopf, J.W., 1999. *The Cradle of Life*. Princeton University Press, Priceton, NJ, 367 pp.

Schopf, J.W. and Packer, B.M., 1987. Early Archaean (3.3 billion to 3.5 billion-year-old) microfossils from Warawoona Group, Australia. *Science*, **237**, 70–3.

Schopf, J.W., Kudryavstev, Agresti, D.G., Wdowiak, T.J., and Czaja, A.D., 2002. Laser-Raman imagery of Earth's earliest fossils. *Nature*, **416**, 73–6.

Schramm, D.N. and Turner, M.S., 1998. Big bang nucleosynthesis enters the precision era. *Rev. Mod. Phys.*, **70**, 303–18.

Schubert, G., Turcotte, D.L., and Olson, P., 2001. *Mantle Convection in the Earth and Planets*. Cambridge University Press, Cambridge, 940 pp.

Schurr, B., Asch, G., Rietbrock, A., Trumbull, R., and Haberland, C., 2003. Complex patterns of fluid and melt transport in the central Andean subduction zone revealed by attenuation tomography. *Earth Planet. Sci. Lett.*, **215**, 105–19.

Sen, G., 1983. A petrological model for the constitution of the upper mantle and crust of the Koolau shield, Oahu, Hawaii and Hawaiian magmatism. *Earth Planet. Sci. Lett.*, **62**, 215–28.

Sen, G., Frey, F.A., Schimizu, N., and Leeman, W.P., 1993. Evolution of the lithosphere beneath Oahu, Hawaii: rare earth element abundances in mantle xenoliths. *Earth Planet. Sci. Lett.*, **119**, 53–69.

Sephton, M.A., Verchovsky, A.B., Bland, P.A., Gilmour, I., Grady, M.M., and Wright, I.P., 2003. Investigating the variations in carbon and nitrogen isotopes in carbonaceous chondrites. *Geochim. Cosmochim. Acta*, **67**, 2093–108.

Shen, Y. and Buick, R., 2004. The antiquity of microbial sulphate reduction. *Earth Sci. Rev.*, **64**, 243–72.

Sherwood Lollar, B., Ballentine, C.J., and O'Nions, R.K., 1997. The fate of mantle-derived carbon in a continental sedimentary basin: integration of C/He relationships and stable isotope signatures. *Geochim. Cosmochim. Acta*, **61**, 2295–307.

Sherwood Lollar, B., Lacrampe-Couloume, G., Slater, G.F., Ward, J., Moser, D.P., Gihring, T.M., Lin, L.-H., and Onstott, T.C., 2006. Unravelling abiogenic and biogenic sources of methane in the Earth's deep subsurface. *Chem. Geol.*, **226**, 328–39.

Sheth, H.C., 1999. Flood basalts and large igneous provinces from deep mantle plumes: fact, fiction or fallacy. *Tectonophysics*, **311**, 1–29.

Shimizu, K., Komiya, T., Hirose, K., Shimizu, N., and Maruyama, S., 2001. Cr-spinel, an excellent micro-container for retaining primitive melts – implications for a hydrous plume origin for komatiites. *Earth Planet. Sci. Lett.*, **189**, 177–88.

Shirey, S.B. and Walker, R.J., 1998. Re–Os isotopes in cosmochemistry and high temperature geochemistry. *Annu. Rev. Earth Planet. Sci.*, **26**, 423–500.

Siebert, C., Kramers, J.D., Meisel, Th., Morel, Ph., and Nagler, Th.F., 2005. PGE, Re–Os and Mo isotope systematics in Archaean and early Proterozoic sedimentary systems as proxies for redox conditions of the early Earth. *Geochim. Cosmochim. Acta*, **69**, 1787–801.

Silk, J., 1994. *A Short History of the Universe*. WH Freeman, New York, 246 pp.

Sleep, N.H., Zahnle, K.J., Kasting, J.F., and Morowitz, H.J., 1989. Annihilation of ecosystems by large asteroid impacts on the early Earth. *Nature*, **342**, 139–42.

Sleep, N.H., Zahnle, K., and Neuhoff, P.S., 2001. Initiation of clement surface conditions on the earliest earth. *Proc. Natl. Acad. Sci. USA*, **98**, 3666–72.

Smith, P.E., Evensen, N.N., York, D., and Moorbath, S., 2005. Oldest reliable terrestrial ^{40}Ar–^{39}Ar age from pyrite crystals at Isua west Greenland. *Geophys. Res. Lett.*, **32**, L21318, doi:10.1029/2005GL024066.

Smithies, R.H., 2000. The Archaean tonalite–trondhjemite–granodiorite (TTG) series is not an analogue of Cenozoic adakite. *Earth Planet. Sci. Lett.*, **182**, 115–25.

Smithies, R.H., 2002. Archaean boninite-like rocks in an intracratonic setting. *Earth Planet. Sci. Lett.*, **197**, 19–34.

Smithies, R.H., Champion, D., and Sun, S.-S., 2004. The case for Archaean boninites. *Contrib. Mineral. Petrol.*, **147**, 705–21.

Smithies, R.H., Van Kranendonk, M.J., and Champion, D.C., 2005. It started with a plume – early Archaean basaltic protocontinental crust. *Earth Planet. Sci. Lett.*, **238**, 284–97.

Soderlund, U., Patchett, J.P., Vervoort, J.D., and Isachsen, C.E., 2004. The ^{176}Lu decay constant determined by Lu–Hf and U–Pb isotope systematics of Precambrian mafic intrusions. *Earth Planet. Sci. Lett.*, **219**, 311–24.

Solomatov, V.S., 2000. Fluid dynamics of a terrestrial magma ocean. In: Canup, R. and Righter, K. (eds) *Origin of the Earth and Moon*. University of Arizona Press, Arizona, pp. 323–38.

Smoliar, M.I., Walker, R.J., and Morgan, J.W., 1996. Re–Os ages of Group IIA, IIIA, IVA, and IVB iron meteorites. *Science*, **271**, 1099–102.

Snow, J.E. and Reisberg, L., 1995. Os isotopic systematics of the MORB mantle: results from altered abyssal peridotites. *Earth Planet. Sci. Lett.*, **136**, 723–33.

Songaila, A., Cowie, L.L., Hogan, C.J., and Rugers, M., 1994. Deuterium abundance and background radiation temperatures in high redshift primordial clouds. *Nature*, **368**, 599–604.

Sproule, R.A., Lesher, C.M., Ayer, J.A., Thurston, P.C., and Herzberg, C.T., 2002. Spatial and temporal variations in the geochemistry of komatiites and komatiitic basalts in the Abitibi greenstone belt. *Precambrian Res.*, **115**, 153–86.

Stachel, T., Harris, J.W., Brey, G.P., and Joswig, W., 2000. Kankan diamonds (Guinea) II: lower mantle inclusion paragenesis. *Contrib. Mineral. Petrol.*, **140**, 16–27.

Staudigel, H., Hart, S.R., Schmincke, H.-U., and Smith, B.M., 1989. Cretaceous ocean crust at DSDP sites 417 and 418: carbon uptake from weathering versus loss by magmatic out-gassing. *Geochim. Cosmochim. Acta*, **53**, 3091–4.

Staudigel, H. and 14 others, 1998. Geochemical Earth Reference Model (GERM): description of the initiative. *Chem. Geol.*, **145**, 153–9.

Stein, M. and Goldstein, S.L., 1996. From plume head to continental lithosphere in the Arabian–Nubian shield. *Nature*, **382**, 773–8.

Stein, M. and Hofmann, A.W., 1994. Mantle plumes and episodic crustal growth. *Nature*, **372**, 63–8.

Stern, C.R. and Kilian, R., 1996. Role of the subducted slab, mantle wedge and continental crust in the generation of adakites from the Andean Austral Volcanic Zone. *Contrib. Mineral. Petrol.*, **123**, 263–81.

Stern, R.A., Percival, J.A., and Mortensen, J.K., 1994. Geochemical evolution of the Minto block: a 2.7 Ga continental magmatic arc built on the Superior proto-craton. *Precambrian Res.*, **65**, 115–53.

Stevenson, D.J., 1987. Origin of the Moon – the collision hypothesis. *Annu. Rev, Earth Planet. Sci.*, **15**, 271–315.

Stevenson, D.J., 1990. Fluid dynamics of core formation. In: Newsom, H.E. and Jones, J.H. (eds) *Origin of the Earth*. Oxford University Press, Oxford, pp. 231–49.

Stevenson, D.J., 2004. Planetary diversity. *Phys. Today*, April, 43–8.

Stevenson, R.K. and Patchett, P.J., 1990. Implications for the evolution of continental crust from Hf isotope systematics of Archaean detrital zircons. *Geochim. Cosmochim. Acta*, **54**, 1683–97.

Strauss, H., 2003. Sulphur isotopes and the early Archaean sulphur cycle. *Precambrian Res.*, **126**, 349–61.

Sumner, D.Y., 2001. Microbial influences on local carbon isotope ratios and their preservation in carbonate. *Astrobiology*, **1**, 57–70.

Sun, S.-S. and McDonough, W.F., 1989. Chemical and isotopic systematics of oceanic basalts: implications for mantle composition and processes. *Geol. Soc. Lond. Spl. Publ.*, **42**, 313–45.

Sutton, J. and Watson, J., 1951. The pre-Torridonian metamorphic history of the Loch Torridon and Scourie areas in the northwest Highlands, and its bearing on the chronological classification of the Lewisian. *J. Geol. Soc. Lond.*, **106**, 241–307.

Sylvester, P.J., Campbell, I.H., and Bowyer, D.A., 1997. Niobium/Uranium evidence for early formation of the continental crust. *Science*, **275**, 521–3.

Takahashi, E. and Scarfe, C.M., 1985. Melting of peridotite to 14 GPa and the genesis of komatiite. *Nature*, **315**, 566–8.

Takano, Y., Kobayashi, K., Yamanaka, T., Marumo, K., and Urabe, T., 2004. Amino acids in the 308 °C deep-sea hydrothermal system of the Suiyo Seamaount, Izu-Bonin Arc, Pacific ocean. *Earth Planet. Sci. Lett.*, **219**, 147–53.

Taylor, S.R., 1987. The unique lunar composition and its bearing on the origin of the Moon. *Geochim. Cosmochim. Acta*, **51**, 1297–309.

Taylor, S.R. and McLennan, S.M., 1981. The composition and evolution of the continental crust: rare earth element evidence from sedimentary rocks. *Phil. Trans. R. Soc. A*, **301**, 381–99.

Taylor, S.R. and McLennan, S.M., 1985. *The Continental Crust: Its Composition and Evolution*. Blackwell, Oxford, 312 pp.

Taylor, S.R. and McLennan, S.M., 1995. The geochemical evolution of the continental crust. *Rev. Geophys.*, **33**, 241–65.

Telesco, C.M. and 10 others, 2005. Mid-infrared images of β Pictoris and the possible role of planetesimal collisions in the central disk. *Nature*, **433**, 133–6.

Thorkelson, D.J. and Breitsprecher, K., 2005. Partial melting of slab window margins: genesis of adakitic and non-adakitic magmas. *Lithos*, **79**, 25–41.

Tice, M.M. and Lowe, D.R., 2004. Photosynthetic microbial mats in the 3,416-Myr-old ocean. *Nature*, **431**, 549–52.

Tiepolo, M., Bottazzi, P., Foley, S.F., Oberti, R., and Zanetti, A., 2001. Fractionation of Nb and Ta from Zr and Hf at mantle depths: the role of titanian–pargasite and kaesutite. *J. Petrol.*, **42**, 221–32.

Tingle, T.N., 1998. Accretion and differentiation of carbon in the early Earth. *Chem. Geol.*, **147**, 3–10.

Tolstikhin, I.N. and Hofmann, A., 2005. Early crust on top of the Earth's core. *Phys. Earth Planet. Interior*, **148**, 109–30.

Tolstikhin, I.N. and O'Nions, R.K., 1994. The earth's missing xenon: a combination of early degassing and of rare gas loss from the atmosphere. *Chem. Geol.*, **115**, 1–6.

Tolstikhin, I.N., Kramers, J.D., and Hofmann, A., 2006. A chemical Earth model with whole mantle convection: the importance of a core–mantle boundary layer (D″) and its early formation. *Chem. Geol.*, **226**, 79–99.

Tonks, W.B. and Melosh, H.J., 1990. The physics of crystal settling and suspension in a turbulent magma ocean. In: Newsom, H.E. and Jones, J.H. (eds) *Origin of the Earth*. Oxford University Press, Oxford, pp. 151–74.

Trampert, J., Deschamps, F., Resovsky, J., and Yuen, D., 2004. Probabilistic tomography maps chemical heterogeneities throughout the lower mantle. *Science*, **306**, 853–6.

Trieloff, M., Falter, M., and Jesseberger, E.K., 2003. The distribution of mantle atmospheric argon in oceanic basaltic glasses. *Geochim. Cosmochim. Acta*, **67**, 1229–45.

Turner, G., Harrison, M.T., Holland, G., Mojzsis, S.J., and Gilmour, J., 2004. Extinct ^{244}Pu in ancient zircons. *Science*, **306**, 89–91.

Ueno, Y., Yamada, K., Yoshida, N., Maruyama, S., and Isozaki, Y., 2006. Evidence from fluid inclusions for microbial methanogenesis in the early Archaean era, *Nature*, **440**, 516–19.

Urey, H.C., 1952. *The Planets, Their Origin and Development*. Yale University Press, New Haven, CT.

Valentine, D.L., Chidthaisong, A., Rice, A., Reeborugh, W.S., and Tyler, S.C., 2004. Carbon and hydrogen isotope fractionation by moderately thermophilic methanogens. *Geochim. Cosmochim. Acta*, **68**, 1571–90.

Valley, J.W., Peck, W.H., King, E.M., and Wilde, S.A., 2002. A cool early Earth. *Geology*, **30**, 351–4.

Van Boekel, R. and 22 others, 2004. The building blocks of planets within the "terrestrial" region of protoplanetary disks. *Nature*, **432**, 469–82.

Van der Meljde, M., Marone, F., Giardini, D., and van der Lee, S., 2003. Seismic evidence for water deep in the Earth's upper mantle. *Science*, **300**, 1556–8.

Van Kranendonk, M.J., 2006. Volcanic degassing, hydrothermal circulation and the flourishing of early life on Earth: a review of the evidence from 3,490–3,240 Ma rocks of the Pilbara supergroup, Pilbara Craton, western Australia. *Earth Sci. Rev.*, **74**, 197–240.

Van Kranendonk, M.J., Webb, G.E., and Kamber, B.S., 2003. Geological and trace element evidence for a marine sedimentary environment of deposition and biogenicity of 3.45 Ga stromatolitic carbonates in the Pilbara Craton and support for a reducing Archaean ocean. *Geobiology*, **1**, 91–108.

Van Zuilen, M.A., Lepland, A., and Arrhenius, G., 2002. Reassessing the evidence for the earliest traces of life. *Nature*, **418**, 627–30.

Van Zuilen, M.A., Lepland, A., Terranes, J., Finarelli, J., Wahalen, M., and Arrhenius, G., 2003. Graphite and carbonates in the 3.8 Ga-old Isua supracrustal belt, southern west Greenland. *Precambrian Res.*, **126**, 331–48.

Van Zuilen, M.A., Mathew, K., Wopenka, B., Lepland, A., Marti, K., and Arrhenius, G., 2005. Nitrogen and argon isotopic signatures in graphite from the 3.8 Ga-old Isua supracrustal belt, southern west Greenland. *Geochim. Cosmochim. Acta*, **69**, 1241–52.

Veizer, J., Hoefs, J., Lowe, D.R., and Thurston, P.C., 1989. Geochemistry of Precambrian carbonates: II. Archaean greenstone belts and Archaean sea water. *Geochim. Cosmochim. Acta*, **53**, 859–71.

Vervoort, J.D. and Blichert-Toft, J., 1999. Evolution of the depleted mantle: Hf isotope evidence from juvenile rocks through time. *Geochim. Cosmochim. Acta*, **63**, 533–56.

Vervoort, J.D., Patchett, P.J., Gehrels, G.E., and Nurman, A.P., 1996. Constraints on early Earth differentiation from hafnium and neodymium isotopes. *Nature*, **379**, 624–7.

Vervoort, J.D., White, W.M., and Thorpe, R.I., 1994. Nd and Pb isotope ratios of the abitibi greenstone belt: new evidence for very early differentiation of the Earth. *Earth Planet. Sci. Lett.*, **128**, 215–29.

von Huerne, R. and Scholl, D.W., 1991. Observations at convergent margins concerning sediment subduction and the growth of the continental crust. *Rev. Geophys.*, **29**, 279–316.

Wachtershauser, G., 1988. Before enzymes and templates: theory of surface metabolism. *Microbiol. Rev.*, **52**, 452–82.

Wade, J. and Wood, B.J., 2005. Core formation and the oxidation state of the Earth. *Earth Planet. Sci.*, **236**, 78–95.

Walker, J.G.C., 1977. *Evolution of the Atmosphere*. McMillan, New York.

Walker, J.G.C., 1983. Possible limits on the composition of the Archaean ocean. *Nature*, **302**, 518–20.

Walker, R.J. and Nisbet, E.G., 2002. ^{187}Os isotopic constraints on Archaean mantle dynamics. *Geochim. Cosmochim. Acta*, **66**, 3317–25.

Walker, R.J., Becker, H., and Morgan, J.W., 1999. Comparative Re–Os isotope systematics of chondrites: implications regarding early solar system processes. *Lunar Planet. Sci. Conf.*, **30**, 1208.

Walker, R.J., Prichard, H.M., Ishiwatari, A., and Pimentel, M., 2002. The Osmium isotopic composition of convecting upper mantle deduced from ophiolitic chromites. *Geochim. Cosmochim. Acta*, **66**, 329–45.

Walraven, F. and Pape, J., 1994. Pb–Pb whole-rock ages for the Pongola supergroup and the Usushwana Complex, South Africa. *J. Afr. Earth Sci.*, **18**, 297–308.

Walter, M.J. and Tronnes, R.G., 2004. Early Earth differentiation. *Earth Planet. Sci. Lett.*, **225**, 253–69.

Walter, M.R., Buick, R., and Dunlop, J.S.R., 1980. Stromatolites, 3,400–3,500 Myr old from the North Pole area, western Australia. *Nature*, **284**, 443–5.

Walther, M.J., Nakamura, E., Tronnes, R.G., and Frost, D.J., 2004. Experimental constraints on crystallization differentiation in a deep magma ocean. *Geochim. Cosmochim. Acta*, **68**, 4267–84.

Wasson, J.T., 1985. *Meteorites: Their Record of Early Solar System History*. W.H. Freeman, New York, p. 267.

Watson, A.J. and Lovelock, J.E., 1983. Biological homeostasis of the global environment: the parable of Daisyworld. *Tellus*, **35B**, 286–9.

Watson, E.B. and Harrison, T.M., 2005. Zircon thermometer reveals minimum melting conditions on earliest Earth. *Science*, **308**, 841–4.

Watson, E.B., Brenan, J.M., and Baker, D.R., 1990. Distribution of fluids in the continental mantle. In: Menzies, M.A. (ed) *Continental Mantle*. Clarendon Press, Oxford, pp. 111–25.

Weaver, B.L. and Tarney, J., 1982. Andesitic magmatism and continental growth. In: Thorpe R.S. (ed) *Andesites*. Wiley, London, pp. 639–61.

Weaver, B.L. and Tarney, J., 1984. Empirical approach to estimating the composition of the continental crust. *Nature*, **310**, 575–7.

Wedepohl, K.H., 1995. The composition of the continental crust. *Geochim. Cosmochim. Acta*, **59**, 1217–32.

Weidenschilling, S.J., Spaute, D., Davis, D.R., Marzari, F., and Ohtsuki, K., 1997. Accretional evolution of a planetesimal swarm. *Icarus*, **128**, 429–55.

Westall, F., de Wit, M.J., Dann, J., van der Gaast, S., de Ronde, C.E.J., and Gerneke, D., 2001. Early Archaean fossil bacteria and biofilms in hydrothermally-influenced sediments from the Barberton greenstone belt, South Africa. *Precambrian Res.*, **106**, 93–116.

Wetherill, G.W., 1990. Formation of the Earth. *Annu. Rev. Earth Planet. Sci.*, **18**, 205–56.

Wever, T., 1992. Archaean and Proterozoic crustal evolution: evidence from crustal seismology. *Comment. Geol.*, **20**, 664–5.

White, R.S., 1988. The Earth's crust and lithosphere. *J. Petrol.* (Special Lithosphere Issue), 1–10.

White, R.V., Tarney, J., Kerr, A.C., Saunders, A.D., Kempton, P.D., Pringle, M.S., and Klaver, G.T., 1999. Modification of an oceanic plateau, Aruba, Dutch Caribbean: implications for the generation of continental crust. *Lithos*, **46**, 43–68.

White, W.M., 1993. $^{238}U/^{204}Pb$ in MORB and open system evolution of the depleted mantle. *Earth Planet. Sci. Lett.*, **115**, 211–26.

Whitehouse, M.J., Kamber, B.S., Fedo, C.M., and Lepland, A., 2005. The importance of combined Pb and S isotope data from early Archaean rocks, southwest Greenland, for the interpretation of S-isotope signatures. *Chem. Geol.*, **222**, 112–31.

Wiechert, U., Halliday, A.N., Lee, D.-C., Snyder, G.A., Taylor, L.A., and Rumble, D., 2001. Oxygen isotopes and the Moon-forming giant impact. *Science*, **294**, 345–8.

Wilde, S.A., Valley, J.W., Peck, W.H., and Graham, C.M., 2001. Evidence from detrital zircons for the existence of continental crust and oceans on the Earth 4.4 Gyr ago. *Nature*, **409**, 175–8.

Wilks, M.E. and Nisbet, E.G., 1985. Archaean stromatolites from the Steep Rock Group, northwestern Ontario, Canada. *Can. J. Earth Sci.*, **22**, 792–9.

Williams, Q. and Hemley, R.J., 2001. Hydrogen in the deep Earth. *Annu. Rev. Earth Planet. Sci.*, **29**, 365–418.

Williams, Q. and Knittle, E., 1997. Constraints on core chemistry from the pressure dependence of the bulk modulus. *Phys. Earth Planet. Interior*, **100**, 49–59.

Wilson, A.H., 2003. A new class of silica enriched, highly depleted komatiites in the southern Kaapvaal Craton, South Africa. *Precambrian Res.*, **127**, 125–41.

Wilson, J.F., Nesbitt, R.W., and Fanning, C.M., 1995. Zircon geochronology of Archaean felsic sequences in the Zimbabwe Craton: a revision of greenstone belt stratigraphy and a model for crustal growth. In: Coward, M.P. and Reis, A.C. (eds) *Early Precambrian Processes*. Geological Society of London Publishing House, Bath, *Spl. Publ. Geol. Soc. Lond.*, **95**, 109–26.

Wilson, M. and Spencer, E.A., 2001. Carbonatite magmatism: constraints on the history of carbonate sediment recycling in the upper mantle since the Archaean. *J. Conf. Abs.* (EUG XI), **6**, 494.

Woese, C.R., Kandler, O., and Wheelis, M.L., 1990. Towards a natural system of organisms: proposal for the domains Archaea, Bacteria, and Eucarya. *Proc. Natl. Acad. Sci. USA*, **87**, 4576–9.

Wood, B.J., 1993. Carbon in the core. *Earth Planet. Sci. Lett.*, **117**, 593–607.

Wood, J.A., 1988. Chondritic meteorites and the solar nebula. *Annu. Rev. Earth Planet. Sci.*, **16**, 53–72.

Wood, J.A., 2004. Formation of chondritic refractory inclusions: the astrophysical setting. *Geochim. Cosmochim. Acta*, **68**, 4007–21.

Woodland, A.B. and Koch, M., 2003. Variation in oxygen fugacity with depth in the upper mantle beneath the Kaapvaal Craton, southern Africa. *Earth Planet. Sci. Lett.*, **214**, 295–310.

Wortmann, U., Bernasconi, S.M., and Bottcher, M.E., 2001. Hypersulfidic deep biosphere indicates extreme sulphur isotope fractionation during single-step mircrobial sulphate reduction. *Geology*, **29**, 647–50.

Wu, F.-Y., Walker, R.J., Ren, X.-W., Sun, D.-Y., and Zhou, X.-H., 2003. Osmium isotopic constraints on the age of lithospheric mantle beneath northeast China. *Chem. Geol.*, **196**, 107–29.

Wyllie, P.J. and Wolf, M.B., 1993. Amphibolite dehydration-melting: sorting out the solidus. *Geol. Soc. Lond. Spl. Publ.*, **76**, 405–16.

Wyllie, P.J., Wolf, M.B., and van der Laan, S.R., 1997. Conditions for formation of tonalites and trondhjemites: magmatic sources and products. In: De Wit, M. and Ashwal, L.D. (eds) *Greenstone Belts*. Oxford University Press, Oxford, pp. 256–66.

Xie, Q. and Kerrich, R., 1994. Silicate-perovskite and majorite signature komatiites from the Archaean Abitibi Greenstone Belt: implications for early mantle differentiation and stratification. *J. Geophys. Res.*, **99**, B15799–812.

Xie, S. and Tackley, P.J., 2004. Evolution of U–Pb and Sm–Nd systems in numerical models of mantle convection and plate tectonics. *J. Geophys. Res.*, **109**, B11204. doi:10.1029/2004JB003176.

Xiong, J., Fischer, W.M., Inoue, K., Nakahara, M., and Bauer, C.E., 2000. Molecular evidence for the early evolution of photosynthesis. *Science*, **289**, 1724–30.

Yamaguchi, K.E., Johnson, C.M., Berad, B.L., and Ohmoto, H., 2005. Biogeochemical cycling of iron in the Archaean–Palaeoproterozoic Earth: constraints from iron isotope variations in sedimentary rocks from the Kaapvaal and Pilbara Cratons. *Chem. Geol.*, **218**, 135–69.

Yang, H.-J., Sen, G., and Shimizu, N., 1998. Mid-ocean ridge melting: constraints from lithospheric xenoliths at Oahu, Hawaii. *J. Petrol.*, **39**, 277–95.

Yin, Q., Jacobsen, S.B., Tamashita, K., Blichert-Toft, J., Telouk, P., and Albarede, F., 2002. A short timescale for terrestrial planet formation from Hf–W chronometry of meteorites. *Nature*, **418**, 949–52.

Yokochi, R. and Marty, B., 2005. Geochemical constraints on mantle dynamics in the Hadean. *Earth Planet. Sci. Lett.*, **238**, 17–30.

Yoshihara, A. and Hamano, Y., 2004. Palaeomagnetic constraints on the Archaean geomagnetic field intensity obtained from komatiites of the Barberton and Belingwe greenstone belts, South Africa and Zimbabwe. *Precambrian Res.*, **131**, 111–42.

Young, E.D. and Russell, S.S., 1998. Oxygen isotope reservoirs in the early solar nebula inferred from Allende CAI. *Science*, **282**, 452–5.

Young, E.D., Simon, J.I., Galy, A., Russell, S.S., Tonui, E., and Lovera, O., 2005. Supra-canonical 26Al/27Al and the residence time of CAI's in the solar protoplanetary disk. *Science*, **308**, 223–7.

Young, G.M., von Brun, V., Gold, D.J.C., and Minter, W.E.L., 1998. Earth's oldest reported glaciation: physical and chemical evidence from the Archaean Mozzan Group (~2.9 Ga) of South Africa. *J. Geol.*, **106**, 523–38.

Yukutake, T., 2000. The inner core and the surface heat flow as clues to estimating the initial temperature of the earth's core. *Phys. Earth Planet. Interior*, **121**, 103–37.

Zahnle, K. and Sleep, N.H., 2002. Carbon dioxide cycling through the mantle and implications for the climate of the ancient Earth. In: Fowler, C.M.R., Ebinger, C.J., and Hawkesworth, C.J. (eds) *The Early Earth: Physical, Chemical and Biological Development. Geol. Soc. Lond. Spl. Publ.*, **199**, 231–57.

Zanda, B., 2004. Chondrules. *Earth Planet. Sci. Lett.*, **224**, 1–17.

Zandt, G. and Ammon, C.J., 1995. Continental crust composition constrained by measurements of crustal Poisson's ratio. *Nature*, **374**, 152–4.

Zartman, R.E. and Haines, S.M., 1988. The plumbotectonic model for Pb isotopic systematics among major terrestrial reservoirs – a case for bi-directional transport. *Geochim. Cosmochim. Acta*, **52**, 1327–39.

Zartman, R.E. and Richardson, S.H., 2005. Evidence from kimberlitic zircon for a decreasing mantle Th/U since the Archaean. *Chem. Geol.*, **220**, 263–83.

Zegers, T.E. and van Keken, P.E., 2001. Middle Archaean continent formation by crustal delamination. *Geology*, **29**, 1083–6.

Zhang, Y. and Zindler, A., 1993. Distribution and evolution of carbon and nitrogen in the Earth. *Earth Planet. Sci. Lett.*, **117**, 331–45.

Zindler, A. and Hart, S., 1986. Chemical geodynamics. *Annu. Rev. Earth Planet. Sci.*, **14**, 493–571.

Zinner, E., Amari, S., Guiness, R., Nguyen, A., Stadermann, F.J., Walker, R.M., and Lewis, R.S., 2003. Presolar spinel grains from the Murray and Murchison carbonaceous chondrites. *Geochim. Cosmochim. Acta*, **67**, 5083–95.

Zolensky, M.E., 2005. Extraterrestrial water. *Elements*, **1**, 39–43.

INDEX

Note: Page numbers from figures are in *italics* while those from boxes and tables are in **bold**